Early Mapping
of Southeast Asia

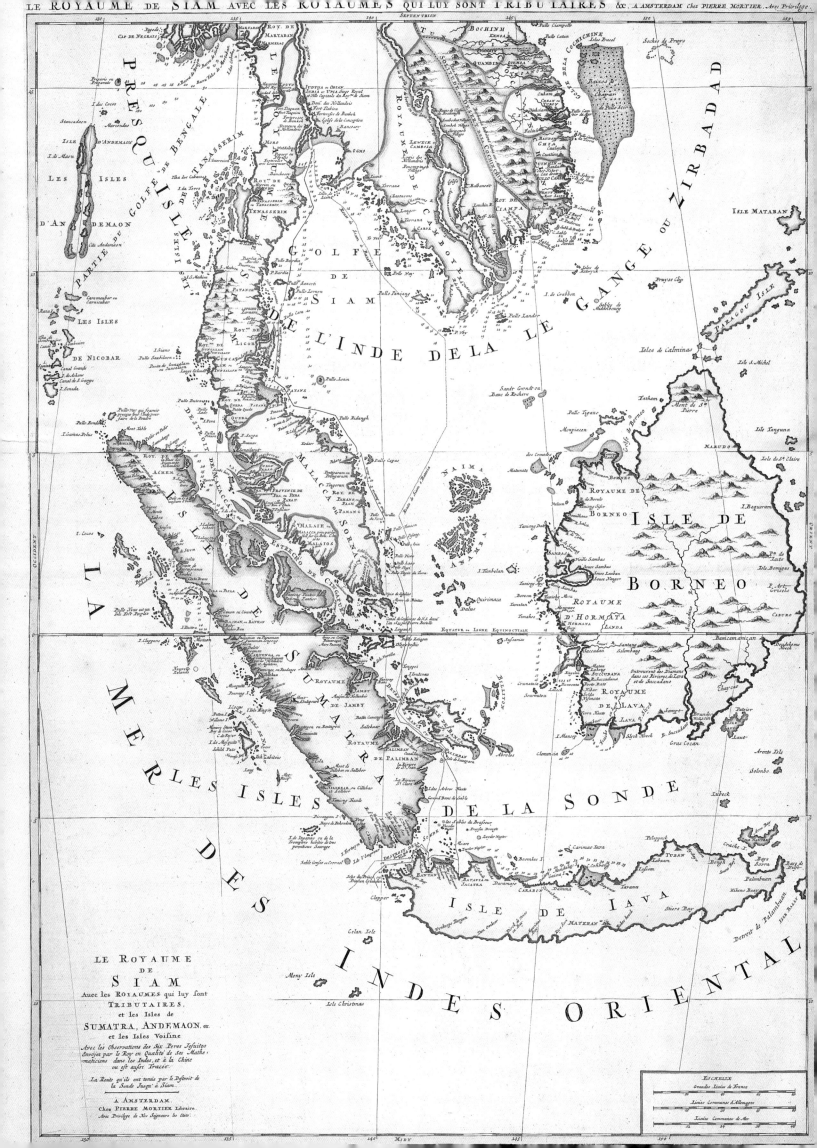

Early Mapping of Southeast Asia

The Epic Story of Seafarers, Adventurers, and Cartographers
Who First Mapped the Regions between China and India

THOMAS SUÁREZ

PERIPLUS EDITIONS
Singapore • Hong Kong • Indonesia

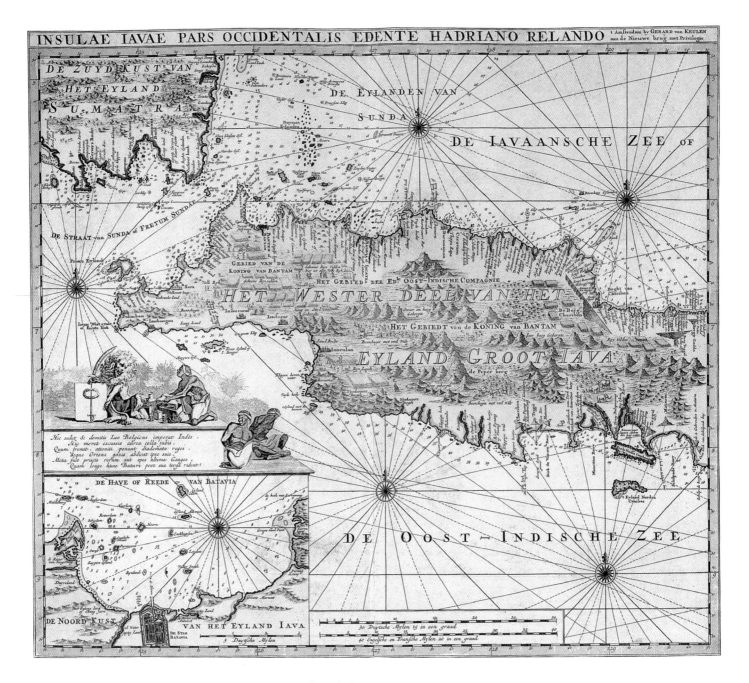

INSULAE IAVAE PARS OCCIDENTALIS EDENTE HADRIANO RELANDO

Published by Periplus Editions (HK) Ltd.

Copyright © 1999All rights reserved

ISBN-10: 962-593-470-7
ISBN-13: 978-962-593-470-9
Printed in Singapore
10 09 08   5 4 3 2

Opening end paper: Gulf of Siam and Indian Ocean, from the *Traiphum*, late eighteenth century. The peninsula in the nine o'clock position is Phatthalung, in southern Thailand; the chain of islands along the bottom is Rama's Bridge leading to Sri Lanka.

Closing end paper: Southeast Asia, from the atlas of the twelfth century Sicilian geographer, Sharif al-Idrisi. A copy of 1553; south is at the top. [Courtesy of the Bodleian Library, Oxford]

Half-title page: Southeast Asia by Pierre Mortier, ca. 1700.

Acknowledgments

Of the many people who generously assisted me while I was researching this book, I would like to acknowledge several in particular. In alphabetical order they are as follows: Deepak Bhattasali of the World Bank, who helped me with Indian history as it relates to Southeast Asia, and with Indian-inspired Southeast Asian cosmology; Richard Casten, who created time which didn't exist to scrutinize the draft and then shared his enormously helpful feedback; Rodrigue Lévesque, whose personal correspondence, as well as his published research on Micronesia, were helpful in sorting through early Pacific voyages and the identities of the islands discovered; Dr. Hans Penth of the Archive of Lan Na Inscriptions, University of Chiang Mai, who took the time to meet with me in person while I was in Thailand, and by mail and fax when I was not; and Dr. Dawn Rooney of Bangkok, who lent her considerable expertise regarding Southeast Asian culture and history, patiently critiqued my draft, and supplied me with difficult-to-find reference material.

I am also grateful to the following people for generously sharing their knowledge with me: Thomas Goodrich, Ambassador William Itoh, Frank Manasek, Greg McIntosh, Professor Günter Schilder, Michael Smithies, G. R. Tibbetts, Lutz Walter, Paul Wheatley, David Woodward, and Michael Wright.

My wife, Ahngsana, was of enormous help with translations and other practical matters. Both she and my daughter, Sainatee, pulled me out of many medieval Southeast Asian jungles in which I appeared to have become lost.

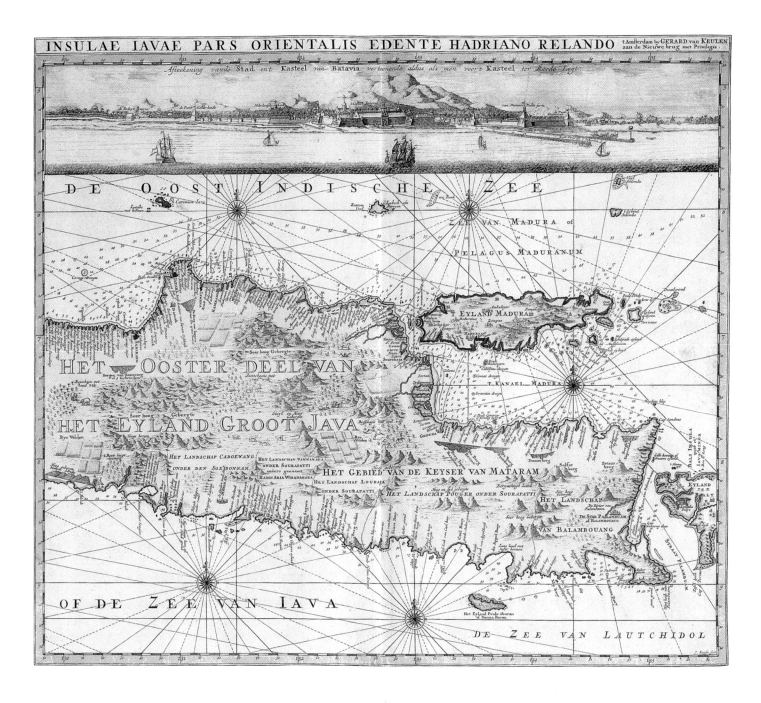

Afteekening vande Stad ent Kasteel van Batavia verteonende alldus als men voor t Kasteel ter Reede Legt

INSULAE IAVAE PARS ORIENTALIS EDENTE HADRIANO RELANDO t Amsterdam by GERARD van KEULEN aan de Nieuwe brug met Privilegie.

DE OOST INDISCHE ZEE

ZEE VAN MADURA of

PELAGUS MADURANUM

'T EYLAND MADURA

HET OOSTER DEEL VAN

HET EYLAND GROOT JAVA

T KANAEL MADURA

HET LANDSCHAP CADOEWANG ONDER DEN SOE'SONNAN

HET LANDSCHAP PANNARAGA ONDER SOURAPATTI

HET GEBIED VAN DE KEYSER VAN MATARAM

HET LANDSCHAP LOUBAJA ONDER SOURAPATTI

HET LANDSCHAP POUGER ONDER SOURAPATTI

HET LANDSCHAP VAN BALAMBOUANG

OF DE ZEE VAN IAVA

Het Eyland Poulo Abaran of Noesa Baron

DE ZEE VAN LAUTCHIDOL

From the institutions whose material I have illustrated, I would like to acknowledge several people in particular: Suwakhon Siriwongwarawat, Director, National Library of Thailand; Joselito B. Zulucta, Director, University of Santo Tomas Publishing House; Jan Werner, Universiteitsbibliotheek Amsterdam; Jim Flatness and David Hsu of the Library of Congress; Elaine Engst and Laura Linke Cornell University; Doris Nicholson, Bodleian Library, Oxford.

Many of the illustrations were lent by antiquarian book and map dealers: Richard B. Arkway (New York), Roderick M. Barron (Sevenoaks, Kent), Clive A. Burden (Rickmansworth, Herts), Rodolphe Chamonal (Paris), Geoff Edwards (Jakarta), Susanna Fisher (Southhampton), Antiquariaat Forum (Utrecht), Mappæ Japoniæ (Tokyo), Robert Augustyn and the firm of Martayan Lan (New York), Edward Lefkowicz (Providence), Jonathan Potter (London), Robert Putman (Amsterdam), and Paulus Swaen Old Maps Internet Auction (Geldrop, Holland).

I was singularly fortunate to have had Julian Davison edit the book. A scholar of Southeast Asian history and ethnography, knowledgeable about maps and a master of page laying out, Dr. Davison polished the book with a precison that no mere editor could have accomplished. Finally, many thanks to Michael Stachels and Eric Oey of Periplus Editions for their expertise and patience in making this book possible. It was my great honor to work with them.

In memory of
Paw U'ee Ma and Phan Ninkhong

Fig. 1 Java, Van Keulen, 1728. (51 x 114 cm)
Following page: Fig. 2 *India Orientalis*,
Hendrik Doncker ca. 1664. [Courtesy Museum of
the Book/Museum Meermanno-Westreenianum,
The Hague]

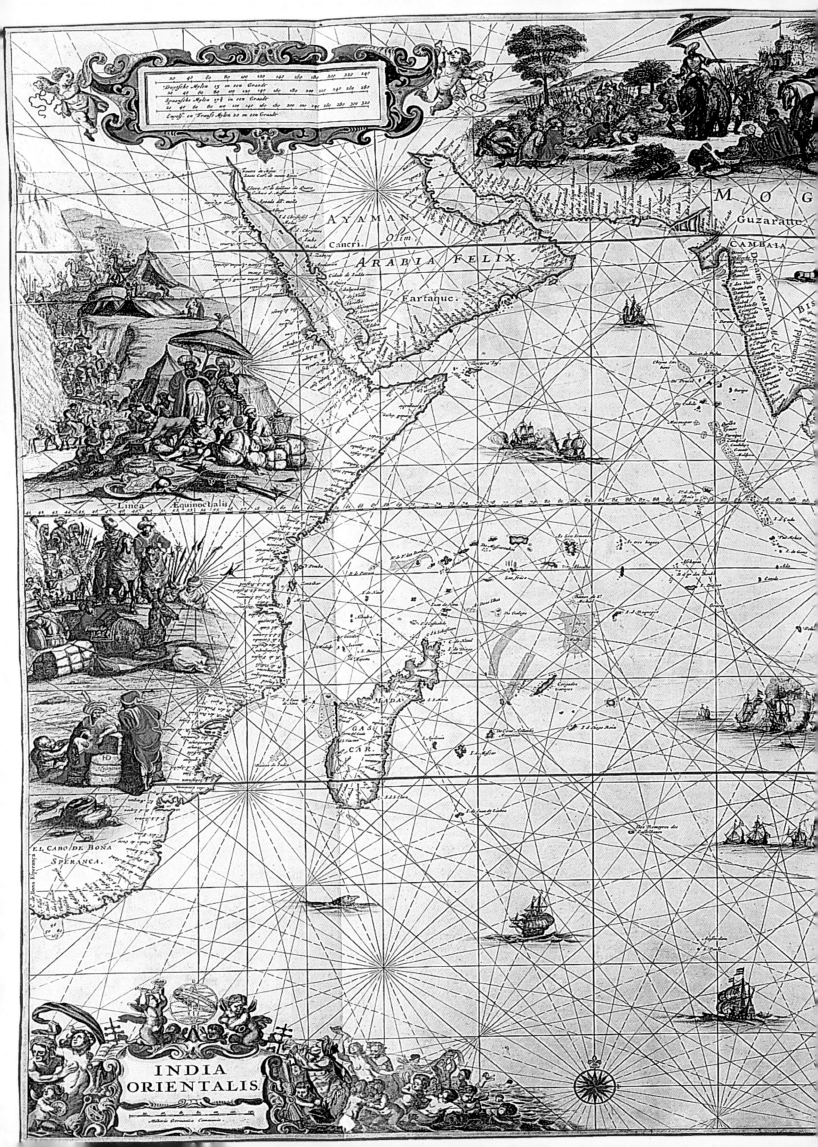

AYAMAN

Cancri. Olim

ARABIA FELIX.

Fartaque.

MOG

Guzaratte

CAMBAIA

Linea   Æquinoctialis

MADA

GASI

CAR.

EL CABO DE BONA
SPERANCA.

INDIA
ORIENTALIS.

Corea
Insula

IAPAN

CHINA

L. BENGALA.   Porto pequeno   Porto Grande   Tropicus Cancri.

ARACAN.

PEGU.

SIAM.

BOR-
NEO.

Celebes

MINDANAO

Ladrones

Iſlas d'

Linea Æquinoctialis

Nova
Guinea

HOLLANDIA

Tropicus Capricorni

NOVA; detecta A.º 1644.

Pascaerte van
OOST INDIEN,
Met de Omgelegen Eylanden; als
Madagascar, Ceylon, Sumatra, Iava,
Borneo, Celebes, Molucco en Banda;
van C. de Bona Speranca tot Iapan.
T'AMSTERDAM,
By Hendrick Doncker Boeckverkoper
en Graadboghmaker, inde Nieuwbrugsteegh.

20	40	60	80	100	120	140	160
Duytsche Mylen 15 in een Graadt							
20	40	60	80	100	140	180	180
Spaensche Mylen 17½ in een Graadt							
40	60	100	140	180	200	220	
Engels. en Fransh Mylen 20 in een Graadt							

Contents

Part IV   Companies and Colonization

Chronological List of Figures

Introduction

One day in the middle 1960s, the Juilliard School of Music, where I was a student, held a rummage sale to dispose of deaccessioned books from its library. While poking through modest piles of worn volumes my eyes were caught by one that looked slightly more dated than the others. I skimmed through the book and came across some pages about Johannes Brahms, a composer who, I truly believed, was the very envy of the gods. I was however confused to find Brahms being spoken of in the present tense, until I realized that when this book was published in the 1890s, Brahms was a *contemporary* composer (he died in 1897). The book was, of course, not especially old. But even as a voice from the recent past, it impressed upon me the fact that Brahms, as well as all the other 'prophets' of my religion called music, were flesh and blood, real people who had walked the earth as do we who are alive today.

Time and place were never the same again. 'Michelangelo' was no longer merely a name affixed to a museum plaque; rather, it was the name of a human being who, like those who composed maps and charted music, laughed and worried and drank wine and just happened to create things that transcended mortal bounds. My feeling for history suffered the same demise, for I was no longer content to navigate the past by mere rote. Regardless of how flawed our record of the past is, the fact is that the past *existed*. Someday it will be we who are the 'past', and people will peer back at us and consider how we lived, loved, and did the best we could in the world that was transiently ours.

Early maps offer a window into worlds which none of us now alive have known. But maps, like music, cannot be understood in a contextual vacuum. Neither one is, as the clichés would have it, a universal language. While a person raised in the highlands of Borneo might find the music of Bach to be aesthetically pleasing, he would scarcely *understand* it any better than I would understand his music, regardless of how enamored I might likewise be of its sounds, for neither of us is literate in the other's musical idiom. As with music, maps which please the senses on the simplest level are those which the casual observer usually finds most appealing, while those which require dedicated attention are often ignored altogether, and those which are 'exotic' risk being appreciated only for their bewitching mystery.

This applies both to the intellectual data of a map as well as to its metaphysical meaning, for there is a subliminal map language which is part of our collective subconscious. But even within a given culture, this language must be learned. How many Western school children, never having studied the grammar of the Mercator-projection world map that hung in their classrooms, grew up believing that Greenland is larger than the United States? Like the music of Sirens, faulty or misunderstood maps can lure travelers astray, cause devastation to ships, instigate wars, and cause untold other miseries.

Early maps present additional challenges. Just as it is not possible for ears which have known Mahler and Bartók to hear Palestrina as Palestrina heard it, it is not possible for eyes which know 'correct' geography and which harbor the modern concept of 'map' — and, indeed, the modern concept of geographic space itself — to see a primitive map as its creator did. Yet to enjoy a medieval *mappamundi* merely for its 'quaintness' is to see it only from the modern language of maps, not the language in which it was created. The challenge of our contemporary selves is to develop a frame of reference with which to see it, just as our ears have established different modes with which to hear Bach, to hear Mahler, and to hear Stravinsky.

Though use of the term 'Southeast Asia' dates only from the Second World War, the recognition of the southeastern part of Asia and its adjacent Pacific islands as a distinct region dates back to antiquity. Indeed, the earth herself created Southeast Asia with its own unique identity, both the mainland and insular regions sharing the same underlying mountain ranges, connected to, yet separate from, neighboring India and China.

The region has been known by many names, in many lands. For China, Southeast Asia lay in the 'Southern Ocean' *(Nanyang)*; for India and western Asia, Southeast Asia was the 'lands below the winds', referring to the region's relation to the monsoons. Southeast Asia was a realm of gold to classical Greece and Rome *(Aurea Chersonesus* and *Chryse)* and to India *(Suvarnadvipa)*. For Ptolemy, and in turn for Renaissance Europe, it was 'India beyond the Ganges' or 'further India' or *'Aurea Chersonesus'*; to many people of medieval Europe and India, Southeast Asia was the 'gold and silver islands'.

The definitions we will use for the regions of Southeast Asia are fairly standard. By the Southeast Asian mainland, we will mean the region comprising what is now the country of Burma to the west, continuing through Thailand, Malaysia, Singapore, Laos, Cambodia, and Vietnam to the east. The areas which are now Bangladesh and eastern India are included since they lie east of the Ganges, the ancient frontier of Southeast Asia, even though they are now culturally distinct. 'Indochina' will refer to the eastern section of Southeast Asia, which includes the countries of Laos, Cambodia, and Vietnam.

Although 'Indochina' is a misleading term coined by French colonizers and may be seen as having become obsolete after the 1954 Geneva Convention, it is a very useful term for which I know of no comparable alternative. Nor are any of the ancient names associated with the region appropriate as a substitute. Until a superior, neutral term is introduced (perhaps '*Mekonia*'?), 'Indochina' will have to suffice.

By 'insular' Southeast Asia, we span the various archipelagos from the Andaman and Nicobar Islands in the west, through Irian Jaya (western New Guinea) in the east, to Luzon in the north, and Timor in the south, encompassing what are now the countries of Malaysia, Indonesia, Brunei, and the Philippines. The only islands which are today divided among nations are Borneo, New Guinea, and — depending on one's map — Timor. Western Micronesia, though properly regarded as Pacific rather than Southeast Asian terrain, is discussed because the exploration and mapping of its islands was closely linked with that of the Philippines. Taiwan, though usually grouped with China rather than Southeast Asia, is touched on because of its role in Southeast Asia's history, being pivotally positioned between the Philippines and China. Finally, Sri Lanka will be included through the sixteenth century because of its long history of confusion with Sumatra; it is not possible to understand the early mapping of one without the other. Having a good modern atlas or map of Southeast Asia handy is recommended in order to follow the text more easily. The index also serves as a glossary for quick reference to the identity of peoples and places.

I have tried to present my subject in a way that will be useful to a wide audience — anyone who would like to explore this most inspired of earth's creations through the looking-glass of the early map. But I would like to note two specific flaws in my writing. Because I attempt to give a broad, generally continuous overview to a story which necessarily combines various epochs, cultures, disciplines, and histories, many themes have become strewn over otherwise unrelated pages. For example, the retracing of the peregrinations of such primary sources as Marco Polo and Antonio Pigafetta is scattered over several maps, according to the cartography of each. Secondly, for this book, the history of the mapping of Southeast Asia has meant for the most part the *European* mapping of the region, which in turn bestows an exaggerated role — at least through the mid-eighteenth century — to Europe's presence there. I accept these deficiencies and hope the reader will understand the book in this perspective.

Note: Wherever applicable, quotations have been changed to modern English spellings, except for place names, which when directly quoted have been left in their original form.

Fig 3 *Terra Java*, from the Vallard atlas, 1547; south is up. [Courtesy of the Henry E. Huntington Library]

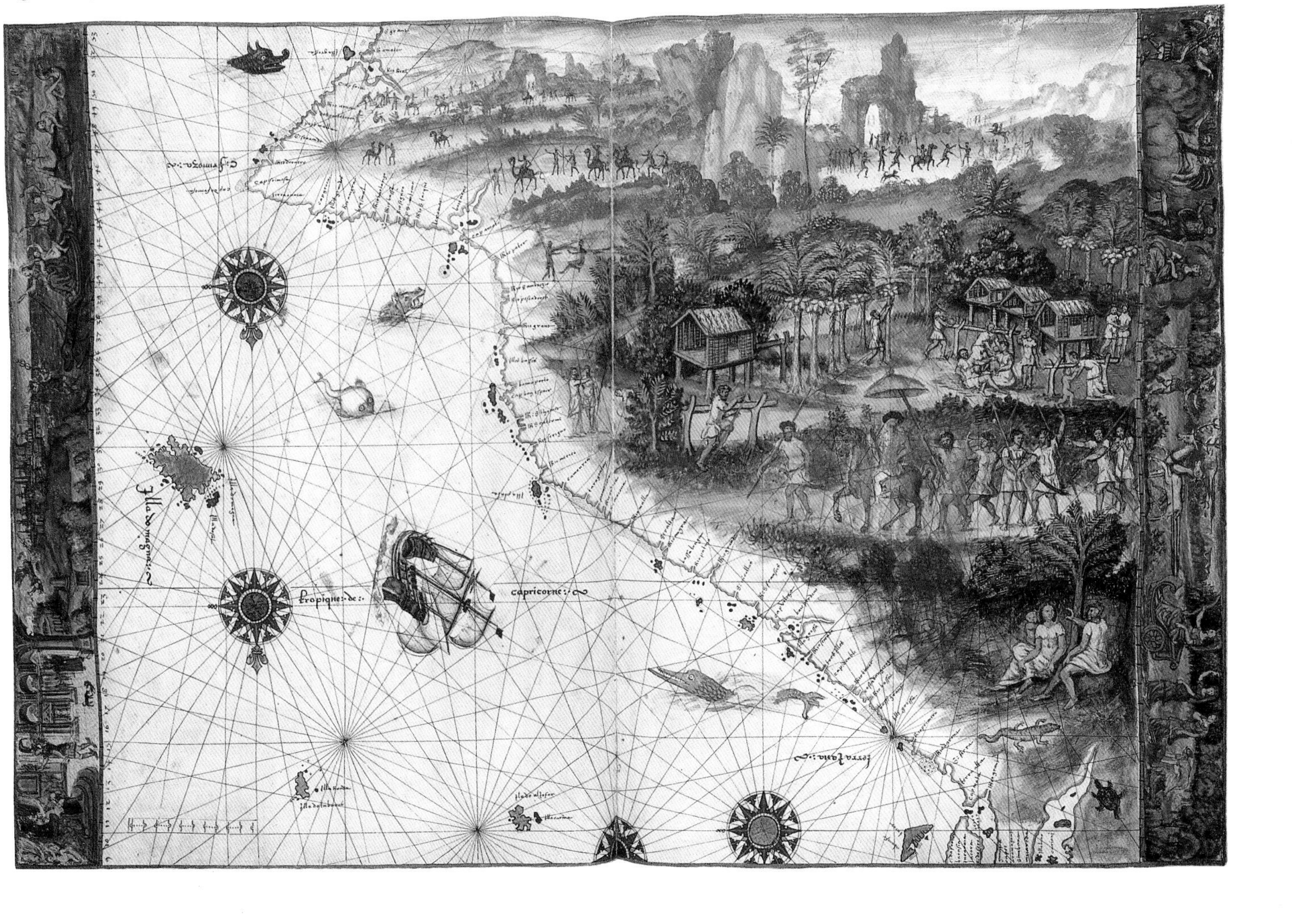

Part I
Southeast Asia

Chapter 1

The Land and Peoples of Southeast Asia

The vast world of islands, peninsulas, rivers, and mountains which is Southeast Asia was begot from the powerful union of the earth's Indian and Pacific Ocean plates. Though the term 'Southeast Asia' usually conjures up images of lush jungles and fertile rice lands, the region varies from the rolling pastoral landscape of northwest Vietnam, to Burma's dry plains, to the lower Himalayas and the snow-covered peaks of Irian Jaya. From the Himalayas sprouts a chain of mountains which forms the spine of continental Southeast Asia and which pierces the southwest Pacific to fashion the many islands of Austronesia. This immense natural threshold between the Indian and Pacific Oceans contributed to the individuality of the Southeast Asian peoples, and also to the rise of India and China as two vastly different and independent civilizations. Thus a parallel can be drawn between land and people — Southeast Asia as a geographic region is brimming with grand diversity, and it has spawned cultures which, though different in their particulars, nevertheless share many fundamental traits. Southeast Asia's geography and topography, and its position relative to India and China, have been fundamental forces in the birth and history of its many civilizations.

Rivers, alternating with mountains or forests, are a primary motif in Southeast Asian geography and in Southeast Asian life. The Sanskrit and Pali words for 'continent' are *dvipa* and *dipa* respectively, which consists of *dvi* = two, and *apa* = water, thus forming the mental image of a landmass as something between two waters, whether seas or rivers.[1] One of the native Vietnamese terms for their country is *Non nuóc*, meaning 'mountain and water.'[2] Because the shapes and courses of rivers, mountains and islands determined the movement of Southeast Asia's people, a map of daily human life in the Southeast Asian mainland and Indonesian islands can be visualized as endless 'vertical' and 'horizontal' lines. For the peninsular and inland people, life was shared with rivers and their generally north-south movement, while the trans-island or coastal travel of the people of insular Southeast Asia was largely east-west. Whereas the islands have multiple river systems, the mainland is dependent upon major river arteries and their fertile plains. The ancient kingdoms of Pagan, Angkor, and Vietnam were all rooted in fertile rice plains (Irrawaddy River plain, Tônle Sap, and Red River basin, respectively). River systems were also, as we

shall see, the arteries of political power, since the nature of Southeast Asia's topography made direct political control over distant populations difficult.

Differences in agriculture have also made their mark upon the map. Some of these differences result from adaptation to environment, such as the farming techniques which hill peoples employ to survive in the mountains of Laos, Thailand, and Burma. Others are largely cultural, producing topographical contrast within the same environment. The Vietnamese and Cambodian sides of the Mekong Delta are an example of this; the eastern side is almost entirely devoted to rice paddies producing two crops a year, while the western side is more varied and wild.

Origins and Influences

The dates, locations, and circumstances of the advent of various skills among the peoples of Southeast Asia have long been debated. It is generally agreed that the domestication of animals, rice cultivation, and pottery making were developed indigenously, and that early Southeast Asian islanders were advanced ship builders and navigators. Although Southeast Asian peoples possessed the ingredients for sophisticated civilizations from early times, urbanized societies did not become the Southeast Asian style until relatively recently.

The Bronze Age in Southeast Asia has been a contentious topic.[3] Bronze metallurgy unearthed in the northeast of Thailand is thought to be among humankind's earliest, and some scholars propose that the earliest bronze working in the world occurred in Southeast Asia. Other researchers trace the beginnings of Southeast Asian metallurgy to the southward expansion of rice cultivation in the Yangtze Valley; the bronzework then was influenced by the region's growing importance in the trading networks of China, India, and the Roman Empire. Southeast Asian Civilization, as it so invariably did with all external influences, modified metallurgical technology to its own requirements.

These early mysteries aside, external influence in Southeast Asia came mostly by sea travelers via the coastal regions, because the eastern Himalayas, which form Southeast Asia's northern frontier, discouraged migration from the north and west. The geography of Southeast Asia also worked against any great consolidation of kingdoms or regions because of the ruggedness of its mountains, rivers, valleys, and rapids, and the vast expanse of ocean over which the various archipelagoes are scattered.

The geography of Southeast Asia insulated the evolving cultures of India and China from each other; at the same time, to varying degrees the civilizations of Southeast Asia borrowed from both, adapting and modifying Indian and Chinese ideas to suit their needs. Chinese influence was strongest in Vietnam, but did not extend to southern Indochina until the eighteenth century. Southern Indochina, comprising Champa and Cambodia, was culturally more similar to the rest of Southeast Asia.

The oft-discussed 'Indianization' of Southeast Asia, which, began as early as the first century A.D. was never a blind embracing of Indian values, but rather the evolution of Southeast Asian culture drawing from the experience and inspiration of Indian civilization. So interwoven were local and Indian beliefs that sociologists have often been at odds over whether a given tradition or god was indigenous, Indian or indigenous but with an Indian name or veneer. Outlying regions, most notably the Philippines, experienced little or no influence from India because of their sheer geographic distance.

Nor was Chinese culture embraced for its own sake; even the traditionally strong influence of China on Vietnam was more limited among common people than royal institutions. For example, while the Vietnamese people adopted the use of chopsticks from China and abandoned elevated houses, raised on posts, Vietnamese music borrowed the Chinese scale structure only within the court's doors. Indeed, the failure of the early elite of the Red River Delta to create a viable state was due in part to their reluctance to incorporate indigenous traditions into their Confucian ideology.[4]

Southeast Asia's history of shaping its own civilization from imported influence is evident in the tradition, recorded by a Chinese ambassador, K'ang T'ai, in the middle of the third century, of the founding of the first known Southeast Asian state, Funan (the name being a Chinese transliteration of the old Khmer *bnam*, modern *phnom*, which means 'mountain').[5] Funan, centered along the Mekong River delta, was said to have been born from the union of a local ruler with a Brahman visitor. The foundation myth records that in the first century A.D., there was a people whose sovereign, a woman named Soma, was the daughter of the water spirit (*naga*), who lived on a mountain. One day a merchant ship arrived from a country which lay 'beyond the seas', carrying a man by the name of Kaundinya. After the man came into the woman's realm and symbolically 'drank water from the land', the two married. While the historical basis of the story will remain a mystery, the legend accurately records two of the most basic truths about the Southeast Asian psyche and Southeast Asian history: firstly, that *water* and *mountains* are a basic fabric of Southeast Asian life (the woman was the progeny of the ruler of the water realm, who lived on a mountain); and secondly, that Southeast Asian civilization, while remaining autonomous (she was the land's sovereign ruler), borrowed freely from external influences (the marriage with a foreigner), though always in its own way and on its own terms. In some versions of the story, Kaundinya was said to have been guided by a dream to board the merchant ship, and to have brought with him a magic bow taken from a temple. The *naga's* daughter was sometimes said to have attempted to raid the visiting ship, but failed. Kaundinya's 'drinking the water of the land' may have also symbolized his importing the technology for Funan's extensive irrigation systems.[6]

The mythology of the civilization that evolved in the north of what is now Vietnam tells a similar tale, acknowledging Vietnam's debt to Chinese civilization, while defying Chinese domination. The myth traces Vietnamese roots to the union of a Chinese woman and a hero, the 'Lac Lord Dragon'. Lac came from the sea to the Red River Plain of what is now northern Vietnam, rid the land of demons, and taught the people "to cultivate rice and to wear clothes."[7] He departed, promising to return if he was needed. When a Chinese monarch came south and tried to subjugate the people, Lac was summoned. He kidnapped the invader's wife, whose name was Au Co, hid her atop a mountain overlooking the Red River, and the Chinese monarch, unable to find her, returned home in despair. A son was born to Au Co and Lac Long Quan, , thus beginning a new dynasty.

A story recorded in Chinese histories originating in the early seventh century records Indian influence by royal marriage in the

Fig.4 The bow being held by this mythical viol player is an early example of a South or Southeast Asian invention being recorded on a European map. The bow reached Europe via the Islamic world and the Byzantine empire in about the tenth century. Although its ultimate Asian origins are not known, the 'spike fiddle' (Indonesian *gending*) was known in Java no later than the eighth century. From a map of the north Atlantic by Ortelius, 1570.

Fig. 5   Javanese musicians at the Banten market, Lodewyckszoon, 1598 (Theodore de Bry, 1599). Lodewyckszoon explains that the gongs are used "to sound the hours, and play their music . . . as well as to summon people in the king's name, which they did when we arrived, to announce that anyone could buy and sell with us." (14 x 17.5 cm)

kingdom of *Langkasuka* (Patani, southern Thailand). During a period of *Langkasuka's* decline, a virtuous man loved by the people but exiled by the king fled to India, where a king gave him his eldest daughter in marriage. When the king of Patani died, the exiled man, accompanied by his Indian wife, was welcomed home as the new sovereign.[8]

Arts and Daily Life

Many generalizations about early Southeast Asian peoples have been made from contemporary accounts left by visitors and from the ethnographic evidence. Most people of Southeast Asia built their houses elevated on posts, whether it was to seek protection against floods, insulate themselves from predators, or to benefit from the body warmth of their livestock, which were stabled underneath during the cooler nights.[9]

Broad generalizations can be made regarding food. Rice was the fundamental ingredient upon which the Southeast Asian diet was based. Throughout most of Southeast Asia, rice was harvested with a finger-knife by women. Fish was a staple, with coastal and inland peoples trading their salt-water and river fish with each other for greater variety. Neither meat nor milk products were substantial

parts of the Southeast Asian diet. Fondness for chewing betel nut, which required the areca nut itself, betel leaf, and lime, was universal in Southeast Asia, and remains so in many areas today.

The music, dance, theatre, and other arts of Southeast Asia are all part of one extended family. That song was a part of everyday life, with the common people singing during their daily tasks, was striking to Europeans. Francisco Alcina, visiting the Philippines in the seventeenth century, claimed that "rarely can a Visayan man or woman be found, unless he is sick, who ceases to sing except when he is asleep." Basic similarities between their musical instruments, the masks and puppets of their theatre and rituals, and in the body movements of their dance were noted, though there were marked regional distinctions in all of these, for example in the scale structure of their music. The gong-chime was common to all peoples of Southeast Asia; neolithic stone slabs, tuned to a seven note scale, have been unearthed in Vietnam, and bronze kettle-drums date back at least two millennia in Indonesia. Although throughout Southeast Asia scales were usually based on a five or seven note system, the actual tuning varied between ensembles, with groups accurately maintaining their own individual calibration as if it were their distinguishing signature.[10] In Indonesia, a particular tuning could be considered rightful property and could not be used by other musicians.

Fig. 6 Javanese dancers, Lodewyckszoon, 1598 (Theodore de Bry, 1599). Lodewyckszoon described how the dancers gently swayed their bodies to the rhythm of the music, their arm and leg movements synchronized, the dancers never leaping in the air. The Europeans also noted dancers' subtle finger movements which are captured in the engraving. The man on the right accompanies them on a metal xylophone or *demung*. (14 x 17.5 cm)

Along with theatre, dance, and music, people's bodies were a medium of art, the splendor of which was an important aesthetic preoccupation. Up until the modern era, a person's hair was felt to be an integral part of his or her physical and spiritual self, and hair styles varied little between the sexes.

Such practical arts as pottery and textile production were at once fairly uniform throughout Southeast Asia in terms of the technologies employed, but individual in terms of specific artistic treatment. Both these vital industries seem to have been exclusively the domain of women and, as with early cultures nearly everywhere, are examples not of sexual inequality in the modern Western notion, but rather of a division of labor and responsibilities according to gender.[11]

Gender

Early European observers did, in fact, remark on common traits regarding the relations between men and women in Southeast Asia and perceived gender roles and relationships — albeit often over simplified and sometimes even patronizing — have been a defining aspect of the region's image in the eyes of European observers.[12]

Southeast Asian daughters were generally welcomed into the world as happily as sons, and were for the most part more empowered than their European, Arab, or Asian sisters, enjoying a status which rivaled, if not directly equalled, that of men. Europeans were surprised to find that Southeast Asian women were assertive in seeking partners, openly enjoyed sex, and even expected men to endure considerable fuss to fulfill their sensual pleasures (most famously the practice, apparently unique to Southeast Asia, of men undergoing the painful implantation of metal or ivory balls into their genitals for the benefit of their partner). Men, rather than women, presented dowries to their fiancées' parents, and newly-weds commonly went to live with the wife's, rather than the husband's, family. Such generalizations even held for the common people of Vietnam, who are so often the exception to the rule in Southeast Asia.[13]

In about 750 A.D., a Mon princess named Jam Thewi became the first ruler of Lamphun (then Hariphunchai), which was the first literate, sophisticated kingdom of Lan Na civilization (in what is now northern Thailand). Tradition records that Jam Thewi was selected by a *rishi* (hermit); when the *rishi* founded Lamphun, he wanted its first ruler to be an offspring of the ruler of Lop Buri but was indifferent as to the person's gender.[14] The civilization of Champa, in the central region of what is now Vietnam, was a matriarchal society. The major entrepôt of Patani on the east coast of the Malayan Peninsula was under the sovereignty of successive queens for over a century (1584-1688), and Bugis kingdoms were often ruled by women, as was the important north Sumatran kingdom of Aceh.

Europeans also noted gender deference. The first Dutch expedition to Java (1596) reported that if the king of Banten sent a male messenger to request the presence of "any subject or stranger dwelling or being in his dominions," the person "may refuse to come; but if once he send a woman, he may not refuse nor make no excuse." When the Portuguese traveler Mendes Pinto was in Banten in 1540, a woman of nearly sixty years of age arrived on a diplomatic mission. According to Pinto, she was paid the highest honors, and it was "a very ancient custom among the rulers of these kingdoms, ever since they began, for matters of great importance requiring peace and harmony to be handled through women." In the seventeenth century, Simon de La Loubère, a Frenchman resident in Siam, reported that "as to the King of Siam's Chamber, the true Officers thereof are Women, 'tis they only that have a Privilege of entering therein", and many Southeast Asian countries celebrated female war heroes in their histories.[15]

Many visitors commented on the authority of women in matters of trade. Chou Ta-Kuan, a Chinese envoy who visited Cambodia in the late thirteenth century, recorded that foreign men who set up residence in Angkor for the purpose of business would find themselves a native spouse as quickly as possible to assist in their commercial affairs. William Dampier, resident in Tongkin (Vietnam) in the late seventeenth century, noted that it was the women who managed the changing of money, and that marriage would establish an alliance between foreign merchants who returned annually and local women with whom they entrusted money and goods.

The Geography of Kingdoms and War

Whereas the modern nation-state is defined by its borders, the traditional Southeast Asian kingdom was defined by its center. This philosophy of geo-political space parallels the *mandala*, or 'contained core', a sacred schematic of the cosmos in Indian philosophy.[16] The kingdom was the worldly *mandala*, defined by its central pivot, not its perimeter. By moving closer to the center of the *mandala*, that is, to the center of the kingdom, one moved closer to the sacred core, traditionally a central temple complex, where spiritual powers and the fertile earth joined. This central "temple mountain" was analogous to a sacred mountain believed to form the center of the world, Sumeru (or Meru), and thus the kingdom was, in effect, a miniature cosmos.

As in feudal Europe, a Southeast Asian kingdom was an array of imprecisely defined spheres of influence, typically consisting of the king's immediate territory, over which he had total control, followed by a succession of further and further removed regions from which he might exact tribute and over which he exerted varying degrees of authority. Beyond these would be outlying regions that had their own monarch but which were not entirely autonomous. These regions might be accountable to one or more larger kingdoms, being obliged to pay tribute and never to act in a manner contrary to the large kingdom's interests.

The further one traveled upriver or upcountry from the pivot-point of power, the more authority faded to various shades of leverage, cooperation, tribute, and influence. There were usually patches of light and dark in the shading; a small, peripheral kingdom might have a particular reason to partially submit to a more distant superpower, most commonly for the protection such an arrangement might offer the petty state against a closer, less benign, power. Insular Southeast Asia's heavily mountainous topography undermined the ability of lowland polities to control upriver regions even more than on the mainland. Perhaps as a result, early insular Southeast Asia was typified by a spread of societies which were roughly 'equal', while the mainland kingdoms tended to see dominant majorities ruling over minorities.

Rulers secured a gradually diminishing degree of hegemony and influence over the upland regions by controlling the lower reaches of river arteries (this was especially true in the case of insular Southeast Asia) or the rice-fertile lowland regions (more so in the mainland than among the islands). A map of the river was, in effect, a barometer of sovereignty. Historians have drawn clear parallels between the control of rivers and irrigation canals, and the rise of despotism in Southeast Asia.

Southeast Asian peoples retained this classical concept of political space until profound European influence on their internal affairs required them to abandon it in the nineteenth century. Sometime prior to this, a new, radical concept of political space and boundary had evolved in Europe in which nations were defined by their perimeter, kingdom *x* lying on one side of an authoritative imaginary line, kingdom *y* on the other. In this new geo-political mind-set, all areas of a kingdom were equally part of that kingdom; the final grain of dirt before the border was as much the property of its kingdom as was its center.

Whereas European eyes presumed that a country's possessions extended as far as its border with its neighboring country, in Southeast Asia there were usually spaces in-between, 'empty' land, which was not part of any kingdom and which sometimes served as a neutral buffer. And while the European boundary formed an invisible wall that was to be guarded lest anyone attempt to violate it, the Southeast Asian border was porous, and was not intended to keep people either 'in' or 'out'. Even a wall built in Vietnam in 1540 to separate rival factions of that country, like the Great Wall of China, was doubtfully perceived as demarcating a precise division of territory.

These differences in the concept of statehood confounded political understanding between Southeast Asian kingdoms and encroaching European powers. Since Thailand alone was never colonized, and therefore Thailand alone negotiated as a sovereign kingdom with the West, clashes of these opposing concepts of political geography are revealed best in nineteenth century British-Siamese relations.

In determining their own political geography, the people of Southeast Asian were, sadly, no different from the rest of mankind. Cultural differences certainly existed — the Hindu-Buddhist Pyus, who lived in the region of what is now Prome (Burma) between the fourth and eighth centuries, appear to have created a truly pacifist society — but, in general, a map of the power struggles in early Southeast Asia would show an ever-changing kaleidoscope of kingdom pitted against kingdom, with fickle alliances of convenience, and the scars of warfare distributed equally among the many nations vying for possession or tribute. The situation was exacerbated by the swords of foreign nations, which altered the map both by outright conquest as well as by manipulating local rivalries between native kingdoms and their neighbors. Contemporary descriptions of local warfare could not be more horrific. So ghastly was inter-kingdom violence that one nineteenth-century British observer commented (though disingenuously) that it was "the absence of pity, which distinguishes the Oriental as opposed to the Occidental."[17]

Religion

Southeast Asian religions represented various combinations of animism, Hinduism, and Buddhism, with Islam reaching many coastal regions by the time the first Europeans arrived. The supplanting of indigenous animism and ancestor worship in Southeast Asia coincided with, and facilitated, the rise of larger social organizations and states. Hinduism, the dominant religion in Southeast Asia until roughly the twelfth century, spread through civilizations along the Gulf of Siam, Cambodia, Vietnam, and parts of Indonesia beginning about the first century A.D., and gradually permeated inland. Ancient edifices in Java and Bali, the most extraordinary of which is Borobudur in central Java (ca. 800 A.D.), demonstrate how far east Indian influences reached. The beginning of Hinduism's decline coincided roughly with the completion of Angkor Wat in the first half of the twelfth century, being gradually supplanted by Buddhism.

The successive waves of religions that came to Southeast Asia never entirely replaced those which preceded them, but rather built layers of combined beliefs. Indigenous animism and ancestor worship can still be discerned throughout Southeast Asia, and elements of Hinduism are still easily visible in regions influenced by Indianization. Bali alone has remained a truly Hindu society, albeit one much modified and informed by Buddhist and local traditions.

Buddhism arrived in Southeast Asia hand-in-hand with Hinduism. Two branches of Buddhism prospered: the older Theravada (or Hinayana) Buddhism, known as the 'Way of the Elders' or 'Lesser Vehicle'; and Mahayana Buddhism, or 'Great Vehicle'. Mahayana Buddhism, which regarded trade more favorably than did Theravada Buddhism, was the first to reach Southeast Asia, brought by the Indian merchants who practiced it. The older Theravada school, however, became the predominate faith, coming later by way of Sri Lanka. Except for Malaysia and southern peninsular Thailand, which are predominantly Muslim, Buddhism is still followed by the overwhelming majority of people in mainland Southeast Asia. Al-Idrisi, a twelfth-century Sicilian geographer, gave an impression of religious life in Indonesia based on the reports of earlier travelers which had reached him:

> The prince is called Jaba, he wears a chlamys and a tiara of gold, enriched with pearls and precious stones. The money is stamped with his portrait. He shows much respect to the Buddha. [The king's temple] is very beautiful and is covered externally with marble. Inside and all around Buddha, can be seen idols made of white marble, the head of each adorned with golden crowns. The prayers in these temples are accompanied by songs, which take place with much pomp and order. Young and beautiful girls execute dances and other pleasing games, before the people who pray or are in the temple.[18]

Islam took root in Southeast Asia through a combination of patience and adaptation. Though extensive Arab contact with Southeast Asia predated Islam itself, and though the Chams of Vietnam adopted Islam in the tenth century, it was not until the end of the thirteenth century — about the time Marco Polo skirted Southeast Asia en route back to Venice — that the faith came to be a major influence in Southeast Asia's coastal regions. It was at first Indian Sufi intermediaries, rather than the Arabs resident in Java and Sumatra, who initiated the acceptance of Islam in the region, which occurred only after certain adjustments had been made in relation to existing cultural orientations. Once established, the spread of Islam accelerated through marriages between the daughters of wealthy Indonesian merchants and Islamic Indian traders, combined with commercial policies which favored Muslims. Like Hinduism and Buddhism before it, Islam accommodated itself to indigenous Southeast Asian values rather than dictating them, and many aspects of Indonesia's pre-Islamic culture, have survived intact to this day.

Christianity, the most recent arrival in Southeast Asia, has had little success on the mainland, even by the Church's own figures, which include token 'converts' in the faith's excellent schools. Even the French failed to convert Indochina to Christianity, or to displace indigenous language and culture, despite their profound and extended presence in the region.

Christianity's unique successes are found in the islands. The Portuguese established Catholicism in East Timor, just as the Spanish did in the central and northern Philippines. Many of the Filipinos combined their former beliefs with the new creed of the Europeans, a practice which the Spanish, who exacted a tax from the islanders to pay for the Church's proselytizing, tolerated.[19] The animism of pre-Spanish central and northern Philippines, involving a deity called Bathala and spirits such as *anitos* and *ninos*, was eventually overshadowed, but not lost.

Conversions were possible in these regions because the indigenous societies had less centralized control, in contrast to those communities which had already embraced Islam. The Spanish, from the beginning, were aware of the distinction: in 1521, Antonio Pigafetta, an observer on the Magellan expedition, wrote that they had burned down a 'pagan' Philippine village whose people had refused conversion, and that they had planted a wooden cross on its site, but that if these Filipinos had been 'Moors', Magellan would have "set up a column of stone . . . for the Moors are much harder to convert than the pagans."

While the failure of Christianity to dominate the map of insular Southeast Asia may be a result of the resolve of the Islamic faith, the tenacity of Buddhism on the mainland appears to have come from its unusual attitude toward religion. The faiths of conquerors and colonizers have after many centuries made only token inroads into these countries, yet these were the very societies that generally did not maintain their beliefs through force. Early European visitors to Southeast Asia frequently commented on the Buddhists' freedom from proselytizing, Mendes Pinto poetically observing that the king of Siam considered himself "master of men's bodies, not their souls." Rather than these people having been the easiest converts, it seems that freedom of choice and non-coercion have kept the religious map of mainland Southeast Asia in this respect unchanged since pre-modern times. Jacques de Bourges, a French missionary in Siam in 1662-63, already understood this paradox, noting that "it is this pernicious indifference [to religious value judgements] which stands as the greatest obstacle to their conversion." When a Siamese man was beaten by a Portuguese for laughing during a Christian ceremony, the king refused any sympathy to his subject, noting only that "he should not another time be so intolerant." Our French observer wrote that he

> sometimes enquired why the King of Siam made himself so lenient in permitting . . . so many religions, since it is a received maxim of the most esteemed politics [i.e., the French Court] that only one be permitted, for fear that, should they multiply, the diversity of beliefs would cause spiritual friction, and so lead to conflict.[20]

He was told that, to the contrary, Siam derives great benefit from the diversity of beliefs, both for the arts and commerce. Further, the Frenchman reported that

Fig 7 Quintessential features of the Southeast Asian landscape: the village of Pilar, on the road from Balanga to the Marivelles Mountains, Philippines. After a drawing by M.E.B. de la Touanne, from the voyage of Baron de Bougainville. Published in Paris, 1828. [From *The Philippines in the 19th Century*, Rudolf J.H. Lietz, Manila 1998. Courtesy of Elizabeth and Rudolf Lietz]

there is another reason for this conduct; this is the view which is held by the Siamese that all religions are good, which is why they show themselves hostile to none.

In Burma in 1710 Alexander Hamilton, a Scottish sea captain who passed much of his life in Southeast Asia, observed that the (Buddhist) monks of Pegu (Burma)

are so benevolent to mankind that they cherish all alike without distinction for the sake of religion. They hold all religions to be good that teach men to be good, and that the deities are pleased with variety of worship, but with none that is hurtful to men, because cruelty must be disagreeable to the nature of a deity: so being all agreed in that fundamental, they have but few polemics, and no persecutions, for they say that our minds are free agents, and ought neither to be forced nor fettered.

Until Siam expelled most foreigners in 1688, the freedom given Christian missionaries to roam the kingdom and preach to the people was in fact misinterpreted as meaning that the king himself was ready to embrace the new faith. The exception was Vietnam, which though officially tolerant did endure infamous incidents of anti-Christian violence in the seventeenth century. Later violence against missionaries in Vietnam contributed directly to France's pretext for intervention.

That Christian missionaries, having left behind a Europe which was tearing itself apart with petty religious rivalries, were routinely perplexed by intractable Buddhist patience is particularly salient in seventeenth-century Cambodia and Laos. The missionaries' message was received with a mixture of tolerance and indifference which was even more impenetrable than the forests they had braved to reach their intended converts. Such an attitude, as modern historians have often observed, afforded the West neither conversion nor martyrdom. The "zeal of the pious and learned missionaries," the French emissary Nicolas Gervaise lamented in Thailand in the late seventeenth century, had failed to negate "the errors that have tainted this people for so many centuries." In our own day, the pressures of totalitarian rule have happily likewise failed to substantially change the metaphysical map of Burma, Laos, and and other parts of Indochina.

Colonialism

Unlike the religious map of Southeast Asia, the *political* map of the region was most certainly shaped by foreign powers. Even Thailand, the sole country in all of Southeast Asia never to have known the colonial yoke, was profoundly influenced by the colonization of its neighbors. Early Southeast Asia was not divided into the major countries we know today, but consisted of numerous smaller kingdoms as in contemporary feudal Europe. But kingdoms which once fought amongst themselves were gradually consolidated according to the interests of their colonizers; the borders established during the colonial era are essentially those that exist today, having changed little even after the treaties drafted to repair the chaos of the Second World War.

The savvy of Thai monarchs has often been credited for their country's success in escaping colonialism and in keeping the forces of Emperor Hirohito 'at bay' during the Second World War. While this is a fair compliment, Thailand's unbroken independence was also the result of sheer geographic luck; as we shall see, the kingdom was spared in part because it buffered the competing claims of England and France in Burma and Indochina, and because the strange course of the mighty Mekong River, which in Europe's eyes defined Thailand's eastern boundary, was wildly misunderstood until the late nineteenth century.

Spanish conquest unified the Philippines as a single country, though the archipelago's many islands encompassed diverse peoples, traditions, and religions; the southern Philippines remain largely Muslim, as they were at the time of Spanish arrival four and a half centuries ago. The various peoples and cultures which today comprise the modern country of Indonesia were first placed under one helm as a corporate entity — for what we know as Indonesia was originally a creation of the *Vereenigde Oost-Indische Compagnie* (the Dutch East India Company, or V.O.C.), not of Dutch colonialism *per se*. By the early twentieth century, Dutch colonial rule, which grew out of the V.O.C.adminstration, encompassed every one of the vast and varied shores stretching from Sumatra to western New Guinea, save for the little Portuguese colony in eastern Timor. Burma is composed of several long-rivalrous kingdoms that were joined together under the British Empire; large minorities to this day do not acknowledge the jurisdiction of Rangoon over their lives. The country of Laos was a French creation, concocted from several petty states caught between occupied Vietnam and independent Thailand. Malaysia was sewn together from various sultanates on the peninsular main-land and from the Bornean territories of the former Brooke Raj (Sarawak) and the North Borneo Company (Sabah). The colonial entrepôt of Singapore became part of post-colonial Malaysia and was then reborn as the independent Republic of Singapore. In a sense, one might see the modern nation of Singapore as the descendent of the Srivijaya empire, which dominated maritime trade through the Singapore-Malacca Straits a millennium ago.

The modern country of Vietnam was also the progeny of colonialism. But, although the tragic war between North and South Vietnam is still fresh in the memory of many of us, bloody intra-Vietnamese conflict dates back to well before French occupation, and the history of Vietnam's divided map began far earlier than the temporary division established at the Geneva Convention of 1954. In fact, the precedent to split Vietnam dates back to 1540, when a wall separating the warring northern and southern regions of the country was constructed at about 17° north latitude by the Chinese. The Japanese occupation of Southeast Asia during World War II, and the Indochina Wars that followed, marked the transition of Southeast Asia out of the colonial era.

Continuing Change

The map of Southeast Asia remains challenged by internal influences. Separatist movements in the southern Philippines and in southern peninsular Thailand, in both cases drawn along religious lines, would like to transform the maps of those countries, while secular ideologies have altered the map in Indochina. The present rulers of Burma have abandoned their country's 'colonialist' name in favor of *Myanmar*, though many people continue to use the older name because the use of 'Myanmar' could be construed as a legitimization of their power (the acknowledgment of sovereignty is, of course, one of the oldest functions of maps and their nomenclature). The island of Timor in Indonesia, which a century ago was split in half to settle competing Dutch and Portuguese claims, finds its niche on the map still being contested. Although western (Dutch) Timor has been part of the country of Indonesia since 1950, East Timor remained a Portuguese colony until 1975, when it was annexed by Jakarta. In 1999, however, economic pressures made Jakarta reconsider whether sovereignty for East Timor might be in Indonesia's interest. On the high periphery of Southeast Asia, China has forcefully asserted its long-standing claims to sovereignty over Tibet. The Paracels, seemingly inconsequential islands which pierce the vast South China Sea to the delight of air travelers with window seats, were seized by China in 1974 at the close of the Vietnam War. As the harvesting, mining, and drilling of the seas becomes as important as that of the land, possession of the Paracels threatens to become a destabilizing issue in the region.

Social and environmental upheavals are also changing the map of Southeast Asia. A dramatic redistribution of traditionally rural societies to urban areas has created mega-cities unprepared to support their people. Poor rural folk learn via modern means of communication and the media that cities, despite their often overwhelming poverty, are also zones of hope and change. This is being exacerbated by radical changes to the topography: the reckless deforestation of vast regions has rendered many local environments unable to sustain their communities. A case in point of this syndrome is the Isaan (northeast) region of Thailand, where land clearance policies enacted decades ago to expand the agricultural frontier, as well as the teak trade, have laid to ruin the forests which since time immemorial had moderated the extremes of the natural dry and rainy seasons. Severe drought, alternating with ruinous floods, is now the norm, leading many of the traditionally poor people of the northeast and other regions to try their luck in the cities, despite the difficulties of finding work and accomodation among the swelling ranks of the country's urban poor.

Some efforts are being made to better distribute population and resources. In a policy reminiscent of the homestead laws enacted in the United States in the 1860s, the government of Indonesia has offered people of Java and Bali a plot of land in Kalimantan (Indonesian Borneo) or Irian Jaya (western New Guinea), complete with a one-way plane ticket and a year's supply of rice. Although this policy of transmigration (as it has come to be known) is done at the expense of further cutting of the rain forest and the cultural isolation of the volunteers, it nonetheless has succeeded in offering a new beginning for many families disenfranchised by the rapid changes of the past few decades. If this redistribution of people is successful, future maps may show much of New Guinea as ethnically Malay. But, for the moment, Irian Jaya is an example of the sometimes awkward incongruity between the political and physical map of Southeast Asia, for many of western New Guinea's indigenous people are only casually aware that their home lies on the map of Indonesia.

Chapter 2

Southeast Asian Maps and Geographic Thought

Moreover, they [the Thais] submitted a map of their own country.
— Ming Annals recording Thai ambassadors in China, 1375.

In order to examine early indigenous Southeast Asian maps, we first need to ask: What is a map? We will use a fairly broad definition: a map is a spatial representation of a place, thing, or concept, actual or imagined. Note that the subject of the map is not restricted. A map can chart the path to a neighboring village, to a successful endeavor or a fortuitous event, or even to the next life; it can illuminate the relationship between various levels of existence or consciousness, or between a previous or future age of the earth. Whether a monk charting the metaphysical, a king illustrating the divine link he shares with the gods, or an ordinary person inspired to scratch out a plan of her paddy in the moist earth for the sheer pleasure of doing so, our definition lays down no parameters for the medium used; a map need not even be of a material nature.

Indian Influence

Until the arrival of Islam, Southeast Asian ideas of man's place on earth were very much influenced by cosmological considerations that ultimately came from India. These external influences, which were themselves the result of a co-mingling of Hindu, Buddhist, and Jain beliefs, conceived the earth as a small part of a vast cosmos, according it a far less central role in Creation than in Christian doctrine. The desire to understand the cosmos was a natural concern since it was simply an extension of knowledge about one's own immediate environment. The nature of the cosmos became even more significant when Buddhism took hold in Southeast Asia, and 'other-worldliness' began to play an increasingly important part in formalized cultural practices. The Buddhist concept of release from the cycle of birth and death, for example, rested on a far-encompassing view of the cosmos.

Jain thought stressed a spatial boundlessness in which the universe contained both known and unknown universes.[22] Hindu traditions stressed the infiniteness of time, illustrated by the concept of *mahakal,* the 'time which lies beyond knowable time', while space was often bounded to reflect the spheres of influence *(kshetra)* of various manifestations of the Hindu concept of god. For followers of the faith, the concept of boundlessness imparted a sense of profound wonder and humility; it may also, on occasion,

have assisted those responsible for preserving and propagating the faith to explain such problematic logistical questions as the whereabouts of gods and deceased souls.

Early Indian instructional texts, compiled over a period of several hundred years and known as the *Purana*, were principally devoted to the mysteries of the creation of the universe, rather than the genesis of humankind on earth. This less mortal-centric perspective probably contributed to a greater emphasis on metaphysical mapmaking in early Southeast Asia than in the West, where a philosophically-engendered geography, as epitomized by 'T-O' maps or *Terra Australis,* used the earth rather than the cosmos as its vehicle.

Indian and Southeast Asian thought regarding the actual Genesis was typical of what appears to have been a nearly universal concept. As with the Judeo-Christian and Islamic traditions which came later to Southeast Asia, Indian cosmogony envisioned the Creation as beginning with a formless seed material; the allegory of an egg, with its yolk surrounded by amorphous embryonic fluid, was a common and natural expression of this thought. This primordial matter gradually assumed the anatomy of the universe (or universes) through the combination of elemental materials formed in the first stage, as well as through the intervention of various agents and natural laws. Eventually, the original fertile substance became articulated with the stars, planets, and all other components of Creation.

In the pre-modern era, such a cosmography, with regional variations and nuances, was accepted both in Asia and in the West. In the West, it gradually lost favor when empirical testing and scientific methods became a new gauge of truth, but in Southeast Asia it remained until the pervasive Western influence of the latter nineteenth century. If the importation of Indian beliefs into Southeast Asia began as a means of legitimizing an Indian-style political system, then Indian influence may have to some extent dampened empirical cartography in Southeast Asia.

Indian-derived Southeast Asian thought envisioned a large cosmos with many universes. Earth and its universe was pivoted on Mount Sumeru, an axis-mountain of fabulous proportions in the Tibetan Himalayas or Central Asia, which we will see more of when we look at Southeast Asian cosmological maps. Water,

Fig. 8 Borobudur, from Thomas Stamford Raffles, *The History of Java* (1817); the cosmic mountain axis, architecturally realised in stone.

mountains, and continents grew from this mountain-axis; the continents were arranged symmetrically like the petals of a lotus blossom, or as concentric circles of alternating seas and continents. The inhabited earth, called *Jambudvipa,* lay at the centre of this scheme of things, or otherwise constituted the southern 'petal', and contained the *Bharatvarsha*, which was the traditional territorial reach of Hindu, Buddhist, and Jain cultures. In the nearest sea, called 'Salt Ocean' (*Lavansagara*), there was a continent, *Angadvipa* (*dvipa* = 'land with water on two sides, or continent'), which may have represented Malaya, and *Yamadvipa*, which may have been Sumatra. Another 'continent' which is frequently cited in popular Hindu, Buddhist, and Jain writings is *Suvarnadvipa,* identified with Indochina, Sumatra, or Southeast Asia in general. In one Thai world view which evolved from this imagery, the earth was a flat surface divided into four continents separated by unnavigable seas, the whole of which was encompassed by a high wall on which all the secrets of Nature were engraved; holy men agilely transported themselves to these walls to learn from its inscriptions.[23]

The sense of humility imparted by concepts of temporal and/or spatial boundlessness nurtured a relatively equitable view of the worth of outside realms. This extended fully from other universes and other worlds, to foreign territories and cultures. In fact, as one moves outward from the *Jambudvipa* [Indianized sphere] of Indian /Southeast Asian cosmology, the various *terrae incognitae* one encounters are not forbidding, but rather become progressively more sublime and idealized (though never reaching the status of the various heavens, which lie vertically 'above'). As a result, the Southeast Asian perspective of other peoples, and even bloody Southeast Asian conquest, were rarely characterized by the chauvinism of the 'Middle Kingdom' notions of the Han Chinese, or of the West's propensity for dividing the world into 'civilized' and 'savage' nations.

Earth and Geography

Throughout pre-modern Southeast Asia, the earth was presumed to be flat. In most regions this belief did not begin to fade until well into the nineteenth century, and in rural areas it could still be found even into the early twentieth century. Acceptance of the earth's sphericity had to overcome religious objections, since Buddhist scriptures contained readily-perceived contradictions to a spherical world that needed to be interpreted anew before the flat earth model could be respectfully discarded. A somewhat ironic parallel might be drawn with Medieval writers in the West who used the Bible to debunk the established Greco-Roman view of a spherical earth, although the flat earth was a minority belief in medieval Europe. Both Christianity and Buddhism rationalized discrepancies between canon and science in similar ways: the scriptures were meant to be taken literally only when it came to matters of spiritual truth; details of natural science are revealed figuratively and allegorically. Perhaps Buddha knew that the people to whom he preached were not yet capable of understanding such fantastic notions as a spherical earth, so it was better that he left such spiritually irrelevant matters aside. To address them would only have distracted his followers from more important metaphysical truths.

The mechanical workings of the universe were rationally envisioned within the context of a flat earth. A representative Thai cosmographical text, the *Traiphum*, described a path between two mountain ranges through which the stars, planets, moon and sun pass "in an orderly fashion", thereby facilitating the calibration of time and the growth of astrological knowledge. In the words of the *Traiphum*, the celestial objects' flight through the valley "enables us to know the years and the months, the days and the nights, and to know the events, good and bad." This

Fig. 9 Northern Thai map combining an itinerary relating to religious sites in India with cosmological concepts.
(No date, but probably a twentieth century copy of earlier maps). [Courtesy of Cornell University Library]

approach to explaining the mechanics of the universe is analogous to a medieval European concept of a mountain which facilitated night and day by forming a partition behind which the sun and moon disappear in the course of their travel.[24]

Indigenous Southeast Asian thought imparted a consciousness of one's physical orientation on earth and in the universe, as regards the cardinal directions, as well as such semi-metaphysical concepts as earth and sky, inside and outside (one's abode), upstream (or upmountain) and downstream (or downmountain), or towards and away from the center of one's kingdom or territory. Later, Indian cosmological precepts were readily assimilated by Southeast Asian peoples, Sumeru becoming the anchor of the cardinal directions and the *mandala* symbolizing the kingdom and its center. As with other Asian civilizations, such concepts were often paralleled with male and female attributes.

For the typical peasant working the fields and harvesting the rivers, the 'world' probably amounted to such immediate concerns as the itinerary from home to field to market — as indeed it still does. The larger view was a more abstract, cosmological and spiritual matter. In this respect, the ordinary Southeast Asian was again, not so very different from his or her European contemporary. Mariners sailing on the open seas, boatmen shuttling up and down a river route, and hill people collecting produce for a central market, all perceived quite a different 'map' of the world. Those societies which were situated on major commercial arteries — for example the cosmopolitan Srivijaya kingdom which dominated trade through the Malacca Strait and along the coasts of Sumatra and Java from about 800-1300 A.D. — were most likely to have a more sophisticated

understanding of the world. At a later date, those regions which accepted Islam might have inherited the Arabic cosmographical tradition, and by the seventeenth century some Islamic courts — notably Makassar — had already solicited and zealously studied European geographical texts and other scientific works. Conversely, other hand island civilizations, such as the Balinese, preserved a very introspective world view.

Extant Buddhist literature from Lan Na illustrates how the world view of peoples of the inland regions was geographically narrow. The authors of these records were primarily interested in the events of their own region, their own village, and their own monastery. Distant places entered into the archives only when specifically relevant to an event at home, for example if the inspiration for the founding of a local school or temple originated elsewhere. Lan Na records — the oldest of which are Mon inscriptions on stone dating from the early 1200s — do not record the history of the local Lawa people, nor any neighboring Burmese kingdom, nor China, nor the Thai kingdoms of Sukhothai or, later, Ayuthaya.[25] On the other hand, Wat Jet Yot, the temple of the seven spires in Chiang Mai, symbolically maps seven places in distant India, because these locations are holy sites that relate directly to the life of Buddha and Buddhist doctrines. Similarly, a map from Lan Na (fig. 9) shows the temple of Bodh Gaya, the place where the Buddha attained enlightenment (represented here by a temple within a box, with several roads branching out from it) and the location of other important Buddhist sites in relation to it. It is both a cosmography and an itinerary rolled into one, with distances marked in either travel time or linear measurements.

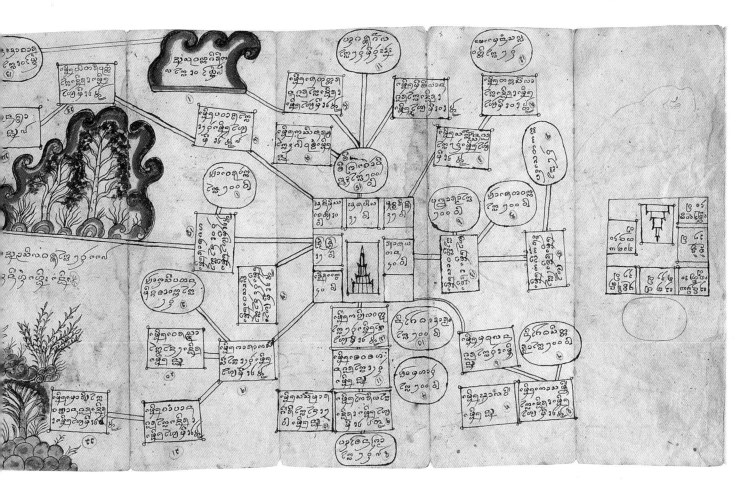

In early Southeast Asia, there was no absolute distinction between the physical, the metaphysical, and the religious; although something might be either predominantly sacred or predominantly profane, essentially abstract or primarily physical, such concepts blurred together at the edges. Southeast Asian geographic thought, like Southeast Asian life, could be at once, both empirical and transcendental. However, as in the case of other seemingly exotic characteristics of pre-modern Southeast Asian cosmographic thought, this oneness of the mundane and the fantastic was typical, not extraordinary, in the medieval world.

Astrology, along with its necessary ingredients, astronomy and mathematics, flourished in ancient Southeast Asia, just as it did in Europe. In turn, the study of mathematics and astrology connected with celestial and cosmographic ideas. For example, in parts of Southeast Asia, the numbers four, eight, sixteen, and thirty-two were considered to be attributes of Sumeru and thus to have special meaning (the sequence of numbers corresponds to increasing powers of two, and to the numbers 10, 100, 1000, 10,000, and 100,000 in a binary system). This idea is developed, for example, in the Thai map in figure 14, which depicts sixteen heavens.

Cosmological and mathematical geography had repercussions in the actual political geography of some kingdoms: in the Malay courts of Kedah and Pahang there were four great chiefs, eight major chiefs, and sixteen minor chiefs; Perak and Malacca once added thirty-two territorial chiefs. During the enthronement ceremonies in Cambodia and Thailand, the new king was surrounded by eight Brahmans, representing the Lokapalas (world protectors in Hindu mythology) guarding the eight points of the Brahman cosmos. In ninth-century Java, the kingdom of *Hi-ling* was ruled by thirty-two high officials.[26]

Allegories were naturally well-suited to many Southeast Asian cosmographies. The people of volcanically active islands of Banda, for example, believed their archipelago to lie on the horns of a great ox, which caused earthquakes when shaking its head. Bali was said to lie on the back of a turtle, Bedawang, who floated on the ocean.[27] In seventeenth-century Mataram (Java), troops were sometimes envisioned as being arranged in the form of a crayfish.[28]

Although Western mapmaking philosophies have now largely supplanted native cartographical traditions, it would be misleading to see the history of Southeast Asian cartography as a gradual incorporation of European values. The modern Western ethic of mapmaking — namely that maps should present geographic data in the most 'accurate', clear and analytical fashion possible — was not the principal aim of Southeast Asian cartographers, just as it was doubtfully a major concern of medieval mapmakers in Europe.

That the assimilation of Western mapmaking and cosmography was not an even process is colorfully illustrated in mid-nineteenth century Siam. Western cosmological principles were by that time already accepted by some members of the Siamese court, yet local mapmakers within its doors evidently felt no need to 'look' Western. Excerpts from an account by Frederick Neale, an Englishman who visited Siam in the 1840s and who was shown a map by the king paint a vivid image of this. The visitor, who found the court to be eclectic and even surreal in its taste, was brought to the palace by a "gorgeously gilded state canoe," and then carried on the boatmen's shoulders to dry land, where he found the palace's courtyard "filled with a strange conglomeration of beautiful Italian statues . . . and of uncouth and unseemly figures of Siamese deities and many-armed gods." Once within the king's chambers, a "self-performing little organ" played music of Mozart as a curtain slowly drew aside to reveal "the corpulent and half-naked body of the mighty and despotic king of Siam."

The king, to illustrate a territorial dispute with Burma, produced "a chart of the two kingdoms which had been drawn by his prime minister." As the canvas map was carefully unrolled on the floor in front of the visitors, the king studied the their faces

> as though he expected that the brilliance of the painting, and the exquisite display of Siamese geographical talent, would have caused us to faint away on the spot, or to go into rapturous fits of delight.

Instead, they thought it such gibberish that they were

> very nearly outraging all propriety by bursting into fits of laughter, and very painful was the curb we were obliged to wear to constrain our merriment.

The map (see fig. 10), indeed, was hardly a 'map' in their eyes:

> [It] was about three feet by two; in the center was a patch of red, about eighteen inches long by ten broad; above it was a patch of green, about ten inches long by three wide. On the whole space occupied by the red was pasted a singular looking figure [the Siamese king], cut out of silver paper, with a pitch-fork in one hand and an orange in the other; there was a crown on the head, and spurs on the heels . . . His Majesty [explained that] such portion of the chart as was painted red indicated the Siamese possessions, whereas the green signified the Burmese territory.

Within the Burmese domain, an ill-formed black figure represented Tharawaddy, the Burmese king, and many disoriented small figures represented his subjects, the whole symbolizing

> what a troubled and disturbed state the Burmese empire was, and what an insignificant personage, in his own dominions, was the Burman King.

Although the European observer mocked the map for its geographic worthlessness, he also understood that its actual purpose was to 'chart' the relative virtue and strength of the Siamese and Burmese sovereigns. This was a map of kingly omnipotence and kingdomly integrity, a propaganda map, with no intended value to a traveler.

Travel, Trade and Statehood

Travel, trade, and statehood promote geographic concepts and are natural precursors to mapmaking. While it is true that the common folk of Southeast Asia stayed, for the most part, within their villages, travel far from home figured naturally in the collective psyche of most societies. A youth might wander about for a few years in search of adventure and profit, and then, having gained at least a patina of glamour, if not riches, return to his natal village. Buddhists and Muslims alike undertook religious pilgrimages and sought teachers of their faith; in the case of Buddhism, the respectability of a life of travel was perhaps set by the religion's foundations, for the life of Buddha was itself largely a series of wanderings. Other traditional (if less respectable) travelers common in Southeast Asia were the troubadours and actors, who led a nomadic existence, meandering

SIAMESE MAP.

Fig. 10 Caricature of a Thai map of Thailand and Burma, Frederick Neale, *Narratives of a Residence at the Capital of the Kingdom of Siam*, 1852. (7.9 x 5.7 cm)

from one village to the next. Chinese texts, for example, record a group of Funanese musicians whose sojourn in China resulted in the establishment of a music institute near Nanking.

Southeast Asia developed sophisticated internal trading patterns, and throughout the seventeenth century — with perhaps the exception of Vietnam's forever awkward relationship with China — these commercial relationships remained more significant than outside influences. Those who were actively involved in trade, needed to be able to repeat each leg of their commercial circuit with confidence, and the images they developed of their itineraries influenced their view of the physical world.

Although the typical Southeast Asian village might be self-sufficient in its production of the all-important staple, rice, other commodities were typically traded up and down river. In many parts of Southeast Asia today, one can still see people from the mountains walking down to the rural markets to sell their modest pickings of odd mushrooms, spices, and other local exotica not found closer to the villages. In earlier times, such trade was an integral part of daily life. Commodities such as the dried fish, fish sauce, and lime produced in the lowlands would be traded for the highlanders' woods, bamboo, herbs, and perhaps lacquer. Most products that were traded across river or mountain did not travel great distances.

Salt was the exception. Of essential commodities in early Southeast Asia, only salt was not obtainable by most people via simple trading channels. Acquired from large pans built at appropriate sites along the coast, salt might typically be part of several trading routes before finally reaching villages far upland. In terms of the complexity and the length of its route from source to final recipient, this most-traveled of essential commodities was surpassed only by prestige goods like Chinese ceramics and other luxuries items.

Commercial intercourse in Southeast Asia received a boost in the first century when the newer Mahayana school of Buddhism, which looked more kindly upon trade than did either Hinduism or the older Theravada Buddhism, eased existing spiritual constraints in India and, in turn, promoted Indian commerce with Southeast Asia. At the same time, larger trading patterns were developing within Southeast Asia following the emergence of major states like Ayuthaya, which started exporting rice to Malacca in the fifteenth century. Southeast Asia also began to develop trade routes which facilitated commerce with India and China. In this instance the people of the region acted as trans-shippers rather than consumers, with Chinese texts recording Malay ships arriving at Chinese ports in the early centuries of the Christian era.

The trade route between Southeast Asia and India was serviced by Indian or Southeast Asian mariners, or a network of the two. Historians had traditionally assigned this role exclusively to Indian vessels, arguing that Southeast Asian peoples were not yet capable of such a voyage. However, scholars have more recently noted that any vessel and pilot capable of reaching Africa from Southeast Asia — as those of Sumatra and Java had done since the first century A.D. — could surely have reached India as well. Malay pilots learned to ride the monsoons, and Malay shipbuilders probably pioneered the balance-lug sail, which were square, pivoting sails set in the front and back of the ship that allowed pilots to sail into the wind by 'tacking'. The technology is related to (and may be the ancestor of) the triangular-shaped lateen sail of the Arab *dhow,* and was in turn borrowed by the Portuguese and Spanish in the design of the caravel.

Migration involved even longer voyages on the open ocean than did trade, and is a recurring motif in Southeast Asian history. During past millennia, groups of various Southeast Asian peoples island-hopped their way east into the unknown ocean sea, settling the various Pacific island archipelagos. Some of these emigrants developed their own cartographic tradition, such as the Marshall Islanders, whose 'stick charts' assisted in navigating the open ocean, or in the instruction of such navigational techniques.

Earlier Southeast Asian seafarers had headed west and south. Roughly two thousand years ago, when European sailors still confined themselves to making coastal passages that kept them within sight of land, Indonesian pilots mastered the long open-ocean voyage to Madagascar and the Cape of Good Hope. Unlike the Southeast Asian settlers of Micronesia and Polynesia, these voyages were commonly made as a round trip. The Indonesian seafarers who settled Madagascar, kept the route between the two distant lands active until the fifteenth century.

Glimpses of craft that Javanese or Sumatrans might have used in the eighth or ninth century have been preserved in stone relief on the magnificent ruins of the 'temple mountain' of Borobudur (see fig. 11), built by the Sailendras of Sumatra. Of the many hundreds of scenes carved on its walls, there are several depicting sophisticated ocean-going vessels with balance-lug sails.

The kingdom of Funan, established on the Mekong River delta by about the first century A.D., is usually regarded as the first known state of Southeast Asia, though the only definite written record of it is from fragments of an account by two Chinese envoys in the third century.[29] The economic success of Funan was largely a result of its strategic location between points east and west, since it was in an ideal position to service merchants and pilgrims who traversed the Isthmus of Kra, or shuttled between India and China by sailing around the Malay Peninsula. Archaeological finds

excavated at the Funan port of Oc-èo, which include Roman and Indian artifacts believed to date from the second and third centuries, indicate that it was a bustling center of maritime trade.[30]

A monarch used maps to record his conquests and to impress upon his people the extent of his sovereignty, to record the lands under his jurisdiction for the levying of taxes from his subjects, and to facilitate the creation of roads and irrigation systems. Two thousand years ago — about the same time that Indonesian people pioneered the voyage to Madagascar and Funan was founded in what is now southern Vietnam — the people of Banaue in northern Luzon constructed monumental rice terraces which have remained largely intact to this day. These stone structures formed vast terraces rising nearly a mile from valley floor to mountaintop. A jewel of ingenuity and engineering, the design created artificial waterfalls, which gently irrigated the terraced crops below. Although the rice terraces in Luzon are the most spectacular of early irrigation systems, sophisticated irrigation complexes with rice-terracing are known throughout Southeast Asia. Funan, and the Cambodian kingdom of Angkor which followed it in southern Indochina, were both highly dependent on their extensive irrigation systems.

War was another situation that provided an incentive for creating maps. An early Western allusion to this kind of map-making comes from Mendes Pinto, who wrote of the queen of Prome (Burma) and her council "mapping out the way in which they were to proceed" with organizing a defense of their city.

Siam boasted about the many neighboring kingdoms over which it claimed suzerainty, and from which it exacted tribute. It was a literate society whose monarch was so attuned to the written word that Nicholas Gervaise, a priest resident in Ayuthaya from 1683-87, wrote that "there is no employment in the royal palace more exhausting than that of the reader" to the king. Although Chinese texts record Siamese geographic mapmaking by the year 1373, three centuries later, when Gervaise wrote of the Siamese king's "eight or ten warehouses, among several others, that are of unimaginable wealth," he made no reference to maps, but simply stated that "it is impossible to say how many precious, rare, and curious things" are in these warehouses.

Simon de La Loubère, who followed Gervaise in Ayuthaya (1687-88), specifically stated that he never saw a Thai map, yet he left us a tantalizing hint of a Thai geographic item of some sort. He wrote that he had hoped to secure a Siamese map of the kingdom, but had to settle for one done by a French engineer, M. de la Mare, "who went up the Menam [Chao Phraya], the Principal River of the Country, to the Frontiers of the Kingdom" (see fig. 128). But since La Loubère believed this map to be inadequate, he had Jean Dominique Cassini, the director of the observatory at the Académie Royale in Paris, "correct it by some Memorials which were given to me at Siam [Ayuthaya]." What were these 'memorials' given to him which helped improve his map of the kingdom? Apparently, they were geographic items of some sort — yet nothing which in his mind befitted the definition of 'map'. La Loubère did, however, offer an insight into Thai cosmography, having studied and recorded the "rules of the Siamese Astronomy, for calculating the Motions of the Sun and Moon."

Writings loosely attributed to Kosa Pan, a Thai emissary who traveled to France in 1686, reveal a conscious interest in maps.[31] In these memoirs, Kosa Pan matter-of-factly requests plans of Chambord Castle and "the great temple Notre Dame". At the Siamese embassy's audience at Versailles, Ambassador Pan "made no secret that our most desired objects were maps of the country, plans of palaces and fortresses [and military images]." A contemporary account of Kosa Pan's visit to France contained in the *Mercure Galant*

Fig. 11 This stone relief is one among many depicting native, ocean-going vessels on the walls of the great temple complex of Borobudur in Central Java, built ca. 800 (see fig. 8). Borobudur was first brought to light in 1814 by Thomas Raffles, who ordered the ruin be cleared of undergrowth and thoroughly surveyed. [Photograph by Richard Casten, 1994]

supports the spirit of these quotations. Independent confirmation of his keen interest in maps and the fact that he did indeed obtain various examples from European sources came four years later, in 1690, when Engelbert Kaempfer visited the ambassador, who was now 'High Chancellor' in charge of foreign affairs, at his home in Ayuthaya. Hanging in "the hall of his house," wrote Kaempfer, were only "pictures of the royal family of France, and European Maps."

The Thai poet Sunthorn Bhoo (1786-1855) speaks of maps quite naturally, writing in a work of fiction that a "ship went out of the way and drifted to an unknown place where nobody could tell where the spot was located on any map."[32]

Extant Southeast Asian Maps

The corpus of extant early maps from Southeast Asia is limited to a relatively small number of geographic and non-geographic maps dating from the past few hundred years, and cosmographic edifices dating as early as the seventh century. Although traces of the missing history of geographic maps can be filled in from references to early Javanese, Vietnamese, and Thai maps in Vietnamese and Chinese chronicles, and from the records of early European explorers, most of Southeast Asia's earlier cartographic history remains mysterious.

Some clues to the very beginnings of humankind's mapmaking have been found in the durable medium of rock, but although proto-cartographic motifs in the form of rock art, or petroglyphs, are extant in neighboring India and China, as well as in Europe, no such record is known in Southeast Asia. Several rock art sites of an uncertain date have been discovered in the Malay Peninsula, Sumatra, Borneo, Sulawesi, Flores, and Timor, but none are known to contain any cartographic elements.[33] Nor do we learn more by moving forward thousands of years to consult ancient texts. Although there are an abundance of geographic and cosmographic descriptions in many old Hindu and Buddhist texts, oddly enough there is no known reference to a 'map' per se, or at least not to what we can now discern as such. The earliest extant textual records of Southeast Asian geographic maps are found in Chinese texts of the fourteenth century (recording a Javanese map of 1293) and fifteenth century (recording a Thai map of 1373). The Europeans who began scouting Southeast Asia at the turn of the sixteenth century were

map-conscious explorers; yet even their record of Southeast Asian maps is frustratingly scant and inconsistent.

Cosmologically-oriented stone structures survive in Cambodia, but no geographic maps. In the middle of the third century A.D., Funan, which was the most important state in southern Indochina before the rise of the Khmer empire, was visited by envoys of the Wu dynasty of China. The Chinese ambassadors described Funan as a place where the people "live in walled cities, palaces, and houses" and have "books and depositories of archives and other things".[34] We do not learn, however, whether or not the 'other things' in their archives might have included maps.

A thousand years later, in 1296-97, the Chinese ambassador Chou Ta-Kuan carefully described many facets of Angkor and Khmer culture, but never mentioned maps. He did, however, provide definite measurements regarding various features of the city, which covered an area of about one hundred square kilometers, making it one of the largest cities in the world at that time. Chou also speaks of astronomers in Cambodia, though does not specifically mention the charting of celestial objects.

"In this country," writes Chou, "there is a hierarchy of ministers, generals, astronomers, and other functionaries"; other passages in his chronicle describe the prediction of eclipses as being among the astronomers' duties. Indeed, some modern researchers have theorized that the making of astronomical observations was part of the purpose of the great temple of Angkor Wat. The building of Angkor Wat in itself might reasonably have involved maps or plans; the height and direction of the edifice's walls, which cover substantial distances, deviate from a theoretically straight line in height and direction by less than one-tenth of one percent. Evidence of the careful recording of the positions of celestial objects has also been noted in Pagan.[35]

Many of the undertakings which modern sensibilities would associate with the making of maps, were accomplished by Southeast Asian peoples without leaving any evidence of mapmaking. There is no indication that charts of any type were made by early Indonesian sailors to assist their monumental voyages. Nor do the Funanese, who must surely have been well-aware of the importance of their pivotal location for trade between the China Sea and Indian Ocean, seem to have created any kind of graphic representation of their land-in-the-crossroads. Nor are there any known plans for any of the ancient irrigation systems or temple complexes of Southeast Asia.

The absence of any evidence of Philippine mapmaking, or even any clear mention of them by early visitors, is perhaps the greatest enigma. There is nothing to show that maps played any role in the bustling intercourse that had developed between the many islands of the Philippines, even though entrepreneurs from Luzon were adventurous enough to have established a trading colony in Malaya before the Portuguese burst on the scene at the turn of the sixteenth century (see page 138). When Thomas Cavendish returned to England in 1588 he brought a map that he had acquired in the Philippines, but it was of Chinese, not Southeast Asian, origin. In the mid-eighteenth century Alexander Dalrymple, the first head of the British Hydrographic Office, reported that a servant from Luzon had given him a map whose bearings generally agreed with Dalrymple's own, but we do not know whether this 'indigenous' map was ultimately based on Filipino or Spanish information.

Whether this silence represents the lack of any substantial geographic tradition in these regions, or simply the failure of local maps to survive the centuries, is disputed. None of the spatial imaging involved with travel, construction, the levying of taxes, or other endeavors necessarily required the making of maps. Various classes of people ranging from nomads to pilgrims to traders and caravaners, routinely traveled confidently without maps. Experience taught the traveler the nature of a route and its itinerary.

Printing, Binding, and the Survival of Early Maps

Indigenous Southeast Asian maps were not, as far as is known, reproduced via printing methods. To put this into perspective, we can note that all but a minute fraction of the impressively large inventory of extant early European maps owe their survival to their mass-production by woodblock or copperplate, which began in the latter part of the fifteenth century, and (to a much lesser extent) to the establishment of chart-producing houses, which made multiple manuscript copies of a given map. Nor did Southeast Asian peoples bind their maps into atlases or other book forms, save for certain, Chinese-influenced, Vietnamese works, and the *Traiphum* and other cosmographic or itinerary-related manuscripts from Thailand. A comparison of extant copies of European maps bound into books as compared to maps produced in similar numbers but sold as loose sheets will demonstrate the enormous effect this has on survival rate. If we were to subtract from the surviving corpus of European maps, those reproduced by printing or by chart-copying houses, and those preserved in books, then the history of European mapmaking would hardly be less mysterious than that of Southeast Asia.

Other factors also limited the chance of a document's survival to our time. In the first place, the climate in most parts of Southeast Asia promotes the decay of organic materials. Vast archives were also lost in war; most famously, the Burmese sacking of Ayuthaya in 1767 is said to have destroyed most contemporary Thai records. The people themselves may have been unconcerned about preserving such documents for the distant future, seeing them simply as temporary creations — the most extreme example of this is in Tibet, where cosmographic *mandalas* are painstakingly created from powdered sand only to be swept away after brief ceremonial use, symbolizing the Buddhist concept of the impermanence of life. Lastly, they may have been deliberately destroyed — there is speculation that Siamese authorities may have periodically purged their archives of older materials which were no longer current. This practice, known as *chamra*, may have become especially common when Siamese authorities adopted Western surveying techniques during the nineteenth century, their new mapping 'language' making the entirety of their past cartographic archives no longer relevant.

Secrecy

Secrecy may also have played an instrumental role in limiting the survival of indigenous maps, just as it did in Renaissance Spain and Portugal, and also in Japan. La Loubère speculated about secrecy in Siam in the seventeenth century, observing that "the Siamese have not made a Map of their Country", or if they had, that they "know how to keep it a secret." But while the experiences of La Loubère does not afford any direct evidence of Thai cartographic secrecy, those of a British ambassador in the early nineteenth century most certainly do. John Crawfurd, who was sent as Britain's ambassador to Bangkok in the early 1820s, actually described his acquisition of geographic data from a Thai mariner. His account of the embassy, published in 1828, included a "Map of the Kingdoms of Siam and Cochin China" by John Walker, compiled from the embassy's own surveys and older sources, as well as from new information obtained (we learn from the book's final appendix) from "a Mohammedan mariner, a native of Siam," whose owns ports-of-call occasionally coincided with those of the British mission.

Crawfurd, it seems, was only able to gather information from his Thai source when far away from Siamese soil and he records how the Thai sailor grew increasingly apprehensive about divulging information the closer they got to Siam. When the embassy met the "Mohammedan mariner" and his colleagues in Penang, the details they revealed in respect of their country "supplied more useful and practical knowledge than all we had before obtained from printed sources." However, as Crawfurd's party "approached Siam they became much more shy and reserved, and now communicated nothing without a strict injunction to secrecy," indicating with gestures that the king would execute them if they were caught divulging such information.[36] Figure 12 illustrates a simple map of Bangkok which was included in the published account of the embassy and which is identified as having been acquired from a native source.

Fig. 12 Plan of Bangkok, taken from a native sketch. Crawfurd, 1828. [Private collection]

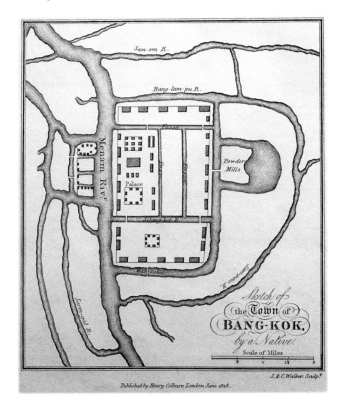

The power structure in Dutch-held Indonesia may also have promoted secrecy regarding the population of the archipelago and its land divisions. The Dutch authorities, in the interest of corporate efficiency, preferred to delegate some administrative responsibilities to local chiefs rather than burden the Dutch East India Company with direct responsibility for the indigenous infrastructure. These chiefs, in turn, would have benefited from keeping confidential particular information about their domain and the size and distribution of its population. Furthermore, traditional seafarers like the Bugis of Sulawesi and the maritime merchants of Java and Sumatra may have been particularly keen to keep their maps and other pilot aids confidential once they witnessed the aggressive commercial spirit of their European visitors.

Mapmaking Media

Although there is only scant record of maps or mapmaking in early Southeast Asia, there is ample reference to writing materials and methods. With the obvious exception of stone edifices, the media used were often volatile, and the climates with which they had to contend promoted decay. Maps for day-to-day affairs may have been drawn on leaves, a common medium for writing in Southeast Asia. Ralph Fitch, a visitor to Burma in the late sixteenth century, witnessed written appeals being presented to the king:

> "supplications [to the king of Burma are] written in the leaves of a tree with the point of an iron bigger than a bodkin. These leaves are an ell long and about two inches broad; they are also double."

This was a description of palm leaves, which maximally measure about 6-7 cm wide by about 55-60 cm long, being used as a writing material. The 'iron bigger than a bodkin' was the stylus which was used to incise the characters. A thick paper, made from the bark of mulberry and other trees, was another common writing medium, and was more durable than leaves. This *khoi* paper was brown or black, but could also be bleached white.

Writing on palm leaves may not always have been immediately obvious to European eyes, unaccustomed to the use of such a medium. The initial 'bruising' of the leaf with the stylus did not always produce a readily apparent image — to heighten the image, the leaf would first be rubbed with a sooty substance and then wiped clean, leaving a black imprint on the rubbed areas of the yellowish leaf.

These leaves were quite practical, as they could be rolled up and carried without concern for their getting wet, which in the torrential downpours of the rainy season and when traveling by river must have been a common occurrence. Although water might wash off the image, leaving the leaf 'blank', the messenger or recipient had only to apply some soot (even from a dirty finger) to restore the image. Thus the writing on a leaf being couriered through the elements might, in such circumstances, not be visible to another person.[37] Our Chinese ambassador in Angkor in 1296-97 was struck by this, writing that when the leaves are "rubbed with something moist, [the characters] disappear."

A more durable medium than palm leaves was described by the same Chinese observer. Chou noted that "for ordinary correspondence, as well as official documents, deer skin is used, which is dyed black" and written on with a type of white chalk. The Thai word for 'book', *nangsü*, derives from *nang* (meaning, 'skin, hide, or bark') and *sü* ('written character').[38] Cloth and cowhide were also used as writing media.

Transient Maps

Maps of an inherently transient nature are documented in the Caroline Islands, also known in the eighteenth century as the 'New Philippines'. The *Philosophical Transactions from the Year MDCC* included a map of the Carolines (fig. 13) copied from an indigenous map consisting of stones arranged to represent islands. The Transactions explains that

> the map was not made by Europeans, for none have yet been upon these islands, but by the islanders themselves, after this manner. Some of the most skilful of them arranged upon a table as many little stones as there are islands belonging to their country; and marked out, as well as they could, the name of each, its extent and distance from the others: And this is the map, thus traced out by the Indians, that is here engraved.

The Body as a Map

Peripheral evidence suggests that the human body might have served as a cartographic medium in Southeast Asia, although this is only speculation. There are two forms this might have taken, one 'permanent', the other temporal: tattooing, and the positioning of the hand and fingers into a map. Tattooing was prevalent in much of Southeast Asia before the dictates of Confucianism, Islam, and Christianity suppressed such body art. A batik map from Java, tentatively dated at about 1800, is known to exist, and tattooing and batik may have a common past.[39] There is no record of tattooed maps in Southeast Asia proper, but what appear to be tattooed cartographic motifs are described in some detail by a Frenchman traveling in the Caroline Islands in the early nineteenth century. According to the visitor, the islanders'

> legs and chest are covered with long straight lines which at first look like striped stockings. They trace the outline of several small fish on their hands, each of about an inch long. It is strange to note that these figures bear the names of various islands. Peseng [one of the islanders], on his left thigh, above the knee, had a certain number of these fish as well as hooks, which represented Lougounor and the neighboring island groups; in addition, each line on his legs and hands was identified with the name of an island, from Faounoupei as far as Pelly. Having accounted for all these islands, there were still a few lines left which he called Manila, Ouon [Guam], Saipan, etc.; and as there were still a few lines left, he named them, chuckling as he went, Ingres [England], Roussia, etc. Perhaps this practice had been introduced to more easily recall the islands of their archipelago. It is a type of geographic rosary . . .[40]

The positioning of the hand and fingers in the form of a map was codified by João de Barros, Lisbon's official historian of Portuguese adventures in the Indies (see page 123, below). Writing in the mid-sixteenth century, Barros told his readers how to 'make' a map of Southeast Asia by placing the fingers of the left hand in certain prescribed positions. Not only were the coastal contours of mainland Southeast Asia reasonably well represented by the fingers, but inland features could be indicated as well, by means of the knuckles and joints. The origin of the idea is not known.

Music as a Map, and the Mapping of Music

In some parts of Southeast Asia, geographic data may have been recorded in an entirely non-material form. Song could preserve the essential knowledge for a voyage or trek in a way that was much

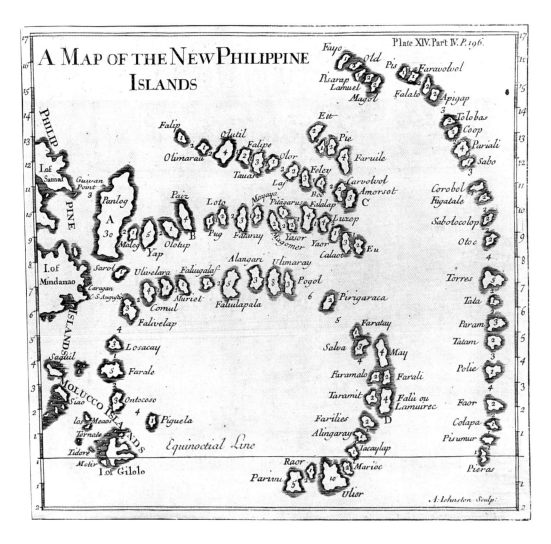

Fig. 13 Map of the Caroline Islands, based on indigenous data, as interpreted by missionaries; from the *Philosophical Transactions*, 1721. The figures within each island indicate the number of days its inhabitants said it took them to sail around it, while the numbers between the islands indicate the duration of a voyage, in days, from one island to the next. (18 x 20.5 cm)

easier to commit to memory than cold data and which was also impervious to material wear and loss, in effect a 'pilot book' bound in the medium of music. The Arab pilot Ahmad ibn Majid, who sailed in Southeast Asian waters, composed a navigational text in 1462 which was written in poetry to facilitate memorizing its instructions and there may have been similar native Southeast Asian examples.

The societies of Southeast Asia proper are not known to have notated their music. The 'mapping' of music is, however, found in Tibet. Possibly as early as the ninth or tenth century, Tibet developed forms of so-called 'neumatic' notation, in which music was recorded with graphic signs representing the pitch movement of a melody — well within our definition of 'map'.[41] Various inflections of a line, looking not unlike the stylized coasts of sea charts, indicated melodic ascent, descent, angular movement, vibrato, and the degrees of each. Forms of neumatic notation were also used in various parts of Europe and western Asia.

Principal Types of Southeast Asian Maps

We do not know how representative surviving Southeast Asian maps are, either in terms of quantity or characteristics, since the durability of the various media used, as well as the inclination of their makers to preserve them, spanned the range from perhaps as little as seconds, to millennia. The extant corpus of Southeast Asian maps probably does not accurately represent the relative numbers of various types of maps in everyday Southeast Asian life, since the survival of maps is heavily skewed in favor of cosmological maps.

Stone edifices, by a great margin the most durable of cartographic media, were used for cosmological and religious cartography, and such cartography did not become obsolete. Geographic maps would more likely have been on far less permanent media, would have been subject to wear and loss, and might be superceded by more current mapping. The sand *mandalas* of Tibet are the exception, since the impermanence of these cosmographic maps was part of their very meaning, but their story is still part of the 'permanent' record of their society.

With these limitations in mind, we can look at the various types of Southeast Asian maps and the dates of surviving examples of each, and then comment on their counterparts in the West. Southeast Asian maps can be divided into four general categories:

1) those which are purely cosmographic (in this context meaning 'metaphysical' or 'spiritual') in nature, or otherwise non-geographic.
2) those that symbolically represent actual geographic features for religious or cosmographic purposes.
3) those that attempt to record true geography, whether by report or empirical observation.
4) itineraries, which might be written, memorized, or committed to song, that served to construct a mental image of time and space, direction and position, topography, and landmarks.

There is not always a fine line separating these categories. Just as daily life in Southeast Asia could entwine the mundane with the magical, Southeast Asian cartographic thought could blur the distinction among the cosmographic, symbolic, and empirical.

Cosmographic Maps

In pre-modern Southeast Asia, mapping one's path to a different level of existence was as important as charting the way to the next valley. Maps representing the intangible or metaphysical world in the form of stone temples or edifices are the earliest surviving examples of Southeast Asian maps.

In order to look at Southeast Asia's cosmological maps, it is necessary to look at the place of mountains in Southeast Asian life. Not only are mountains are a primary feature of the Southeast Asian physical landscape, but they are also an integral part of the region's spiritual landscape too. A king's sovereignty was often inextricably linked with a mountain, either actual or symbolic. Mountains could represent the embodiment of higher states of existence and at the same time be the dwelling place of gods. This reverence for mountains was indigenous, but was complemented by outside influences. The most famous of spiritual mountains was Sumeru, believed to lie to the north, in the center of the world, that is, in the Tibetan Himalayas or the Pamir Mountains of Central Asia.

When Hinduism spread through Southeast Asia, beginning in about the first century A.D., it brought with it the idea of Sumeru and other Indian traditions about mountains, which subsequently merged with indigenous animist religions. The Hindu god, Indra, was sovereign of Sumeru, and dwelled in a place called *Trayastrima* at the mountain's summit. The Khmers of Cambodia had their own equivalent of Sumeru, which they called Mount Mahendra; when Hindu beliefs reached Cambodia, they merged with the religious aspect of Mahendra with early Khmer rulers becoming identified as the earthly incarnations of the deities of this cosmic mountain.[42] Siva, the most powerful of Hindu gods, was the 'Lord of the Mountain', and the veneration of Siva as a mountain deity existed in Cambodia as early as the fifth century.[43] In Burma as well, the imported Hindu veneration of mountains assimilated indigenous spirits.[44] The concept of sacred mountains continued with the arrival of Buddhism; early thirteenth century Chinese annals record that in a country to the west of Cambodia "there is a mountain called *Wu-nung* [probably from the Malay word for mountain, *gunong*]" from which one enters Nirvana.[45]

The island of Bali, according to local legend, was originally a flat, mountainless land. However, when Islam supplanted Hinduism throughout most of Java, the Hindu gods, having elected to resettle on neighboring Bali, first needed to create on their new island home, mountains that were high enough to be their abode. In another version, the mountains were moved from eastern Java. The most exalted of these god-dwellings, the Balinese 'Sumeru', was Gunung Agung, on the east of the island.[46] Bali was the world, and Mount Agung was its 'navel' *(puséh)*. Representations of Sumeru, which are in effect cosmological maps, are invariably found in the temples of even the most humble Balinese villages.

Mountains were the source of spiritual life; they were also the origin of the other 'element' of the world, rivers, and thus were the source of physical life as well. In this last respect, the perfect silhouette of an Indonesian mountain, towering above the valley floor or rising above the seas, was sometimes compared to the image of a breast.

The fertility of the land was part of the iconography of a type of symbolic map which is known in the form of stone carvings from Champa (central eastern Vietnam).[47] These had their origins as magical "stones of the soil," in which the stone embodied the god of the earth; unlike imported beliefs, in which gods *inhabited* earthly entities, the indigenous view may have regarded the earth *itself* as a god. Such mystical objects were analogous to stones or earth mounds worshiped by Vietnamese and Chinese. The Champa monoliths were in the form of *lingas,* Hindu phallic symbols representing the god Siva, or the reproductive power of nature. They were placed at the center of a village and appear to have formed a multi-faceted microcosm of the territory, the embodiment of the kingdom or feudal group of which the *linga* was the center. The top of the stone was the emblem of Siva; Siva was accorded the apex of stone both because he was 'Lord of the Mountain' (Meru) and because political power in early Champa and Cambodia was legitimized by the identification of the king with Siva. An octagonal section under this cap represented the god Vishnu, and a four-sided section below it symbolized Brahma, the overseer of all things sacred (another instance of the special significance of these numbers). A lowermost section represented the earth, that is, the kingdom. Thus the stone was at once a schematic of the order of the universe, a symbol of the divine authority of the king and his singular embodiment with Siva, and a map of the powers that bestowed fertility to the land and people. This 'map' of power, fecundity, and earth probably served as a sacred tool in fertility rituals. Other Cham stone carvings were semi-cadastral in content, and will be mentioned in connection with empirical maps (see page 38).

Sumeru was symbolically replicated throughout Hindu-Buddhist Southeast Asia. Hindu temples centered on shrines representing Sumeru, the axis of the universe, date from the seventh century in Java, and edifices with Buddhist architectural symbols based on Sumeru survive from the eighth century. The epitome of these is the great temple at Borobudur (ca. 800 A.D.). Borobudur and neighboring Hindu-Buddhist temples which chart the various states of existence, are the earliest surviving maps in Southeast Asia. The lower levels of Borobudur depict mundane matters and represent lesser states of spiritual awareness: as one rises to the upper levels the subjects become increasingly elevated and metaphysical, with the very top symbolizing the achievement of blissful enlightenment. Taken as a whole, the structure maps the oneness of the cosmos.

Similar motifs, though employing different media, are recorded in early Vietnam. By the late tenth century, about the time Vietnam regained independence from China after nearly a thousand years of vassalage, Vietnamese artists mapped cosmological beliefs by constructing a 'mountain' of bamboo in a river, and in the following century by building a brick edifice with adjoining ponds.[48]

In Cambodia, the magnificent temples at Angkor (early twelfth century) also constitute a cosmological map, charting both space and time. Modern researchers have discovered that the distance between elements of the temple, going east to west, corresponds to the number of years in the present — final — age of the earth, according to Hindu belief, measured in *hat* (1 *hat* = approximately 0.4 meters).[49]

Figure 14 is a Thai map of the cosmos, or *cakkavala*.[50] The bell-shaped enclosure looking like a walled city contains the sixteen lower heavens which rise above Sumeru. Each of the sixteen heavens is represented by one row of stupas (dome-like mounds

Fig. 14 **Thai map of the cosmos, or *cakkavala*. The walled enclosure contains the sixteen lower heavens which rise above Sumeru. The moon is to the left of the stupas, while the sun is on the right, with the various stars of the cosmos sparkling overhead. On the left and right are the mansions of the world guardians in the four cardinal directions, and in between is the mansion of Phraya Yom, god of the infernal or nether regions. The mortal world lies at the bottom of the heavens. Fish from the ocean surrounding Sumeru are identified in the bottom line of text, and lotus which purportedly grow in these ocean waters are named in the text near the moon. [National Library of Thailand]**

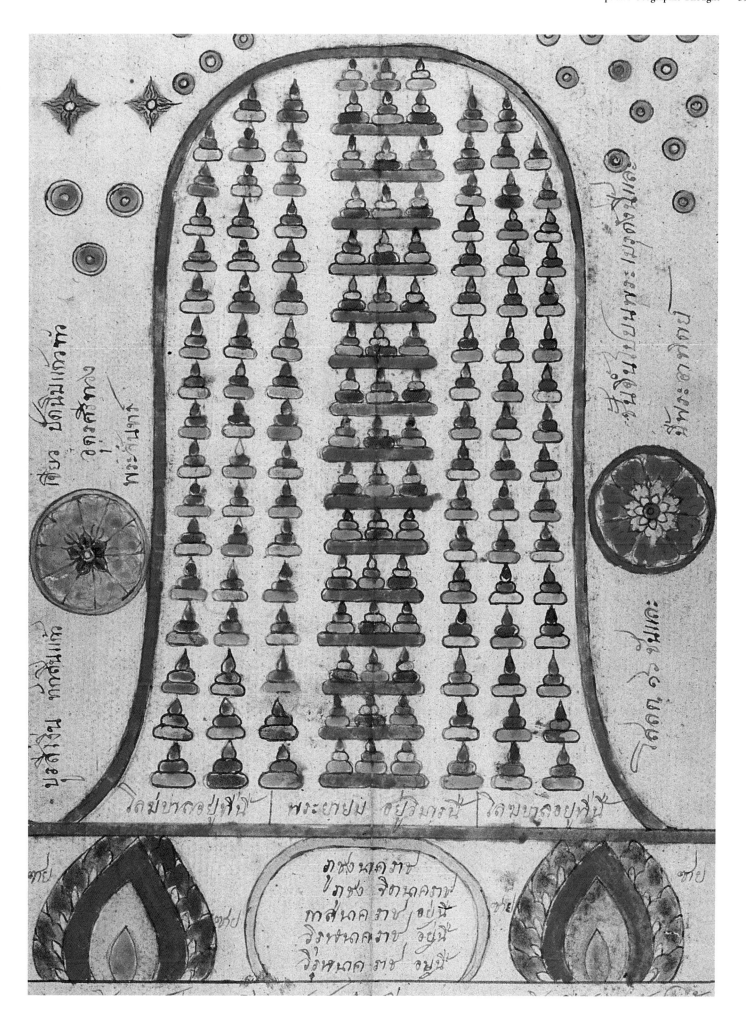

Fig, 15 World map, Antonio Saliba /de Jode, ca. 1600 (Jollain, 1681).
(71 x 61.5 cm [map], 71 x 111 cm [with text])

Fig. 16 World map (detail), Antonio Saliba /de Jode, ca. 1600 (Jollain, 1681).
Note the connection of Chiang Mai Lake with a reservoir deep within the earth,
as well as the unusual focus on islands of the Western Pacific — *Nadadores,
Barbudos, dos Hermanas, Arecifes, Saprovechada* (*Desaprovechada*), and others.

containing a shrine) in three groups of three stupas each. Two extra stupas are placed on the top left and right to complete a play on numbers: 16 x 3 = 48 stupas; 3 x 48 = 144 stupas; the extra four stupas brings the total to 148 stupas. Our mortal world lies at the bottom of the heavens. On the left and right are the mansions of the world guardians in the four directions, and in between is the mansion of Phraya Yom, god of the infernal or nether regions. To the left of the stupas is the moon, to the right is the sun, with the various stars of the cosmos sparkling overhead. The oval-shaped enclosure in the lower center is the residence of various *naga* (serpent) kings. The upper of the two lines of text on the bottom seems to explain that the three mountains which support Sumeru themselves rest on a diamond slab (the figures on either side of the *naga* residence may be two of these stones), and the diamond slab itself rests on a podium. Several of the fish which are said to swim about the immense ocean surrounding Sumeru are identified in the bottom line of text. A few of the kinds of lotus which purportedly grow in these ocean waters are named in the text near the moon.

When Buddha stepped back onto earth after achieving enlightenment, his feet made an imprint in the various places he stepped, the first and most venerated of these being the supposed imprint atop Adam's Peak in Ceylon. This not only added to the Southeast Asian veneration of mountains, since it was via Ceylon that Theravada Buddhism spread to Southeast Asia, beginning in the late twelfth to early thirteenth century, but also led to the use of

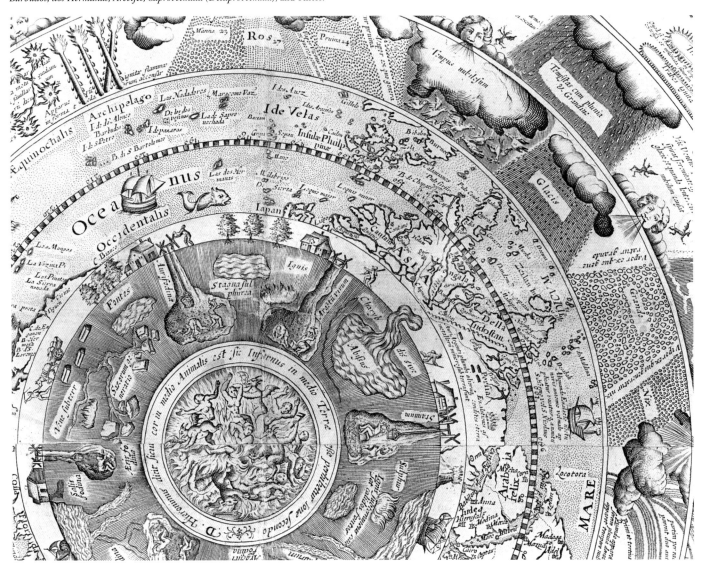

Buddha's footprint itself as a cosmological framework. For a follower of the faith, the imprint of Buddha's footprint upon the earth is roughly comparable to symbolic significance of the crucifix for Christians. Just as the cross assumed a cosmographic significance in the iconography of medieval Europe, the image of Buddha's footprint sometimes served as the framework for maps of Creation in Southeast Asia. Known as *Buddhapada*, these depict the cosmos in a similar, though sometimes more elaborate, format as the map in figure 14.[51]

Western Parallels of Cosmographic Maps

Representations of a multi-dimensional universe, such as the temples that emulate Sumeru and its spectrum of states of existence, are a hallmark of Asian cartographic thought, but have analogies in the West as well. Homeric mythology speaks of the heights of Mount Olympus as the abode of the gods, a region of great tranquility, and this idea was perpetuated in medieval Europe by such authors as Pomponius Mela and Solinus. Cosmological renderings of the earth in the context of the elements and religious/celestial spheres reflected many Europeans' view of Creation. Had Dante known of Borobudur, he would have found its symbolism eminently natural — like Dante's *Commedia,* it uses a mountain to stratify good and evil, heaven and hell.

A map by Antonio Saliba is one example of a Western map depicting the earth in such a cosmographical context (see rendering by de Jode, figs. 40 & 41). The map, originally published in 1582, depicts an earth-centered universe in cross-section, with several concentric circles symbolizing various levels of creation. At the very center are the bowels of the earth, depicted as an inferno. The writhing agony of condemned souls and the glee of satanic demons make clear that this is the Judeo-Christian Hell, analogous to the bottom level of Borobudur, not merely a geological feature. Moving out from this base level of the cosmos we reach the elements, mineral mines, and water sources of the earth, and then the earth's surface, in the form of a terrestrial map. Above the earth we ascend through the different levels of the sky, passing meteorological phenomenon, celestial bodies, and in the outermost ring, the fire of the sun.

The Portuguese poet Luis vaz de Camões, who himself spent many years sailing about the Indies, described a similar metaphysical geography in his epic poem, *The Lusiads.* In imagery that would be as readily understood at Borobudur as in Lisbon or Rome, Camões tells of Vasco da Gama, the Portuguese conqueror of the sea route to the Indies, being led by a nymph up a mountain to a "lofty mountain-top, in a meadow studded with emeralds and rubies that proclaimed to the eye it was no earthly ground they trod." On this fabulous mountain, which a Buddhist or Hindu could easily mistake for Sumeru, "they beheld, suspended in the air, a globe" which replicated the "the mighty fabric of creation, ethereal and elemental."

Next, in words that could equally well describe the symbolism of Borobudur's summit, the nymph explained that the highest sphere "rotates about the other lesser spheres within, and shines with a light so radiant as to blind men's eyes and their imperfect understanding as well." This highest level, where "the souls of the pure attain to that Supreme Good that is God himself," seems interchangeable with the state of *moksa* in Hinduism and Jainism, and the Buddhist concept of Nirvana. Camões' nymph then described the various spheres below, which include the stars, sun, and planets. Beneath these, in the center, is situated "the abode of mankind", and below that lies the realm of Hecate, goddess of darkness. Thus in European metaphysical thought, as in Southeast Asia, the spatial coordinates 'high' and 'low' provided a metaphor for higher and lower states of existence which in turn were linked to the notion of good and evil.

Fig 17 'T-O' world map, 1472. The Biblical Ham, Shem, and Japhet are marked on the continent where each propagated humanity after the Deluge. (6.5 cm. diameter) [Sidney R. Knafel]

Sacred Geographic Maps

Some early Southeast Asian maps symbolically depict the geography of spiritual events, thus forming a genre intermediate between the cosmographic and geographic. This kind of representation of sacred geographic space, found in several stone temples, constitutes the earliest surviving non-cosmographic Southeast Asian maps. Burmese temples in Pagan (thirteenth century) and Pegu (fifteenth century), as well as Lan Na temples in Chiang Rai and Chiang Mai (both fifteenth century and now part of Thailand), replicate sites around Bodh Gaya in northeast India, where Buddha reached Nirvana, or enlightenment.[52] Burmese chronicles record that artisans were sent to the temple at Bodh Gaya (thought to have been made in the third century B.C.), to draw up plans for the Pegu rendering of the temple. The Thai temple of Wat Jet Yot, in Chiang Mai, has seven spires representing the various sites where Buddha tested his tenets after attaining enlightenment. The Lan Na manuscript illustrated in figure 9 is part a religious itinerary map, and part cosmography. It is probably typical of spiritually-oriented geographic maps made for hundreds of years, although this surviving example is most probably a twentieth century copy.

Western Parallels of Sacred Geographic Maps

As with cosmological maps, there are abundant Western analogies for such maps. Representations of the path of the Buddha, which we have already seen in the arrangement of temple spires and will find again in the *Traiphum* manuscript, parallel European maps which illuminate the peregrinations of saints and prophets, or the exodus of the Jews from Egypt. A more subtle example is found in the medieval 'T-O' map (fig. 17), which depicts the earth as a circle with a superimposed 'T' shape, dividing the circle into Asia (the area above the 'T'), Africa (lower right), and Europe (lower left). The 'T' itself, while providing the rudimentary divisions of the continents, may have been derived from the Greek *tau,* an ancient form of the cross that was adopted by early Christians as a

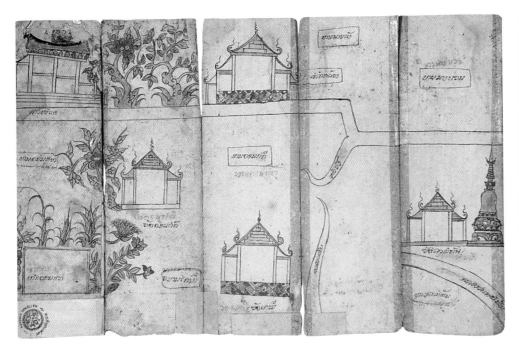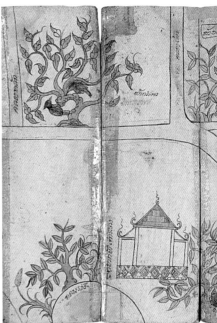

Fig. 18 Parts of a long itinerary map from southern Thailand dating from the late seventeenth or early eighteenth century. According to a label accompanying the manuscript, it pertains to the history of Nakhon si Thammarat, and was given to the National Library by Prince Damrong. [National Library of Thailand]

clandestine symbol of their faith. Thus, the 'T' symbolically superimposed Christ upon the entire earth, and, in fact, some medieval *mappaemundi* literally superimposed the figure of Christ over the world. The letters 'T' and 'O' may, for some authors, have also denoted *Orbis Terrarum,* that is, the 'sphere of the earth'.

Empirical Geography: Indigenous Southeast Asian Mapping
The mapping of Southeast Asian soil, in a most rudimentary sense, may be found in semi-cosmographic, semi-'terrestrial' stone carvings which were placed at the center of some Cham villages or territorial groups.[53] Such monoliths were a microcosm of a well-defined territory and were ritually linked to stones located on the territory's boundary. The all-important central stone was in part an elementary cadastral symbol, a 'map' of the land; at the same time, it was a guarantor of fertility and an object of religious veneration. Since territorial statues and religion were entwined, and since religion partly served to legitimize rights to land, the monolith's cadastral and religious aspects worked together in harmony.

Apart from the purely speculative evidence cited earlier to suggest a long history of indigenous mapmaking, the known record of Southeast Asian cartography begins with textual references to maps from Vietnam, Java, and Thailand. Although none of these maps survive, they antedate the semi-geographic Chiang Mai and Chiang Rai temple maps. The Vietnamese maps originated with the needs of government, which are of course a major impetus for the making of maps, both for the protection of boundaries and the assessment of population and taxes. In 1075, according to a Vietnamese history book of 1479, a map was made of Vietnam's southern border with Champa; the same book records a map made in the 1170s that was the result of "a royal inspection tour of the coasts and the frontiers."[54] Given the long Chinese tradition for fastidious record-keeping and the strong influence of China on Vietnam's governmental affairs, it is not surprising that it is in Vietnam where we find the earliest surviving record of 'practical' Southeast Asian maps.

Vietnam initiated a more thorough cadastral mapping of its land and people in the second half of the fifteenth century, when its court imported Chinese notions of social infrastructure and a stronger, more centralized government. Equipping the government with better control over its land and people, mechanisms were established for a methodical, thorough survey of the entire country. In 1467 the king ordered the (then) twelve provinces of Vietnam to map their respective regions. The maps were to record topography, man-made and natural features, travel routes, and other relevant data. The court collected the individual surveys to which they added extra statistics about the populace and the newly-annexed Champa, which Vietnam had conquered in the 1470s. The combined information was declared the official atlas of the kingdom in 1490, just before the European onslaught, though the earliest surviving Vietnamese maps, date from the seventeenth (or possibly sixteenth) century.[55] The nineteenth-century Vietnamese atlas illustrated in figure 163 probably bears stylistic similarities to such earlier works.

Chinese records demonstrate that mapmaking had a part in the Thai court by the later 1300s. The Ming Annals record that in 1373 the Ming court received an embassy from the country of *Hsian-Lo* (believed to be the region of Sukhothai, Ayuthaya, and Lop Buri, central Thailand). According to the Ming Annals,

> . . . envoys were sent [from Hsian-Lo] to congratulate on the New Year Festival of the next year and to present native products. Moreover, they [the Thais] submitted a map of their own country.[56]

An interesting example of a map apparently intended to record or facilitate a person's travels is illustrated in figure 18. Seven sections, out of a total of about 41 are shown. The map comes from the southern part of Thailand, and was probably made in the late Ayuthaya or early Bangkok period (latter part of the eighteenth century). Temples, houses, landmarks, fauna and plants passed *en route* are depicted and labeled. The several lines of continuous text appear to relate to the circumstances and events pertaining to a

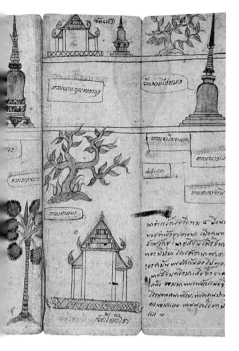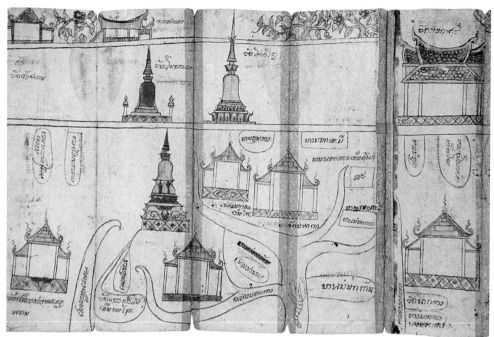

person's travels, speaking of someone who lives beyond the rice paddy, describing meetings with a man of high rank, a story involving a dowry, and the like.[57]

Another geographical map from southern Thailand has been preserved among the papers of Henry Burney, a British captain active in the mapping of Siam in the 1820s. The map (fig. 19) depicts roughly 250 kilometers of the west Malay coast, from the Thai-Burmese border, south to Phuket (the large rectangular island at the bottom of the map). Captain Burney wrote on the map (lower left) that this is a "Chart sketched by a Siamese Priest at Pungah" (Phang-nga, the mainland region directly above Phuket).[58] Another British envoy, James Low, also recorded his pleasure at the quantity and quality of indigenous maps that he had acquired in Penang in the 1820s.

The earliest known reference to a map from insular Southeast Asia is found in a Chinese account of the Yuan invasion of Java in 1292-93. Compiled in 1369-70, the *Yuan shi* (*History of the Yuan*) records that in 1293, Raden Vijaya, a leader of the Javanese state of Kediri, presented a map and census record to a Yuan military commander, thus symbolizing submission to Chinese rule. This event suggests that mapmaking had already been a formal part of governmental affairs in Java.

In 1505, the Bolognese traveler, Ludovico di Varthema, mentions a chart used by the pilot of the vessel on which he sailed to Java from either Buru or Bornei.[59] Boarding the local vessel,

we took our way towards the beautiful island called Giava, at which we arrived in five days, sailing towards the south. The captain of the said ship carried the compass with the magnet after our manner, and had a chart which was all marked with lines, perpendicular and across.

A more impassioned record of Indonesian sea charts came from Alfonso de Albuquerque, founder of Portugal's empire in Southeast Asia. In a letter to King Manuel of Portugal, dated April 1st of 1512, Albuquerque wrote that he had acquired

a large map of a Javanese pilot, containing the Cape of Good Hope, Portugal and the Land of Brazil, the Red Sea and the Sea of Persia, the Clove Islands, the navigation of the Chinese and the Gores [people of Formosa], with their rhumbs and direct routes followed by the ships, and the hinterland, and how the kingdoms border on each other. It seems to me, Sir, that this was the best thing I have ever seen.

He claimed to the king that the Javanese chart revealed

the course your ships must take to the Clove Islands [Moluccas], and where the gold mines lie, and the islands of Java and Banda, of nutmeg and mace, and the land of the king of Siam.

Albuquerque assured his king that the map was a

very accurate and ascertained thing, because it is the real navigation, whence they come and whither they return.

If Albuquerque was accurate in reporting that the Javanese chart included Portugal and the 'Land of Brazil', which clearly must have come from European sources, then the sharing of geographic data was already a reciprocal affair.

These or similar Javanese maps probably supplied some of the data for charts drawn by the Portuguese pilot Francisco Rodrigues in about 1513. Although the Javanese original is not known to have survived, an extant manuscript has been identified as being the Rodrigues derivative.[60] Rodrigues, together with António de Abreu and Francisco Serrão, penetrated insular Southeast Asia as far as Banda with the assistance of Malay pilots in 1512. His charts record these and other islands, and were probably compiled on the basis of Javanese information.

Yet there remains the enigma of why Indonesian charts, which were so readily acquired in the early years of Portugal's presence in Southeast Asia, would subsequently have become so mysterious. Willem Lodewijcksz, a member of Cornelis de Houtman's expedition to Java in 1596, reported that the Javanese, although transporters of

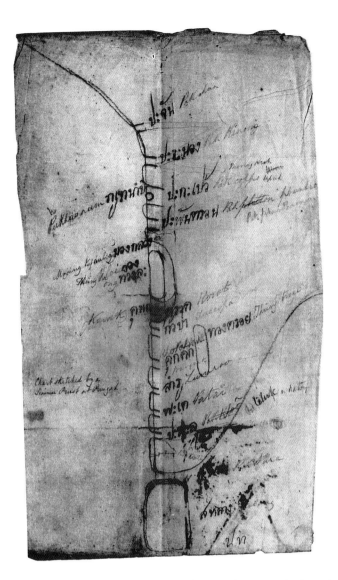

Fig. 19   Thai map of the west Malay coastline from the Thai-Burmese border south to Phuket, "sketched by a Siamese Priest at Pungah [Phang-nga]." From Henry Burney's papers.[By permission of the Syndics of Cambridge University Library]

Burma enjoys a relatively rich cache of surviving geographic maps, or copies of such maps. In 1795 Francis Hamilton sought out maps while traveling through Burma, and his analysis, along with engraved reproductions, was published in Edinburgh in the early nineteenth century.[63] But with the act of examining something comes the risk of altering it, and indeed Hamilton's wonderful zeal to acquire indigenous maps led him to commission them, which in turn may have influenced his Burmese cartographers. Hamilton himself makes this bluntly clear, noting that the Burmese map-makers he employed were "wonderfully quick in comprehending the nature of our maps [and] very soon improved their plans." Thus Burmese maps often form exceptions to generalizations which might be made about indigenous Southeast Asian maps, having been influenced by Western prototypes.

Although surviving Southeast Asian geographic maps present a wide diversity of characteristics, some generalizations can be made. Cartography was doubtfully a profession in itself. As with pre-Renaissance Europe, Southeast Asian cartographers rarely left any mark of their authorship, and maps were rarely dated. Maps generally lacked a uniform scale and were not constructed on any particular projection. No known Southeast Asian map has any grid to represent latitude or longitude, though both Varthema and Albuquerque alluded to rhumb lines on Javanese charts.

Southeast Asian cartographers tended to stylize features and exaggerate waterways. Chinese art and cartography are similar in this regard, and the stylization of water systems can be found on medieval European maps as well.[64] Stylized maps have their uses and are still employed when simplicity and clarity are of a higher priority than accuracy of scale or direction, as is the case when mapping subway lines or depicting electronic circuits.

The more stylized and metaphorical nature of early Southeast Asian maps, as opposed to European maps, may be said to be consistent with their approach to art and language in general. The French envoy to Siam, La Loubère, certainly oversimplifying, correlated this to climate:

> In cold countries, where the imagination is cold, every thing is called by its Name . . . it is not the same in hot Countries, [where] the briskness of the Imagination employs them in a hundred different ways, all figurative . . . to them it seems that an exact Imitation [in art] is too easie, wherefore they overdo every thing. They will therefore have Extravagancies in Painting, as we will have Wonders in Poetry . . . the Secret is, to give all these things a Facility, which may make them appear Natural.

Itineraries

Travel in early Southeast Asia was probably more dependent on itineraries than maps *per se.* The sorts of geographic aids used by Southeast Asian peoples may well have amounted to instructions, not so different from the types of aids a pilgrim or sailor may have used in medieval Europe. The village to which one was traveling would be 'mapped' as being so many days by *perahu* upriver, after which the traveler finds a bend in its course and a certain landmark, at which point one then leaves the boat and crosses the mountains just to the left of where the sun rises, and after two more days following sundry other topographical clues, one reaches one's destination.

In such itineraries, two different methods could be used to measure distances. A journey could be recorded either in units of linear measurement, or else in units of elapsed time. In Thailand, for example, linear distance was measured in *wa* (fathom), a unit which probably equaled about 1.8 meters, but is today fixed at 2 meters.[65]

goods among the islands, did not use maps for sailing, nor did they have the compass until they acquired it from the Portuguese. Buginese (Sulawesi) mariners' charts from the eighteenth or nineteenth centuries are extant, but these show a strong Western influence.[61]

After the Portuguese reference to tapping indigenous charts in the early sixteenth century, there is little mention of European reliance on native Southeast Asian maps for two and a half centuries. In the middle and latter eighteenth century some mapmakers dedicated to surveying Southeast Asian shores, notably Alexander Dalrymple, openly acknowledged appropriating Malay charts for their own (page 238, below). Was the practice of European cartographers incorporating Southeast Asian sources unusual? Or was Dalrymple simply being more forthright than most?

In the northern periphery of Southeast Asia, evidence of indigenous surveying is found in early European accounts of their attempts to map the imposing expanses of mountains and plateaus in Tibet. Missionaries made a large-scale survey of Tibet in 1717 in which they took maps "drawn on the spot from the report of the natives, by the Lama Mathematicians". The map explains that since "the Lamas made no Astronomical observations in the Course of their Survey, the Missionaries have corrected this Map with their own."[62]

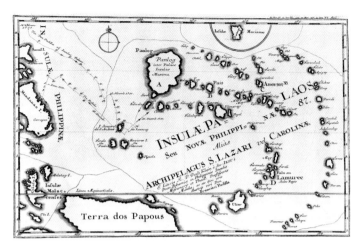

Fig. 20 Part of Mindanao and the Carolines, based on indigenous data contained in figure 13. From a Jesuit report by Joseph Stöcklein, 1726. (22.4 x 35.3 cm)

According to oral tradition, one of the people in the king's retinue would carry a 'wa-stick', a stick cut to the exact length of one wa. It would be that person's responsibility to flick the stick over and over with his wrist, as they walked, to record the number of wa traveled.

Elapsed travel time, however, was the more common gauge of distance for longer journeys, and in Thailand this was usually measured in khrao (overnight stages). The association of physical space with time, rather than literal distance, is perhaps one reason why indigenous maps generally appear to have no uniform scale — scale may have been consistent with the elapsed time of travel rather than linear distances *per se*.

A surviving itinerary from sixteenth century Lan Na, *The Chronicle of Chiang Mai*, records the travels of King Mä Ku from Chiang Mai to Chiang Rai, Thöng-Phayao, Phrä-Lampäng, Lamphun, and back to Chiang Mai, in 1559. This text records distances both in linear terms (wa, or fathoms), as well as in units of elapsed time (khrao, or overnight stages). An excerpt follows:

> From Wang Kham he [King Mä Ku of Chiang Mai] went to sleep [i.e., one overnight stay, or khrao] at Pa Sieo, 12,000 fathoms [i.e., the surveyor had recorded approximately 12,000 flicks of the wa-stick]. From Pa Sieo he went to Phræ, 12,000 fathoms [another 12,000 flicks of the wa-stick], and he stayed in Phræ for 12 days. On the fourth waning of the seventh moon he installed Phraya Chiang Lüak to rule Phræ. From Phræ, he went to sleep at Khrao Ton Hua, 5,000 fathoms [5,000 flicks of the wa-stick]. From Khrao Ton Hua, he came to sleep at the Nam Ta, 9,000 fathoms.[66]

Interpreting route lists was an art — as was the interpretation of itineraries and pilot guides in medieval Europe (it is no coincidence that the title of a great Spanish navigational treatise, published in 1545, is *Arte de Navigar*, literally 'Art of Navigation', implying an acquired skill rather than a precise science). As with European and Arab pilot books, it is likely that even if a map was drawn to illustrate a journey, the guiding data would be the itinerary. The southern Thai map in figure 18 is an example of itinerary text illuminated with a map.

Caroline Islanders memorized the relative distance between their islands in terms of travel time on the sea. Spanish missionaries recorded this data in the early eighteenth century and entered it on a map engraved for the *Philosophical Transactions* of 1721 (fig. 13). The *Transactions* explains that the numbers on that map placed in, and between, the islands are travel times as related by the islanders:

> The figure in the midst of every island, shows how many days' sail it is in circumference. The figure between each island, shows how many days are required to pass from one to the other,

This local information was subsequently converted into European linear scales, such as leagues of 3 Italian miles on a map compiled by Father Cantova (fig. 20).

Arab navigators also frequently used units of elapsed time rather than linear distance in their navigational texts; the *zam*, for example, equalled one watch at sea, or three hours.

In the later seventeenth century, the French traveler La Loubère observed that the Thai have a linear measurement, "their Fathom, which equals the French Toise within an inch," This unit was used in the surveying of land, "and especially in measuring the Roads, or Channels, through which the King generally passes." Along the road from Ayuthaya to Lop Buri, which lies roughly 70 kilometers to the north, "every Mile is marked with a Post, on which they have writ the number of the Mile."

The Traiphum

Literally meaning 'three worlds', the *Traiphum* (figs. 21 & 22) is a Thai text and map which gives an account of the Creation from the Theravada Buddhist perspective, although the work is ultimately part of Southeast Asia's larger Hindu-Buddhist heritage. An all-encompassing cosmography, the *Traiphum* charts everything from the mechanics of the universe to the wanderings of Buddha, from earthly geography to relative states of desire. The *Traiphum* is a map of Existence itself, a cycle of genesis and apocalypse which charts the journey that an individual, or indeed any living creature, may, over many lifetimes, take. Our 'real' world is but one of thirty-one states of existence, in three different worlds, which beings inhabit according to their level of accumulated merit. Mount Sumeru is the central axis of the earth and of the sun, moon, and stars, and accounts for changes in the seasons. Various subjects from the *Traiphum* are still found in the murals which decorate temple walls throughout Thailand.

The *Traiphum* is believed to have originated in the mid-fourteenth century in Sukhothai, Siam's capital before Ayuthaya. The text has been modified and expanded since then, and the earliest surviving fragment dates from the sixteenth or early seventeenth century. There was a renaissance in *Traiphum*-making after the Burmese pillaging of Ayuthaya in 1767, during the court's decade and a half sojourn in Thonburi before settling in Bangkok. After the centuries-old court was laid to ruins, the remaining court archives were brought to Thonburi, and the deteriorating documents copied. The first king of the new capital in Thonburi had the *Traiphum* text reconstructed, and both manuscripts illustrated here date from this transitional period during the latter part of the eighteenth century.

A journey through the *Traiphum* illustrated in figure 21, from left to right, begins with sacred Buddhist sites in India, such as revered temples, the bo tree under which Buddha was born, his mother's town, and the bo tree under which he reached enlightenment.[67] The greatest concentration of sacred places is found in this section. Moving to the right, just before the mid-point of the map, we descend from the previous, more metaphysical realms, to 'tangible' places within the immediate sphere of Siam, although there is never a clear break between the empirical, the mythological, and the sacred.

This central portion of the *Traiphum* identifies a number of towns in Lan Na and northern Thailand which are clearly recognizable today, including Chiang Mai and Sukhothai. Strong Lan Na influence is not surprising since Chiang Mai had been important in Siam's affairs for centuries, with close communication between north and

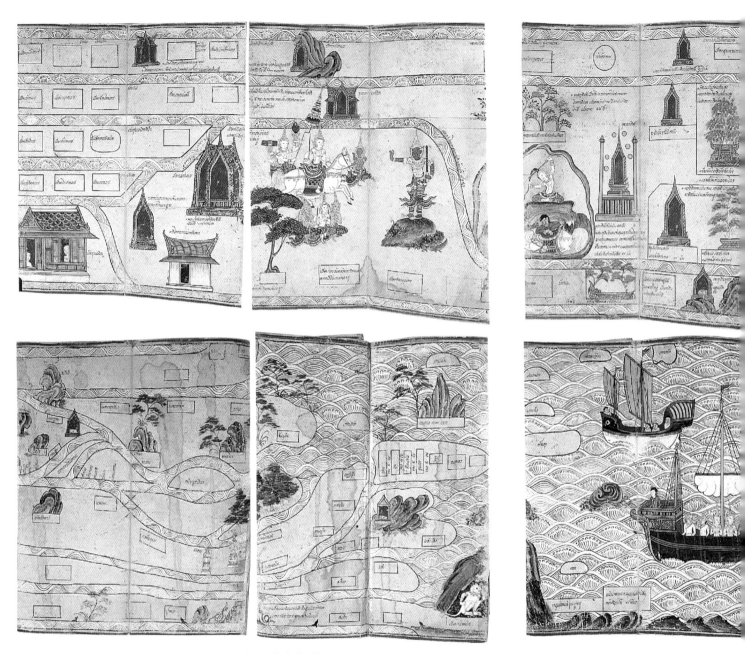

Fig. 21 The *Traiphum*, manuscript, 1776. [National Library of Thailand]

south at this time as the two kingdoms set aside their animosities to join forces against Burma, their perennial nemesis. No doubt it was this close contact between Thonburi and Chiang Mai (whose court moved to a camp south of Lamphun in 1775) which resulted in this particular *Traiphum's* pronounced emphasis on northern features.

Continuing to the right, the mapping of actual towns continues into the territory of Siam, with Ayuthaya prominent (even though it then lay in ruins). Reaching Bangkok and the southern Thai peninsula, the nature of the map changes yet again, now marking the number of days' travel between towns. But the itinerary nature of this section soon fades. To the west (the bottom mainland portion), in Burma, two *nagas* (waters spirits, represented as serpents) appear in the Narai River, just north of Rangoon. They ask Buddha to leave his footprint upon the land, which would be a holy and fortuitous occurrence.

Nearby, a monkey is perched at the edge of the mainland, over-looking a chain of little islands which runs along the bottom of the map and leads to a large island, which is Ceylon. We have now entered the realm of the Indian epic, the *Ramayana,* which tells how

Rama rescues his wife from captivity on Ceylon by running along a 'bridge' linking it with the mainland; the monkey at the northern end of the bridge is Rama's knight.

Buddha used the bridge as well, stepping from the mainland over Rama's Bridge to Ceylon after he had attained enlightenment. In doing so he left his footprint upon the summit of Adam's Peak, which is of course the imposing and precipitous mountain on the far right. There is a note referring to this most venerated of all Buddha footprints atop the mountain.

'Rama's Bridge' is in fact a line of minuscule sand banks dotting the waters between the island and India. The *Traiphum*, however, depicts a massive chain stretching out to what would seem to be the southern part of the Malay Peninsula. This is probably simple license on the part of the artist, since scale and orientation are fluid on the map, allowing us to journey from the Indian features on the far left to Malaya, Burma and Rama's Bridge. This image of Rama's Bridge and Ceylon might, however, be alternatively seen as evidence of European influence — for a long time Sumatra was identified as

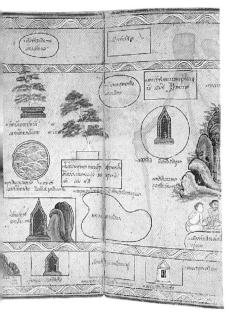
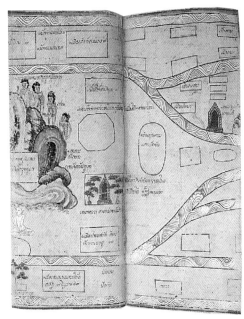

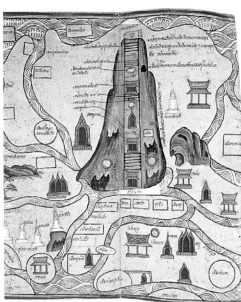

Ptolemy's island of *Taprobana* (Ceylon) on Western maps. If so, the stepping stones of Rama's Bridge then become the Andaman and Nicobar Islands, with '*Lanka*' placed in the position of Sumatra.

The *Traiphum's* representation of Ceylon itself is clearly based on indigenous sources. The four great stupas of Anuradhapura are marked in the southwest of the island, as are a couple of points along the pilgrimage route to Adam's Peak; Europeans are unlikely to have bothered to map such features, even if they had known of them. The one conspicuous omission in the depiction of Ceylon is the inland city of Kandy. Although Kandy is an ancient city, it gained importance when it became the last (albeit brief) retreat of the Ceylonese kings after the Portuguese took control of the kingdom of *Kotte* (Colombo) in 1539. The absence of Kandy on the *Traiphum* in figure 21 could either have been a mistake in the copying process, or because the image was transcribed, without alteration, from a much older model.

The smaller island to the right of Ceylon is identified as an isle of 'naked people', a characterization generally associated with the Nicobar islands. Though seemingly far displaced from the Nicobars'

true location to the north of Sumatra, mapping it as the final island on the extreme right fits logically into the *Traiphum's* fluid scale and panoramic orientation.

Figure 22 illustrates a markedly contrasting *Traiphum* map, now located in Berlin. More geographic than cosmographic, it depicts Asia from Japan to India, with east and south at the top. The map's scale and orientation is wildly inconsistent, beginning with China and Japan on the left, and ending with Ceylon (the yellow, triangular island) on the right. The latter clearly lies off the coast of India here, lacking the ambiguity of the previous *Traiphum.* The six white islands and the white corner of the adjacent foreshore are all Japan, with China occupying the rest of the lefthand mainland. The cluster of eight islands to the right of Japan are the 'black *farang*' islands, *farang* meaning both 'foreigner' and 'guava'. The fat peninsula is Indochina and the river presumably the Mekong; the long peninsula is Malaya; and the peninsula on the extreme right is India. Lastly, the square box at the bottom center is Ayuthaya, though the city already lay in ruins when the map was made. A ship enters the Gulf of Siam above it.

Chapter 3

Asian Maps
of Southeast Asia

Southeast Asia had a place in much of the literature and cosmography of her continental neighbors. Some of these references were direct cartographic records, while others were cosmographic concepts in which Southeast Asia played a significant role. Most often, however, Southeast Asia is found in textual entries. These include literary allusions, and the substantive content of travel records, as well as the itineraries of the pilots who sailed to the 'lands below the winds' or the 'southern ocean'.

Arab and Indian pilots relied on itineraries and sailing directions rather than charts. Although Marco Polo and other early European travelers in the Indian Ocean mentioned their pilots' 'charts', no such Arab or Indian navigational maps of the region are known. Detailed lists of places, latitudes, and relative compass bearings contained in some Arab navigational texts could in theory be used to construct maritime maps of the seas and oceans, but there is no firm evidence to suggest that any such charts were ever employed.

Marco Polo, making the trek westward across the Indian Ocean in the latter part of the thirteenth century, twice mentioned seeing maps. Once, in an apparent reference to sea charts and pilots' books used by his vessel's pilots, Polo stated that "it is a fact that in this sea of India there are 12,700 Islands, inhabited and uninhabited, according to the charts and documents of experienced mariners who navigate that Indian Sea."

Polo's other testimony to his Indian Ocean pilots' use of maps is especially important, because in it he unknowingly left us one detail which corroborates his story. He explains that although Ceylon has a circumference of

> 2,400 miles . . . in old times it was greater still, for it then
> had a circuit of about 3,600 miles, as you find in the
> charts of the mariners of those seas.

Polo's explanation of the size accorded Ceylon on the chart was that the chart's geography originated at an earlier time before much of the island had been submerged. In fact, what this passage indicates is that the chart followed the Ptolemaic model with its characteristic reversal of the relative proportions of Ceylon and India. Yet Ptolemy's *Geographia,* and maps constructed from it, were virtually unknown in Europe at this time, even among academics, and remained so until a century after Polo's return. Thus Polo clearly did not fabricate this key Ptolemaic error, which he himself did not understand. Ptolemy's

Geographia was, however, known to Arab scholars, and had profoundly influenced the Arab conception of Southeast Asia. But the fact that the map seen by Polo retained such an incorrect dimension for Ceylon supports the view that native pilots guided their vessels by navigational texts, and did not refer to the charts themselves.

Another important European witness to south Asian sailing was Nicolò de' Conti. In the first half of the fifteenth century, Conti mentioned that Arab and Indian sailors steered their vessels for the most part by the stars of the southern hemisphere, and made a statement which has commonly been interpreted as meaning that they were not acquainted with the use of the compass. In fact, he merely said that they did not *rely* on the needle for navigation.[68]

At the very end of the fifteenth century, Vasco da Gama was purportedly shown a chart of India by a 'Moor of Guzarat', just before his crossing of the Arabian Sea, but this is only mentioned retrospectively by João de Barros in the 1540s, and is not reported in earlier accounts of the voyage. Barros wrote that this chart was "of all the coast of India, with the bearings laid down after the manner of the Moors, which was with meridians and parallels." This is reminiscent of Ludovico di Varthema's claim that his Southeast Asian (presumably Malay) pilot consulted a chart marked with coordinates (1505). Barros described the map seen by da Gama as containing "bearings of north and south, and east and west, with great certainty, without that multiplication of bearings of the points of the compass" which typified Portuguese charts.

India

India's record of Southeast Asia is an enigma. Despite the profound influence of Indian civilization on much of Southeast Asia, there remains hardly any trace of Indian voyages to the east. No Indian maps of Southeast Asia whatsoever are known, nor geographic treatises detailing the itineraries and commerce of Indian sailors and traders. How is the contradiction between the undeniably extensive Indian presence in Southeast Asia and the utter void in cartographic and historical evidence reconciled?

India never 'colonized' Southeast Asia. Contact was not organized on any large scale, nor did Indian culture have the sense of posterity which led the Chinese to keep meticulous records of the world as they knew it. With the exception of military expeditions sent by the Chola emperors to Malaya and Sumatra in the eleventh century, India did not undertake a conquest of Southeast Asia. Rather, Indian influence was probably the result of successive individual initiatives as merchants sailed east to find their fortunes among the fabled isles of gold. No doubt many perished, but others established themselves in coastal communities where they married the daughters of local chiefs and assumed some degree of influence. These same local rulers, noting the legitimacy to a king's power afforded by Indian religion and political thought, were receptive to adapting the foreign ideas for their own ends, and similarities between indigenous and Indian traditions made this assimilation all the more natural and fluid. Indians who became respected citizens on Southeast Asian soil eventually returned to their homeland, where others in their family or village, on hearing their story, elected to join them when next they ventured east.

Although the sort of small-scale peregrinations which seem to have characterized Indian contact with Southeast Asia did not leave any formal histories or maps, what they did foster were references to Southeast Asia in Indian literature. Early traces are found in India's *Jataka* fables of popular Buddhist lore, which originated well over two thousand years ago but assimilated stories about Southeast Asia as Indians returned and shared their adventures. These tales became associated with Mahayana Buddhism and its affinity for common folk, for trade, and in turn, travel.

Some of the legends describe Indian merchants who sailed to Southeast Asia on trading expeditions. We hear, for example, of a Prince Mahajanaka, who joined a group of merchants bound for *Suvarnabhumi,* the Land of Gold, representing either Sumatra or Southeast Asia as a whole. Similarly, in the tale of *Kathasaritsagara,* a Princess Gunavati, while *en route* to India from *Kataha* (possibly Kedah, Sumatra), is shipwrecked on the coast of *Suvarnadvipa* (Golden Island or Golden Peninsula). Clear references to Southeast Asia are also found in the *Ramayana,* the classic epic poem about the abduction of Rama's wife by the king of Ceylon and Rama's attempts to rescue her. These stories record seven kingdoms on the 'Gold and Silver Islands' beyond Ceylon.

China

Chinese cartography, which dates back to ancient times, influenced Vietnamese mapmaking, but was not a major cartographic influence in the rest of Southeast Asia (and the West, in turn, was not as much of an influence on Chinese mapmaking as once was assumed).[69]

In China, as in Southeast Asia, the earth was generally believed to be flat. Chinese cosmography, however, held that the flat earth was not level. The plane of the earth was believed to be tilted, that is to say, inclined to the mountainous northwest and falling away to the southeast. The incline made the waters of the earth flow via rivers 'downward' from northwest China, emptying into the ocean sea, which itself leaned to the south and east. Southeast Asia figured importantly in this tilted flat earth concept, since it was in the 'low' southeast corner of the earth, the vast sea world of Austronesia, that all the earth's waters ultimately accumulated. Chinese seafarers, heading south to the lands of the 'barbarians', may thus have envisioned their course as literally 'down'. Mendes Pinto, exploring Southeast Asia and China in the mid-sixteenth century, noted that Sumatra, Makassar, and the other Indonesian islands are "referred to as 'the outer edge of the world' in the geographical works of the Chinese, Siamese, *Gueos* [a purported Southeast Asian nation of cannibals] and Ryukyu [the chain which includes Okinawa]."

Yet the idea of a spherical earth, literally and poetically, co-existed along side this scheme of things. At least as far back as the second century B.C., Chinese astronomers had written about the sphericity of the earth.[70] Taoist cosmography, philosophically describing a spherical earth, held that heaven, after which man was modeled, 'revolved' from left to right, while earth, after which woman was conceived, did so from right to left.[71] The traditional concept of *yin* and *yang* was also applied to the Chinese world concept. *Yin,* the passive power, was associated with the colder north, while the active power, *yang,* was associated with the hotter climes of southern China and the southern realms of Southeast Asia.

In Taoist creation myths, the emperor of the South Sea (that is, Southeast Asia) was *Shu* (Brief). *Shu* periodically visited the central region, *Hun-tun,* which was conceived as a cosmic egg or gourd, where he met with *Hu* (Sudden), the emperor of the North Sea.[72] Interestingly, the analogy of an egg yoke for the earth floating in the heavens was used both in ancient China and ancient Greece.

China and Southeast Asian Trade

Chinese awareness of India, of the Roman Empire, and of the possibilities of trade with both, was heightened in the latter part of the second century B.C. (Han Dynasty) as a result of the adventures of an explorer/diplomat by the name of Chang Ch'ien. Chang made two expeditions, the first in 128 B.C. to Central Asia, during which he was taken prisoner for a decade by the Hsiung Nu (Huns) in the Altai Mountains, and again in 115 B.C. to western China. On his first expedition he found cloth and bamboo in Bactria and Fergana (north of modern Afghanistan), which in turn had been acquired from India, but which Chang recognized as being ultimately of southern Chinese origin. This was to prove eventful for both China and Southeast Asia, since it opened China's eyes to the possibilities of more direct trade with lands to the west, and it set the stage for the role that Southeast Asia would play as a facilitator of this trade. On the second expedition, Chang had his envoys continue further west, bringing gold and silks to Persia and the eastern periphery of the Greco-Roman world. Chang's endeavors led to the birth of the Silk Road along whose length there subsequently flowed not only trade but also an improved knowledge of the world. The latter was shared between Rome and China, and the lands that bordered the route; China learned of Burma and other neighbors in Southeast Asia.

The Southeast Asian mainland, however, was not itself an important destination for the earliest Chinese traders. What little it offered them in terms of indigenous resources could be obtained in ample quantities from sources farther north. It was, rather, itineraries to the west that first lured Chinese seafarers into the Indian Ocean. Thus for early Chinese sailors, Southeast Asia was an impediment as well as a destination, in the same manner that America was initially seen as an obstacle to Europeans sailing west in quest of Asia. Similarly, both the Europeans in America and the Chinese in Southeast Asia sought short-cuts across isthmuses. Many Chinese and Indian traders may have opted to cross the northern neck of the Malay Peninsula at the Isthmus of Kra rather than undertake the arduous voyage around the peninsula and through the Malacca Strait, just as European sailors experimented with crossing Central America at Darien to avoid the lengthy route around South America and through the Magellan Strait. Yet another parallel can be found between the Gulf of Siam and the vast mouth of the Rio de la Plata in Brazil; both must surely have tricked pilots into believing that they had reached the end of the continental obstruction, only for them subsequently to discover that they still had the full Malay Peninsula and the whole of South America, respectively, to round.

Fig. 22 *Traiphum,* 1776 (anonymous).
(Section illustrated measures 51.8 x 138 cm) [With permission of the Staatliche Museen zu Berlin - Preußischer Kulturbesitz Museum für Indische Kunst]

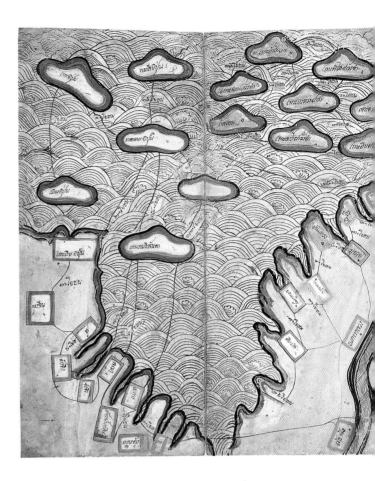

The fragmentation of the celestial kingdom resulting from the fall of the Han dynasty between 190 - 225 A.D., expedited the beginning of Chinese intercourse with Southeast Asia. As a result of the dynasty's demise, most of the territory south of the Yangtze River became part of the kingdom of Wu which, though isolated from countries to the west, controlled the long southern Chinese coastline and thus was in an ideal position to trade with Southeast Asia. In order to exploit this window of opportunity looking on to the countries that lay to the south, an embassy was dispatched, in the third century A.D., to southern Indochina under the guidance of K'ang-T'ai, a senior secretary, and Chu-Ying, who was in charge of cultural relations. Although the original accounts of this enterprise are lost, much of their content has been passed on to us by way of the many later Chinese documents that quote directly from them. These extracts are often confusing and have probably been corrupted by copyists, but nonetheless they constitute much of what is known about Southeast Asia at the time of the early Christian period, and they have provided us with the only clear record of the kingdom of Funan.

Chinese maritime contact with Southeast Asia probably began over two thousand years ago. According to the *Han Shu* (*History of the Han Dynasty*), Chinese vessels were visiting Sumatra, Burma, Ceylon, and southeastern India during the Western Han Dynasty (206 B.C. - 8 A.D.).[73] The scholar and official, Jia Dan (730-805), described the sea route from Canton (Guangzhou) to Baghdad, via Singapore and the Malacca Straits, the Nicobar Islands and the Indian Ocean, Ceylon and India, and finally the Arabian Sea and the Euphrates, at which point the journey was completed by land.

Chinese vessels began regularly to make the round trip to Southeast Asia in the eleventh century, during the Song Dynasty.[74] Although the Song era is remembered as being primarily a period of intellectual strength, and a time when advances were made in printing techniques, it was also one in which curiosity about the outside world was not deemed respectable. Confucian philosophers, in particular, sought to discredit both the accuracy and the merit of knowledge about distant realms.

Some Song government bureaucrats did, however, chronicle the reports they heard about the lands to the south and two texts have survived with details pertaining to Southeast Asia. One was written in 1178 by Chou Ch'ü-fei, an official of the maritime province of Kuang-hsi, the second a half century later, in 1226, by Chao Ju-kua, the Commissioner of Foreign Trade at Ch'üan-chou (coastal province of Fukien). Chou Ch'ü-fei explained that

> The great Encircling Ocean bounds the barbarian countries [Southeast Asia]. In every quarter they have their kingdoms, each with its peculiar products, each with its emporium on which the prosperity of the state depends. The kingdoms situated directly south [of China] have [the Sumatran maritime state of Srivijaya] as their emporium; those to the south-east [of China] have *She-p'o* [Java].[75]

Referring to Indochina, Chou states that although

> it is impossible to enumerate the countries of the South-Western Ocean . . . we have to the south [of *Chiao-chih* = Tongkin] *Chan-ch'eng* [Annam], *Chen-la* [Cambodia], and *Fo-lo-an* [?].

To the west of Cambodia (in present-day Thailand?) lies

> the country of Teng-liu-mei. Its ruler wears flowers in his hair, which is gathered into a knot. Over his shoulders he wears a red garment covered with white. On audience days he ascends an open dais, since the country is wholly without palace buildings of any kind. Palm leaves are used as dishes in eating and drinking; spoons nor chopsticks are used in eating which is done with the fingers.

Another kingdom is *Tan-ma-ling,* probably in the region of Ligor. Around the city of *Tan-ma-ling*

> there is a wooden palisade six or seven feet thick and over twenty feet high, which can be used as a platform for fighting. The people of Tan-ma-ling ride buffaloes, knot their hair behind and go barefoot. For their houses, officials use wood while the common people build bamboo huts with leaf partitions and rattan bindings.

Among the products of *Tan-ma-ling* are bee's wax, various woods including ebony, camphor, ivory, and rhinoceros horn.

Langkasuka

Six days and six nights' sail from *Tan-ma-ling* is *Langkasuka,* one of the most enduring names of early Southeast Asia. *Langkasuka* was centered in the vicinity of Patani, on the east coast of the Malay Peninsula, and is amply recorded in Chinese, Arabic, and Javanese history (fig 121). Probably founded in about the second century, *Langkasuka* experienced the eclipses and renaissances of any long-lived state; it still appeared on a Chinese map compiled from early fifteenth century data known as the *Wubei zhi* chart (fig. 23), but seems to have disappeared just before Portuguese familiarity with Malaya began in the early sixteenth century. So prevalent is the

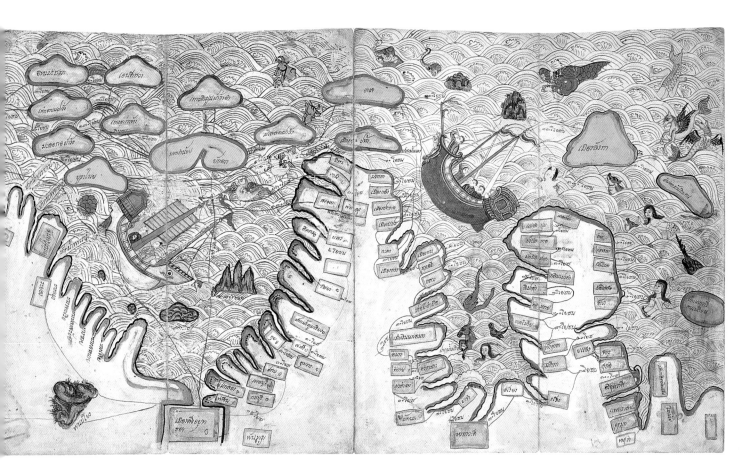

kingdom in early annals that its name was considered for that of independent Malaysia after the Second World War.

Chinese records of *Langkasuka* date back as early as the seventh century. It is described as a kingdom in the Southern Sea, covering an area thirty days' journey east-to-west, and twenty north-to-south, lying 24,000 *li* from Canton. Its climate and products are similar to those of Funan. The capital

> is surrounded by walls to form a city with double gates, towers and pavilions. When the king goes forth he rides upon an elephant. He is accompanied by banners, fly-whisks, flags and drums and he is shaded with a white parasol. It is customary for men and women to go with the upper part of the body naked, with their hair hanging dishevelled down their backs, and wearing a cotton sarong.

Langkasuka was also mentioned by Chinese monks making the pilgrimage to India. I-Ching records one such visit in the late seventh century, and was clearly impressed by the warm hospitality of the inhabitants of *Langkasuka*. On one voyage, three pilgrims "let hang the mooring ropes" from their port on the Gulf of Tongkin

> and weathered innumerable billows. In their ship they passed *Chen-la* ['Funan' in the text] and anchored at Langkasuka, [where] the king [bestowed] the courtesy appropriate to distinguished guests.

Another Buddhist pilgrim *en route* to *Langkasuka*

> buffeted through the southern wastes in an ocean-going junk, [passing] *Ho-lin*g [Java] and traversed the Naked Country [Nicobars]. The kings of those countries where he stayed showed him exceeding courtesy and treated him with great generosity[76]

The Islands

Chou Ch'ü-fei wrote that "to the south of [Srivijaya, i.e., Sumatra] is the great Southern Ocean, in which are islands inhabited by a myriad and more of peoples." Then the concept of a flat earth comes into play, so that "beyond these one can go no further south." The Chinese belief that the flat world is angled becomes especially important to the east of Java, for here "is the great Eastern Ocean where the water begins to go downward."

Chou Ch'ü-fei described the relative importance of these trading itineraries to the south: "of all the wealthy foreign lands which have great store of precious and varied goods, none surpasses the realm of *Ta-shih* (Arabia)." He believed that trade with Java *(She-p'o)* was second in importance, and Sumatra *(San-fo-ch'I)* ranked third. Sumatra, however, because of its position, "is an important thoroughfare on the sea-routes of the foreigners on their way to and from [China]."

Chao Ju-kua also recorded an active role for the Sumatran intermediaries in trade via Southeast Asia, noting that "because the country [Sumatra] is an important thoroughfare for the traffic of foreign nations, the produce of all other countries is intercepted and kept in store there for the trade of foreign ships." He compiled information about twenty-eight countries from discussions with both Chinese and foreign sailors and his book, entitled *Chu-fan-chih* (*Description of the Barbarians*), records information about various countries in Southeast Asia, South Asia, and as far west as Africa and even the Mediterranean. Although the reports from his first-hand interviews form the principal value of the work, it is supplemented with information from older records.

From Chao Ju-kua we also learn a curious lesson about how Southeast Asian nomenclature could be deliberately manipulated. The Chinese appetite for Javanese pepper was such that the Chinese court, alarmed about the considerable exodus of copper

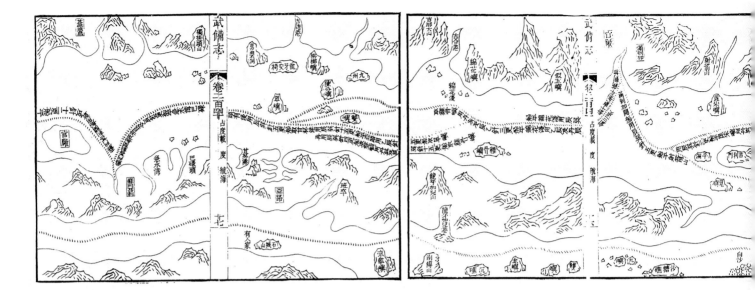

Fig. 23 The section of the *Wubei zhi* Chart covering Southeast Asia, 1621. [Courtesy of the Library of Congress, Washington, D.C., E701.M32.1]

coinage to Java to pay for it, banned trade with the island. The Javanese traders, in order to circumvent the trade ban, simply re-named their island; the Chinese traders were now buying their pepper from a land called *Sukadana (Su-ki-tan)*.[77]

Religion was another reason for Chinese incursions into the Indian Ocean, and another reason why Southeast Asia benefited from being on the crossroads between two great civilizations. By the first century A.D. Buddhism had reached China, and by the third century it was established along the delta of the Red River in Vietnam. Soon, some of its more devoted adherents began to undertake pilgrimages to their Holy Land, India. Monks traveled to India by both a land route through Burma and the sea route via the Malacca Strait. The earliest surviving record left by such a pilgrim is that of Shih Fa-Hsien, who, inspired to make an accurate Chinese transcription of Buddhist texts from the original Sanskrit, traveled overland from China to India in 399 A.D., returning by sea in 413-14. On the way back home to China from Ceylon, Fa-Hsien's ship went aground off the coast of Java, and he was lost at sea for seventy days before finally reaching China. This is the earliest record of a return to China from southern Asia via the maritime route.

As a result of this sea route, Buddhism was well established in Sumatra by the seventh century. I-Ching, who was in India and the southern seas between 671-695 and compiled records of sixty pilgrims' journeys to India, mentioned a multi-national community of a thousand monks (Mahayana Buddhist) in the Sumatran state of Srivijaya in 671.

China and the Philippines
Chinese commercial interest in the Philippines dates back at least to A.D. 982, when an Arab ship carrying goods from Mindoro is recorded as having reached Canton.[78] Direct Chinese trade with islands to the east began by the twelfth century. In 1127 A.D., the Song rulers were forced south of the Yangtze River, and a southern capital was established at Hangzhou, from whence ceramics and other commodities were exported to the Philippines. Chinese sailors became increasingly familiar with northern Borneo and the Philippines, and trade links were established as far afield as the Spice Islands which were reached via the Sulu Sea. These trading networks probably elevated Filipino knowledge of their islands as

well, since they precipitated an elaborate system of trading centers to gather the forest goods sought by the Chinese and to distribute the wares acquired from them. In 1226, Chao Ju-kua referred to the Philippines by the general term *Ma-yi,* and to the Visayan islands as *San-hsü* (three islands). He also used the term *Lin-hsing,* which probably referred to Luzon. Interest in Philippine commodities is evidenced by a Chinese writer in 1349, who noted that "Sulu pearls are whiter and rounder than those of India," and that they commanded a high price.[79] Embassies from Luzon were sent to China in 1372 and 1408, bringing such gifts as "small but strong" horses, and returning with Chinese silk, porcelain, and other goods. Chinese trade with the Philippines was evident to the earliest Europeans to reach the islands; the lords of Cebu had Ming porcelain when Magellan reached there in 1521.

Although Chinese interest in Southeast Asia was traditionally commercial, the Philippines were briefly the target of an emperor's conquest. In about 1405-1410 Yung Lo, second Ming emperor, sent an imposing fleet under Zheng He (Cheng Ho) in a bid to establish a foothold in the Philippines. He was unsuccessful.

Zheng He
Zheng He, however, had considerable success in opening China up to much of southern Asia and parts of eastern Africa. In the years between 1405 and 1433 — ironically, the very period that Portugal was beginning to flex its muscles and push ever further around Africa — this Chinese navigator, who became known as the 'three-jewel eunuch', led seven expeditions to the southern seas, following the Southeast Asian coast into the Indian Ocean and along the eastern shores of Africa, possibly reaching as far as Kerguelen Island in the southern Indian Ocean. The scale of these undertakings was fantastic. The first expedition purportedly boasted 62 large vessels, 225 smaller ones and a crew in excess of 27,000 men; it touched on the shores of Sumatra on the outward voyage, and Siam and Java on the return. By the seventh voyage, Zheng He had won China commercial and diplomatic ties with 35 countries in the Indian Ocean, Persian Gulf, and eastern Africa. Fra Mauro, on his world map of 1459, records Chinese naval junks off the east coast of Africa — probably those of Zheng He — which came from Arabic sources.

A chart based on Zheng He's voyages (fig. 23) is found in a printed work entitled *Wubei zhi* (*Treatise on Military Preparations*)

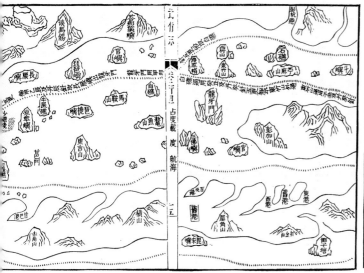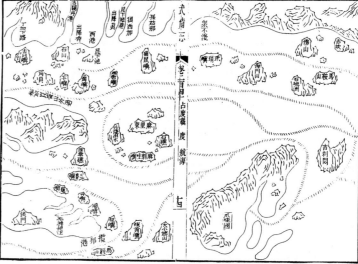

completed by Mao Yuanji about 1621 (the date of the book's preface), and presented to the throne in 1628. Mao does not name the source of the map, but there is little question that it is based extensively on Zheng's voyages. "His maps," Mao states, "record carefully and correctly the distances of the road and the various countries and I have inserted them for the information of posterity and as a momento of [his] military achievements."[80] We know from the text of the *Treatise* that maps and information were collected before each of Zheng's voyages, and that charts were compared and corrected for compass bearings and guiding stars, with copies made of drawings of the configuration of islands, water bodies, and the land.[81] The map, as it has survived, however, appears to have been constructed in part from textual sources.

Originally, the *Wubei zhi* chart was probably a long, single piece, stored as a scroll, though for the book it was divided into a series of strips. One consequence of the strip format (whether in scroll or segmented) is that orientation is not consistent. In addition, scale is stretched and compressed according to the amount of detail included in a particular section. The cluttered and dangerous coastal area of Singapore, for example, is drawn on a scale more than three times larger than that of the east coast of Malaya and two-and-a-half that of the west coast.[82]

The map included sailing instructions that modern scholars have found to be fairly accurate.[83] Sailing the Malacca and Singapore Straits from west to east, the pilot guide states that (using modern compass bearings)

> having made the Aroa Islands, setting a course of 120° and then 110°, after 3 watches the ship is abreast of [South Sands]. Setting a course of 115° and then 120° for 3 watches the ship comes abreast of [Cotton Island]. After 10 watches on a course of 130° the ship is abreast of [Malacca]. Setting sail from Malacca on a course of 130°, after 5 watches the ship is abreast of [Gunong Banang]; after 3 watches on a course of 130° the ship is abreast of [Pulau Pisang] . . .[84]

From Pulau Pisang, a course of 135° brings them to Karimun, and from there, 5 watches of 115° and then 120° and the ship makes Blakang Mati, passing out through Dragon-Teeth Strait. With 5 watches at 85°, the vessel then reaches Pedra Branca (Pulau Batu

Puteh), and after passing Pedra Branca, sets a course of 25° and then 15° for 5 watches, which brings the ship abreast of 'East Bamboo Mountain', one of the two peaks of Pulau Aur. Finally, setting a course of 350° and then 15°, the ship passes outside of Pulau Condor.

The budding commercial empire pioneered by Zheng He was short-lived. After the death in 1424 of the emperor who sponsored him, Yung-lo, and the death of Zheng He himself a decade later, Chinese authorities rejected any further forays in the southern seas. Commerce in the Indian Ocean trade was once again relinquished to networks of Muslim and Southeast Asian traders on the eve of the Portuguese penetration of eastern waters.

Japan and Korea

Although Japanese vessels were plying Southeast Asian seas from about the turn of the fifteenth century, no Japanese charts are known prior to the arrival of a substantive European presence in the region. Japanese traders were already well familiar with the South China Sea when Europeans first appeared in those waters, and are mentioned, for example, by Spanish sailors reaching the northern Philippine islands of Luzon and Mindoro in the 1560s, yet Japanese mapmaking of Southeast Asian waters is not known until after Europeans introduced their chartmaking techniques into Japan in the latter sixteenth century. These charts of maritime Southeast Asia were known as *nankai karuta* (south-sea charts), as differentiated from charts of their own coasts, *nihon karuta*.

'India' appears in early symbolic Japanese and Korean maps of the world which were inspired by Buddhist pilgrimages to holy sites in the land where Buddha was born. One such world view which may have had a place in China, Japan, and Korea was the so-called Buddhist world map, or *Gotenjiku* ('Five Indias'). Inspired by the travels of a Chinese monk in the Tang Dynasty, named Xuanzhuang (602-64), the world is depicted here in the shape of an egg, with north, the larger end of the egg, uppermost. As in many other world views which have their origins in Buddhist and Hindu cosmologies, Mount Sumeru lies at the center. Southeast Asia is not recorded on this map as such; a 'Mt. Malaya' in the south is another mythical mountain on whose summit sits the 'Castle of *Lanka*'.

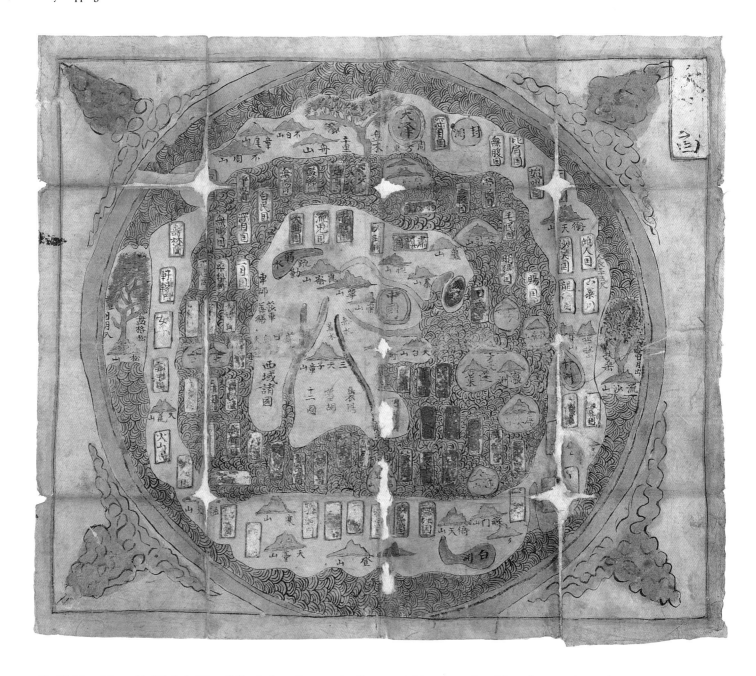

Fig. 24 Map of the world, *Ch'ŏnhado* ('*Map of all under heaven*'), manuscript, Korean, probably seventeenth or eighteenth century. Among the countries in the inner ocean ring are the 'islands' of Siam and Cambodia. (26.8 x 32.4 cm) [Martayan Lan, New York]

Southeast Asia is, however, recorded in a popular Korean view of the world known as the '*Map of all under heaven*', or *Ch'ŏnhado* (fig. 24). This particular representation of the world probably originated in the late sixteenth or early seventeenth century, and was copied with little or no change into the nineteenth century. The *Ch'ŏnhado* did not reflect the best geographic knowledge available to the society which created it, but, as with the more enduring naive maps in Europe, it is doubtful whether this was its authors' intention. The map depicts a central landmass comprising China, Korea, and several historical and mythical countries. In the inner of the two encircling oceans lie the world's island countries, which in addition to Japan and the Ryukyus, include the 'islands' of Cambodia and Thailand. Mainland Southeast Asia was a long sea voyage from Korea, and it is understandable that Cambodia and Thailand would have been treated as islands like Japan and the Ryukyus (just as many Western mapmakers of the sixteenth and seventeenth centuries believed Korea to be an island). Reinforcing

the idea of Cambodia and Thailand's insularity was the cultural 'sea' separating them from China and Korea, the mapmaker envisioning them as lands of a different tradition and people.

The ocean sea which harbors these 'islands' is itself ringed by land. While the central landmass and its neighboring 'islands' embrace both the real and the fictional, this outer ring of land records exclusively fictional places and features. The *Ch'ŏnhado* remained popular through the late nineteenth century.

The Arab View of Southeast Asia

Arab pilots began searching the shores of Southeast Asia for spices and drugs in the early seventh century. The information by these sailors and traders, who were primarily interested in commerce rather than in science for its own sake, subsequently appeared in travelers' accounts, and in the geographical, historical, and medical (herbal) treatises which, in turn, drew on these accounts. The earliest known compilations of Arab travelers' tales dealing with Southeast Asia, date from the mid-ninth century, though some of the stories which appear in them are older.[85] Some authors interviewed their informants directly, but most — in particular the later geographers — simply copied and elaborated upon earlier writings. This material is the source of most of what is known of the early Arab experience in Southeast Asian waters.

Most of the places mentioned in the various Arabic geographies and navigational tracts cannot be identified with certainty, nor can they even be presumed to have always represented the same place to each sailor or author. Although most of the locations commonly mentioned were originally based on fact, many gradually assumed a mythological status, slowly becoming the progeny of sailors' lore. Some of the fanciful places may derive from the Alexander Romances, or perhaps from attempts to reconcile Ptolemaic geography with the Qur'an.

Arab geographic knowledge of Southeast Asia after about 1000 was beset by two major problems. First, geographers failed to tap reports of more recent voyages to the region, if such information was available. Instead, the centuries-old stories, by now largely mythicized, continued to be recycled. The great Arab geographer, al Sharif al-Idrisi, based in Sicily in the twelfth century, certainly dedicated himself to gathering information from travelers, yet his Southeast Asian material remains the product of lore dating back three centuries. Not even the account of Ibn Battuta's journeys in the early fourteenth century, the period's unique surviving first-hand travel narrative, appears to have been utilized by geographers.

Secondly, the introduction of Ptolemaic theory into the Arab world view created irreparable confusion. Geographers such as Khwarizmi (tenth-century) attempted to incorporate elements of Ptolemy's *Geographia* into their conception of Southeast Asia, juggling their data to make it 'work'. This was exacerbated by the Ptolemaic idea of an enclosed Indian Ocean which led to a mixing up of information from Southeast Asia *(al-Zabaj)* with material from East Africa *(al-Zanj)*, even though Arab geographers did not accept the existence of the land bridge itself (see figs. 27 & 28).

When the Portuguese reached Southeast Asia at the turn of the sixteenth century, the Arab geographical concept of Southeast Asia had stagnated for three centuries, and Persian and Turkish geographers were beginning to rise to prominence. For example, 'Abd al-Razzaq, a Persian historian of the fifteenth century, already cites *Tenasserim* (Burma) and *Shahr-I Naw* (Siam) as being among the destinations of sailors leaving Hurmuz. Arabic mapping of Southeast Asia began to utilize European sources in the late seventeenth or early eighteenth century.

Southeast Asian Landfalls in Arab Navigational Texts

Navigational texts offer better insight into the Arab geographic conception of Southeast Asia than the geographers' treatises, though even they are often ambiguous and inconsistent. The better Arab geographical texts, such as that of Ibn Sa'id (d. 1274), also record detail about topography, and quote information derived from earlier navigational tracts.

The earliest surviving Arab navigational treatises detailing travel in Southeast Asian waters are relatively late, dating from the latter part of the fifteenth or early sixteenth century, but incorporate data from previous centuries of navigation. They were written by two fifteenth-century sea captains, Ahmad ibn Majid, whom tradition tenuously ascribes as the pilot Vasco da Gama hired to conduct his fleet's crossing from northeast Africa to India, and Sulaiman al-Mahri. Ibn Majid's work was written in poetry to facilitate

committing its instructions to memory. These pilot books seem to have been the culmination of at least four centuries of such texts; ibn Majid, in fact, cites predecessors to his work dating back to the early twelfth century, and a travel narrative from the beginning of the eleventh. These navigational works, however, had no obvious impact on Arab mapmaking.

The pilots' tracts cite compass bearings, and latitudes, measured in *isba'* (approximately 1° 43′) from a given star. A sample passage about Sumatra from Ibn Majid's text gives a flavor of these works:

> Sumatra begins on the north with mountain of Lamuri at 7⁷/₈ *isba'* against the Little Bear, but according to some 7³/₄ *isba'*; and it ends in the south with a place called Tiku Tarmid. People disagree as to the latitude of this place and there are three opinions: the first that there the Little Bear is at 4 *isba'* and most of the Indians lean toward this; the second is not quite 4 *isba'* and the Arans and some of the Cholas prefer this; while according to others, it is 3¹/₂ *isba'*.

Ibn Majid believed the last of these to be the most accurate.

Sulaiman al-Mahri's navigational tract reads similarly. He notes islands off the west coast of Siam (Malaya)

> which are called Takwa, and are from 5 *isba'* - 2 *isba'* Pole Star. The first of them is the island of Fali. This is a large island, the northern point of which is 5 *isba'* Pole Star, and the southern one, 4 ³/₄ *isba'* Pole Star. Next, to the south in line with it is Fali Kara, the northern point of which is 4¹/₂ *isba'* Pole Star, and facing it on the east is the island of Lamamand and the estuary of Markhi. After it on the south is another island nearby called Awzamanda, having the appearance of the large sail [of an Arab boat]. The Pole Star there is 4¹/₄ *isba'*. . . [Another is] the island of Pulau Lanta, an inhabited island, whose inhabitants are permanently settled. Fruits are found there and it is at 2³/₄ *isba'* Pole Star . . . Now the island of Urang Salah [*Junk Ceylon*, = Phuket] is a large, long island [whose cape is] at 2³/₄ *isba'* Pole Star.

The most important destination covered by these navigational texts, and one whose identity is not in question, is Malacca, which had risen as the region's principle trading center for Arab navigators during the fifteenth century (see extract, page 104). Singapore, parts of Sumatra and Java, and China were also focal points of these sailing instructions.

The coasts of Burma and the Andaman-Nicobar Islands were well-frequented, and are described in the pilot books. The texts group the shores of Malaya with Siam, and the mainland to the east with China, though *Zaiton* (Quanzhou) is considered the threshold of the kingdom of China proper. Along the route from Singapore to China, the ports described included *Shahr-I Naw,* referring to the Thai metropolis of Lop Buri or Ayuthaya. The term *Shahr-I Naw* (new city) was later borrowed from Arab acquaintances by Europeans such as Mendes Pinto, and incorporated by some mapmakers (see fig. 80). Beyond it, along the eastern shores of Indochina, the navigators frequented Champa *(Shanba),* and the port of *Kiao-Chi* in Annam. Borneo was visited, the Arabs' *Barni* referring to the sultanate of Brunei; other parts of Borneo are probably recorded by some of the unidentified place-names, the sailors who visited them not realizing that they were parts of the same island. To the northeast, Arab merchants sailed as far as Formosa, which was known as *Likiwa* (Liu-ch'iu) or *al-Ghur* and was described in Ibn Majid's navigational text of 1462. The Portuguese chronicler, Tomé Pires, writing in 1512-15, had heard both names, noting that "the *Lequeos* are called *Guores.*" Timor, Sulawesi and Banda were the southeastern limit of the Arab pilots' traditional ports of call. This passage from al-Mahri illustrates the limits of Arab navigational confidence to the east, as well as the texts' tendency to offset the Indonesian islands to the south:

> Know that to the south of the island of Jawa [Sumatra or Java] are found many islands called Timor and that to the east of Timor are the islands of Bandam, also a large number. The latter are places for sandal, aloeswood, and mace. The island called the Isles of the Clove [i.e., the Moluccas] as east of Jawa; they are called Maluku.

Since many of the islands described in the navigational and travel texts have not been identified with any confidence, the extent of Arab sailors' familiarity with the region is not known. For example, although Arab traders were apparently transporting merchandise from the Philippine island of Mindoro to Canton at least as early as 982, there is no reference to the Philippines in the pilot books which is clearly recognizable to us now. Some of the unidentified places mentioned in the pilot books are probably parts of the Philippines. For example, the island of *Fariyuq*, which is described in a navigational treatise by the fifteenth-century captain, Sulaiman al-Mahri as a "large, inhabited island to the southeast of the ports of China", may be Palawan.

Geographical Treatises and Travel Narratives

Whereas the authors of navigational texts had little interest in recording anything that was not specifically of value to the seafaring merchant, the authors of geographical treatises and travel narratives filled their pages with descriptions of fauna, vegetation, strange peoples and lore about the region, which they compiled from existing books, or from the testimony of sailors and traders who claimed to have been there. These works are the most numerous extant Arab texts about Southeast Asia.

The one first-hand travel narrative of a credible voyage to Southeast Asia is the classic *Rihlah* (*Travels*) of the most famous of Arab travelers, Ibn Battuta. Battuta's odyssey, which began in 1325, came relatively late in the history of Arab voyages to Southeast Asia, and although his text is certainly based to a large extent on the author's own extensive travels, whether or not even he actually reached Southeast Asia is disputed. The only earlier extant account which purports to be first-hand is the tenth-century work *Meadow of Gold* by al-Ma'sudi, who claimed to have traveled to Southeast Asia and China in the pursuit of knowledge, but the veracity of his voyage is doubted, and his work was probably compiled from the stories he heard in the port of Siraf.[86]

One of the most important Southeast Asian destinations, according to Arab texts, was the port of *Zabaj*, which was said to belong to the empire of a Maharaja and was located along the sea route to China, though far to the south, perhaps on the equator. It was believed to be closer to China than India, but was considered part of latter. There is no definite identification for *Zabaj*; though it was certainly an Indonesian island. Probably it was Java or the east coast of Sumatra; *Zabaj* may be related to the term Srivijaya, the great Indonesian maritime empire of a thousand years ago.[87]

Some of the same texts which describe *Zabaj* also refer to the island of *Jaba*. As with medieval European maps which depict an island of '*Java*', *Jaba* may have been either Java or Sumatra, or both, depending on the author. From *Jaba* it is said to be about fifteen days' sail to the Islands of Spice — clearly Java, assuming the texts' 'spice islands' to be the Moluccas — yet it is also described as being situated on the sea route to China, which makes Sumatra more likely.

Volcanoes are mentioned in connection with both *Zabaj* and *Jaba*. Akbar al-Sin wa 'l-Hind (mid-ninth century), wrote that

> near Zabaj is a mountain called the Mountain of Fire, which it is not possible to approach. Smoke escapes from it by day and a flame by night, and from its foot comes forth a spring of cold fresh water and a spring of hot water.

The report conjures up an image of a monumental volcano such as Krakatau or one of the smaller volcanic Indonesian islands known as '*Gunung Api*' ('fire mountain'); a tamer one is described at about the same time by Ibn Khurdadhbih as being on *Jaba*:

> There is in Jaba a small mountain with fire on its summit stretching for the distance of a hundred cubits but having only the height of a lance. One sees its flames at night but only smoke during the day.

The island of *Ramni* is clearly Sumatra, usually described as being the first to be reached after leaving Ceylon, and washed by two seas — *Harkand* (Bay of Bengal) and *Salahit* (Malacca Strait). In the mid-ninth century, Akbar al-Sin wa 'l-Hind wrote that on the island of *Ramni* there are plantations called *fansur,* the latter being the name of a Sumatran kingdom later visited by Marco Polo. By about the year 1000, there are references in Arab texts to an island named *Lamuri,* another of Polo's Sumatran ports-of-call, as well as of Friar Odoric and other early European visitors. In the fourteenth century we find the island referred to by its modern name, *Sumutra,* in the text of Rashid al-Din (d. 1318) and *Samutra* in the travelogue of Ibn Battuta (dictated in 1355).

Another Southeast Asian port frequented by Arab sailors in the route from India to China was *Kalah*. Reference to *Kalah* is found as early as 650 in accounts of Nestorian Christians, and appears in Arab writings two centuries later. It lay roughly six days' sail from the Nicobars, and ten days from Tioman, an island off the southeast coast of Malaya. In about 940, Abu Dulaf stated that it was at the limit of Chinese sea-faring. *Kalah's* people, or at least its trading community, were dressed in sarongs, the men and the women alike. Abu Dulaf describe the town as

> very great, with great walls, numerous gardens and abundant springs. I found there a tin mine, such as does not exist in any other part of the world.

The thousand-year-old lore of Sindbad described *Kalah* as a great empire bordering on India, in which there are mines of tin, plantations of bamboo, and excellent camphor." Scholars have proposed that *Kalah* lay on the coast of *Mergui* (northwestern Malaya), or Kedah, or one of the islands which lie on the western waters of Malaya, such as Phuket.[88]

Champa is described by many texts, which refer to the kingdom as *Sanf*. It is said to lie along the sea route to China between *Qmar* (Khmer, i.e., Cambodia) and *Luqin* (Lung-p'ien, a port at the mouth of the Red River). Akbar al-Sin wa 'l-Hind (ca. 850) noted the kingdom's export of aloeswood, and that mariners en route to China next called at an island with fresh water named *Sundur Fulat* (logically Hainan, but possibly Pulo Condore).

As with the *Ramayana* and India, the legendary voyages of Sindbad record some distant memory of Arab voyages to Southeast Asia. Sindbad traded extensively in India (the 'Sind' of Sindbad denoting a region of India, now part of Pakistan) and traveled to Ceylon and beyond. In the course of his first voyage, he visited the island of *Kasil*, describing it in terms matching the island of *Bartayil* of the geographers. On his third voyage, Sindbad reaches *Salahit,* a place cited by several Arab geographers which probably lies in Sumatra. During the course of the fourth voyage, he went to

Naqus (Nicobar Islands) and *Kalah* (western Malaya?), and the fifth voyage he reached islands with spices and aloeswood. The people of these last islands "love adultery and wine and do not know about the proper methods of praying" — complaints commonly made by Arab sailors about some Southeast Asian islands.

Waq-waq: the Life of a Myth

One of the most recurring and enduring place names recorded in Islamic tradition which is associated with Southeast Asia is the island or archipelago of *Waq-waq*. Various solutions have been suggested to the riddle of the island's identity — Sumatra is most often proposed, but so is Madagascar, since the integration of Ptolemaic geography into the Arab world view had created confusion between the islands of eastern Africa and those of southern Asia. The only realistic answer is that *Waq-waq* was some undetermined Indonesian island, and that the popular lore associated with it eventually transformed it into a mythological place just out of reach of the conventional trade routes. Different sailors may well have identified different landfalls as *Waq-waq*.[89] The island's most distinguishing characteristics were its namesake tree, which bore fruit of girls (figs. 25 & 26), and its abundance of gold.

The metamorphosis through which *Waq-waq* was transformed from a real place to a legend by the progressively fanciful lore surrounding it, offers an ideal chance to examine the life of a Southeast Asian geographic myth. We will look at several descriptions of *Waq-waq* and its tree, in chronological order, and see how an unassuming description of local features was gradually transformed into the magical land of *Waq-waq*.

A geography by the Persian author Hudu al-'Alam, dated 982 relates that

> "east of [China] is the Eastern Ocean; south of it, the confines of Waq-waq, the Sarandib mountain, and the Great Sea . . Waq-waq belongs to the hot zone . . . its capital is M.qys, which is a small town (where) merchants of various classes stay."[90]

Another text, the *Masalik wa'l Mamalik* attributed to Kurtubi (d. 1094?), states that "occasionally due to the strength of a violent wind some ships reach the island and through the perseverance of the sailors land on [it]."[91]

Our metamorphosis begins in about the year 1000, with the geographical treatise of Aja'ib al-Hind. Al-Hind wrote that

> Muhammad ibn Babishad told me that he had learned from men who had landed in the country of Waq-waq, that there is found a species of large tree, the leaves of which are round but sometimes oblong, which bear a fruit similar to a gourd, but larger and having the appearance of a human figure. When the wind shook it there came from it a voice . . . A sailor seeing one of these fruits, the form of which pleased him, cut it off to bring it back, but it immediately collapsed and there remained in his hands [only a flabby thing] like a dead crow.

There are no mysterious or mystical claims here. The author had simply heard from second-hand sources that the island had a curious tree which bore a gourd-like fruit that resembled the human figure. The fruit made a sound with the wind, and it lost its hardness when removed from the branch. Many retelling the story of the tree, however, fell to the temptation to elaborate upon the curious description of the tree's fruit. Shortly after Al-Hind's text, another author, Biruni (973-1048), was already trying to stop the wild stories that had sprouted up about *Waq-waq* :

> The island of al-Waq-waq belongs to the Qmair islands [Qmar]. Qmair is not, as common people believe, the name of a tree which produces screaming human heads instead of fruits, but the name of a people the color of whom is whitish. They are short in stature and of a build like that of the Turks.

Similar repudiation came from al-Idrisi (d. 1165), who sifted through texts and travelers' reports in Sicily: "[In *Waq-waq*] there is the tree about which Mas'udi [Arab geographical author, d. 956] tells us unbelievable stories which are not worth telling."

Nonetheless, the writings of another twelfth-century author, Kitab al-Jughrafiya (Spanish), reveal how embellished and horticulturally precise the myth had become:

> In the part of the land of China which is in the sea . . . the largest and most important island is Waq-waq. It is so called because there are great tall trees there [which] bear fruit in the month of Adar [March], and they are [at first] fruit like the fruits of the palm-tree. These fruits end in the feet of young girls, which project from them; on the second day of the month the two legs protrude, and on the third day the two legs and thighs. This continues so that a little more protrudes each day until they have completely emerged on the last day of the month of Nisan [April]. In the month of May their head comes out and the whole figure is complete. They are suspended by their hair. Their form and statures are most beautiful and admirable. At the beginning of the month of June, they begin to fall from these trees and by the middle of the month there is not one left on the trees. At the moment of falling to the ground they utter two cries: "Waq, Waq". When they have fallen to the ground, flesh without bones is found. They are more beautiful than words can describe but are without life or soul.

The writer explains that *Waq-waq* lies at the end of the inhabited world to the east, where the coast touches the Greater Sea. A century later Qazwini (d. 1283), much of whose material was recycled from older writings, adds an Amazonian twist to the island. He repeats the story that the *Waq-waq* fruit looks like women hanging from the branches by their hair, and notes that

> al-Mubarak of Siraf claimed to have been into the island and seen the queen seated upon a throne, completely naked and with a crown on her head, surrounded by four thousand young virgin slaves, also naked. Others say that these islands are called this because there is found a species of tree, having fruit, which produces a noise "Waq-waq!" The inhabitants of this island understand this noise and draw disagreeable omens from it.

Kurtubi informs us that man does not live on *Waq-waq*, but rather that

> there is a kind of great tree whose fruit, which grow among its blossoms and boughs, are always lovely women such that those who see the beauty of their shape and the grace of their body are astonished. The breast and vulva of each one are like [those of] other women, and in the branches of the tree they are suspended from the branches by their heads [hair] like a kind of fruit. Sometimes they all make the sound vak-vak. Therefore they call the island Vak-Vak.

Fig. 25 The *Waq-waq* tree, said to grow on the island of *Waq-waq*, somewhere in insular Southeast Asia. Ottoman, ca. 1600. [Topkapi Palace Museum, Istanbul]

Fig 26 A Thai adaptation of the *waq-waq* tree, the landmark of an Indonesian island described in Arabic geographic texts. The tree was incorporated into Thai poetry as "a garden of magical fruits [whose] trees bear beautiful women as fruits," and the artist who executed this image, labeled it as a tree with "women mango fruits." From a southern Thai manuscript, probably late eighteenth century.

Friar Odoric, a European who reached China via Southeast Asia in the early thirteenth century, was told a version of the story of the *Waq-waq* tree, and a Chinese writer, Tu Huan, noted one rendering of it as well. The fabulous *Waq-waq* tree is depicted in Islamic art with both male and female 'fruit', but the tree was described in the Pali language as *nari phala* ('girl-fruit tree'). In the early nineteenth century the Thai poet Sunthorn Bhoo borrowed the idea, writing of "a garden of magical fruits [whose] trees bear beautiful women as fruits," and the image of the tree is found in Thai art as a tree with 'woman mango fruit' (fig. 26). Another recurring motif of *Waq-waq* is the plenitude of gold to be found there. Gold is considered to be so abundant in *Waq-waq* that it is not highly esteemed, and is even used for dog collars.

The corroborative evidence for *Waq-waq* led al-Idrisi to include it as a group of islands in his atlas of the world (fig. 28), though he dismissed the fantastic lore associated with it. *Waq-waq* remained a prominently mapped feature into the middle of the seventeenth century, when the island is found in the position of Sumatra on the map of the world by Indo-Islamic cartographer Sadiq Isfahani.

Bartayil

Bartayil was an island which was associated with the beating of drums, musical instruments, and the sounds of dancing. From this island "one continually hears the noise of the drums, flutes, lutes, and all sorts of musical instruments, a sound soft and agreeable, and at the same time dancing steps and the clapping of hands." This story was repeated by various writers with minor embellishment (the size of the ensemble grew in the later renderings). The island was also believed by sailors to be the dwelling place of the Antichrist. Some sources noted that cloves are purchased there from "invisible people", probably referring to the common system of silent barter. An animal, looking like a horse but with a mane of such length that it trailed on the ground, came onto the island from the sea. Sailors were said to be too frightened to investigate the island or its music. The island was visited by the legendary Sindbad, who called it *Kasil*.

As with *Waq-waq*, there is no reliable identity for *Bartayil*. Sailors' imagination probably made the association of *Bartayil* with the Antichrist from the appearance of the island's people (they were said to have faces resembling shields of leather, with split ears) and the unearthly effect of its incessant music with no visible source. The idea that the Antichrist would be found in the east probably grew from the tradition of the apocalyptic monsters, Gog and Magog, while *Bartayil*'s strange horse may have evolved from the lore of the Chinese triple-headed goddess, Kwan-yin, who was sometimes said to assume the form of a horse, and who was

Fig. 27 World map of al-Idrisi, mid-twelfth century, a copy bearing the date 960 AH (1553 A.D.). Oriented with south at the top, Africa wraps around the southern Indian Ocean. The Arabic title of al-Idrisi's geography translates as *The Recreation for Him Who Wishes to Travel Through the Countries.* (Approx. 23 cm diameter) [Bodleian Library, Oxford, ms. Poc. 375]

supposed to have been born in the southern ocean from a father whose empire extended from India to Southeast Asia. The purchase of cloves does not mean that the island was itself a source of the spice (which at that time would have meant the Moluccas), but simply that it was a market place for them — as the sixteenth-century Portuguese historian, João de Barros, noted, Portuguese and Malay merchants preferred to conduct their trade in the Banda group, rather than sail all the way to Ternate itself.

The Islands of Spice
'Islands of Spice' are, of course, presumably references to the Moluccan and Banda Islands, but many allusions to spice islands in Arab texts are vague and semi-mythical. The principal question regarding the identification of the Arab texts' reference to "Spice Islands" with the Moluccas is whether the sailing time they quote for reaching the islands is sufficient, and the answer to that question depends on the identity of the points of departure. As early as about 850, Ibn Khurdadhbih stated that the "Islands of Spice" are said to be reached after fifteen days' sailing from "Jaba, Salahit, and Harang." If, as some scholars believe, these three places lay at the southern end of the Malacca Strait, then the fifteen days would doubtfully have taken a vessel further than eastern Java.[92] Other authors repeat the same figure, among them, interestingly,

the Italian traveler Nicolò de' Conti. Conti stated that he sailed fifteen days from Sumatra and Java to *Sanday,* a source for nutmeg, and *Bandam* (Banda), a source for cloves. In fact, Banda was the source for nutmeg, and only served as a market for cloves which were brought there from Ternate and Tidore in the Moluccas.

Even as late as the fifteenth-century, al-Mahri's navigational text is little more specific than to acknowledge Banda as the source of nutmeg, that "the islands of Cloves are called Maluku and are four islands [actually five]," and to place the islands to the southeast of *Jawa* (they actually lie to the northeast, but Arab pilots tended to envision Indonesian islands aligning to the southeast). He states that the latitude of these islands "is certainly unknown, although people of some knowledge have suggested their latitudes." Islands of spices of less definite identity are described in earlier Arab writings, recorded by different names and not placed in clearly identifiable places, but probably all refer to the Moluccas, even if based on second-hand data.

Cloves were said to be purchased on these islands by silent barter. Traders, upon reaching the shore, would spread out leather sheets and place upon them their purses with *dinars* corresponding to the amount of cloves they wished to purchase, then return to their ship for the night. The next day they would return to the shore, in the anticipation that the islanders will have replaced the

Fig. 28 Southeast Asia from al-Idrisi's atlas, mid-twelfth century, a copy bearing the date 960 AH (1553 A.D.). Six pages of the atlas have been joined together in this illustration; south is at the top. (Each section is approx. 21 x 30.5 cm) [Bodleian Library, Oxford, ms. Poc. 375, fol. 33v-34r, 38v-39r, 42v-43r, 77v-78r, 81v-82r, 84v-85r]

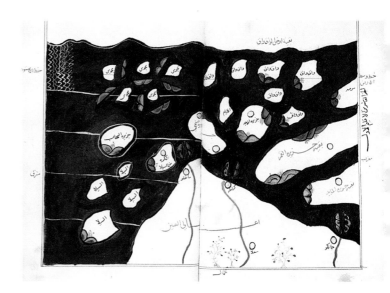

money on each leather sheet with cloves. If the trader was satisfied with the exchange, he would gather up the cloves, thereby consummating the purchase. If not, he left the cloves on the leather sheet, returning the next day to find his *dinars* fully replaced in his purse. Such exchanges were said to be undertaken with confidence, without fear of injustice.

The Island of Women

In contrast to the Islands of Spice, the long and colorful tradition of an island of women somewhere in Southeast Asia defies identification and probably belongs to the region's purely mythical landscape. One source dating back about a thousand years, the geographer 'Aja'ib al-Hind, relates the story of a ship in the Sea of *Malayu* (Malaya), *en route* to China, which after encountering a storm landed on an unknown island. As the sailors were disembarking from their ship,

> a party of women arrived from the interior of the island, the number of which God alone could count. They fell on the men, a thousand or more to each man. The women carried them off to the mountains and forced them to become the instruments of their pleasure.

All but one man died of exhaustion, the survivor, a Spaniard, being hidden by one of the woman. Together they escaped in a boat filled with gold that she had discovered on her island, and the two safely reached the port from whence the sailor had come.

Other authors note that the island of women "is situated at the limit of the sea of China," and repeat the various traditions that its inhabitants make themselves pregnant by facing the wind, by eating the fruit of a particular tree, or by beholding their own image (see Ortelius, fig. 86). As we have seen, the island of *Waqwaq*, as a result of its fabled tree, also became the object of Amazonian lore. In the fifteenth and sixteenth centuries, Western mapmakers recorded such islands in Southeast Asia (see, for example, Bordone & Gastaldi, pages 120 & 143, below).

The Island of the Castle

Some early Arab authors wrote of "a white castle which stands on the sea and appears to the sailors before the dawn." Alexander the Great is said to have reached the island with the castle, whose inhabitants were clothed in leaves. Although one story noted that sailors "rejoice when they see the island, for it assures them of safety, profit, and good luck," stepping foot on the island was also said to induce insanity, which could only be cured by eating a particular fruit which grew there. Other stories associated the island with misfortune. Some accounts give a specific location to the island, such as "in the sea off Champa (central Vietnam)," though the tradition is generally of a place somewhere in the outer ocean.

The Motionless Sea (Sea of Darkness)

Sailors venturing into the eastern periphery of Arab voyages risked drifting into a 'Sea of Darkness' *(Bahr al-Muzlim),* in which sailors are tossed about forever.[93] Ibn Battuta, in Southeast Asia in the early fourteenth century, reported that after 34 days' sail from *Mul-Jáwa* (Java) he and his party "came to the sluggish or motionless sea

[in which] there are no winds or waves or movement at all in it, in spite of its wide extent" (see page 105, below).[94] Friar Odoric, his European contemporary, repeated a similar story when describing the seas near Java. Odoric refers to an island called *Panten,* or *Tathalamasin,* whose king counts many islands under his dominion. "By this country," according to Odoric, is a sea called *Mare mortuum* (dead sea), which flows continually to the south, and into which "whosoever falleth is never seen again." These images perhaps tie in with the Chinese idea of a flat world tilted 'down' to the southeast, to where the earth's water incessantly flows.

Sharif al-Idrisi

The work of the twelfth-century Sicilian geographer al-Idrisi is the earliest surviving Arab view of Southeast Asia. Al-Idrisi's atlas was divided into a world map (fig. 27) and seventy regional maps, of which six covered southern and Southeast Asia (joined together in figure 28). His map is a composite of sources. Some of his data can be traced back as far as the mid-ninth century text, *Akhbar al-Sin,* and the geographer, Ibn Khurdadhbih; much of it is rooted in intervening geographers; and some of it is new, or at least newly-interpreted. The maps illustrated are from a fine copy of al-Idrisi's atlas, bearing the date of 960 AH (1553 A.D.).

The landmass running along the top (south) is Africa, which al-Idrisi, influenced by earlier Arab geographers, shaped around a

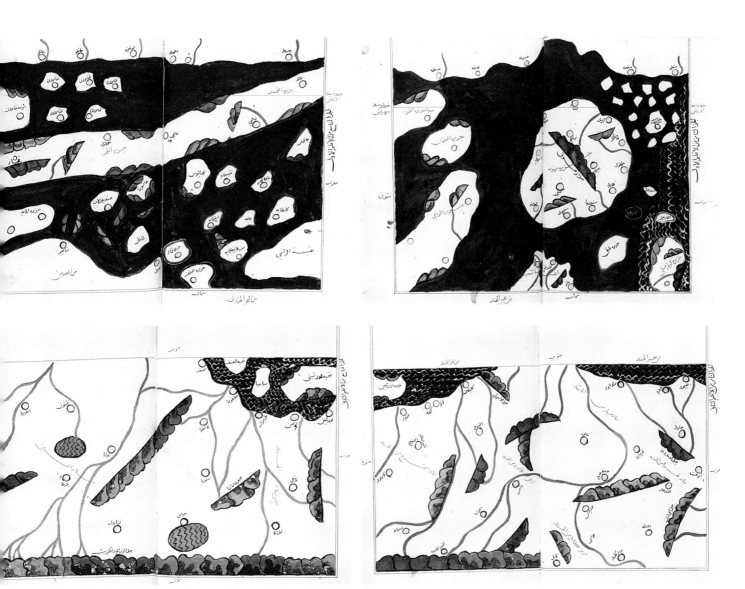

landmass which Ptolemy believed connected Southeast Asia with Africa (see al-Idrisi's world map, figure 27). This concept, Arab geographers' one dramatic concession to Ptolemy, was influential, being found in Western sources such as the 1320 *mappamundi* of Sanudo (fig. 3). But contrary to Ptolemy, al-Idrisi followed the overwhelming Arab consensus in leaving the eastern (left) end of the Indian Ocean open.

The southeasternmost archipelago (upper left) comprise the islands of *Waq-Waq,* a colorful land of plentiful gold and trees with blossoms of girls. These isles lie in the sea off China, which is the mainland coast below. To the left of the mainland cape lies the archipelago of *Sila,* which is identified with Korea. Skipping to the upper right (west), the large round island in the corner is Ceylon, and the large, fat island near the mainland to the lower left (northeast) of Ceylon is *Ramni,* a kingdom of Sumatra for which the island was sometimes named. Various small 'Java' islands lie to east (left) of *Ramni,* including *Salahit,* visited by Sindbad and probably part of Sumatra, and *Harang,* one of the departure points for the Spice Islands and probably part of Sumatra or Java. *Sanf,* which is Champa in Indochina, is the island directly between eastern *Ramni* and the mainland. Among the mainland coastal cities near the eastern end of *Ramni* is *Cattigara,* an emporium noted by Ptolemy whose location was eagerly sought during the Renaissance.

Now we reach the long, prominent narrow island which

dominates the Indian Ocean. al-Idrisi identifies this as *al-Qumr,* which is Madagascar, the island lying in its 'proper' place off the wildly mis-located African coast. Al-Idrisi notes that the same island is also called *Mala'i.* This could refer to a placename on Madagascar, or to the *Malayu* of other Arab geographers, which is part of Malaya or Sumatra. In addition, one of the locales on the large island is *Qmar,* which is Cambodia (i.e., Khmer). Thus the island may be both Madagascar and part of the Southeast Asian mainland and Sumatra.

Champa *(Sanf)* was erroneously depicted as an island because little was known about it except that it was a port-of-call *en route* to China. A more dramatic example of the pitfalls of charting lands based on inexact textual references can be found in three archipelagoes off eastern 'Madagascar'. They are *Ma'id,* the large island between 'Madagascar' and the mainland which is evenly divided between the upper-left and upper-central sheet; *Muja,* one of the islands off the southeast coast of 'Madagascar'; and *Qamrun,* the group of five small islands caught between 'Madagascar' and the Ptolemaic African coast. All three places are described in the mid-ninth century text, *Akhbar al-Sin,* as lying between India and China, which al-Idrisi interpreted as meaning that they lie on the *sea route* between the two. In fact they lie on the overland route: *Qamrun* is Assam, while *Ma'id* and *Muja* are kingdoms along the border between Burma and Yunnan.

Part II

The Early Mediterranean and European Record

Chapter 4
Asia and Classical Europe

Of the various peoples inhabiting Europe in ancient times, only the dominant Greek and Roman civilizations were in a position to seriously consider who and what lay far to their east. It is doubtful if those who lived in western and northern Europe, whatever their sophistication and aspirations, learned anything about the East which had not been sifted through those Mediterranean powers.

Greece

The earliest known cartographic allusion in Greek civilization, is the so-called 'shield of Achilles' from the *Iliad* of Homer (before 700 B.C.). This was a cosmological device not intended to show specific geographic features; neither Asia as an entity, or even the divisions of the continents, were reflected in this view. Achilles' shield depicted a 'River Ocean' surrounding the cosmological whole, which foreshadowed the idea of an ocean sea encircling the continental land. The entire earth, as well as the seas, sun, moon and stars, all formed part of the center of the disk. True geographic divisions were drawn in the sixth century B.C. by Anaximander, a theoretician who, like Homer, lived in what is now Turkey. The Greek world, itself blessed with a wealth of islands, probably envisioned islands dotting much of the ocean sea.

The Name 'Asia'
'Asia' is an ancient name. In the fifth century B.C., Herodotus was already speculating about its origin, noting that most Greeks believed the continent to have been named after the wife of Prometheus. Other accounts describe Asia as the mother, by Iapetus, of Atlas, Prometheus, and Epimetheus. In any event, Asia was an Oceanid, a sea-nymph daughter of Oceanus and Tethys, similar to a Nereid.

Greek Excursions to the East
In about 513 B.C., the region of the Indus valley was taken by Darius of Persia. A Greek officer, Scylax of Caryanda, was sent into the newly-acquired territory, sailing much of the Indus River and helping to push the Persian — and ultimately the Greek — world view eastward into Asia. Scylax, however, reported that the Indus flowed to the southeast, an error that, as we shall see, helped stunt even a hypothetical expansion of the continent farther east (although segments of the river do indeed flow to the southeast, the overall orientation is flowing to the southwest). Scylax also initiated the durable tradition of concocting strange quasi-human inhabitants for the Orient, such as the people whose feet are so large that they use them as umbrellas.

The ancient world never reached any consensus regarding the extent of Asia. About 500 B.C., Hecataeus of Miletus espoused the notion of a disc-shaped earth with Asia and Europe of roughly equal size. Herodotus (ca. 484-425 B.C.), remembered fondly as the 'father of history', argued instead that Europe was larger than Asia, and questioned whether the inhabited world was entirely surrounded by water. "I cannot help laughing," Herodotus mused, "at the absurdity of all the mapmakers — there were plenty of them — who show Ocean running like a river round a perfectly circular earth, with Asia and Europe of the same size." He believed Europe to be as large as Asia and Africa combined. Among his early critics was Ctesias of Cnidus (ca. 400 B.C.); whereas Herodotus vastly understated the size of Asia, Ctesias exaggerated it.

A proponent of experience-founded empirical geography, Herodotus reviewed a great number of writings and reports about Asia. In his opinion, "Asia is inhabited as far as India; further east the country is uninhabited, and nobody knows what it is like." He dismisses reports furnished by his countryman Aristaeus[95] of the sundry people said to live far across Asia, flatly stating that "eastward of India lies [an uninhabitable] desert of sand; indeed, of all the inhabitants of Asia of whom we have any reliable information, the Indians are the most easterly."

Alexander the Great
Thus by the time Alexander the Great (356-323 B.C.) pushed eastward in his attempt to conquer the world, Greek civilization had diverse ideas regarding the eastern end of the world. Some of Alexander's soldiers probably believed that they would need only to subdue India before meeting the ocean sea which ringed the earth. But Alexander had been tutored by the great philosopher Aristotle (384-322 B.C.), who postulated a vast Asian continent, extending much farther to the east than Alexander's exploits. Aristotle believed that the circumference of the earth was 400,000 *stades,*

which exceeded the correct figure by at least 50 percent, and that the expanse of sea separating the eastern shores of 'India' from Spain was not great. Thus he envisioned the Asian continent extending through, and well beyond, our Southeast Asia. Alexander never learned for himself what lay past the Indus, since his troops refused to advance past the river. "Those who accompanied Alexander the Great," noted Pliny in the first century A.D., "have written . . . that India comprises a third of the whole land surface of the world and that its populations are uncountable."

Alexander's conquests, and his decision to split his people into three segments for a return via inland, coastal, and ocean routes, greatly expanded Greek knowledge of the East and of the western Indian Ocean region. With his conquests began regular trade in India's ivory and spices. Despite Alexander the Great's failure to reach Southeast Asia, two thousand years later some 'credible' European writers would claim that the extraordinary Cambodian edifices at Angkor were built by him.[96] As absurd as such an assertion was, it would lend legitimacy to European designs on Southeast Asian soil.

Following the death of Alexander, the various factions of the Greek empire fell into ruinous warfare among themselves, dampening any energy for exploits into India until Seleucus Nikator consolidated power in about 300 B.C. Seleucus, realizing that he could not overpower the Maurya rulers east of the Indus, instead negotiated safe passage for an ambassador named Megasthenes to cross India to the Maurya court in *Pātaliputra* (Patna), along the valley of the Ganges. Megasthenes was, indeed, on the frontier of 'India beyond the Ganges', Southeast Asia. His account of India, *Indica,* was the paramount Greek record of 'eastern' Asia. It was Europe's first notice of Ceylon, of Tibet, of the origin of the Ganges in the Himalayas, the shape of the Indian subcontinent, and of the monsoons, that great natural engine which would facilitate travel across the Indian Ocean.

The various reports gained from Alexander's exploits and Megasthenes' embassy were examined by Eratosthenes (ca. 276-ca.196 B.C.), a brilliant librarian at Alexandria, born about a half century after Alexander's death. Eratosthenes, who is best remembered for having calculated an uncannily accurate figure for the earth's circumference, learned of Ceylon and the peninsular nature of India from Megasthenes' reports.

But Eratosthenes, perhaps extrapolating from the erroneous southeasterly flow of the Indus assigned by Scylax, concluded that the Indian subcontinent was oriented to the southeast, leading him and his contemporaries to believe that India formed the southeastern threshold of the Asian continent. It is true, of course, that the Ganges River was now known, suggesting that 'India beyond the Ganges', our Southeast Asia, was there by implication, since something obviously had to form the eastern banks of the Ganges. But the Ganges was instead adapted to fit the existing world view, flowing to the east into the ocean sea rather than south into the Indian Ocean. As a result of their misconceptions, 'beyond the Ganges' meant *north* of the Ganges. In the mid-sixteenth century Gerard Mercator, under entirely different circumstances and for unrelated reasons, reached the same conclusion, deciding that the Ganges was in reality the Canton River in southern China (see page 101, below).

With the decline of the Greek empire, Greek civilization and the Hellenic world effectively lost touch with the world east of the Hindu Kush mountain range in Central Asia, dampening any dreams of further expansion. Intermediaries seized the opportunity for trade, and Indian merchants may have reached Alexandria directly by the second century B.C. In 146 B.C., the remnants of the Greek states were effectively subjugated by the next emerging European superpower, Rome.

Rome

Contact between Rome and entrepôts far to the east is recorded in 26 B.C., in which year Augustus received envoys from Sri Lanka. A white elephant exhibited in Rome during Augustus' time may have originated in Southeast Asia, in the general region of what is now Thailand (the 'white' elephant in this case was probably an albino animal rather than the so-called 'white elephant' which is revered in Siam and Burma — the latter is identified by a number of physical traits, of which color is of relatively minor importance).[97] In about 166 A.D., a group of Roman musicians and acrobats traveled to China via Burma; a Roman lamp dating from this era has been unearthed in P'ong Tük, on the northeast part of the Malay Peninsula, along what may have then been a trans-peninsular shortcut.[98]

Hippalus

The consolidation of the Roman Empire helped circumvent many of the middle-traders who had come into the Indian trade after Alexander. But until the very early Christian era, Mediterranean merchants had either to rely on the monopoly of traders in southwestern Arabia, or sail the lengthy, dangerous route along the coast of Persia and the Arabian Sea. This limitation, however, was broken in the middle of the first century, when a Greek sailor named Hippalus revealed one of nature's great secrets: Hippalus learned how to harness the seasonal winds.

Hippalus determined that the prevailing winds in the Indian Ocean reversed direction seasonally, and thereby theorized that one could sail to India via open ocean, out of sight of land, and return when the winds reversed. In about 45 A.D., he successfully sailed from the mouth of the Red Sea to the delta of the Indus River, and thence to ports along the Malabar Coast. The spring and summer monsoons upon which he had gambled then safely returned his vessel, now filled with Indian goods, to Egypt. A record of his voyage was preserved in the writings of the Roman historian Pliny, as well as in an anonymous document written in the first century, the *Periplus of the Erythrean Sea,* (i.e., Indian Ocean), a guide for mariners wishing to trade with India.

Both the *Periplus* and Pliny helped relay the new reports about India. The *Periplus* described the coasts from the Indus to the Ganges delta, as well as information about Sri Lanka based on secondary sources. Pliny records information clearly based on actual navigation of the waters east of India, noting that "the sea between Taprobana [Sri Lanka] and the Indian mainland is shallow, not more than 18 feet deep, but in certain channels the depth is such that no anchors can rest on the sea-bed." He also noted the voyage to Ceylon from *Prasii* being made in seven days "in boats built of papyrus and with the kind of rigging employed in the Nile."

The new sailing method brought India much closer to Europe and the Mediterranean, and this, in turn, brought Southeast Asia that much closer as well. With Roman sailors frequenting the eastern coast of India by the second century A.D., Asian commodities were reaching the Mediterranean world in generous quantities, silks generally coming by land via Antioch (Syria), and spices by sea via Alexandria. Merchandise being sent on to Italy was generally shipped to Pozzuoli, near Naples.

Though profit rather than knowledge was the driving force (Pliny remarking that "India is brought near by lust for gain"), more extensive commerce provided new channels through which geographers could enrich their storehouse of data. The fact that products from eastern Asia reached Rome meant that ideas, information, and perhaps even geographic data, could make that journey as well.

The range of marketable Asian goods increased as well. Diverse commodities were imported, from 'exotic' Asian animals to asbestos (used for oil lamp wicks) and slave women — a far greater range of valuables than just silks and spices. An ivory statue of an Indian goddess of fortune, Lakshmi, unearthed in the twentieth century in the remains of Pompeii, was clearly acquired before the city was buried by Vesuvius in 79 A.D. Chinese earthenware and bronzes have been excavated in Roman cities, and Roman traders appear to have reached China in the second and third centuries A.D. Roman coins, directly or indirectly, reached the kingdom of Funan in the Mekong River delta as early as the second century.

In our day of cheap, plentiful spices, it is difficult to imagine the fuss made about spice-producing regions in ancient times, and indeed up until relatively recent times. Two thousand years ago, Pliny wondered similarly:

> It is amazing that pepper is so popular. Some substances attract by their sweetness, others by their appearance, yet pepper has neither fruit nor berries to commend it. Its only attraction is its bitter flavor, and to think that we travel to India for it! Both pepper and ginger grow wild in their own countries, yet they are purchased by weight as if they were gold and silver.

Chryse and Argyre

Gold and silver, in fact, characterize the earliest extant specific Western reference to Southeast Asia. Pomponius Mela (*fl.* 37-43 A.D.), a Roman geographer and native of southern Spain, largely carried on the Greek tradition about the East, perpetuating stories about Amazons, people without heads, griffins, and other such characters, but adds two lands which lay to the east of India. One was *Chryse*, said to boast soil of gold, the other *Argyre*, said to have soil of silver:

> In the vicinity of *Tamus* is the island of *Chryse*; in the vicinity of the Ganges that of *Argyre*. According to olden writers, the soil of the former consists of gold, that of the latter is of silver and it seems very probable that either the name arises from this fact or the legend derives from the name.

Mela was quoting earlier, unknown sources and he goes on to vaguely mention the possibility of a Southeast Asian peninsula:

> Between *Colis* [southeastern tip of Asia] and *Tamus* [China?] the coast runs straight. It is inhabited by retiring peoples who garner rich harvests from the sea.

Pliny also alludes to a Southeast Asian peninsula. Noting that the *Seres* [Chinese] wait for trade to come to them, he lists three rivers of China, which are followed by "the promontory of *Chryse*",

Fig 29. World map after Strabo/Mela. An anonymous sixteenth century manuscript. (30.5 x 43.5 cm) [Sidney R. Knafel]

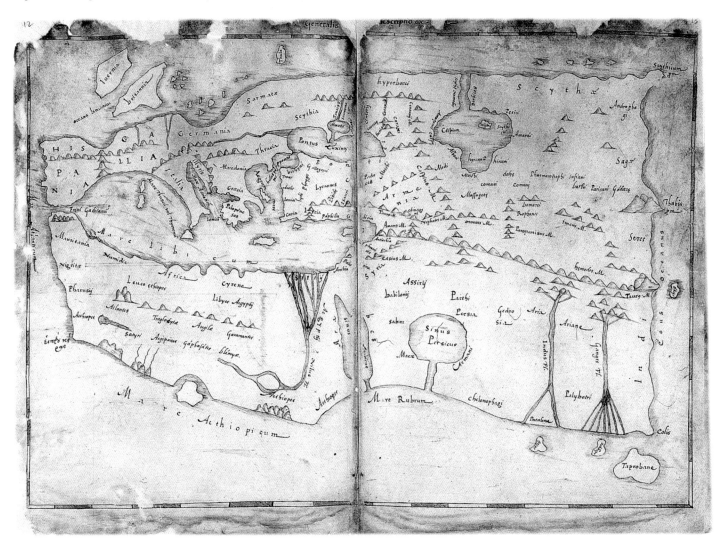

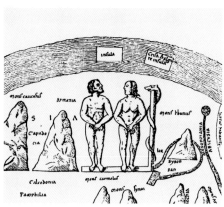

Fig. 30 The 'Turin' world map, twelfth century. Of the two islands in the ocean sea immediately above Adam and Eve, the righthand one represents the Southeast Asian realms of *Argyre* and *Chryse*. Reproduced from the facsimile in Nordenskiöld, *Facsimile Atlas*, Stockholm, 1889.

Fig. 31 'Turin' world map (detail). Adam and Eve, the Serpent and the island of *Chryse* and *Argyre*.

and then a bay. Elsewhere in his *Natural History,* however, Pliny refers to *Chryse* as an island. The discrepancy probably results from his having compiled news of the 'land of gold' from contact via land (peninsula) and sea (island). It was more often mapped as an island in medieval *mappaemundi.*

Chryse most likely represented Malaya, while *Argyre* was probably Burma, perhaps Arakan. Both are seen as islands in the world map after Mela (fig. 9), *Chryse* being the island off the east Asian coast, and *Argyre* the island at the Ganges delta next to *Taprobana.* On the twelfth-century 'Turin' world map (figs. 30 & 31), they appear as a single island in the easternmost ocean sea, the right-hand isle of the two immediately above Adam and Eve (the left-hand isle is simply designated *insula,* and thus may have been intended for either *Chryse* or *Argyre*).

Mention of *Chryse* is also made in the *Periplus of the Erythrean Sea,* which describes *Chryse* as "the last part of the inhabited world toward the east, under the rising sun itself," a land from which comes "the best tortoise-shell of all the places on the Erythræan Sea." The work's anonymous author then described the land of *This* (China) and city of *Thinæ,* from which raw silk, silk yarn, and silk cloth, acquired through silent barter, were brought overland to India.[99] Isidorus Hispalensis (Isidore of Seville, ca. 560-636 A.D.), in his *Etymologiae,* one of the most popular cosmographies of the Middle Ages, also placed the lands of *Chryse* and *Argyre* in the southeastern extreme of the world, along with *Taprobana* and *Tyle* (Tile, an island near India).

Interestingly, *Chryse* and *Argyre* are reminiscent of some aspects of Buddhist cosmology where the waters that pour forth from Sumeru flow into four canals separated by four mountains, of which one is gold, another silver, and the other two, precious stones and crystal.[100] The image of four canals separating four landmasses, can also be compared with a view of the Arctic region found in a medieval European text and used in later world maps by the Renaissance cartographers Ruysch (1507, fig. 56), Fine (1531, fig.48) and Mercator (1569).

The search for gold also promoted intra-Asian maritime trade in the Indian Ocean during the first century A.D. As a result of the disruption by internal disorders of the traditional routes through the steppes of Central Asia to Siberian gold reserves, new sources for the metal, a medium of exchange between various Asian peoples, were sought. Rome decreased the gold content of its coins and introduced measures to halt their exportation. At the same time, new ocean-going vessels and navigational techniques made it more feasible for Indian merchants to pursue the 'Islands of Gold' to their east.

The association of Southeast Asia with gold was so strong that Josephus, in his *Antiquities of the Jews* (second half of the first century), wrote that Ophir, the land from which King Solomon had fetched gold, is now known as *Aurea Chersonesus* (Golden Peninsula, i.e. Malaya). Josephus thus began the recurring idea that the Ophir of the Bible was in Southeast Asia, a belief that can be found in earnest through the latter nineteenth century. Various places were believed to have been the site of Ophir, from Malaya to Indochina, Sumatra, and the Pacific Ocean.

Marinus and Ptolemy

At about the same time as Josephus, a 'Golden Peninsula' in Southeast Asia was described by a geographer from Tyre by the name of Marinus. Tyre lies on the eastern Mediterranean coast, in what is now Lebanon. Today it is a peninsula, but in ancient times it was an island blessed with two bustling ports. Tyre had been the capital of Phoenicia from the eleventh to sixth centuries B.C., but was captured by Alexander in the fourth century B.C. It subsequently came under the control of Rome, remaining in Roman hands until the seventh century A.D. Tyre was ideally situated for gathering information about the East from traders and seafarers, and Marinus used his city's advantageous location to expand the world map from their reports.

Marinus' text served as the foundation for the work of Claudius Ptolemy, the most famous of ancient geographers. Ptolemy, who can be seen as representing the culmination of the Greek cosmographic

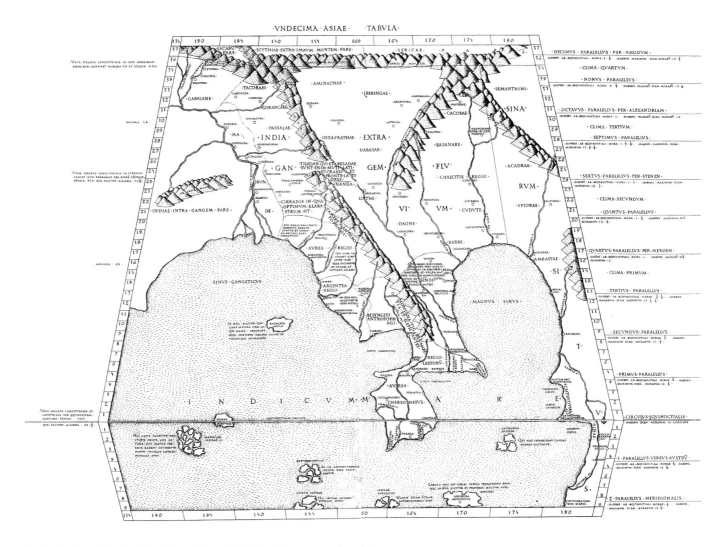

Fig. 32 Southeast Asia, Claudius Ptolemy, Rome, 1478. (37 x 42 cm + external words)

tradition, was an Egyptian who headed the library of Alexandria between about 127 - 150 A.D. His *Geographia,* largely a critical adaptation of Marinus' work, pushed the frontiers of the world map to include a 'true' rendering of Asia, though Ptolemy's map, as normally reconstructed, did not record the continent's eastern shores.

Perhaps the most extraordinary part of the story of Ptolemy is that his writings effectively lay dormant for over a thousand years, and then galloped into European consciousness as fresh and revolutionary in the fifteenth century as they has been in the second century A.D. In truth, although Ptolemy was unknown in Western Europe for the first one and a half millennia of Christendom, the *Geographia* was reviewed by Byzantine academics in the tenth or eleventh century, and it is possible that some of what we know as Ptolemy's work may in fact have originated with, or at least may have been emendated by, those scholars. Whatever the secrets of this most famous of ancient cosmographical texts, for our purposes we will consider it in its Renaissance incarnation rather than as a closing chapter of the science of ancient Greece (page 82, below).

Roman Geography

The Roman approach to cosmography was more empirical and practical than that of Greece. The quest for a comprehensive cosmographical overview was overshadowed by the need for more immediate knowledge of places actually traveled by Roman citizens and controlled by the Roman military. Thus despite apparent early

Greek knowledge of a peninsula and islands in Southeast Asia (that is, presuming Ptolemy=s eleventh map of Asia is not a Byzantine addition), the Roman world view traditionally terminated Asia along the eastern bounds of India (see fig. 29).

Julius Caesar (100 - 44 B.C.) ordered that a survey of the world be compiled, dividing the task into four sections according to the cardinal directions, with 'east' designating all the land east of Asia Minor. Caesar's project was carried out during the reign of his successor, Augustus (reigned 27 B.C. - 14 A.D.), under the direction of Augustus' son-in-law, the general Marcus Vipsanius Agrippa. The completed map of the known world was prominently displayed for the Roman public, in the belief that it would inspire awe for the Roman Empire, encourage trade, and foster colonization. Although neither Agrippa's displayed map, nor any of the copies of it that were probably disseminated throughout the Empire, are extant, the basic prototype may have been a forerunner of the medieval archetype seen in figures 34 and 39 (*Rudimentum Novitiorum*).

Practical Roman cartographers also produced road maps that one could follow fully from England to the southeasternmost part of Asia, which for them meant the Indian shores opposite Ceylon. One such map has survived by way of a copy, made in the twelfth or early thirteenth century, of a mid-fourth century original. Known as the 'Peutinger' map after the man who inherited and studied it in the early sixteenth century, the map is six and three-quarters meters long but only thirty-four centimeters high, and was

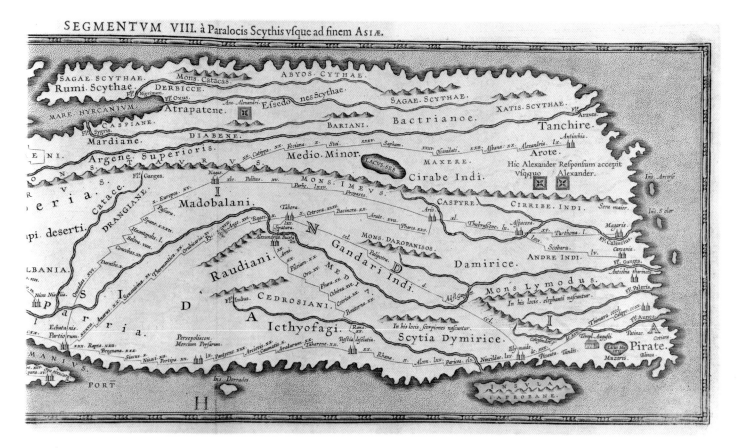

Fig. 33 The easternmost part of Asia, from the anonymous 'Peutinger' map. The map is probably typical of the kind of map a Roman traveler would have used as early as the first century A.D., and survives only by virtue of a twelfth-century copy of a fourth-century original. Abraham Ortelius made this first engraved rendering in 1598. Although in the broadest terms the map runs from the west on the left to the east on the right, there is no consistent orientation. Nor is there a fixed scale, and although travel distances are given, they are not always expressed in the same units of measurement. Most are in Roman miles, but in Gaul they are expressed in leagues, in Persia they are marked in *parasangs*, and on the easternmost section illustrated here, they appear to be in an Indian unit of measurement. But the map's purpose was to show significant places and natural features which a traveler would encounter on any major road from England to India and in this it was an efficient, accurate tool.

oriented solely for the needs of the traveler on roads, and — like many Asian maps — made no attempt to maintain direction or scale. The easternmost section, in an engraved copy made by Ortelius in 1598, is illustrated in figure 33. As with Strabo and Mela, the author of the Peutinger map perceived India to be the southeast tip of the world, with Ceylon lying off its southeast coast. The Indus empties just above Ceylon, and the Ganges flows into the eastern ocean sea, just below the silk-producing region of China (*Sera maior*). Approximately 104,000 km of roads are shown, with staging posts, spas, rivers, and forests. The staging posts facilitated a transportation system organized by Augustus, the *cursus publicus*, in which couriers traveled about fifty Roman miles (74 km) a day. Distances between staging posts are marked, usually in Roman miles, but in Gaul they are in leagues, in the region of Persia they are in *parasangs*, and in the southeast of Asia they appear be in an Indian unit of measure.

There were also Roman geographers who carried on the more theoretical cosmography of the Greeks. Strabo (ca. 63 B.C. - ca. 21 A.D.) was a geographer, historian, and philosopher from Amasia in the ancient country of Pontus, by the Black Sea. Straboclaimed to have traveled widely in the Roman world reaching as far east as southwestern Asia. The information gathered during his travels, he later set down in his *Geography,* completed in about 18 A.D. Strabo acknowledged Crates of Mallos, a geographer who had lived in Asia Minor about one and a half centuries earlier, as one of his sources.

Although Strabo never reached Southeast Asia, his text figured importantly in European designs on the Indies fifteen centuries later. It described the shipping routes between the Mediterranean and India, but most importantly, it postulated that the ultimate source of precious spices C the Indies C could be attained by sailing west from Spain. He encouraged the use of globes to avoid the distortion of flat maps, recommending that they be large (the equivalent of three meters in diameter or greater) to be able include sufficient detail. His ideas were introduced to the West at the Council of Florence in 1459, whose participants included Toscanelli, who in turn influenced Columbus.

Another important figure who was largely inspired by the writings of Crates was Ambrosius Theodosius Macrobius, a geographer, astronomer, and philosopher who flourished ca. 399 - 423 A.D. But as with Ptolemy and Strabo, Macrobius influenced the mapping of Southeast Asia in a very distant future, when his depiction of an open Indian Ocean encouraged the efforts of Europeans attempting to reach India — and ultimately Southeast Asia — by sea. Like Ptolemy, he figured prominently in the European Renaissance; unlike Ptolemy, he was studied, rebuked, and admired during the European Middle Ages (page 68, below).

Chapter 5

Medieval Europe

To medieval Europe, the East was the source of silks, spices, and other exotica. It was the environs of Paradise, the place of the original Garden but also of the original Sin. It was the horizon from whence the sun rose, the point from which humankind dispersed throughout the inhabited earth, and the subject of much philosophical speculation. The East invoked montages of splendorous images of strange people with exotic customs, the demoniac realm of Gog and Magog, and the holy lands of Prester John, the apostle Saint Thomas, even Paradise itself. The belief that Paradise lay at the eastern bounds of the world made the Far East special in the minds of Europeans.

When examining the various ideas espoused by geographers and cosmographers of the past, it is important to remember that those who were 'right' about a given question were not necessarily wiser or better-informed than those who were wrong. And many profound blunders were the result of careful, informed, intelligent deduction. Similarly, the fears and misconceptions of medieval travelers and sailors were not necessarily the result of ignorance or superstition. It is often pointed out, for example, that an accurate estimate of the earth's circumference existed in ancient Greece; true, but grossly inaccurate estimates existed as well, and we only know in hindsight which one was correct. Many medieval sailors may have feared that the tropical sea was impassable from heat, or that the ocean sea could not be traversed, but these 'wrong' ideas were not unfounded, and indeed were espoused by some of the very same classical texts which we now commend as having been correct on other matters. The image of the hapless, superstitious sailor venturing into the ocean while fearing sea monsters or the edge of the earth, is a modern-day chauvinism.

Posterity remembers the first millennium of Christian Europe as one in which society's mores and values were narrowly dictated by the Church, and the quest for furthering the understanding of the physical world was largely abandoned. Though judging those centuries through modern biases may be unfair, Europe in the Middle Ages seemed content with crude maps similar to those known in the ancient Greek world. Nearly two millennia after they were developed in classical Greece, simple zonal maps thrived in medieval Europe. It is true, of course, that later medieval Europe also produced sophisticated portolan charts of Mediterranean

waters, as well as *mappaemundi* resourcefully drawing from travelers' reports, and thus was more riddled with contradictions than was ancient Greece; but as late as the fifteenth century, mainstream European academia was still largely a receptive audience for primitive maps and the fanciful travel lore that often accompanied them.

Throughout medieval times, the sphericity of the earth was commonly accepted, and thus Southeast Asia was seen as the logical far equatorial shores of the Western Ocean. These were the shores that Christopher Columbus would have reached had America not rudely been in his way. But the nature and extent of the southeastern corner of the world remained mysterious for the first one and a half millennia of Christian civilization.

Southern Asia on Medieval European Maps

In the eyes of most medieval geographers, that southeasternmost point was still India. In the fifth century, Orosius wrote that "Asia begins in the East where lies the mouth of the River Ganges, facing the Eastern Ocean."[101] That the Ganges was 'facing the eastern ocean' meant, in other words, that what we now refer to as India was the end of the continent, not just that 'India' was being used in its general sense as 'Asia'. Orosius' view of a truncated Asia was repeated by numerous Western writers through the twelfth century. Further inhibiting the eastward expansion of the map of Asia was the fact that many early scholars — a notable exception being Ptolemy — believed that the Caspian Sea emptied into the ocean sea through northern or eastern Asia. Not until the thirteenth-century traveler, William of Rubruquis (see page 76), was this proven erroneous by actual observation.[102]

Zonal Maps and the Question of Equatorial Heat
Several types of elementary maps were common in Medieval Europe. Simple diagrammatic maps that divided the earth into climate 'zones' showed which parts of the earth were habitable and which were not. Oriented with north at the top, the upper- and lowermost regions were uninhabitable because of their severe cold, while the middle, equatorial region, whether land or an 'ocean

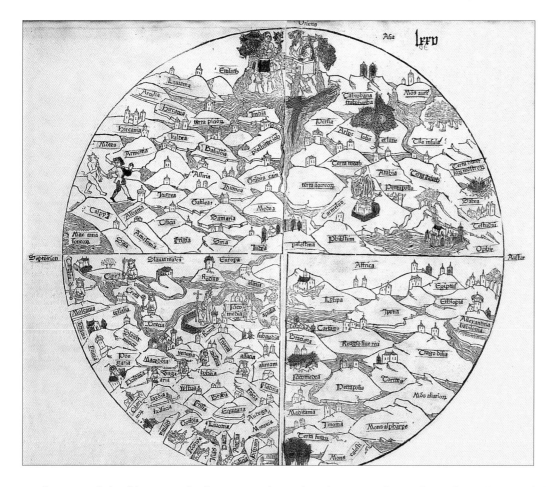

Fig 34 World map from the *Rudimentum Novitorium*, meaning "an elementary book for beginners", printed in Lübeck, in 1475. Its unknown author was conscious of his place in the ongoing revolution of printing, writing that his book was created with the aid of the art of printing, newly invented by the grace of God to the redemption of the faithful. The map itself is essentially a woodblock adaptation of the *mappamundi* which typified many medieval European's image of the world. It is based on the 'T-O' format, with Asia occupying the top half of the image and Europe and Africa the left and right lower quadrants respectively. (38 cm diameter)

river', was uninhabitable as a result of its extreme heat. Thus the southern Malay Peninsula, as well as Sumatra, Java, Borneo, and all the other near-equatorial islands would in theory be lifeless, parched landscapes.

The concept dates back to antiquity. Aristotle (384-322 B.C.) had theorized that the equatorial region was an uninhabitable continent, while Macrobius, whose work *In Somnium Scipionis Expositio* (early fifth century) was widely copied in medieval Europe, believed the equatorial region to be an ocean which was "scorched with the heat of the sun" and thus impassable. There were, however, some classical sources, for example, Polybius and Posidonius, who challenged the belief that it was uninhabitable on account of first-hand reports from expeditions that had actually penetrated the torrid zone.

The bleakest medieval assessment of the torrid zone went beyond mere temperature. Gervase of Tilbury (fl. 1200) deduced that the heat of the equatorial region would cause such a profound evaporation of its seas — and thus their dramatically increased salinity — that equatorial oceans would be so viscous as to render them nearly as treacherous to navigate as the frozen waters of the north.[103] On the other side of the debate were such people as the thirteenth-century geographer Albertus Magnus, who noted that cities are reported to exist in southern India and in near-equatorial Ethiopia. By the sixteenth century, 'torrid' having no further practical meaning, Peter Martyr noted that Magellan has crossed the Pacific following "what philosophers call" the torrid zone.

Sebastian Münster's *A Treatyse of the Newe India,* translated into English by Richard Eden in 1553, still considered the views of ancient authors on the matter. Pliny "doubted whether habitable regions may be found under the Equinoctial line"; Eratosthenes "is of the opinion that the air is there very temperate"; Polybius believed "that the earth is there very high, and watered with many

showers"; Posidonius "supposed that there are no mountains under the Equinoctial."

Some classical texts actually envisioned the equatorial earth as a raised ocean which, being at a high elevation, was nearer to the sun. This idea can be found, albeit in an in allegorical form, as early as Homer. As Münster explains,

> Some thought that the Equinoctial was extended beyond the earth over the main Ocean Sea: which thing the poet Homer seemeth to insinuate, where he [says] that the horses which draw the chariot of sun, drink of the Ocean Sea, and the sun itself to take his nourishment of the same.

Even in the later seventeenth century, long after zonal maps were mere historical curiosities, the torrid zone was rationalized by the French missionary Gervaise. Resident in Thailand, he wrote that the country "would certainly be uninhabitable, as it was long believed to be, were not the excessive heat of the sun tempered both by the number of rivers that irrigate the land and by the abundant rainfall that refreshes it."

The Tripartite Map

The tripartite map was the most basic type of medieval representation of the world after the elementary zonal map. Many medieval texts, in particular that of Isidorus, were illuminated by the so-called 'T-O' map, which depicted the three known continents by placing a 'T' shape within a circle (fig.17). With east at the top, the vertical line of the 'T' represented the Mediterranean Sea, while the horizontal line represented the Nile River and the other known seas which were roughly in line with the eastern end of the Mediterranean. Asia occupied the area above the horizontal line of the 'T', Europe the lower left quarter, and Africa the lower right quarter. With the horizontal line of the 'T' placed at the middle of the circle, the

Fig. 35 Macrobius, world map, ca. 1500; one of several very similar woodcut renderings used at the turn of the sixteenth century. This view of one hemisphere illustrates Macrobius' ideas about terrestrial quadrants and an equatorial ocean. Renaissance interpreters of the map have given the known world (upper land mass) recognizable form by loosely adapting the Ptolemaic model to Macrobius' larger view. Note that all of Africa lay north of the equator, implying that an easy sea route existed between Europe and the Indian Ocean, if the ocean were navigable. In concept, Macrobius' western antipodes foreshadowed the discovery of America.

In some renderings of the Macrobius map, such as this one, east and west are reversed, placing Southeast Asia on the left and Europe on the right. This was probably due to expedient making of woodblocks, the wood cutter not cutting in mirror-image. (14.4 x 14.2 cm)

proportions of the continents agreed with medieval writers who stated that Asia was of equal size to Europe and Africa combined. Although Europe is in fact only about one third of the area of Africa, the belief that the combined areas of Europe and Africa equalled the size of Asia was quite accurate. Some, such as Gervase of Tilbury, noted that Shem, the Biblical progenitor of humankind in Asia after the Deluge, was the oldest of Noah's sons, and therefore would have been given the largest continent. If scale were taken literally, T-O maps vastly overstate the north-south breadth of Asia, while depicting the east-west breadth as being roughly equivalent to that of the Mediterranean, placing equatorial eastern Asia at only about the longitude of the Maldive Islands.

More sophisticated encyclopedic world maps were based on the same tripartite skeleton as the T-O map, but actually located real geographical places and topographical features. The greatest extant example of such a *mappamundi* is that in Hereford Cathedral, while printed descendants of the genre date from as early as the woodcut world map in the 1475 *Rudimentum Novitiorum* (fig. 34).

Virtually all such *mappaemundi* were oriented with east at the top (hence the very term 'oriented'), with Southeast Asia thus falling at about the one o'clock to two o'clock position. East was the preferred orientation, perhaps in part because the 'T' of the T-O may have been derived from the Greek *tau*, an ancient form of the cross, and thus needed to be upright. The *tau* had been adopted by early Christians as a clandestine symbol of their faith, and it was later incorporated in maps to symbolically superimpose a Christian framework over the entire earth — a devise with parallels in such Southeast Asian maps as the Thai *Buddhapada*, which employs Buddha's footprint in much the same way. Some medieval maps even used the literal figure of Christ in place of the 'T', with his head in Paradise, at the top of the map.[104] Further, since the Judeo-Christian Paradise was believed to lie at the eastern 'end' of the world, the East most definitely needed to be accorded a position at the 'summit'.

Macrobius

One of the most interesting cosmographical texts, with an accompanying map, which a motivated European sailor, merchant, academic or dreamer may have been aware of at that time, was the work of the early fifth century theoretician, Macrobius (fig. 35).

Macrobius' ideas were inspired by Crates of Mallos, a geographer who headed the library at Pergamum, a highly cultured city in the northwest of what is now Turkey, in about 150 B.C. His ideas are expounded by way of a commentary on Cicero's *Dream of Scipio*, which was part of the *De Re Publica* and was to a certain extent copied from Plato's *Vision of Er* from the *Republic*.

Despite the fact that aspects of Macrobius' text contradicted Church teachings and brought cries of heresy from some, it was immensely popular, being read, for example, by the mathematician and astronomer, Pope Sylvester II, at the end of the tenth century. Macrobius believed the earth was girdled by two ocean belts, one equatorial, the other along opposing meridians, forming four roughly equal quadrants of land, of which the known world (comprising Europe, Asia, and a truncated Africa) was one. This geography meant, correctly, that the Indian Ocean was an open sea, but it also vastly abbreviated the southerly breadth of Africa. Africa, according to his map, lay entirely north of the equator, and would therefore be easily circumnavigable were it not for what Macrobius believed to be the impenetrable heat of the equatorial ocean. The belief in a wide equatorial ocean complemented the idea that Africa could only be a narrow sliver of land, which had been mistakenly deemed a separate continent only because it was separated from Europe by a sea (the Mediterranean).

By the middle of the fifteenth century, Portuguese expeditions had proven the fallacy of the concept of a hot equatorial ocean, and so Macrobius' geography, if correct, would mean that there existed an easy sea route to the Indies. In truth, not only was the sea route to the East to prove far more difficult than Macrobius' geography suggested, but a great coincidence would fool European pioneers to the Indies into believing — for a brief time — that Macrobius' sea route was correct. In Macrobius' view, southern Africa lay between the Tropic of Cancer and the equator, a latitude which by coincidence corresponded to the true southern coast of *West* Africa. So, when Portuguese mariners first sailed east along the southern coast of West Africa, they probably believed they had indeed reached the bottom of the continent, and had thus 'confirmed' Macrobius' geography and his scenario of an easy sea route to India and beyond.

There were others who supported a dramatically abbreviated Africa. Orosius included a chapter describing the countries of the world in his fifth-century work *Historiae adversus paganos.* This text, which was widely disseminated and plagiarized by others throughout medieval times, stated that Africa, despite the sea separating it from Europe, could simply be considered part of the continent of Europe "because of its [Africa's] small size." A similarly truncated Africa can be seen in the world map after Mela (fig. 29). Given the varied sources promising a foreshortened African continent, the turn of West Africa's coastline to the south at Cameroon must have come as a staggering disappointment to Portuguese mariners and cosmographers alike.

Macrobius, Pliny, and Buddhist Cosmography

In a curious quirk of history, Macrobius' basic concept of the earth is strikingly similar to that of some Buddhist cosmography. Just as Macrobius believed the earth to be composed of four quadrants of land, each separated by a belt of ocean, so too did Thai cosmographers, who, according to Gervaise and other seventeenth-century observers, deem that "the earth is divided into four equal parts called *tavîp*" (*thavip* = continent), which are "separated from each other by the same sea that covers the *Càu pra Someratcha*" (Khao Phrasumane = Mount Sumeru). Furthermore, Macrobius believed that these ocean belts were impassable, so that the people of each of the four continents would forever remain ignorant of the people in the other three. This is precisely what Gervaise tells us about his hosts' cosmography as well: "as this sea is not navigable anywhere," the Thai cosmographers told him, "it is impossible for the inhabitants of one part to have any communication with those of another."

A further comparison can be made with the philosophical context of Macrobius' map and Buddhist thought. Macrobius' world view, like that of eastern cosmographers, transcended worldly details of geography. Macrobius viewed the Earth as itself an insignificant, impermanent, and minute point in the cosmos. Thus, he believed, any person's pretense of fame and glory would be a self-deluding indulgence. Moreover, the earth was regularly subject to "floods and conflagrations", so that even if the facade of fame and glory were attained, it would be only a momentary boast. We must, rather, be content with what we have, for our happiness will come only from within. While such comments regarding the transient nature of life could well have been summarized from Pali texts written by Buddhist monks in Southeast Asia, they are in fact paraphrased from Macrobius' commentary on the *Dream of Scipio*.

Pliny spoke similarly. Regarding "this dot in the universe," as he called the Earth, "we fill positions of power and here we covet wealth and put mankind into a turmoil repeatedly and fight wars — even civil wars — and empty the land by killing one another." Pliny sarcastically sums up "the outward madness of nations" by observing that "we drive out our neighbors and dig up and steal their turf to add to our own, so that he who has marked his acres most widely and driven off his neighbors may rejoice in possessing an infinitesimal part of the earth."

Paradise and Ophir

The Judeo-Christian Paradise, where Man and Women were created and from which they were banished after the Original Sin, was believed by most medieval commentators to have been in easternmost Asia. The Biblical account of Genesis alludes to this position, stating that "the Lord planted a garden in Eden, to the east, and there He put the man [Adam] whom He had formed." Though usually placed at the 'top' of the continental land, as in the twelfth-century map in fig. 1, there were also medieval maps which placed Paradise on an island in the ocean sea. Dissenting opinions on Paradise's location in easternmost Asia were generally based on the premise that Alexander the Great (the story of whose conquests had, over the centuries, been transformed into epic fiction) had conquered India fully to the end of the earth, but clearly had not stumbled across the primeval Garden.

Several popular cosmographies of the Middle Ages 'confirmed' that the far eastern end of the world was indeed Paradise. Perhaps most influential was Isidorus of Seville, whose *Etymologiae* (between 622-633), an indiscriminate and uncritical compilation of diverse texts which Isidorus himself sometimes did not appear to fully understand, asserted that God placed Eden in the East. An anonymous mid-seventh century text known as the '*Ravenna Cosmography*' reiterated much of the same, judging Paradise to be "a country in the farthest East, beyond all known land." Similarly, a popular encyclopedic work of uncertain authorship, *De Imagine Mundi* (ca. 1100), taught that "the first place in the East is Paradise, a garden famous for its delights."

We have John of Hesse (fl. ca. 1389) going beyond mere faith on the matter, claiming actually to have seen Paradise from a distance in the Far East. Dante placed Paradise as the summit of a mountain (the lower part of the mountain being Purgatory), and located it in the southern hemisphere antipodal to Jerusalem. While in truth this spot lies in the South Pacific Ocean, south of Tahiti and the Marquesas, in his mind he almost surely considered this to be in Southeast Asia, because of his concept of the earth. John Marignolli claims to have been assured by the people of Ceylon that Paradise lay about forty miles from Adam's Peak, and that it was sometimes possible to hear the water falling from the river which "went out of Eden to water the Garden." The popular medieval text of Gervase of Tilbury was more dramatic, informing its readers that Eden was separated from our inhabited earth by a long tract of land and sea and elevated so high that it reaches the sphere of the moon.

The virility of the lush, southern latitudes was a perfect stage for the creation of Man and the garden which was Paradise; in addition, the *antithesis* of Creation, the apocalypse, would come in part from the Scythian — northern — reaches of Asia, wrought by the terror of Gog and Magog when freed from their incarceration behind the Caspian Gates.

An interesting example of an early map locating Paradise in Southeast Asia proper is the so-called 'Beatus' type, contained in a codex dedicated to Bishop Etherius of Osma (fig. 36). In this view dating from the early thirteenth century, the southeast corner of Asia is *Paradisus,* with the four rivers traditionally believed to flow from there, being the Ganges, Nile, Tigris, and Euphrates.[105] In the small corner of Southeast Asia outside of Paradise itself is *edem,* the Garden of Eden. Just north of Paradise is the apostle St. Thomas, who later became associated directly with Southeast Asia on the Martellus map of ca. 1489 and several Renaissance texts. Of the islands lying in the Beatus map's ocean sea, the two directly off the coast of Paradise are *Argire* and *Crise,* our *Argyre* and *Chryse,* the Southeast Asia of early Roman lore. Some rendering similar to this map was probably among the several cosmographies and world views which were in the mind of the Genoese navigator Christopher Columbus as he crossed the Atlantic.

Even 'authentic' travelers' accounts could reinforce paradisiacal images of Southeast Asia. Traveling south from China in the early fourteenth century, Friar Odoric "arrived at a certain country called Melistorte," where there were two mountains enclosed by a wall.

> Within this wall there were the fairest and most crystal
> fountains in the whole world: and about the said
> fountains there were most beautiful virgins in great
> number, and goodly horses also, and in a word, everything
> that could be devised for bodily solace and delight, and
> therefore the inhabitants of the county call the same place
> by the name of Paradise.

"By certain conduits," an aged man who had constructed the walls "makes wine and milk to flow abundantly," and, as if he were the keeper of the Gates, "when he saw any proper and valiant young man, he would admit him into his paradise."

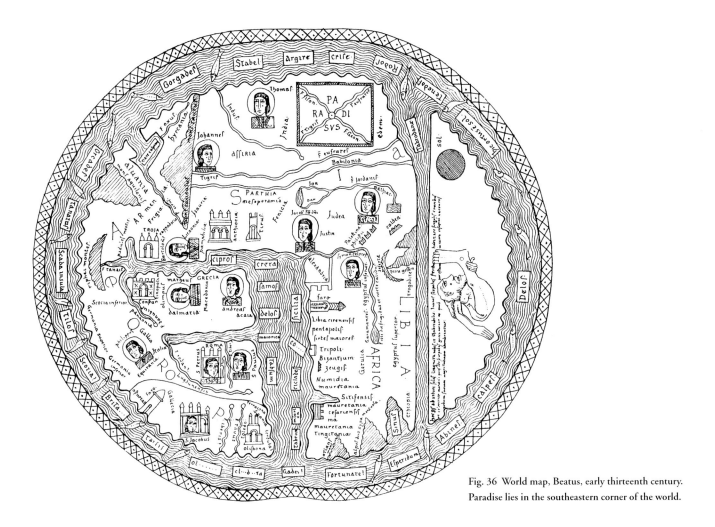

Fig. 36 World map, Beatus, early thirteenth century.
Paradise lies in the southeastern corner of the world.

Fig. 37 World, from M. Elucidarius, *Von allerhand Geschöpffen Gottes* . . . 1595.

Similar images of the Far East continued to be published even in the latter part of the sixteenth century. A German woodcut map of the world from 1595 (fig. 37), shows the earth divided into four parts by the four rivers of Paradise. This was consistent with Biblical descriptions of of the Garden of Eden which stated that "a river went out of Eden to water the garden, and from there it parted and became four riverheads." The idyllic and fruitful nature of the two eastern (uppermost) portions of the world contrast sharply with the portrayal of Europe and Africa below, which are, quite literally, in the shade. Adam and Eve stand in the principal eastern sector of the world, beside the source of the four holy rivers, while idealized and fanciful fauna in the 'Southeast Asian' quarter to their right, identify this region as the original Garden of Eden.

Similar ideas had been promoted, at a much earlier date, by the influential Isidorus of Seville. Writing in the seventh century, Isidorus noted that "from the middle of the Garden, a spring gushes forth to water the whole grove and, dividing up, it provides the sources of four rivers." Nine hundred years later, serious Renaissance scholars such as Abraham Ortelius continued to allude to Asia as a paradise even on 'modern' maps. For example, on his 1567 map of Asia, Ortelius observes that

> The whole of [Cathay and China], together with the neighboring islands, is and always was most celebrated and fortunate . . . as regards the numbers of peoples, their wealth, character, the magnificence of things, in all necessities, and also various luxuries, it seems to surpass Europe and the other parts of the world, so that it can be deservedly called a paradise on earth.[106]

Etas prima mundi   Folium VII

Fig. 38 Woodcut depicting the Garden of Eden and the Expulsion from Paradise, from Hartman Schedel's *Liber Chronicarum*, the so-called 'Nuremburg Chronicle' of 1493. From Eden, portrayed as a garden filled with tropical foliage, flow the four rivers of Paradise. The typical Judeo-Christian image of paradise was of a tropical setting in easternmost Asia. parallels can be drawn between this European view and the Southeast Asian idea of a cosmic mountain, Sumeru, around which the world was pivoted. The 'Nuremberg Chronicle' marked the maiden use of printing to produce a monumental, lavishly illustrated work conceived for a mass-market audience, and indeed the author, a humanist, expressed his desire to make the work available "for the common delight". (26 x 23 cm)

Columbus, Southeast Asia, and Paradise

Christopher Columbus harbored ideas of divine purpose, perhaps in anticipation of the coming of the Antichrist and the end of the world.[107] He was thus the perfect figure to go forth and 'discover' the foothills of Paradise in the southeast of Asia, which was his intended destination. When the Genoese navigator reached the shoulder of the South American continent on his third voyage, he conceded that they were in fact near "the Earthly Paradise . . . where no man may go, save for by the grace of God." Columbus believed that Paradise lay more to the south than east — that is to say in Southeast Asia rather than in China. In his copy of Pierre d'Ailly's *Imago Mundi*, an influential geographical treatise which was written in 1410 and published in about 1483, Columbus underscored passages alluding to the temperate nature and material wealth of the equatorial regions. Had he been in Asian waters upon reaching the Caribbean (as he claims to have believed), his northern coast of South America would have been at about the latitude of the southern coast of Indochina.

Ophir

The Judeo-Christian Bible recorded that King Solomon's fleet sailed for three years to find Ophir, load its treasure, and return to Jerusalem to build Solomon's temple. Whereas Paradise was almost universally believed to lie in eastern Asia, Ophir's location was more ambiguous, being commonly envisioned in Africa as well, until the Portuguese failed to find any trace of it there in the course of their fifteenth-century voyages. The *Imago Mundi* placed Ophir as an island off the coast of Southeast Asia; Columbus probably expected to find the island where d'Ailly had stipulated, in his anticipated destination in the South China Sea. By 1498, according to letters written by Columbus to Ferdinand and Isabella, he concluded that the 'Southeast Asian' island of Hispaniola was Solomon's Ophir.[108]

Within a decade, once America was recognized as being a distinct landmass, the Pacific Ocean became the logical site for Ophir. In the 1520s, Rodrigo de Santa Ella, founder of the University of Seville, wrote that Ophir lay in the middle of the Pacific, a view shared by many Spanish pilots and cosmographers but scoffed at by others.

Despite increasing skepticism, Ophir is known to have been a Southeast Asian goal for pilots like Sebastian Cabot, in his aborted attempt to reach the Moluccas in 1526, and Ruy López de Villalobos in his crossing from Mexico to the Philippines in 1542. In the sixteenth century, Mendes Pinto, searching for Southeast Asia's "elusive Isle of Gold", claimed that some people in either Malaya or Sumatra "tell of reading in their chronicles that the queen of Sheba once had a trading post there which supplied her . . . with a large quantity of gold, which she contributed to the temple of Jerusalem when she visited King Solomon." The placement of Ophir in both Malaya and Sumatra was not contradictory, since many people during the Renaissance believed that the Malacca Straits which separates Sumatra from Malaya did not exist until after Ptolemy's time (second century). Sumatra, in this view, was the southern region of the Golden Peninsula in Biblical days.

When the French ambassador Choisy reached Ayuthaya in 1685, he speculated that an alloy of gold and copper used by the Siamese for various statues might actually be the *electrum* mentioned by Solomon.[109] Even in the late-nineteenth century, some French explorers of the Mekong River still equated Indochina with Ophir.

'True' Southeast Asia on Medieval Maps

The first inklings of insular Southeast Asia are difficult to pin-point because the tradition of a scattering of islands in the ocean sea are prevalent in early cosmographical writings. Archipelagoes in the ocean sea were commonly, but arbitrarily, represented in many world views. Nonetheless, the many allusions to 'islands of gold' that occur in Indian literature, in the *Ramayana,* for example, together with similar references from early Western authors such as Pomponius Mela, must surely represent some actual knowledge of Malaya, Sumatra, and other Indonesian islands.

Columbus' discovery of various Caribbean islands fitted nicely into medieval thought, since the belief that innumerable islands dotted the ocean sea was one that cut across all cosmographical and philosophical camps. This set the stage for the incorporation of the many East Indian islands into the world picture, defying the most orthodox Christian view which, citing the dogma that only Noah and his people survived the great Deluge, held such islands to be uninhabited because the entirety of post-deluge humanity must lie within land accessible to the known continents — a belief which, in turn, later became an argument supporting the idea that eastern Asia and western North America were joined.

Early references to the mainland of Southeast Asia are easier to separate from generic myth. The Malay Peninsula is recorded by Ptolemy as the Golden Chersonese, and is found in other Western writings as early as Martianus of Heraclea in the fifth century. Martianus writes in his *Periplus of the Outer Sea* that

> in Trans-Gangetic India is the Golden Chersonese, and beyond is the Great Gulf, in the middle of which is the frontier between Trans-Gangetic India and the Sinai. Then come the Sinai and their capital, called Thinai [China]. It is the boundary between lands known and unknown.

In a subsequent book, the same author elaborates on Southeast Asia, offering specific measurement of its size and a compendium of its topography and people. He states that

> Trans-Gangetic India has a maximum length of 11,650 stades [approximately 1800 kilometers], and a maximum breadth of 19,000 stades [3000 kilometers].[110] It comprises 50 tribes or satrapies; 67 towns, important villages, or

markets; 18 important mountains; 5 noteworthy capes; 3 notable ports; one large bay [Gulf of Siam?]; and 30 important islands.

The lands of *Chryse* and *Argyre* were also remembered during the early Middle Ages. Isidorus repeats older writings, stating that "Chryse and Argyre, islands situated in the Indian Ocean, are so rich in mines that, according to most writers, their soil is of gold and silver — whence are derived their names." Isidorus believed that these islands had dragons and griffins that guarded mountains of gold, as well as flowers that were always in bloom. This fabulous Southeast Asian 'mountain of gold' is found in many medieval *mappaemundi,* such as that which accompanied the *Rudimentum Novitiorum* (fig. 39). *Chryse* and *Argyre* are also mentioned by the unknown author of the *Ravenna Cosmography* (mid-seventh century), who informs us that "in the same part of the same ocean there is the Island of Chryse, that is to say, the Golden." However, repeating an idea that dated back at least to Herodotus, he believed that beyond India lay an impassable desert.

Interestingly, the unknown author of the *Ravenna Cosmography* acknowledges several sources for his text's information, of which Isidorus and Ptolemy could account for these allusions to Southeast Asia. His work is but one hint that during so-called 'Dark Ages' of the first Christian millennium, the intellectual inquisitiveness about the outside world may not have been as uninspired as our lack of memorials might suggest. Occidentals still traveled east and Orientals still journeyed west; sailors and traders continued to exchange stories; and truths about distant lands could be discerned in amongst the fantasy of exotic tales. The West was in fact already trading in goods that passed through Indonesian intermediaries. From the seventh to thirteenth centuries the Srivijaya empire dominated trade between the China Sea and Indian Ocean, handling commodities of diverse origin. Western items, such as the forest products considered precious and medicinal by the Chinese, were exchanged for manufactured goods, including porcelain, lacquer, and silk. The Srivijaya leg of this commerce could not always have been invisible to Europe — that we have any written Western record of Southeast Asia at all from the Middle Ages suggests that an interested individual of the day could glean more of its place on the map than simply its position to the south of Paradise.

Ample records remain of inquisitive minds from that era. By the early eleventh century, the cathedral school of Chartres, known for its independent, intellectual attitude, was well-established. In the first half of the twelfth century, Chartres' Adelard of Bath eloquently questioned the blind acceptance of the Church's authority and espoused the 'universal judge' of reason as being part of, rather than in conflict with, God's will. His contemporary William of Conches was more eccentric, complaining of the 'wretched ones' who explain phenomena merely by stating that God made it so.[111] People such as Adelard and William were of course the exceptions, and William was ultimately forced by the ecclesiastical elite to recant many of his opinions; but they demonstrate that these centuries were not as 'dark' as we sometimes assume.

The Spanish city of Toledo, so strongly influenced by Muslim culture, was also a center of scientific research. In about 1080, a Muslim scholar in Toledo, Az-Zarqali, wrote a commentary on astronomical tables that had been composed by a group of Muslim and Jewish scholars; the work, known as the *Toledo Tables,* contains some geographical information derived from Ptolemy's *Geography.* It was translated into Latin in the following century by Gerard of Cremona (d. 1187), a prodigious translator of scientific texts.

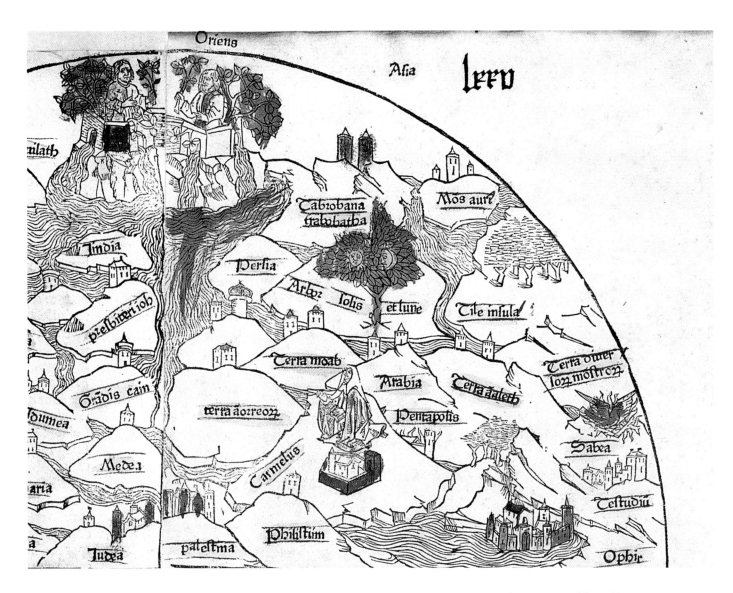

Fig. 39 World map from the *Rudimentum Novitiorum*, Lübeck, 1475 (detail). The walled garden on the mountain at the eastern 'top' of the world is
Eden, and is the source for four major rivers. At one o'clock we find the island of *Tabrobana* (Sri Lanka), and at two o'clock, *Mons auri*, 'mountain of
gold', representing the Classical and medieval concept of Southeast Asia as a peninsula or island of gold. Isidorus, for example, described Southeast Asia as
mountains of gold guarded by dragons and griffons. After this lies *Tile insula*, the island of *Tile* near India. (38 cm diameter)

Another region with strong Arabic influence was Norman Sicily.
The twelfth-century was a remarkable period in Sicily, one in which
Arabs, Greeks, Italians, and Jews cohabited in peace and in which the
arts and sciences flourished. Even the consolidator of this empire,
Roger II, purportedly

> gave himself up to [questioning travelers about distant
> parts of the earth and collecting Arabic geographical
> treatises] for fifteen years, never ceasing to examine
> personally into all geographical questions, to search for
> their solution and to verify facts . . .[112]

This compliment came from a Sicilian subject and one of the
foremost geographers of the Middle Ages, al Sharif al-Idrisi. King
Roger provided al-Idrisi with all the necessary facilities for making
maps (see figs. 27 & 28), and the two collaborated together, choosing
"certain intelligent men, who were dispatched on travels and were
accompanied by draughtsmen." Southeast Asia was far too distant to
benefit directly from this effort, however, and al-Idrisi's cartography
of the region is largely based on Arabic textual sources which were
already three centuries old.

Although al-Idrisi's work had little direct influence on medieval
Europe, the sum of Sicilian culture did. Navigational charts are
found early on in Sicily; the arts of navigation (and perhaps the
seeds for Genoese portolan charts?) were exported from Sicily to
Genoa in the thirteenth century, and thence to Portugal. Arabic
influence on Italian thought is abundantly evident as early as the
Marino Sanudo world map of about 1320, one of the most
important of medieval *mappaemundi* (fig.44). Sanudo's mapping of
Africa, the Indian Ocean, and Southeast Asia, are directly adapted
from Arabic data.

Chapter 6

European Pioneers

'I am a geographer,' said the old gentleman.

'What is a geographer'

'A geographer is a scholar who knows the location of all the seas, the rivers, the cities, the mountains and the desserts.'

…'Your planet is very beautiful. Are there any oceans?'

'I have no way of knowing' said the geographer.

…'And any mountains?'

'I really couldn't tell you that either,' said the geographer.

'But you are a geographer!'

'Exactly,' said the geographer, 'but I am not an explorer.

I have no explorers on my planet.'

— from *The Little Prince,* by Antoine de Saint-Exupéry

Scant record remains of direct contact between Europe and the Far East during the earlier Middle Ages. *Mappaemundi* of the period which attempted any detail within Asia did so mainly based on theological and Biblical concerns, or on the traditions inherited from the Romans. Travel to the farthest East was seen by some as being improper; typical of this attitude was the unknown mid-seventh century author of the *Ravenna Cosmography,* who noted that such a journey would be blasphemous, since the Scriptures state that no mortal may venture to the Paradise of God, which is the most eastern part of the world. Other images of the East, including those of religious context, were far less forbidding. Even Cosmas Indicopleustes (ca. 540), whose cosmography was molded from theological deductions, unimpassionately described the sea route from the Indian Ocean to China.

The earliest surviving map specifically of Asia accompanied the works of Jerome, who passed part of his life in Constantinople and the Levant, and is best remembered for having translated the Bible into Latin. Although the map may have originated with Jerome himself (latter fourth to early fifth century), the only known example is a copy dating from the twelfth century.[113] Like most of its classical predecessors, it dramatically abbreviates the breadth of Asia, if in fact its author intended any relative scale. It depicts the Indus flowing due south, with Ophir on its southeast bank, and the Ganges emptying through the continent's eastern coast, with five islands, including *Taprobana, Argyre,* and *Chryse,* lying in the southeastern corner. The map appears to have been a straight-forward attempt to record the continent based on whatever empirical data was available.

During the Middle Ages, Europe gradually lost its dependency on Asian traders for silks, but not for spices. A monk is supposed to have smuggled silkworm eggs into Constantinople by the year 553, from which domestic European production began. By about the eighth century, silk making had spread to Greece, Sicily, Spain, and, by the time of the Crusades, to Italy. Spices, however, could not be grown in Europe. Venice entered the spice trade in the tenth century, adding another middle-man to the trade in eastern goods rather than breaking the trading monopoly of the Levant.

Europe Returns 'Home' to Southeast Asia?

To what extent the European Middle Ages was a period of self-perpetuating ignorance, or a formative society finding its own way and doing as well as it could, given its values and its resources, is today a matter of historical debate. But the perception of medieval Europe as a 'dark' age was pivotal to humanists and explorers of the Renaissance whose sense of 'manifest destiny' legitimized their claiming distant lands as their own. Their words suggest that newly discovered or newly accessible shores — and those of easternmost Asia in particular — were not part of a foreign map, but in truth part of their *own* map, which had been denied them during a long era of ignorance and repentance.

Art and science, according to one text of about 1530, had now "gloriously rearisen after twelve centuries of swoon"[114] — a rather sweeping judgement on the twelve hundred years between the decline of the Roman Empire and the newly-achieved conquest of the seas. One mid-sixteenth century commentator who had a particular interest in the Far East, João de Barros, was more explicit: he judged that Europe had for so long been kept ignorant of the greater part of the world as punishment for the sins of man, a veil having obscured the original creation just as the Tower of Babel had cast a veil over the original language. Probably echoing the belief of many Christians, Barros believed that the mission of explorers and mapmakers alike was thus the very shedding of this 'veil'. Similarly, Simon Grynaeus, who in his preface to the 1532 Basle edition of Huttich's *Novus orbis,* declared that the navigators of Europe were "invincible and truly divine souls . . . who have reclaimed the empire of the earth and the sea as the heritage of their first parents." Significantly, the work in which these comments appeared contained a map of the world which depicts Ludovico di Varthema striding across Southeast Asia a quarter of century earlier (fig. 68) and reaching a building with classical columns. This foreshadowed a tendency to attribute ancient Asian stone edifies to classical European sources, thereby reinforcing Western precedent in the region. When missionaries found the ruins of Angkor in the later sixteenth century, there was a widespread belief that the city had been built by Alexander the Great or the Romans.

It was this profound belief in the West's rightful place in the Indies that led the great Portuguese poet Camões to create his epic masterpiece, *The Lusiads*, which elevates Vasco da Gama's discovery of the sea route to India to the heights of Virgil's *Aeneid*. Peter Martyr, writing in the early sixteenth century, was moved to compare Iberian prowess to classical feats, bragging of Magellan that "if a Greek had [made the circuit of the globe], what would not the Greeks have written about his incredible feat!" Martyr described the seductive lure of the Spiceries with the mystique of a *femme fatale,* remarking that

> from the date of Rome's luxury [the Moluccas] have, so to say, glided into our ken, not without serious consequences to us; for characters soften, men become effeminate, virtue weakens, and people are seduced by these voluptuous odors, perfumes, and spices.

In the mid-seventeenth century an Italian engraver of maps, Antonio Lucini, wrote that "the same nature which gave laws to man, shall receive them back." Whereas previously, man had been controlled by the elements, he would now control them, as though the earth were a temptress who had been conquered. With the East Indian archipelagoes likely in his mind, Lucini added that the "many islands thrown by nature far away in the boundless sea" would enrich Christendom while themselves being bettered by it.[115]

Thus we see that those who demystified the map of Further Asia could also be perceived as having secured the salvation of the children of Eden, who had now repented long enough to retake the world as their birth-right. To whatever extent Europe believed Eden to have been in the tropical Far East, her quest for Southeast Asia may well have meant a *return* voyage, a voyage home. A letter written by Matthew Paris in 1237 speaks of the sovereign whose "prelacy extended over India the Greater and the Kingdom of Prester John and other realms still nearer the rising sun."[116] Paris' other lands, "still nearer the rising sun", would be Southeast Asia. An inscription by Martin Behaim, on his globe of 1492, states that "all this land, sea and islands, countries and kings" in the vicinity of Southeast Asia "were given by the Three Holy Kings to the Emperor Prester John." This historically confused statement (it has the mythical Christian king already on the throne at the time of Christ's birth), demonstrates the belief that Southeast Asia was rightfully Christian turf. At the turn of the sixteenth century, with the Portuguese standing on the threshold of Southeast Asia, some mapmakers identified Indochina with the mythical 'Seven Cities' founded by Christian kings (fig. 58).

There was yet another dimension to the special status of the eastern limits of Asia. The sun rises in the east, the sun sets in the west; the original Garden of Eden and the Creation of Man occurred in the east, from whence mankind dispersed westward. As early as the fourth century there was a theory that Man, created at the very rising of the sun (Southeast Asia), would reach the end of all ages upon his arrival at the ends of the earth in the west and the setting of the sun.[117] That age was nigh, of course; and, although the turn of the sixteenth century was certainly a period of great hope and excitement, there existed at the same time a fear of apocalypse among some. The '*Nuremberg Chronicle*' *(Liber Cronicarum),* a great history of the world published in 1493, is sometimes found with seven blank leaves at the end, supposedly, it is believed, for the book's owner to record the seven remaining years of mortal existence before the end of the world in 1500. Perhaps for some, the prospect of returning to the rising of the sun, from whence we began — that is, sailing to the Far East and its islands — helped portray the conquest of the Indies as a return to the Garden, a subliminal return to the womb, a route to salvation. Christendom had completed its circuit of the world, returning to the place from whence it had come.

Southeast Asia During the Early European Renaissance

What was going on in Southeast Asia in the mid-to-later thirteenth century that European travelers might have discovered had they been there to see it? A flourishing civilization in Cambodia, based around Angkor, extended through much of what is now Thailand. Its sophisticated irrigation systems harnessed the region's alternating six months of torrential rain and infertile drought, to produce three rice harvests a year and feed more than one million Cambodian people. The magnificent temple at Angkor, whose ruins remain the largest religious structure in the world, was a century old, and a new temple-mountain, the Bayon, was already completed. Angkor's magnificence was rivaled by that of Pagan, in what is now Burma, though that splendid metropolis was destroyed by the Mongols in 1287. In the north of what is today Thailand, Thai people were reaching the area of Lan Na, and in 1296, within a year of Marco Polo's return to Venice, Lan Na's principal city-state of Chiang Mai was founded. The great maritime trading empire of Srivijaya, based around Sumatra and Malaya, had flourished for half a millennium, but was now waning with the advent of direct Chinese trade across the Southeast Asian threshold was beginning. In the jungles of central Java, the imposing complex of Borobudur, which had inspired awe-struck pilgrims for generations, already lay abandoned and was being reclaimed by the forest. The various peoples of the Philippine archipelago remained aloof from the Hindu-Buddhist sphere of the mainland and Indonesian islands; although Chinese and Arab merchants frequented their coastal communities, little is known of the Philippines from this period.

The Predecessors of Marco Polo

One of the earliest Europeans known to have penetrated far into Asia was Benjamin of Tudela, a rabbi from Spain who lived in the twelfth century. Benjamin journeyed for thirteen years, traversing Europe and the Middle East to Persia and beyond, reporting about the country of '*Zin*' (China) and describing an eastern sea where violent winds blow — a reference to the monsoons of Southeast Asia or the typhoons of the China Sea.[118] In 1177, Pope Alexander III sent his physician, Philip, to deliver a letter to Prester John, 'king of the three Indias', but there is no record of Philip having returned.

In the mid-thirteenth century, Marco Polo's immediate predecessors benefitted from the rise of the Mongol Empire, which facilitated greater intercourse between the Far East and Europe and was accommodating to trade. This relative freedom of movement lasted for about a century, until China regained control of their kingdom in the middle of the fourteenth century, and the Ottoman Empire came to power in the western part of Asia. During this century of relatively easy trans-Asian travel, accounts left by travelers began to chip away at the mythification of the Far East and the underestimating of the size of the Asian continent. One of the first of these chronicles was that of Friar John of Plano Carpini, a friend and disciple of St. Francis of Assisi. In 1245-47 Carpini journeyed into Mongolia as an ambassador of Pope Innocent IV. His mission was to convert the Khan to Christianity and enjoin his help in fighting the Muslims, their common enemy, as well as to record the history and customs of the Mongols. By having kept a careful log of his outbound progress, Friar John gave those who read his report a better sense of the vastness of the Asian continent, for despite his lengthy journey he never reached the ocean sea. He makes only passing reference to the various demons or oddities rumored to inhabit that part of the earth, and so did not reinforce the medieval lore which dampened European courage to travel east. His chronicle, *The Book of the Tartars,* was included in the work of Vincent of Beauvais (ca. 1250). In 1253, William of Rubruquis, a Franciscan Friar sent by Louis IX on the same mission as Carpini, crossed Central Asia and met with the Mangu Khan in Karakorum, Mongolia. His *Journey to the Eastern Parts of the World* is an intelligent, unimpassioned narrative, which included a brief description of the Far East based on second-hand sources. William mentions (but did not reach) Tibet, and tells of "great Cathaya, the inhabitants whereof (as I suppose) were of old time, called Seres, for from them are brought most excellent stuffs of silk" (see figure 29, for China as *Seres,* and a reference to silk).

Roger Bacon, who had a keen interest in the Far East, was among those who read the account of William's journey. A founder of what we now consider the scientific method, Bacon harbored many 'dangerous' and 'novel' ideas for which he was imprisoned (ironically during Marco Polo's trip to China). These ideas, which included gunpowder, and perhaps spectacles, may in part have originated from his contacts with Asia and with Arab writings. Not always able to discern reported fact from fancy, Bacon gave new life to the myth of Gog and Magog, demons in the Far East who were held behind the Caspian Gates, which were commonly believed to lie on the 'other' side of Paradise, the northeast of Asia.

Nicolò and Maffeo Polo, the father and uncle of Marco, followed soon after John of Carpini and William, purportedly reaching what is now Beijing in their journey of 1260-69. They set off for the East again in 1271, this time accompanied by young Marco, then seventeen. Contact between East and West was two-way; while the three Polos were in China, their host, the Kublai Khan, sent a Nestorian monk who had been born in Beijing, Rabban Bar Sauma, to Europe as an ambassador.

Marco Polo and Subsequent Travelers

The figure who most revolutionized Europe's early knowledge of Southeast Asia was the Venetian Marco Polo, who reached China by land through Central Asia. Polo may have skimmed the northern frontier of Southeast Asia when, in employ of Kublai Khan, he traveled far to the southwest and reached (or at least learned of) the country of *Mien* which, many scholars believe, was Burma. A kingdom of *Mien* still appears in western China on the Ortelius *Indiae Orientalis* of 1570 (fig.86), to resurface in Burma on maps based on Martini (e.g., the 1655 map of China by Blaeu), and in Cambodia on various late-seventeenth century maps by Vincenzo Coronelli. Polo tells us that *Mien* lies "towards the south on the confines of India", and that after fifteen days' journey through "very inaccessible places and through vast jungles" one reached the capital city of *Mien.* Polo's *Mien,* may in fact be the ancient Burmese kingdom of Pagan, whose tragic destruction he records in 1287:

> the king of Mien and Bangala, who was a very puissant
> prince, with much territory and treasure and people; and
> he was not yet subject to the Great Khan, though it was
> not long after that the latter conquered him and took
> from him both the kingdoms I have named.

But this excursion aside, Marco Polo might have never explored Southeast Asia, or even have returned to his homeland to relate his saga, had it not been for fateful events early in the last decade of the thirteenth century. Envoys of Persia's King Arghun reached Kublai Khan on horseback with news of the death of Arghun's Queen,

Bulagan. The queen, whose family was from the Khan's domain, had stipulated in her will that only a woman of her own blood should succeed her, and so a relative of seventeen named Kokachin was selected to be the new Persian queen. A party set out to escort the young woman to her new life in Persia. In the interim, however, war had broken out amongst the Tartar kings, rendering the roads to western Asia impassable, and so after a futile eight months of travel the party returned to the Khan's country. Rather than wait for a cessation of hostilities or risk the same route again, it was decided that Kokachin should go by sea. The Polos were chosen to accompany the bride-to-be, as part of her entourage; and so began Polo's journey through Southeast Asia and his ultimate return to Europe.

Marco reached Venice in about 1296, ending what posterity claims as among the most fantastic of human odysseys.[119] But surely there were many others, from all parts of the world, who had adventures as extraordinary as his, back when the world was a far 'larger' place than it is today. What separates Marco from those lost voices is that he returned safely to a homeland eager for news of the world, and that his story was widely disseminated in highly-readable, entertaining prose. Had a romance writer by the name of Rusticello not been a fellow prisoner of war with Polo during a period of confrontation between the city-states of Genoa and Venice, and had Rusticello not undertaken the project of laying Polo's memories to paper, Polo might have survived as merely a curious footnote in history, an eccentric and boasting importer of Eastern goods who claimed to have traveled to the edge of the continental world and into the sea of islands surrounding it.

In his day, Marco was accused of cavalier exaggeration and fabrication. Both he and his story earned the name *Il Milione* for his addiction to speaking in the grandest terms, always quantifying things as 'millions'. For this he may be fairly criticized, but the basic integrity of his observations and reporting have stood modern scrutiny reasonably well; for example, a Chinese document discovered earlier this century corroborated the names which Polo had given of the leaders of the voyage from *Zaiton* to Persia. Polo falters when he reports things he had heard second-hand (such as Arabian Nights-inspired reports of birds in Madagascar which could carry away elephants), and where the pages concern less 'exotic' places closer to to home which were perhaps spiced up a little, by Rusticello, to make them more 'interesting'.

There are, nonetheless, some serious problems with the text which modern historians have labored to reconcile. The very truth of Polo's travels to China has been questioned by some historians, citing, for example, problems of chronology, his failure to note the Great Wall and Chinese foot binding, and the lack of Chinese records confirming that Polo held the official posts he claims to have held.[120] According to this theory, Polo could have learned about the Far East while in western Asia, probably Persia. Others counter that foot binding was not generally practiced at the time Polo was in China, that the Great Wall did not assume its present form until Ming times, well after Polo, and that the leaders of Mongol China did not maintain records in the manner of the Chinese. Nor has any alternative source for Polo's information ever been identified.

Even scholars who suspect that Polo was an imposter still acknowledge the value of his text, for it reflects, if not his own observations, much factual data about Asia gleaned from his hosts in the Middle East that was not previously available to Europeans. The conflict, at least in part, stems from the fact that Polo's book has been erroneously presumed to read as a travel log, providing an itinerary from which his travels can be reconstructed. The very title of the book as it is commonly known in English, the *Travels,* has contributed to this misunderstanding, creating the false assumption that the details of its itinerary should always fall into logical sequence. The nature of the book is better described by its original title, *Devisement dou Monde,* a description of the known world. The narrative is woven around his travels, but contains frequent digressions, which distort the continuity of time and place.

Marco came home to a cosmopolitan Venice, eager to enter into trading relations with the East. In 1291, only a few years before his return, Venice's rival city-state, Genoa, had financed an expedition led by two brothers, Ugolino and Guido Vivaldi, who were to try to find an ocean passage to 'India'. Many people must have believed that they would soon be returning, their ship laden with cargo from the Indies, to the pleasure of Genoa and the consternation of Venice.

Since Polo's story was set to paper while he was a prisoner of Genoa, one can easily imagine Genoese officials taking advantage of his confinement to question him in the hope that he might have heard something about the progress of the Vivaldi enterprise. Enough time had elapsed since the Vivaldis' departure for them to worry, but certainly not so much as to lose hope. With the Vivaldi expedition in mind, Genoese geographers probably tried to sort out the sundry lands Polo spoke of. The mental 'map' image Polo had of his Indian Ocean journey probably resembled the "*mappemundi* of the mariners of this sea" he saw *en route*. Whether he ever made a simple cartographic sketch of his itinerary for his often disbelieving peers, and whether the later maps which incorporated his travels did so by copying such a draft or entirely through extrapolation from his text, is not known.[121]

Friar Odoric

Following closely after Polo was the Franciscan friar, Odoric of Pordenone. Bent on exporting Christianity rather than importing goods, Odoric left for the East between 1316 and 1318, traveling in the reverse direction to Polo. He reached Southeast Asia by sea from India and continued on to China in 1322, returning to the West overland, via Central Asia. As with Polo, some historians have questioned the veracity of Odoric's trip. Most scholars accept the basic truth of his story, although he at times seems to rely on standard clichés of travel lore and to have had a gullible ear for wild stories. Odoric's account is nonetheless valuable — he reached Java (which Polo had not), filled in some gaps in Polo's picture of China, and brought Europe its first description of Tibet since ancient times.

Whatever truth lies behind the journeys of Polo and Odoric, many cartographers nonetheless believed the essence of their travels, and were prepared to risk committing their reports to maps. The earliest surviving map of certain date to draw from Polo's or Odoric's travels is the fabulous Catalan Atlas (fig. 4), a map of the world made about 1375 and attributed to the Catalan chartmaker Abraham Cresques (1336-87), a Spanish Jew from Majorca who was cartographer to the King of Aragon.[122] Cresques possessed copies of both Marco Polo's *Description of the World* and Odoric's *Description of the Eastern Regions*. A brilliant illustrator and resourceful mapmaker, he depicted India clearly as a peninsula for the first time, probably from Arabic sources, and marked sixty-five places in the Far East. There is no Southeast Asian peninsula as yet, but there is a plethora of islands in the region. Most are arbitrary, merely representing the belief that archipelagoes dotted that corner of the world.

There are specific islands as well. Cresques' group of 7,448 islands, taken directly from Polo, arguably represents the first depiction of the Philippines on a Western map (see page 107, below). *Jana* is also a 'real' island, either Sumatra or Java. '*Trapobana*',

the dominant feature in insular Southeast Asia, heralds the beginning of confusion between Ceylon and Sumatra. *Taprobana* is Ceylon, but explorers and mapmakers of the fifteenth and sixteenth centuries usually identified it as Sumatra.

During the first part of the fourteenth century, the journey from Europe to the Far East had probably become routine, with travelers enjoying the relative safety of roads that were under Mongol protection and along which an occasional Franciscan missionary might be found. When Friar Odoric returned from China in 1322, he indicated that many Venetians had previously been there before him, while a *Merchant's Handbook* written in Venice in 1340 advised (not knowing that the situation was already changing) that the road from the Black Sea to Cathay was "perfectly safe, whether by day or night".[123]

But commencing about the time of Odoric's adventure, the climate for travel to the East began to deteriorate. With the conversion of much of the region to Islam, beginning with Persia in 1316, came a growing hostility to Western travelers, and by about 1340 the traditional routes through Central Asia were essentially closed. Even if a Westerner could safely cross to the Far East, he would be less welcome in China than Polo or Odoric had been, since the Ming Dynasty rulers, who had succeeded the Mongols in in 1370, were less receptive to foreigners.

Ibn Battuta

Ibn Battuta (1304-1377), a native of Tangier in Morocco, was the well-educated son of a Muslim judge who set off, at age 21, on a pilgrimage to Mecca. That pilgrimage evolved into a quarter-century of travels through Africa and Asia. He reached Southeast Asia in 1346-47, getting as far Sumatra, Java, and other islands. Upon his return to Tangier, the local wizir provided him with a secretary, Ibn Juzayy, to set his memories of his journey to paper. The text is the only surviving original Arab travel account of Southeast Asia.

Religion was a paramount concern in both Battuta's and Odoric's narratives of their travels, Battuta hoping that Islam would be more widely embraced and more carefully followed, Odoric hoping that Christianity might gain a foothold in the region. Like Odoric, Battuta frets over various peoples' mores, lamenting, for example, his inability to get local women to wear clothes. In great contrast to Odoric, Battuta marries wives as he travels, and receives gifts of slaves. Battuta's is a more lively and detailed account; but neither is especially good at geographic descriptions and itineraries.

Although Ibn Battuta's book had no impact on either Arab or European mapping of Southeast Asia, we will refer to his text when it helps illustrate Southeast Asian travel in the late medieval period.

Sir John Mandeville

Sir John Mandeville is a curious case. He was doubtfully a knight, he probably did not travel beyond Western Europe (though it is possible that he visited Egypt and the Levant), and he may not even have been English. Some scholars believe that 'Sir John Mandeville' was a pseudonym for either Jean de Bourgogne or Jean d'Outremeuse, both of Liege. Others argue that the name was genuine, that a writer named John Mandeville was born in St. Albans in the late thirteenth century and passed much of his life on the Continent, and that his book was only intended as a travel romance rather than as a factual account. The book, which may have been completed around 1356, was in any event commonly construed as a legitimate travel log and was consulted by many geographers attempting to map Southeast Asia. In the 1560s, Gerard Mercator still put his faith in Mandeville's text.

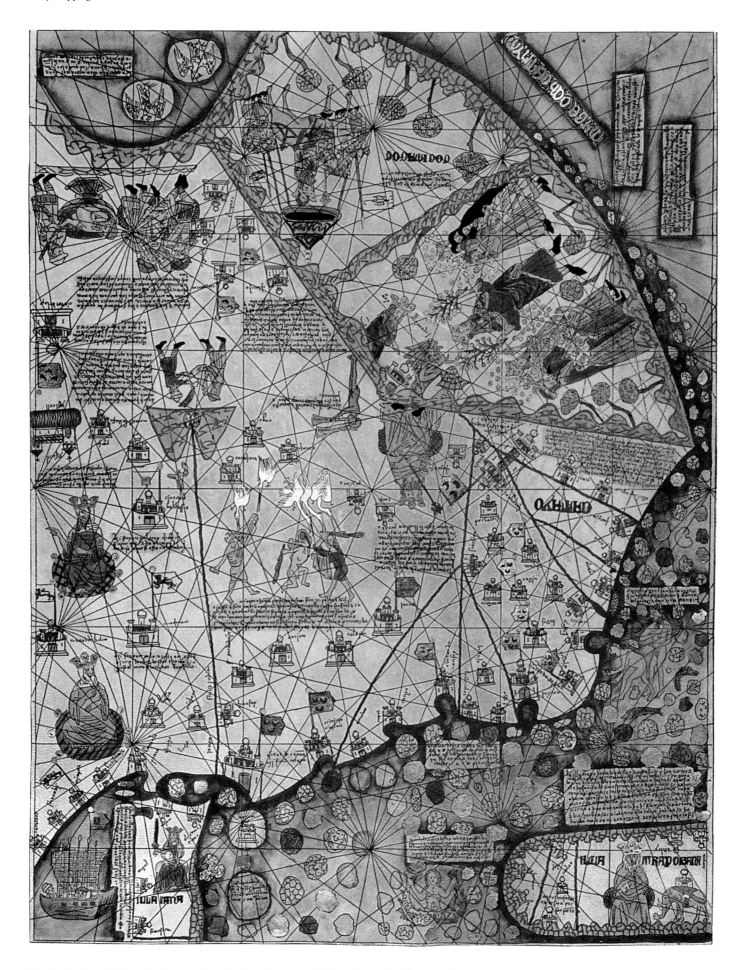

Fig. 40 Southeast Asia from the *Catalan Atlas*, Abraham Cresques, 1375 (from a facsimile). The original is manuscript on vellum and comprises 12 sheets. (Each sheet approx. 25 x 65 cm) [Library of Congress]

Nicolò de' Conti

No European is known to have penetrated far into the islands of the Indonesian Archipelago until more than a century after Polo and Odoric, when a merchant from the Venetian island of Chioggia, Nicolò de' Conti, set off on an extraordinary journey to the East. In about 1419, Conti made his way to Damascus, where he remained for several years, learning Arabic. Conti then went to Persia and from there to India. He subsequently visited the Andaman Islands (*Andamania*), the Irrawaddy River region of Burma (*Panconia*), Malaya, Sumatra (to him *Taprobana*), Java, and may even have reached as far as Sumbawa and Banda. In all, Conti passed about twenty-five years in the East, marrying an Asian woman and rearing four children *en route*. During the return journey to Venice, his wife and two of their children died of the plague. When he arrived back in Italy in around 1444, he was carefully questioned by Poggio Bracciolini, a humanist and papal secretary, who took down his words and left us with a first-hand account of Conti's time in the Indies.

During the course of his travels, to facilitate the freedom of his movement and his safety, Conti 'renounced' Christianity and 'converted' to Islam. Tradition has it that Bracciolini's setting to paper of his experiences was the penance imposed upon Conti by Pope Eugene IV for this impropriety. The account is preserved in the fourth volume of Bracciolini's *De Varietate Fortunae* (*On the Varieties of Fortune*), dating from about 1447.

Posterity has perhaps slighted the accomplishments of Conti, as compared to his fellow Venetian, Polo. Both passed about a quarter-century in Asia, and were perceptive observers. But the narrative of Polo is more detailed, and was disseminated in printed form as early as 1477, while Conti's story was not widely available until the middle of the sixteenth century, when it was included in Ramusio's *Delle Navigationi et Viaggi*.

Although Conti's account of the East did not enjoy the wide circulation of Polo's, his reports were shared among intellectual circles, and data from Conti's journey appeared on maps as early as 1457. An anonymous *mappamundi* of that year known as the 'Genoese' world map shows *Java Major* and *Java Minor*, as well as Conti's *Sanday* and *Bandam,* probably Banda, these representing the first appearance of any of the legendary Spice Islands on a Western map. Conti, who was probably the first European to reach these islands, reported that

> sailing fifteen days [beyond Java and Sumatra] towards the Orient, you come unto two other Islands, the one is named Sanday, where there is Nutmegges and Al maxiga or Masticke. The other is called Bandam, where Cloves grow, and from thence it is carried unto the Islands named Clavas.

Two years later, in 1459, the reports of both Conti and Polo were used by Fra Mauro, a cartographer working in a monastery on the island of Murano, Venice, for a map commissioned by King Afonso V of Portugal[124] It was with Fra Mauro's map that the identity of *Taprobana* was formally transferred to Sumatra. Conti had reported visiting a city on the island of '*Taprobanam*', which he claimed the natives call *Shamathera;* and so Fra Mauro identifies '*Siamatra*' with *Taprobana,* and clinches the error by noting that "Ptolemy, professing to describe *Taprobana,* has really only described Saylan [Ceylon]."

Perhaps the most important opinion Bracciolini gleaned from Conti was that the Indian Ocean was an open sea, contradicting (as we shall see) the writings of Claudius Ptolemy, which Conti may well not have been acquainted with. Fra Mauro, taking his lead from Conti, depicts an entirely open Indian Ocean, strewn with myriad islands.

Ludovico di Varthema

The Bolognese traveler Ludovico di Varthema sailed through the Middle East, India, Burma, Malaya, Siam, and Indonesia between the years 1502-1508. Like Conti, he pretended to be a Muslim in order to travel freely and safely. But Varthema's pretensions to Islam failed to convince the sultan at Aden, who had him arrested as a Christian spy.

After escaping (according to his account, he escaped as the result of a love affair with one of the sultan's wives), Varthema went to Persia, where he befriended a wealthy merchant. During the years 1504-05, the pair meandered through Ceylon, Malaya, Burma, Siam, Sumatra, Java, and possibly as far as the Moluccas and Sulawesi. Varthema's account, published in 1510 under the title *Itinerario,* was tapped by mapmakers for many years to come; its cartographic influence persisted as late as Gerard Mercator, who quoted it on his landmark world map of 1569 as part of the explanation for his juggling of Southeast Asian lands and *Terra Australis.*

Tomé Pires

Tomé Pires was a Portuguese apothecary who reached India in 1511 at the age of 40. While in Malacca between 1512 -1515 he put to pen whatever he could learn about the vast world of the Indies, and entitled his work *Suma Oriental.* Pires was then sent by Afonso de Albuquerque to Sumatra and Cochin-China, and afterwards was appointed Portugal's first ambassador to China. The *Suma Oriental* is the most remarkable record of Southeast Asia at the dawn of the European period. This truly extraordinary compilation of data remained unpublished until Ramusio included brief excerpts of it in his *Navigationi.* Since it was essentially unknown until after more current reports had been circulated, the *Suma Oriental* had no direct effect on published maps.

Forgotten Voices of the Past

Posterity has been kind to such travelers, whether legitimate or fanciful, as Polo, Odoric, Ibn Battuta, Mandeville, and Conti. But the cartographic record demonstrates that information about Southeast Asia reached European geographers from other sources as well. We have seen, for example, evidence of Arabic influences in the rendering of Southeast Asia in some fourteenth and fifteenth century European maps, and we will encounter place names, such as *lanna* (Lan Na), for which there is no known source but which are too specific to easily be dismissed as chance similarities. Martin Behaim, for example, cites a Bartolomeo Fiorentino as a source for Southeast Asian data on his globe of 1492. Fiorentino purportedly traveled to the East in 1400-1424, and gave an account of his adventures to Pope Eugene IV. Whether or not Fiorentino's account is credible (it may be no more factual than Mandeville's), information from otherwise unknown sources must account for some pre-Portuguese Southeast Asian features recorded by early Renaissance mapmakers.

Part III
The View from the Deck: Early European Maps

Chapter 7

Europe's Quest for a Sea Route to the Indies

The course of late medieval and early Renaissance Europe was partly shaped by an obsession with attaining a sea route to the Indies. The hunger for a sea route to Asia was first felt by the northern Italian city-states who had suffered most directly from the various problems in trans-Levant commerce. As a result it was they, not the Iberian powers, who made the first stabs at a sea route to the Orient. As we saw earlier, such an attempt is known to have been made while Marco Polo was still in the Far East. In 1291, perhaps motivated by the papacy's imposition of a ban on trade with Egypt in that year following the fall of Acre and Tyre to the Egyptians, the brothers Ugolino and Guido Vivaldi of Genoa sailed through the Strait of Gibraltar bound for 'India'. They followed the west African coast at least as far as Cape Juby, in line with the Canaries, but after that they were never heard from again.

Coincidentally, at this same time that Venice and Genoa were searching for ways to maintain trade through the Levant, the great Srivijaya maritime empire based around eastern Sumatra and western Malaya was suffering a similar fate, losing its grip as the facilitators of commerce between China and points west. In the centuries to come, the Portuguese, the Dutch, and the English would take their turns at being the controlling presence in this strait.

Europe's obstacles to a sea route to the East were of two types: technological limitations of the sailing vessels and navigational techniques, and the geographical realities that such a voyage would entail. The technological status might change; the geography of unknown regions, whatever it was, would not. Thus academics in Italy and elsewhere during the fifteenth century were eager to discover the truths of a newly-available text, the *Geographia* of Claudius Ptolemy, a scientist of Alexandria who flourished ca. 150 A.D. If the nature of the Indian Ocean espoused by the *Geographia* were accurate, the quest for a sea route to the Indies might prove to be a futile folly.

The *Geographia* of Claudius Ptolemy

By 1406, Ptolemy's *Geographia* had been translated into Latin, and manuscript copies had become available to fortunate academics and interested authorities throughout Europe by the time Conti returned from the East in about 1444. Printed editions, without maps, appeared by 1475, and the first issue illuminated with the

full complement of world and regional maps, appeared only two years later. By 1500, editions were available with printed maps that were reproduced using either copperplate or woodblock printing techniques. Some of these editions included improved renderings of certain regions, but only with the strictly Ptolemaic view of Southeast Asia.

A key tenet of the *Geographia* held that the Indian Ocean was a closed body of water, not part of the ocean sea, which meant that Southeast Asia was not accessible by sea from Europe. Certainly many fifteenth-century theorists in Europe, having secured a copy of the *Geographia*, must have believed they now understood the fate of the Vivaldi brothers over a century earlier. The expedition, it must certainly have seemed, was abruptly foiled when it reached the mysterious outer shores of a great landmass that isolated the Indian Ocean from the ocean sea (see figs. 41 & 42).

Ptolemy divided Asia into twelve parts, of which Southeast Asia and *Sinae* (China) are the eleventh, and *Taprobana* (Ceylon) the twelfth portion. A perusal of a Ptolemaic Southeast Asia map, for example the one which appears in the 1478 Rome edition of his *Geographia*, or the 1511 edition by Bernard Sylvanus (figs. 32 & 43, respectively), shows a peninsula in the eastern Indian Ocean which is a rudimentary approximation of Malaya and Indochina. So what are the identities of the many cities, rivers, and mountains of Southeast Asia recorded in the *Geographia*? Ptolemy gave precise coordinates for approximately eight thousand places on the earth that he catalogued, and it is tempting to try and correlate those which occur in Southeast Asia with modern-day locations on the map. It seems reasonable that one would need only compensate for Ptolemy's various errors to reconstruct Southeast Asia of two millennia past (or at least the Southeast Asia of the intervening Byzantine scholars who may be responsible for parts of the *Geographia*). The high esteem with which Ptolemy has been regarded, combined with the meticulous appearance of his coordinates, would give the impression that there was real meaning to be harvested from his figures.

One would first need to adjust for Ptolemy's understating the circumference of the earth. Ptolemy had accepted the figure of 180,000 *stadia* proposed by Posidonius (ca. 135-50 B.C.), rather than the more accurate figure of 252,000 *stadia* proposed by Eratosthenes in the third century B.C. This error was cumulative

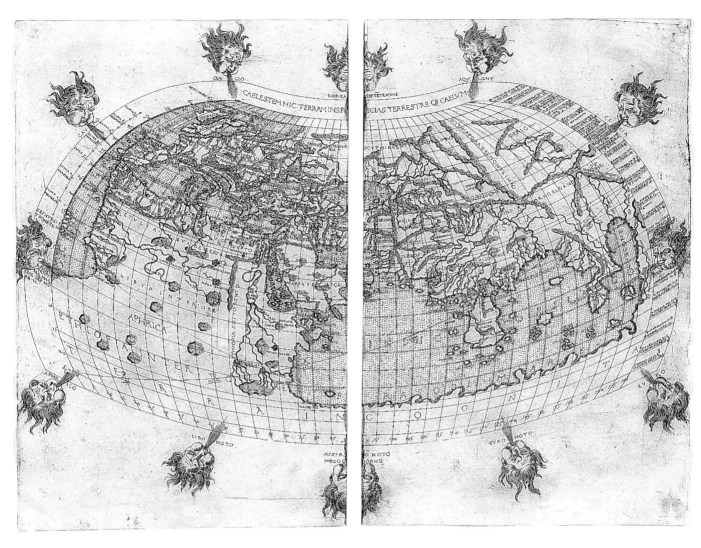

Fig. 41 World map from the *Geographia* of Francesco Berlinghieri, Florence, 1482. Note the *Terra Incognita* landmass which wraps around the Indian Ocean from southern Africa to Southeast Asia, making the Indian Ocean a closed sea and rendering the East inaccessible by sea from Europe. This strictly Ptolemaic image of the world was accompanied by Berlinghieri's adaptation of the text of Ptolemy's *Geographia*, set to Italian verse. [Courtesy of Martayan Lan, New York]

from his prime meridian, and as he had chosen the Fortunate Islands (Canaries) as his prime, the Far East was the most affected by it. But one could compensate for this, as well as for Ptolemy's misplacement of the Canaries themselves. One could even recalibrate Ptolemy's coordinates to adjust for the fact that his equator was only a theoretical one based on calculations, using his undervalued degree, from the Tropic of Cancer.

However, in the final analysis, the true locations and identities of Ptolemy's catalogue of Southeast Asian place names can today serve only as fodder for intelligent speculation. His coordinates were for the most part based on the reports and itineraries of travelers. Astronomical observation may have played a limited part in determining the positions of places within the immediate realm of the Greek and Roman civilizations, but doubtfully had any role whatsoever in the Indian Ocean where reports from travelers and the dead-reckoning of pilots provided all the available data. Indeed, Ptolemy himself warns the reader of the limitations of his sources and his system. In the *Geographia* he describes the various empirical methods available to him to compile his map of the world, and he divides them into two categories. One of these he defines as being "whatever relations to fixed positions can be tested by meteorological instruments for recording shadows," and he describes this as "a certain method . . . in no respect doubtful." But all the other

methods, he warns, are subject to error. These include "the history of travel, and the great store of knowledge obtained from the reports of those who have diligently explored certain regions." It was solely on these kinds of fallible data that Ptolemy was obliged to rely upon when constructing his map of Southeast Asia. Given the questionable nature of his materials, the relative accuracy of Ptolemy's Malaya testifies to the extent and sophistication of sailing in the Indian Ocean, at least at the time when the *Geographia* was reviewed a thousand years ago, if not in the age of Ptolemy himself.

The *Geographia* correctly ordered the continents according to their relative size: "of the three parts of the world Asia is the largest, Africa is next in size and Europe is the smallest." But in contrast to the typical medieval view, which abbreviated Asia's easterly extent, Ptolemy overestimated it, and stretched the known world (i.e., from the western coast of Iberia to China) to occupy 180° of longitude, fully half the earth's circumference. This worked in favor of a westward crossing to Asia, and indeed was part of the evidence used by Christopher Columbus to support his Enterprise of the Indies. Nor was the *Geographia* alone in overestimating the relative breadth of the Asia; long before Ptolemy's text was available in Europe, the thirteenth century treatise *De sphaera* of Sacrobosco also did so. However, since Ptolemy undervalued the degree, his erroneously high longitudinal figure did not oversize the continent in real terms.

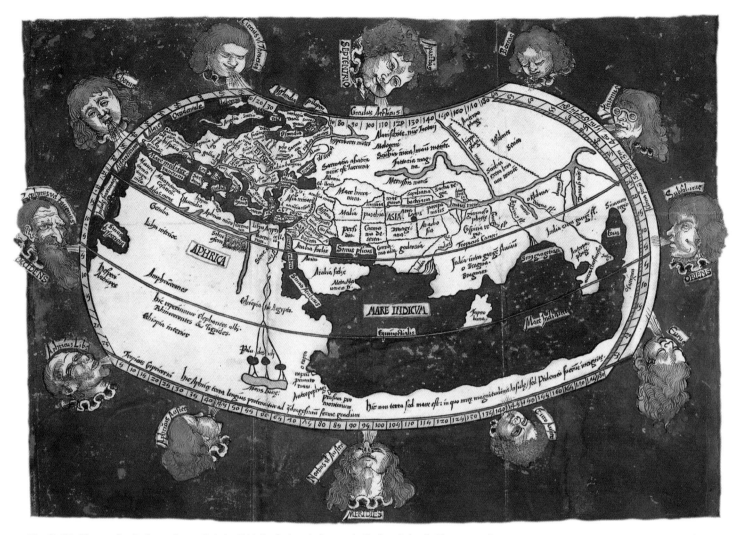

Fig. 42 World map after Ptolemy, Gregor Reisch, 1503, In the inscription on the Ptolemaic landbridge connecting Southeast Asia and Africa, Reisch explains that the region may in fact be islands. (28 x 41 cm)

Southeast Asia figured in Ptolemy's world as 'Further India', what he termed *India beyond the Ganges*. "India beyond the Ganges is terminated on the west by the Ganges River," Ptolemy relates in Chapter II of Book Seven of the *Geographia,*

> on the north by the accessible parts of Scythia and Serica;
> on the east by the Sinae region along the meridian line
> running from the border of Serica as far as the bay called
> the Great Bay.

Various features of Ptolemy's Southeast Asia can be seen in the copperplate rendering prepared for the Rome edition of 1478 (fig. 32) and also the Sylvanus edition of 1511 (fig. 43). After crossing east of the Ganges (whose delta is on the left), we enter *Aurea regio,* a kingdom of gold, which is roughly located where Burma begins today. Above it lies *Cirradia,* from where, Ptolemy tells us, comes the finest cinnamon. Further down the coast one comes to *Argentea Regio,* a kingdom of silver, "in which there is said to be much well-guarded metal." *Besyngiti, w*hich is also said to have much gold is situated close by. The region's inhabitants are reported to be "white, short, with flat noses." Here the *Temala* River, because of its position and because it empties through a southerly elbow of land, appears to be the Irrawaddy. If so, the *Sinus Sabaricus* would be the Gulf of Martaban, whose eastern shores begin the Malay Peninsula, and the *Sinus Permimulicus* would be the Gulf of Siam.

That active commerce was conducted between Malaya and points west during the period in which the *Geographia* was compiled is suggested by the *Geographia's* reference to the Indian port of *Alosygni* (probably in the Godavari delta) as a "place of embarkation for those who set sail for the *Golden Chersonese.*" The Golden Chersonese is generally accepted to be Malaya, which Ptolemy considered to be an important place:

> of the most noted islands and peninsulas, the first is
> Taprobana [Sri Lanka], the second the island of Albion
> [England] . . . the third is the Golden Chersonesus . . .

We see it here in the shape of a leaf, connected by its stem to the mainland north of the peninsula. The southern emporium of *Sabana* is placed at a latitude of 3° south, in agreement with the coordinates recorded in the *Geographia*, which means that the Malay Peninsula is extended well to the south of its true limits at 1° 15′ north.

Ptolemy's 'leaf', may in fact be Sumatra, portrayed here without the strait that separates it from Malaya. If this is so, it would explain both the shape of his southern Malay Peninsula and also its extension into the southern hemisphere. During the Renaissance, this observation led some scholars to argue that Malaya and Sumatra were once joined in some earlier geological age, rather than conclude that Ptolemy was simply unaware of the Malacca Strait. As Abraham Ortelius explained in a legend on his 1567 map of Asia:

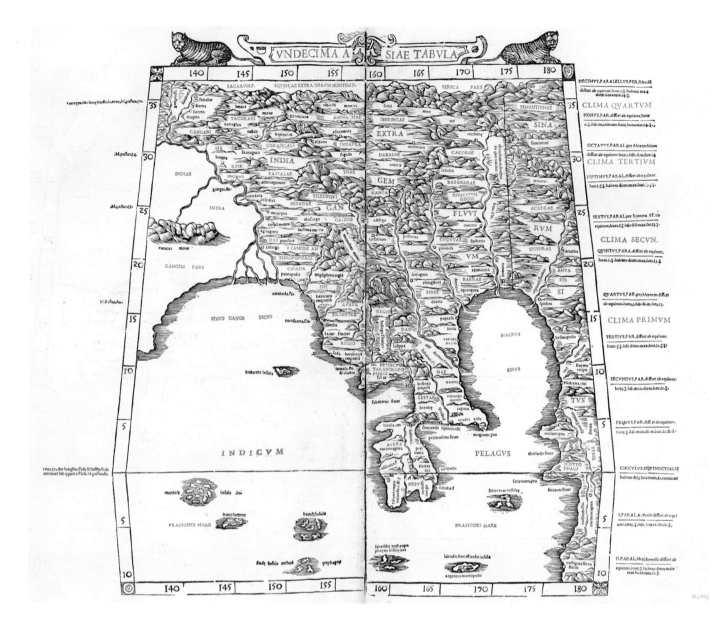

Fig. 43 Ptolemaic Southeast Asia from Bernard Sylvanus' rendering of the *Geographia*, Venice, 1511 (40 x 46cm). This woodcut image was the first map of the region to be *printed* in color (rather than colored by hand later). A second register was used to print the important words in red. [Courtesy of Martayan Lan, New York]

It is true that Samotra is not now a peninsula, but it is very likely that it was torn from the continent by the force of the Ocean after Ptolemy's time. Moreover, if you imagine Samotra being joined to Malacca with an isthmus, it will agree very well with the shape of the Golden Chersonese as described by Ptolemy.

In 1597, Ortelius made a map depicting his image of classical Asia, *Erythraei sive Rubri Maris Periplus,* which depicts Sumatra connected to the Golden Chersonese by a narrow neck. Other writers who promoted this idea in the sixteenth century were the Portuguese historian Barros, the Portuguese poet Camões, and the Portuguese-influenced Dutch adventurer Linschoten. Linschoten, writing in his *Itinerario* of 1596, stated that some believe Sumatra to have been the *Chersoneso Aurea* of old, and that "others affirme it to be Ophir, from whence Salomon had his Golde, as the Scriptures rehearseth," and that "in times past it was firme land unto Malacca," that is, that Sumatra used to be joined to Malaya. Camões explains in his *Lusiads* that the "noble island of Sumatra was said in ancient

times to be joined to the mainland, until ocean breakers eating into the coast drove a wedge between."

Many of the coastal cities in Ptolemy's Southeast Asia are marked as 'emporiums'. These include the ports of *Baracura, Barabonna,* and *Bsyga* in the region of what would now be Burma. Further along the Malay Peninsula one comes to the emporium of *Tacola,* with *Sabana* at the southernmost tip, in the position of modern day Singapore. Scholars have tried to identify these places, but there is no convincing consensus. The *Sinus Perimulicus* is probably the Gulf of Siam (if not the *Sinus Magnus*), with *Regio Lestoru* being lower Siam and Cambodia. A spine of mountains running north from there, in what is now northern Thailand and Burma, is according to Ptolemy a "habitat of tigers and elephants," as well as "lions and robbers and wild men who live in caves, having skins like the Hippopotamus, who are able to hurl darts with ease."

In the waters west of the peninsula, in the relative position of the Nicobar Islands, we find *Bazacata Insula,* where there "are many shellfish, and the inhabitants of the island," who are called *Agmatae,* "are said at all times to go without clothing." In 673 A.D. the

Chinese pilgrim I-Ching, sailing northward from *Chieh-ch'a* (Kedah, Malaysia), reached what he called the "Kingdom of the Naked People," which has been identified as the Nicobars. Arab vessels called in on the Nicobars as well; one of the earliest references to them comes from Ibn Khurdadhbih (ca. 850), who stated that

> If one wishes to go to China one leaves . . . Sirandib [Ceylon] on the right and goes to Alankabalus [Nicobars or Andamans], which is situated at the distance of ten to fifteen days from Sirandib. The inhabitants of this island are naked and live on bananas, fresh fish and coconuts. Iron is a precious metal to them. They are visited by foreign merchants.

Similarly, an 'island of naked people' which appears in the seas to the east of Ceylon on the Thai *Traiphum* illustrated in figure 21 (extreme right), probably represents the Nicobars.

In the mid-sixteenth century, Gerard Mercator wrongly determined that *Bazacata* was actually the Philippine island of Palawan. This was a logical conclusion for Mercator, because he mistakenly believed *Taprobana* to be Sumatra (rather than Ceylon), and judged Palawan to be in the same position relative to Sumatra as Ptolemy's *Bazacata* was to *Taprobana*.

Ptolemy's enticing imagery of fresh seafood and nude people on *Bazacata* is a sharp contrast to the islands which lie to the south. The latter include the island of *Bonae Fortunae*, three called *Sindae*, five called *Brussae*, three more called *Sabadicae*, and a group of ten called *Maniolae*. They are the islands of the Anthropophagi, the eaters of human flesh, a people mentioned as early as the ancient historian Herodotus in the fifth century B.C. Ptolemy warns that vessels constructed with nails should not approach near the last of these archipelagoes, the *Maniolae* (southwesternmost islands on the map), "lest at any time the magnetic stone which is found near these islands should draw them to destruction." To protect against this, boats are strengthened with beams of wood. Indigenous vessels in the Indian Ocean were in fact made with coconut fiber coir, rather than nails, a fact commented upon by early European visitors such as Polo and Conti. This belief persisted, Bordone noting in his *Isolario* of 1528 that the magnetic stone of this island would pull out metal nails out of an approaching vessel until it sinks. Off the southeastern coast of the Malay Peninsula is a group of three islands called the *Satyrorum* which, as the name implies, is inhabited by people who are "said to have tails such as they picture satyrs having." In this myth, Ptolemy was probably influenced by Ctesias (ca. 400 B.C.), who concocted this and other absurdities about the East which were later taken up by Pliny and Solinus.

Some later mapmakers, no longer able to reconcile all of Ptolemy's Southeast Asian lands with current discoveries, moved the satyr islands and other Southeast Asia locals far to the north. Abraham Ortelius, for example, on his *Indiae Orientalis* of 1570 (fig. 86), placed *Satyrorum* at a latitude of about 60° north, in the waters above Japan. Perhaps contributing to this transposition is the fact that the story of boats with metal being drawn to peril, such as was said of *Maniolae*, was later commonly associated with the Arctic region.

A similar transformation occurred in the case of the 'Silver Islands' which probably originated in mainland Southeast Asia in the vicinity of Burma. Travelers in the first half of the sixteenth century, among them Mendes Pinto, searched without success for these islands of silver in Southeast Asia. In the end, Abraham Ortelius, who was probably influenced by the claim of the Jesuit, Francis Xavier, that silver mines would be found in Japan but not in Southeast Asia, re-invented the silver isles as a large island directly to the north of Japan on his *Maris Pacifici* of 1589 (fig. 90).

Ptolemy's Land Bridge and the Great Gulf

Now we return to what is arguably the strangest part of Ptolemy's entire world picture, the mischievous geographic quirk which most worried Europeans interested in Southeast Asia: the enclosed Indian Ocean. Ptolemy's landlocking of the Indian Ocean posed a sobering problem for the seafaring nations of Europe. If Ptolemy were correct, if indeed the Indian Ocean were a closed sea, it would mean that no maritime route to the East existed as an alternative to the cumbersome, monopolized routes through the Middle East. For much of the fifteenth century, Portugal sacrificed many lives and a great deal of resources on the gamble that Ptolemy was wrong. We shall see that Ptolemy's 'Great Promontory' lived on in maps long after the myth of the closed Indian Ocean was debunked, being variously mistaken for the Malay Peninsula, an extra Asian peninsula, and lastly for Central and South America.

Ptolemy placed *Sinae* (China) in easternmost Asia, "terminated on the north by the accessible parts of Serica," that is to say, it lay just south of the silk-producing region. The cities of *Sinae* are said to be walled towns, "but none deservedly renowned," and the region is located on a great promontory which forms the eastern shores of a 'Great Bay', the *Magnus Sinus*.

What was this huge bay Ptolemy believed to lie east of the Malay Peninsula? The simplest explanation would be that it is either the Gulf of Siam, or the Gulf of Tongkin and South China Sea, or a combination of both of these. But then what is his 'Great Promontory'? In clarifying this, the text of the *Geography* raises larger questions. Ptolemy states that Southeast Asia is bordered "on the south by the Indian sea and a part of the Parassadis sea which extends from Menuthiadae island as far as the opposite shore of the Great Bay," meaning that the land forming the eastern shores of the bay is contiguous with the coast of Africa. Therefore, his bay is bounded on the east by a neck of land connecting southern China to sub-equatorial Africa, wrapping around the lower perimeter of his world map at roughly 15° south of the equator and rendering the Indian Ocean completely landlocked. Southeast Asia is inaccessible by sea from the world's oceans. If there is any ambiguity about the meaning of this passage, elsewhere the *Geographia* states quite bluntly that ". . . the Indian sea . . . with its gulfs, the Arabian, the Persian, the Gangetic, and that which is called the Great Gulf, it is entirely shut in . . . by land on all sides."

Various theories have been advanced to explain this strange geographic anomaly. Cursory knowledge of the shores of Sumatra may have been confused with a continental coast (though northwest Sumatra was probably the southern 'leaf' of Ptolemaic Malaya). Or perhaps Ptolemy wished to keep his map compatible with the ancient precept of a river sea encircling the earth, and since the Indian Ocean as he knew it could not be part of that sea, some landmass must separate the two. Yet another possibility is that Ptolemy of Alexandria and Marinus of Tyre, who were both from cities which had benefited enormously from trans-Levant trade in the Mediterranean, deliberately invented (or sustained) the myth of a land bridge to create the delusion that no alternative sea route existed into the Indian Ocean around Africa for Mediterranean entrepreneurs to circumvent their control over trade with the East. Lastly, scholars since the Renaissance have advanced the theory that Ptolemy's land bridge represented the western coast of South America, and that the 'Great Gulf' was the Pacific Ocean.

One of the first to equate Ptolemy's 'bay' with the Pacific was Peter Martyr, who in the early sixteenth century wrote that "the great Gulf of *Cattigara*" is where the Moluccas lie (*Cattigara*, as we shall see, was the pre-eminent port on the Great Promontory). Later that century, Abraham Ortelius also believed that Ptolemy's 'Great

Bay' was the Pacific, writing that "Ptolemy falsly termes [the Pacific] Sinum Magnum, a great bay; whereas he should have nam'd it Mare Magnum, a great sea." Ortelius' statement was more radical than Martyr's, since Ortelius envisioned a larger Pacific and a wholly autonomous America. Ptolemy, however, would have been surprised to see his *Magnus Sinus* equated with such a vast ocean, for he states in the *Geographia* that it is the earth's third largest gulf after the Bay of Bengal and Persian Gulf. Yet he does also acknowledge that his charting of the 'Great Promontory' is nearly arbitrary.

Dispelling the Myth of a Closed Indian Ocean

The voyages of Polo, Odoric, and Conti had not altogether disproved Ptolemy's land bridge. They had simply proved that there was a strait which passed through it in the east, a conclusion which had earlier been reached by many Arab geographers (fig. 27). But there was a long tradition of an open Indian Ocean, predating Europe's knowledge of Ptolemy. Herodotus recorded the claim that Phoenician seamen, sent by king Neco, "sailed on a westerly course round the southern end of Libya [Africa], [and] they had the sun on their right — to the northward of them," skeptically recording what appears to have been a circumnavigation of Africa in about 610-595 B.C., which in effect proved that the Indian Ocean was an open sea. In recording the exploits of Darius of Persia, Herodotus ambiguously noted that "all Asia, with the exception of the easterly part, has been proved to be surrounded by sea."

The writings of the early fifth century scholar Macrobius, whose map (fig. 35) depicted the Indian Ocean as an open sea, was popular among the sort of intellectual humanist who might have chatted with Nicolò de' Conti or read Marco Polo's narrative. Another classical author who believed that the Indian Ocean was open, but whose text was not known to late medieval Europe, was Cosmas Indicopleustes. Cosmas, who lived a century or so after Macrobius (ca. 540), was a less sophisticated writer, but one who (unlike Macrobius) had first-hand experience in the Indian Ocean. Originally a merchant-trader in the Red Sea and Indian Ocean, Cosmas converted to Christianity and passed his remaining years in a Sinai cloister discrediting classical writings. Though his ideas were based largely on theological considerations, he clearly indicates an open Indian Ocean by stating that a vessel bound for China must, after sailing sufficiently far east, turn north and continue "at least as far as a vessel bound for Chaldea would have to run up the Straits of Hormuz to the mouth of the Euphrates,"[125].

Early Arab geographers tapped Ptolemy's *Geographia,* but knew that the Indian Ocean was not landlocked. One of the earliest whose text is extant, Khwarizmi (tenth century), modified his essentially Ptolemaic views to open up the Indian Ocean. Another tenth-century Muslim scholar, Al-Mas'udi, theorized Africa as sea-girt and separated by a narrow strait from an austral landmass, while at the same time acknowledging that he did not know the true shape of Southeast Asia.[126] The following century a Muslim geographer by the name of Biruni, who was primarily interested in northern India, proposed that the Southern Sea communicates with the waters of the north via a gap in the mountains of southern Africa.[127] A little later, a student of Plato of Tivoli's translation (ca. 1140) of the earlier Arabic scholar Al-Battani's *Astronomy,* would have deduced that the Indian Ocean was open, because according to the text, a branch of that ocean extends to the furthest point of 'India', at which point one has reached China *(Thiema).*[128] Al-Battani's 'branch' of the Indian Ocean through which one reaches China was, of course, the Malacca Strait and South China Sea.

An open Indian Ocean was also favored by theorists like Roger Bacon, Pierre d'Ailly, and Aeneas Piccolomini (who was Pope Pius II). In the mid-fifteenth century, Giovanni da Fontana postulated a semi-landlocked Indian Ocean, which was bound by land on the south and east as in Ptolemy' account, but with a passage connecting it to the waters of the outer ocean sea. Da Fontana probably created the strait to accommodate the reports of Polo, Odoric, and Conti, since he notes that "those who owe their information to true experience and the fruits of wide travel and diligent navigation, have found beyond the equinoctial circle on its southern side a notable habitable land uncovered by water together with many famous islands."[129] And, finally, there was Nicolò de' Conti, who himself had sailed to the Spiceries, and who believed that the Indian Ocean was an open sea.

Thus Ptolemy's Southeast Asian land bridge had already been challenged well before Portugal reached the Indian Ocean, and the mapmakers were not far behind. The *mappamundi* of Marino Sanudo and Petro Vesconte (ca. 1320) is a remarkably early example (fig. 44). Based largely on Arabic sources, it depicts Africa in a wide, truncated fashion typical of Arabic-influenced maps, with a passage into the Indian Ocean and Southeast Asia which is roughly 1500 kilometers wide (extrapolating from the size of other features on the map). In 1448, the land bridge was breached by a narrow strait on a map of the world by Andreas Walsperger, and by a wider strait on a world map of Giovanni Leardo. Nine years later, in 1457, the anonymous 'Genoese' world map did away with the land bridge altogether, and noted that "in this southern sea they navigate without sight of the northern stars", indicating Arab sources. The 1459 map of Fra Mauro also disposed of the old land bridge.

The final two solutions to the land bridge issue followed in the wake of the Portuguese penetration of the Indian Ocean in 1488, when they successfully round the Cape of Good Hope and sailed into eastern waters for the first time. They are rooted in the manuscript maps of Henricus Martellus, a German mapmaker working in Florence in about 1489-1490, and the portolan chart known by the name of the Italian diplomat who smuggled it out of Lisbon, the Cantino map of 1502. The Martellus maps modified the Ptolemaic view only just as much as the new Portuguese discoveries required, while the Cantino versions tapped new (Arabic) sources. Both had a profound influence on the early crop of printed maps. We will visit these two maps and their printed progeny in the next chapter.

Technology and Theory

Although Italy had for centuries enjoyed her ideal perch for trans-Levant trade with the East, it was Portugal who was best situated for launching vessels into the Atlantic. Soon after the ill-fated Vivaldi enterprise of 129, Genoese merchants and mariners were increasingly residing in the ports of Lisbon and Oporto. Portuguese navigators learned from their Genoese visitors, and by the early fifteenth century they had supplanted the Italians as the pioneers of Atlantic exploration. Thus by the time Prince Henry led the pivotal conquest of Ceuta (1415), Portugal was primed to begin a series of voyages to the south and east around Africa, culminating with the voyage of Bartolomeu Dias around the Cape of Good Hope (1488), and of Vasco da Gama to the Indian coast at Calicut (1498). With their successes, Portugal's faith in a trans-African sea route to the markets of Asia was vindicated, and the myth of Ptolemy's closed Indian Ocean was finally laid to rest.

But what lay to the east of India and Malaya? Portugal's navigators had proven that the Indian Ocean was one and the same waters as the ocean sea, but had not entirely settled the question of

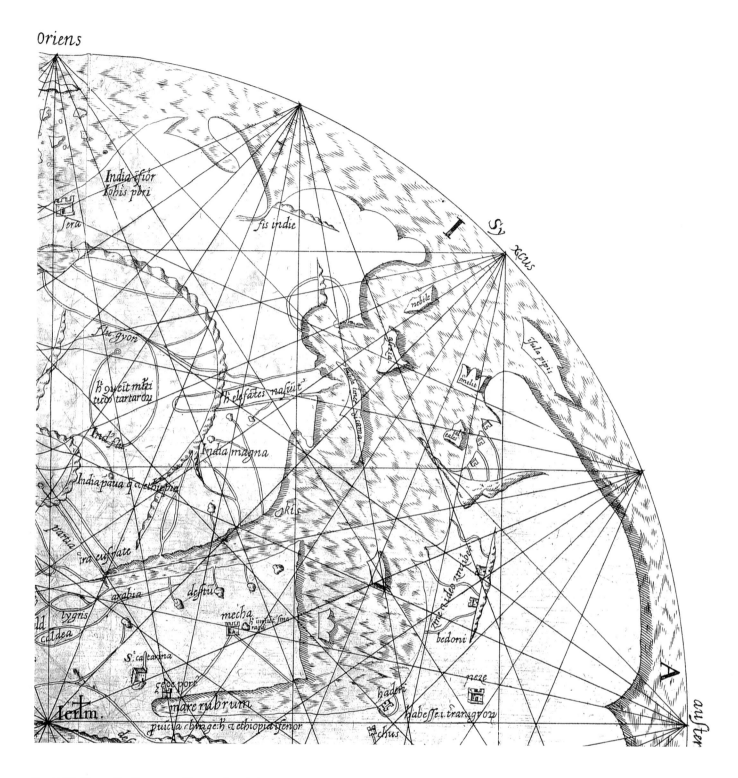

Fig. 44 **World map** (detail), ca. 1320, Vesconte-Sanudo (Bongars, 1611). (32.5 cm diameter)

the 'Great Promontory'. The first landfalls on Malayan shores may, in fact, have 'proven' that the eastern segment of this huge landmass did indeed existed. The story of the Southeast Asian land bridge does not, therefore, end with the Portuguese in India, nor even with the arrival of the Portuguese in Malacca.

At the same time that these epic voyages were being undertaken, revolutions were also taking place in technology and theory. The development of printing in the West, followed two decades later by the printing of maps, had an effect whose profundity can scarcely be overstated. The earliest known printed maps of Western origin which can be definitely dated are a T-O map of 1472 (fig. 17), and

a *mappamundi* of 1475 (figs. 34 & 39), which is based on the same tripartite skeleton as the T-O map but which records the various realms of the world. Once the development of printing allowed a dramatic increase in the quantity of maps produced, paper (rather than vellum) became the common medium. The technique for making paper had been imported from China, reaching the Middle East in the eighth century, and Iberia perhaps by the ninth century.

Though printed maps were in themselves revolutionary, the medium was slower than manuscript maps to incorporate new data from travelers' and sailors' reports. The first known printed map to reflect Dias' rounding of southernmost Africa was a separately published world map made in Florence by Francesco Rosselli in 1492 or '93. Maps in printed books were more conservative still. A world map in a 1482 Venice edition of Pomponius Mela reveals some knowledge of Portuguese voyages around West Africa, but is otherwise Ptolemaic in form; this same configuration was used eleven years later for the world map in the '*Nuremberg Chronicle*' of Hartmann Schedel. With the exception of Macrobius (whose *Dream of Scipio* first appeared in print in 1483), it was not until 1503 that a map reflecting any knowledge of an open Indian Ocean, was finally appeared in a printed book, and even then this was speculative. The map in question is found in Gregor Reisch's *Margarita Philosphica* (fig. 42). It retains the Africa-Asia land bridge but qualifies it with a legend stating that "here there is not land but sea, in which there are such islands not known to Ptolemy." From a modern perspective we may wonder whether the author of this statement had in mind the islands of Southeast Asia or the islands of the Caribbean, but at that time the Caribbean and the South China Sea were essentially synonymous and it is unlikely that the map's author himself made any distinction between the two island worlds.

The first, and one of the few Ptolemaic maps to open up the Indian Ocean appeared in a 1511 edition of the *Geographia* by Bernard Sylvanus. Most subsequent issues of the *Geographia*, however, reverted back to the strict Ptolemaic rendering for their classical map. But this was academic, since beginning in 1507 most editions of the *Geographia* included a 'modern' world map in addition to the Ptolemaic one, the Sylvanus edition among them.

Rapid changes were also occurring in the world of shipbuilding and navigation. Mediterranean and Iberian vessels borrowed a number of ship-building principles from Arab vessels in the fifteenth century, including the balance-lug sail, which may have originated with Malay seamen. Vasco da Gama had hired an Arab navigator to guide his vessel through the waters of the Indian Ocean to Calicut, and he returned to Lisbon with newly-acquired navigational techniques as well as several *kamals,* instruments for observing the altitude of the stars. The Europeans were excellent students and by the turn of the sixteenth century they had surpassed their Arab and South Asian mentors both in navigation as well as in shipbuilding techniques.

The goal of India having been solidly attained, Portugal's eyes turned yet again eastward. A Venetian spy, who was sent to Lisbon to learn about Portuguese meddling in the spice trade reported in 1506, that even greater rewards awaited whomever might sail further east and return.[130] This 'look' eastwards culminated in Albuquerque's conquest of Malacca in 1511, and at the close of that year Malay navigators tutored Portuguese pilots on their maiden penetration of Indonesian seas. The Portuguese did not create new trade routes or displace old trading networks, but rather became new participants in established commercial circuits.

Chapter 8

A Confusion of Peninsulas and Dragon Tails

But the captain [Ferdinan Magellan] knew that he had to sail through a very hidden strait,
as he had seen in the archives of the king of Portugal,
on a map drawn by a worthy man named Martin of Bohemia.

— Antonio Pigafetta, *The Voyage of Magellan*.

By the early years of the sixteenth century Portugal had proven that Southeast Asia could be reached by sea and thus discredited the western extent of Ptolemy's land bridge connecting Southeast Asia and sub-equatorial Africa. This was a step beyond the earlier odysseys of Marco Polo, Friar Odoric, and Nicolò de' Conti, which had demonstrated that a strait existed between the Southeast Asian mainland and Sumatra, but had not actually proven the east-west breadth of the land bridge to be fallacious. In response to the Portuguese successes, mapmakers at first simply disconnected the promontory from Africa, sometimes leaving bits of it lying about in the Indian Ocean as unidentified islands, and keeping the eastern portion of it intact, forming a mammoth, phantom Southeast Asian peninsular subcontinent. This set the stage for one of the most extraordinary of cartographic adventures: for the next half century, cosmographers juggled the remains of Ptolemy's 'Great Promontory' to accommodate discoveries in Southeast Asia and America.

A map on a polar projection by Peter Apianus, 1524 (fig. 19), is a good illustration of the land bridge vestige with part of its east-west extension still retained; the large island at 130° longitude is Ceylon, the peninsula at 160° is Malaya, and the huge '*India Merid[ionalis]*' is the old land bridge, newly reborn as an Asian subcontinent approximately as large as the rest of Asia. The peninsula's claw-shaped southern appendage is a vestige of the southern land which had formerly connected to Africa. Since this Southeast Asian subcontinent extends much farther south than India or Malaya, Apianus referred to it as *India Meridionalis* ('Southern India').

The Fate of Cattigara and the Great Promontory

Cartographers' assimilation of the new subcontinent born from Ptolemy's 'Great Promontory' to the rapidly changing view of Southeast Asia occasioned by successive voyages of discovery can be traced by following the history of its great southern emporium, *Cattigara*. Ptolemy made quite a fuss over the precise location of *Cattigara*, which he attempted to determine based on an account of a voyage by the navigator Alexander, as related by Marinus of Tyre. Alexander purportedly left the Golden Chersonesus (Malay Peninsula) and sailed for twenty days to *Zaba*, a place found on typical Ptolemaic renderings (e.g., figs. 32 & 43) as an eastern promontory of Malaya, which some sixteenth century observers, such as Barros, believed to be Singapore, but which most modern historians believe to have been in Indochina. From *Zaba*, he reached *Cattigara* by sailing "southward and toward the left" (i.e., east) for a period of time described by Alexander as "some days," re-interpreted by Marinus as meaning "many" days, and then evaluated again by Ptolemy. The journey from the "capital Sina" (China) to the "gate of Cattigara" runs, according to Ptolemy, to the southwest, while Marinus reported that their meridians coincide. Ptolemy tried to determine the precise coordinates of *Cattigara* through an utterly comical (if unintentionally so) attempt to extrapolate a precise reckoning from the non-existent 'data' hidden in the words 'some' and 'many'. His final verdict on the location of this pivotal southerly emporium of the Great Promontory was that it lay 177° east of the Canary Islands, and 8° 30′ south of the equator.

The cartographical transformation of the 'Great Promontory' in sixteenth-century maps can be summarized in three steps:

1) The world map of Martellus (ca. 1489) deletes the western half of Ptolemy's land bridge, but retains *Cattigara* on the resulting Southeast Asian peninsula. (For a slightly later example of this, see Waldseemüller's Malay Peninsula in his world map of 1507, fig. 45; *Cattigara* is the uppermost city visible on the peninsula). Quite possibly, this configuration was based on earlier models which are no longer extant (perhaps, for example, one purportedly acquired by a son of Prince Henry in 1428; see page 95, below).

2) Some geographers, however, held the view that the new Southeast Asian subcontinent was in fact America. The sketch map by Alessandro Zorzi and Bartolomeo Columbus (brother of Christopher) (1506?/1522, fig. 51), adapts the old land bridge to function as Central America. Believing that *Cattigara* would be found on the promontory, and knowing that Ptolemy located it well below the equator, they consequently extended Central America much too far to the south. Oronce Fine (1531 and 1534; figs. 52 and 71) also decided that what Ptolemy called the Great Promontory was in fact the New World — his delineation of America is much improved over that of Zorzi, and *Cattigara* now 'properly' lies in South America.

3) Sebastian Münster completed the transition, as can be seen on his maps of Asia and America of 1540 (fig. 72). Münster makes North and South America an autonomous continent, completely severed from Asia, but retains *Cattigara* in its 'correct' latitude on the Peruvian coast. Münster's radical solution to the riddle of *Cattigara* was the logical conclusion to a half-century of adapting Ptolemy's 'Great Promontory' to the new Portuguese and Spanish discoveries.

In addition, some portolan charts abandoned the old promontory altogether and located *Cattigara* on an island in the East Indies. The Genoese chartmaker Battista Agnese (fig. 47), working in Venice in the mid-sixteenth century, identified his South China Sea with Ptolemy's Great Gulf, and consequently located his *catigara'* in the Moluccas, sometimes as an indefinite landmass, at other times (such as in fig, 47) as a small island. This was a sensible solution to the puzzle; it simply restaged Alexander's voyage in terms of an open Indian Ocean, and loosely honored Ptolemy's statement that *Cattigara* was part of *Sinae*, or southern China.

What was *Cattigara,* and why was it so important? Of the several metropolises Ptolemy placed on the 'Great Promontory', Renaissance cartographers were most reluctant to abandon *Cattigara* because of the significance accorded it in the *Geographia* as a major emporium marking the most southeasterly point in the known world.

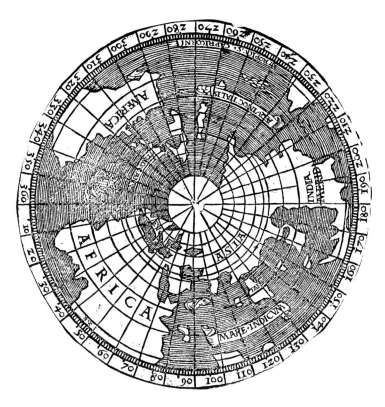

Fig. 45 The Malay Peninsula, from the world map of Martin Waldseemüller, 1507 (from a facsimile). Portions of two of the woodcut map's twelve sheets are shown. (Full map 132 x 236 cm)

Fig. 46 World map on a north polar projection, Peter Apianus, 1524. The map was part of a volvelle, in which moving paper pieces, pivoted at the center with a string, rotated around the map to make various calculations. (11 cm diameter)

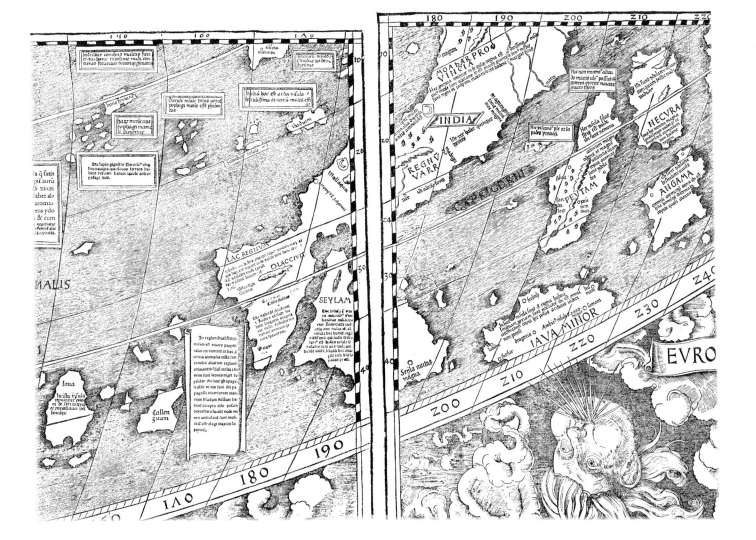

Fig, 47 Map of the world by Battista Agnese, ca. 1544, recording the route of the Magellan circumnavigation. The island to the lower left of the wind head at the 3 o'clock position is identified as *Cattigara*. Despite Agnese's advanced depiction of Southeast Asia and America, he defers to Ptolemy in reversing the relative proportions of the Indian subcontinent and Ceylon. (Manuscript on vellum, approx. 16 x 23 cm) [Library of Congress]

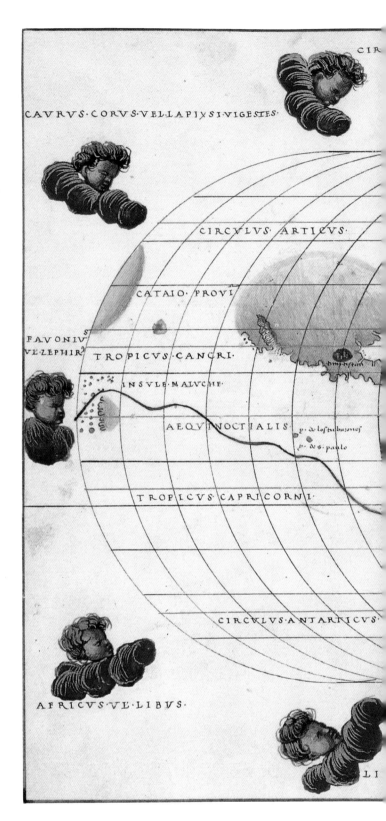

This strategically-positioned entrepôt of *Cattigara* was specifically sought by European pioneers in Southeast Asia. Magellan, for one, hoped to reach it after crossing the Pacific. But some early observers soon deduced that *Cattigara* lay much further north than Ptolemy's coordinates indicated. Antonio Pigafetta, recording the course of Magellan's fleet across the Pacific, explained that

> we changed our course to west by south, in order that we
> might approach nearer to the land of Cape Cattigara, but
> that cape, with the pardon of cosmographers, for they
> have not seen it, is not found where it is imagined to be.

Maximilian of Transylvania, who chronicled the Magellan voyage from interviews with survivors, similarly noted that when they had reached the Moluccas, "the Spaniards . . . agreed that [the ship *Victoria*] should sail to the cape of Cattigara." However, "though they went as far as 12° south, they did not find Cattigara, which Ptolemy considered to lie considerably south of the equator."[131] Peter Martyr implied that it was the inaccuracy of Ptolemy's figures that caused Magellan's crew to miss *Cattigara*. Martyr wrote that although Magellan had intended "as much as were possible, to approach to the cape called of the old writers Cattigara," this geographical feature "is not found as the old cosmographers have described it, but it is toward the north about xii degrees as they afterward understood."[132]

Of the various theories then in circulation regarding the riddle of *Cattigara*, Peter Martyr's corresponds best to modern appraisals. Scholars are generally in agreement that *Cattigara* actually either lay along the Red River Delta, near what is now Hanoi, or else it was in the southernmost part of China, the Canton or *Zaiton* of its day. *Cattigara* has also been identified with Funan, the ancient kingdom pivoted around southern Vietnam and the Mekong River delta, which flourished by the time the *Geographia* was composed in about 150 A.D.

Although exploration had rendered Ptolemy's *Geographia* a matter of mere academic curiosity by the early sixteenth century, there still remained a stubborn propensity to try to rectify and rationalize its basic precepts. New editions of the work were common up until the very last years of the sixteenth century, and it continued to be published, though less frequently, well into the eighteenth. As late as about 1786 one finds a map included in the *Philosophical Transactions* which still tries to come to terms with the enigma of *Cattigara*, of the Great Bay, and of the fateful words 'some' versus 'many' quoted by Alexander and Marinus. According to a lengthy inscription accompanying this map, it was "the difficulty of doubling Cape Siam in sailing out of the bay" that was responsible for the "inconsistent information which Ptolemy received," and this was also "the Chief Cause of his Changing his opinion concerning a proper allowance for the Word *Some*." The authors of this map believed the 'Great Bay' of Ptolemy to have been the Gulf of Siam, and 'the Bay of the *Sinae*', Ptolemy's *Sinarum Sinus* ('China Gulf'), to be the little double inlet to the southeast of Chantanburi in southeast Thailand. The enigmatic *Cattigara*, in this scheme of things, was said to lie 3° west of 'Cambodia City' (Phnom Penh or Lovek).

Classical Foundation for the Southeast Asian Subcontinent
Two observations from classical times may have encouraged Renaissance geographers to believe that there was still a huge Southeast Asian subcontinent extending far to the south and toward Africa, even if Ptolemy was wrong about its landlocking the Indian Ocean. One instance was Pliny's record of an embassy from *Taprobana* (Ceylon) which had visited the court of Emperor Claudius. According to Pliny, the eastern ambassadors were surprised to find that in Rome their shadows fell to the north, whereas in their land it was the sun which was to the north. Pliny also spoke of an Indian island named *Patale*, with a celebrated

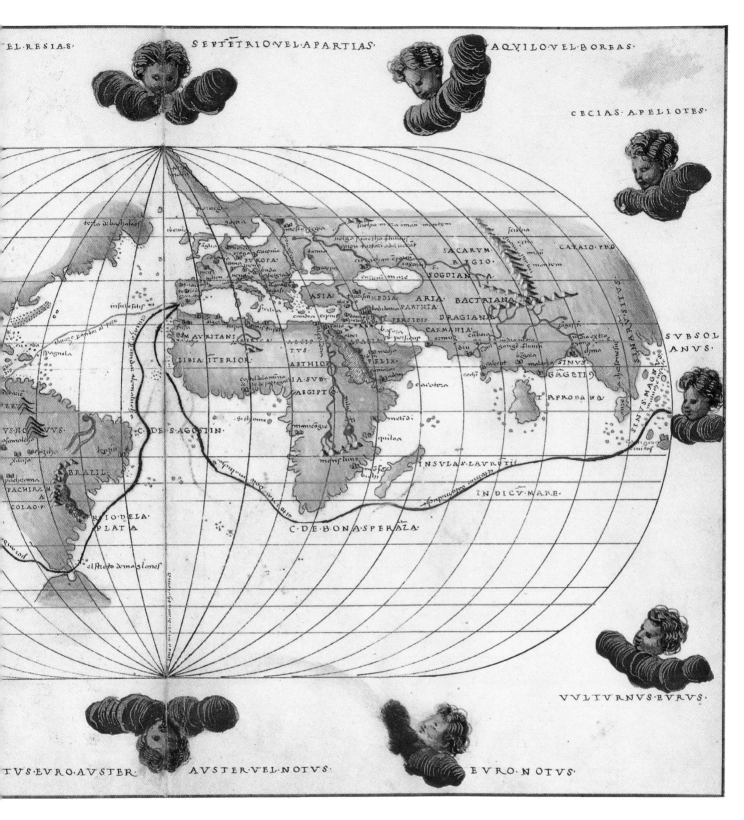

port, where the shadows always fell to the south. Most Renaissance geographers mistakenly associated *Taprobana* with the island of Sumatra, which helped to push both Sumatra and the neighboring mainland peninsula well south of the equator. Aristotle may have helped preserve the bend of the vestigial land bridge toward Africa with his observation that since elephants inhabited both India (which by now had become synonymous with the remnants of Ptolemy's Southeast Asian land bridge) and Africa, the two must lie correspondingly close to each other. Seneca, in fact, had written that one could sail from Africa to Asia in just a few days with favorable winds.

The 'Three Indias'

Medieval thought commonly envisioned 'three Indias', which probably arose from the Biblical account of the three eastern kings who visited the infant Christ. The myth of the legendary Christian king, Prester John, no doubt reinforced the idea. "Our magnificence dominates the Three Indias", Prester John told the leaders of western Europe in the famous forged letter allegedly 'written' by the oriental Prester and circulated among various European rulers in about 1165.

This notion of 'three Indias' took literal form on world maps in the late fifteenth and early sixteenth century. No doubt Renaissance mapmakers felt they had made better sense out of the 'three Indias'

Fig. 48 Southeast Asia from the globe by Martin Behaim, 1492. Globes, by their nature, are particularly vulnerable to the ravages of time. Although textual references leave no doubt that terrestrial globemaking in Europe dates back to at least the thirteenth century (and probably earlier), Martin Behaim's *erdapfel* ('earth-apple'), made in Nuremberg 1492, is the earliest surviving example. Behaim's configuration for Southeast Asia is closely related to that of Martellus. (50.7 cm diameter) [From a facsimilie by Greaves & Thomas, London]

than their predecessors. Soon afterwards, the belief that Asia possessed three major peninsular subcontinents took on a life of its own, culminating with the belief that America was the easternmost 'India'.[133]

The Life and Death of the Phantom Peninsulas

There were two distinct patterns established for the proto-Malay Peninsula subcontinent born from the old 'Great Promontory' of Ptolemy. In both cases, their extremities veer to the southwest, which is contrary to the true contour of southern Malaya, but consistent with their ancestry in the old land bridge.

The Martellus 'Dragon Tail' Model
The first pattern for the phantom peninsula, rooted in the Martellus world map of ca. 1489, is by far the larger. It retains the Southeast Asian portion of the defunct land bridge as a huge subcontinent partitioning the Indian Ocean from the China Sea, like an imposing Oriental screen or the tail of a dragon, reaching southwards beyond the Tropic of Capricorn. The Ptolemaic Southeast Asian mainland is retained as well, though it is dwarfed by the false peninsular subcontinent. Martellus places various Southeast Asian islands known from Marco Polo and other travelers in relation to the bogus peninsula rather than true Ptolemaic Malaya. He also inscribes the new subcontinent with Polian comments about the Malay Peninsula, reinforcing the original error. The Martellus map (or its unknown prototype) thus gave birth to an error that would have a profound impact on the mapping of Southeast Asia, as well as on the mapping of America, for the next half century.

Although the Martellus map is the oldest extant representation of this model, it may in fact be based on earlier, unknown proto-types. The basic concept of a Southeast Asian 'India' which leans toward Africa is found in earlier literary records. Pierre d'Ailly, for example, writing in about 1410, believed India to extend south to the Tropic of Capricorn, and to reach close to Africa. An inscription on the map in his *Imago Mundi* states that "according to some, the southern front of India extends to the Tropic of Capricorn [and] extends to near the limit of Africa," which is clearly evocative of the Martellus

geography. Such a map would have portrayed an open Indian Ocean based on Arab influence; the success of Portuguese voyages could not account for the 'dragon tail' much before the Martellus map.[134]

The most intriguing hint of a Martellus prototype is left to us by António Galvão, a Portuguese adventurer in Southeast Asia who was governor of the Moluccas in the late 1530s, and who devoted his later years to writing a history of the Spiceries and Portugal's voyages to the East. In a tantalizing reference to a map which could well have been a precursor to the Martellus model, Galvão wrote that

> in the year 1428, it is written that Don Peter, the King of Portugal's eldest son, was a great traveler. [In Rome or Venice] he bought a map of the world, which had all the parts of the world and earth described.

He added that by this map, "Dom Henry, the king's third son, was much helped and furthered in his discoveries." Most interesting of all, he notes that "the Strait of Magellan was called in it The Dragon's Tail."[135]

The Strait of Magellan, *per se,* could not, of course, appear on a map of 1428.[136] From Galvão's perspective, however, the Strait of Magellan was a passageway around a mammoth Southeast Asian 'American' subcontinent, not necessarily a strait around a separate New World. A Southeast Asian 'dragon's tail' such as that depicted on the Martellus map or Behaim globe is precisely what Magellan imagined himself skirting when he rounded the South American mainland at the end of 1519. Pigafetta confirms this when he remarks that Magellan "knew that he had to sail through a very hidden strait, as he had seen in the archives of the king of Portugal, on a map drawn by a worthy man named Martin of Bohemia" (see Behaim globe, fig. 48).[137] The monstrous Southeast Asian peninsula of Martellus and Behaim does in fact resemble the tail of a dragon, and had Galvão seen a prototype of the map or heard a description of it, the tip of the dragon's tail would have formed a proto- Strait of Magellan.

The Cantino Model
The second model of a Southeast Asian peninsular subcontinent was established by the so-called 'Cantino' planisphere of 1502, a Portuguese portolan chart of the world which was apparently smuggled from Lisbon to Italy by Alberto Cantino. Cantino, who

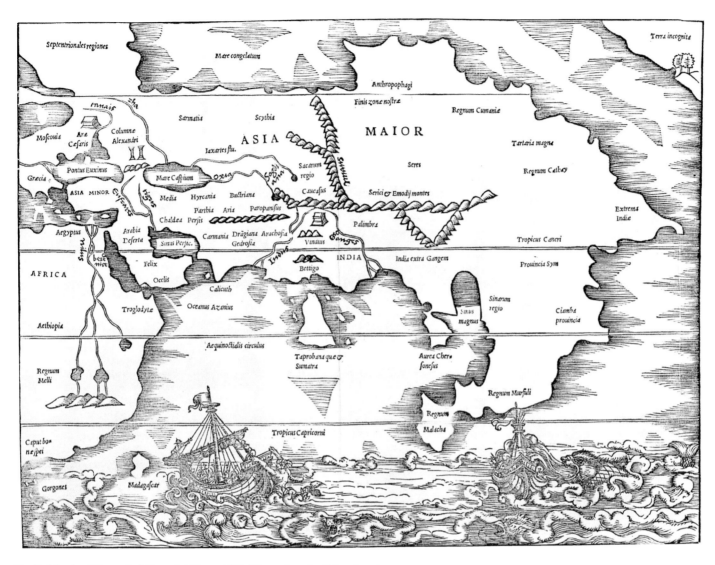

Fig. 49 Sebastian Münster, in Solinus, *Polyhistory*, 1538. [Mappæ Japoniæ, Tokyo, Japan]

might be best described as a spy, was a diplomat in the employ of Duke of Ferrara, a member of the wealthy Este family which was undoubtedly uneasy over Portugal's imminent access to the Eastern markets. Two or three years after the Cantino map was completed, very similar geography was used in a chart of the world by Nicolo Caveri (or Canerio).

The Cantino peninsula is smaller than that of Martellus (though still vastly oversized), and has been straightened from the contours of the land bridge. Unlike the Martellus peninsula, the Cantino peninsula replaced (rather than supplemented) Ptolemaic Malaya. Though at a glance it might appear to be nearly as 'wrong' as the Martellus peninsula, this subcontinent is in fact a first attempt to map 'true' Malaya and Indochina, and was probably extrapolated in large part from Arabic textual sources. Some of the nomenclature on the map agrees with Islamic navigational texts, and its use of Pole Star altitudes might also be derived from such documents.[138] The prominently mapped island of *Timonia* (Pulau Tioman), for example, is not known to have been in the Portuguese repertoire in 1502, but was a routine stopover point for Arab and Malay mariners; we will visit Tioman on the Cantino-based Waldseemüller map of Asia (page 111, below). Some scholars have suggested a similarity between the Cantino map's depiction of Southeast Asia and that of the twelfth century Arab geographer al-Idrisi (fig. 28), but any parallels are extremely vague at best: a better comparison is found in a map

compiled from the text of Ibn Sa'id, whose figures were ultimately derived from al-Idrisi and which bears a slight similarity to the Cantino coastline and classical toponymy of one or two place names.[139]

In summary: while the Martellus model had two Southeast Asian peninsular subcontinents (the Ptolemaic Golden Chersonese and the vestigial Great Promontory), the Cantino pattern had one; whereas the Martellus peninsula supplemented Ptolemaic Malaya/Indochina, the Cantino peninsula actually replaced it; and while Martellus fitted the peninsula with terminology from Marco Polo, the Cantino version introduced nomenclature previously unknown in Europe. Prominent examples of maps modeled on the Martellus lineage include many of the early printed 'modern' world maps, such as those of Contarini-Rosselli (1506), Martin Waldseemüller (1507, fig. 45), Sylvanus (1511), Apianus (1520), and Sebastian Münster (1532 & 1538, figs. 68 & 49). Waldseemüller used the Cantino pattern for the map of Asia in his 1513 edition of the *Geographia* (fig. 57), and again on his great *Carta Marina* of 1516. He also employed a unique variant for his world map in the 1513 *Geographia* (fig. 60).

Martin Waldseemüller

Waldseemüller's world map from the 1513 atlas was most likely prepared in about 1507. Its unique configuration combines all three of the non-Ptolemaic Asian subcontinents so far invented: 'modern' India, the Cantino peninsula, and the 'fake' peninsula of

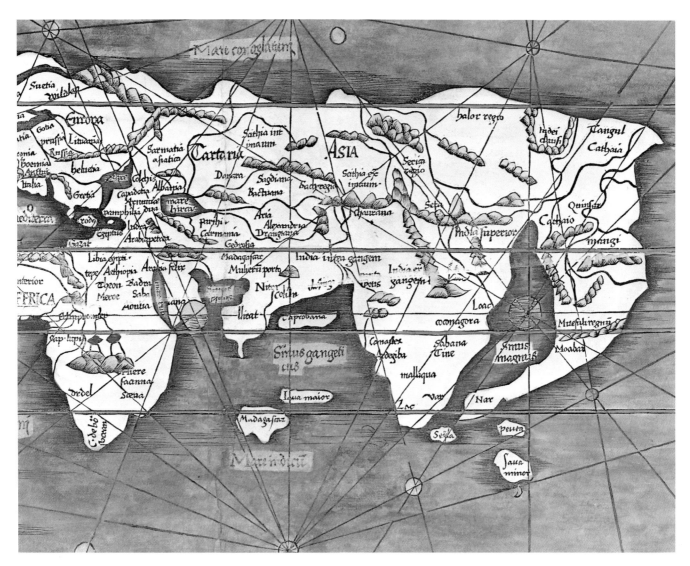

Fig. 50 World map by Lorenz Fries, 1522 (detail). (31.5 x 48 cm)

Martellus. This was the only map to employ both the Cantino and Martellus sub-continents. In the same atlas, Waldseemüller also included a conventional Cantino-based map of Asia, which we will examine as well. At that time, it was not unusual for contradictory maps and information to appear side-by-side in the same publication. The discrepancies between different versions reflect changing geographical opinions during the period that the atlas was being prepared, or else the desire or pressure to endorse a particular theory, or even, quite simply, uncertainty on the part of the cartographer.

A strange variation of Waldseemüller's 'three Indias' pattern occurred in 1522, at the hands of Lorenz Fries, who substituted an entirely new subcontinent for the Cantino peninsula (detail, fig. 50). This new 'India' is, in fact, roughly shaped like the real India, except for its gargantuan proportions, being about the size of Africa. Most Southeast Asian nomenclature has been allocated to this vast new land. The monumental bay which separates it from the Martellus peninsula to the east is identified as Ptolemy's Great Bay, and extends northward well beyond the Tropic of Cancer. Although Sumatra *(Java minor)* correctly lies to the south of 'Indochina' and near the island of Bintan *(Peuta),* Java *(Iava maior)* is grossly misplaced, being positioned due south of Sri Lanka and nearly on top of Madagascar which occupies the middle of the Indian Ocean.

The Martellus Dragon Tail 'Becomes' America

The most resourceful solution to the puzzle of Ptolemy's Great Promontory was the theory that Southeast Asia and the newly-discovered American coastline were its opposite shores. The Martellus 'Dragon Tail' peninsula thus became the mainland of Southeast Asia and the New World, its southern tip forming the Strait of Magellan with *terra australis.* That this Amerasian subcontinent was the vestige of the old promontory can be seen by following the mapping of Ptolemy's great emporium, *Cattigara,* which first was placed on the western shores of this hybrid continent, and then subsequently on the west coast of an autonomous South America.

Balboa and the 'South Sea'
When Vasco Nuñez de Balboa beheld the 'South Sea' (as he called the Pacific Ocean) in 1513, after crossing the Central American isthmus, it is unlikely that he realized he was gazing out at a vast new ocean. Rather, he believed that he had crossed the mountains of Ptolemy's Great Promontory, and that *Cattigara* lay somewhere to the south. Thus for Balboa, the *South Sea* was none other than the 'Great Gulf' region of the Indian Ocean, truly an *austral* sea. The oft-cited explanation for the term 'south sea' — that the isthmus at Darien is oriented east-west, which meant that Balboa

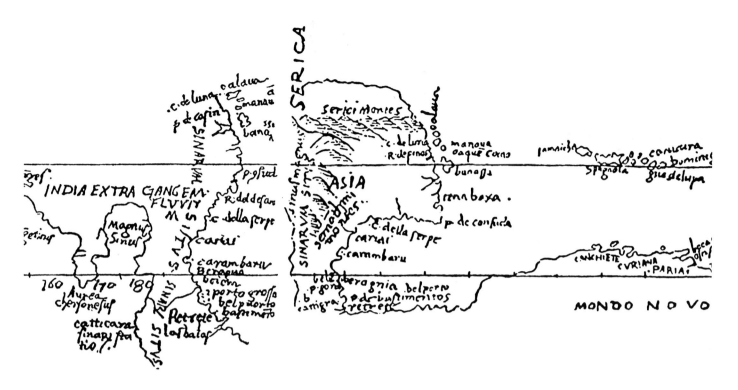

Fig. 51 Two sketches, joined, of Asia and the Pacific, Zorzi /B. Columbus, 1506-22. [R. F. v. Wiesner, Die Carte des Bartolomeo Colombo]

was heading south rather than west when he reached the ocean — merely confirmed his view of the Pacific as an Indian sea.

Ancient Greek civilization had made the same comparison. Herodotus, writing in the fifth century B.C., referred to the Mediterranean as the '*Northern Sea*', and the Indian Ocean as the '*Southern Sea*'. Peter Martyr, writing in 1516, suggested that Balboa's 'south sea' was where the Spiceries lay. Even as late as 1548, in the map of Southeast Asia by Gastaldi (fig. 73), the Indian seas to the southwest of Malaya and Sumatra are called *Oceano Meridionale* — 'South Sea'.

Columbus and St. Thomas

A letter written to Columbus a few years after his discovery of America demonstrates the pervasive belief that Southeast Asia and the New World were one and the same:

> very soon you will be by Divine grace in the Sinus Magnus [Ptolemy's 'Great Gulf'], near which the glorious Thomas left his sacred body [i.e., India].[140]

In the eyes of the correspondent, Columbus had indeed reached islands off a Martellus-type promontory. Upon crossing or circumventing this promontory, he would, according to this scheme of things, enter Ptolemy's 'Great Gulf' and shortly afterwards reach the land where St. Thomas died — India. Clinching this scenario is the fact that many maps (e.g., Fries, fig. 62) started to place Maabar, the region of India traditionally associated with St.Thomas' grave, in Southeast Asia. The tomb of St. Thomas as a link between America and Asia continued into the early sixteenth century, when the Spanish historian Bartolomé de Las Casas reported that Brazilian natives preserved the memory of the saint, and had showed Portuguese missionaries his footprint on the bank of a river.[141]

Columbus believed that he had reached Asian soil when he made his landfall on the shores continental South America and was not deterred even by the fact that it was a southern land with a northern coast, rather than vice-versa. One of the possible

explanations that Columbus is likely to have considered was that he had reached the shores of an island which Marco Polo had said was the largest in the world — Java — an entirely sensible solution, given the prevailing geographical preceptions of the day.

Peter Martyr, who was privy to the inner circles of Spanish exploration, wrote in 1511 that the Venezuela coast is "said to be part of the Indian [i.e., Asian] Continent." Martyr's observations did indeed make sense in the context of some world maps he must surely have seen, for example the chart of the world attributed to Juan de la Cosa (1500-08?).[142] Nor did better knowledge of the New World necessarily dispel this assumption in the eyes of geographers, since the apparent contradiction of Southeast Asia having northern continental shores was easily reconcilable. It was, as we will see, eminently logical for such mapmakers as of Zorzi, Schöner, and Fine.

Natural History Used as Evidence

For many observers, similarities between the people, plants and fauna of the New World and those of Asia provided further evidence that the New World and Southeast Asia were the same. The Florentine merchant Giovanni da Empoli, who sailed to India in 1503, followed by Malacca in 1509, Sumatra in 1515, and thence to China, wrote in a letter in 1514 that

> the land of Santa Cruz ['land of the Holy Cross', as South
> America was then known], called Brasil; which land is not
> yet fully discovered, because the Antilles of the king of
> Castile and also the land of Corte Real [North America] is
> presumed and judged and is all made one land with
> Malacca because the people, the animals and all other
> things are similar.[143]

Although Da Empoli's letters remained unpublished until Ramusio included them in his *Navigationi* in the mid-sixteenth century, they were nevertheless influential being circulated in manuscript form among interested scholars.

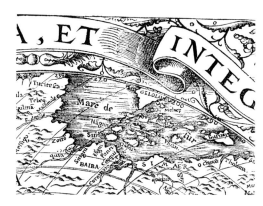

Fig. 52 (*above*) Detail of Southeast Asia, from the world map of Oronce Fine, 1531, showing Halmahera [*Gelolo*] and the Moluccas lying off the coast of Mexico [*Meffigu*].

Fig. 53 (*left*) World map, Oronce Fine, 1531. (29 x 42 cm)

The 'Southeast Asian' Subcontinent of America

The geography of America, seen as Ptolemy's Great Promontory, is espoused on the sketch map attributed to Alessandro Zorzi and Bartolomeo Columbus of ca. 1506/22 (fig. 51). This map shows South America connected by way of a hybrid isthmus (Central America/Ptolemaic land bridge vestige) to China, with the Caribbean islands off the China coast. Along the western shores of the neck connecting South America with China lies our familiar emporium of *Cattigara*. A small woodblock world map in hemispheres by Franciscus Monachus, dating from about 1527, updates the theory to reflect Magellan's voyage around South America.[144] Monachus' subcontinents are *India* (India proper), *Alta India* (Malaya), and then North America, which is separated by a narrow strait from South America.

The most refined and explicit depiction of America as Ptolemy's Great Promontory was provided by of Oronce Fine, on a world map with a double cordiform projection, which appeared in 1531 (figs. 52 & 53) and again on a world map using a single, 'true' cordiform projection which came out in 1534 (fig. 71), On both occasions, Fine used a second (middle) 'India', similar to that of Fries' world map of 1522, to form Malaya and Indochina, discarding the Martellus and Cantino peninsulas while concluding that Mexico, Central America, and South America represent a third 'Indian' peninsula in the east. Fine, unlike Monachus, does not break North and South America with a strait, and shows far greater detail.

At about the same time, the eminent geographer Johann Schöner invoked Magellan's voyage as proof that America was an eastern subcontinent of Asia. In his Opusculum geographicum, which was published in 1533 but probably begun a decade earlier, he states that

> very lately, thanks to the recent navigations accomplished . . . by Magellan, towards the Molucca islands, which some call *Malaquas*, which are situate in the extreme east, it has been ascertained that [America] is the country of Upper India, which is a portion of Asia.

Among the regions of easternmost Asia referred to by Schöner in his text is the place name *Bachalaos*, which was given to North America on account of the cod fish so abundantly caught in its waters (see, e.g., fig. 113, Gemma Frisius) Other places named include Mexico, Parias (Columbus' landfall on the shoulder of South America), and Darien (the narrow isthmus of Central America).

The opinions expressed by Schöner explain much of Fine's reasoning. For instance, Schöner bluntly states that Mexico is China:

> By a very long circuit westward, starting from Spain, there is a land called Mexico and Temistitan in Upper India, which in former times was called Quinsay [China]; that is, the City of Heaven, in the language of the country.[145]

Criticizing an opinion attributed (though perhaps wrongly) to Vespucci that South America was insular, Schöner added that "hydrographers . . . have found that [America] is a continent, which is Asia," and that explorers have reached "as far as the Molucca Islands in Upper India."

This solution, as strange as it may seem to us today, was entirely logical, neatly adapting Ptolemaic geography to the new discoveries. America had proven to be a mammoth curtain partitioning the oceans in a fashion not so dissimilar to Ptolemy's 'Great Promontory', which meant that the old promontory need not be discarded. The belief in a massive *terra australis* bridging the southern part of the globe, rekindled by Magellan's discovery of Tierra del Fuego, fitted nicely into the theory as well, since it explained the southern, east-west extension of the old land bridge. In this light it seemed that Ptolemaic geography was not so much mistaken as simply in need of fine-tuning. Ptolemy's land-locked Indian Ocean was, in effect, nothing more than a misconstrual of America and *terra australis* — Fine's combined Indian-Pacific Ocean *was*, indeed, nearly sealed, only accessible from the west by a narrow strait below South America (the Strait of Magellan) and from the east by a wider strait between South Africa and *terra australis.*

These considerations explain how Fine managed to square his Southeast Asian, American and antipodean geography with that of Ptolemy, but it does not explain why he saw no contradiction between his map and the vast Pacific Ocean described in the published accounts of the Magellan expedition, with which he certainly was familiar. We will see, however, that Fine, in fact, sized his Pacific correctly, according to the one published version of Pigafetta's journal which would have been available to him in Paris ar that time (page 125, below).

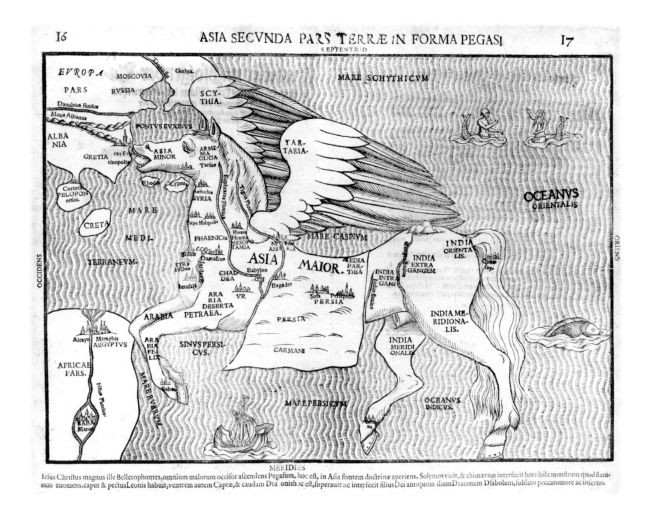

Fig. 54 Asia as Pegasus, Heinrich Bünting, 1581 [Martayan Lan, New York].

In 1538, young Gerard Mercator made a world map based on Fine's. Contrary to Fine, Mercator believed that America was an autonomous continent, and that it therefore could no longer serve as the third Asian peninsula. He pried the New World and all its nomenclature away from Southeast Asia, leaving behind a new, far smaller peninsula, a rudimentary Indochina.

Münster and Cattigara

Two years later, in 1540, Sebastian Münster took the story of the Great Promontory to its final and logical conclusion (fig. 72). Like Fine and Mercator, Münster believed the Great Promontory to be America; but Münster created a 'true' Pacific from the Great Gulf, that is, a large, autonomous ocean stretched between the unconnected continents of Asia and America. As a result, *Cattigara,* that pivotal emporium of Ptolemy's Southeast Asia, lies on the coast of Peru.

We saw earlier how Ptolemy and Marinus tried to determine the location of *Cattigara* by reconstructing the voyage to the emporium by the ancient navigator Alexander (page 90, above). Alexander reported that he had sailed "some days" from *Zaba* (probably Indochina) to *Cattigara.* Marinus, according to Ptolemy, believed that Alexander had used the word 'some' days "because the number of days [sailing to *Cattigara*] were too many to be counted." Ptolemy rejected this interpretation and placed his *Cattigara* a 'few' days from *Zaba* (his rendering of Alexander's 'some'). Münster, however, decided that Alexander had truly gone the 'many' (!) days imagined by Marinus, concluding that he had crossed the Pacific Ocean and returned.

Modern Southeast Asia

On his world map of 1546, Giacomo Gastaldi offered a far more advanced depiction of Southeast Asia than had Fine, Mercator, or Münster. Although Gastaldi connects Asia and America along the North Pacific, America is not an Asian subcontinent. His map is free of the ghost of the Great Promontory and its associated confusion, though his configuration for Malaya still bears a distant resemblance to the Golden Chersonese of Ptolemy. Gastaldi's Southeast Asian geography was copied two years later for a separate map of the region created to supplement his edition of the *Geographia* (fig. 73, and page 130, below).

Later editions of printed maps by cartographers like Bordone, Huttich, van Watte, and Honter, continued to clutch at the Martellus or Cantino model for Southeast Asia into the second half of the sixteenth century. At the end of the century, the double Southeast Asian peninsula was briefly revived for poetic use by the theologian Heinrich Bünting, who designed a map allegorically depicting Asia as the horse Pegasus (fig. 54), with the horse's hind legs formed from the two peninsulas.

Chapter 9

Printed Maps Through 1538

The reports of Southeast Asia that reached the desks of cartographers in the early sixteenth century contained a mass of new information in the form of both nomenclature and geographic features. There was often a large element of guesswork involved in the interpretation of this data and also a certain amount of speculative reasoning as to how these new discoveries could be reconciled with older, even ancient, sources.

Taprobana, Java Major, and Java Minor

These three names have been the source of much confusion and consternation which to this day has never been satisfactorily sorted out.

Taprobana

Taprobana is apparently a word of Sanskrit origin (*tamraparni*) meaning 'copper leaf'. News of this island was first brought to Europe via Megathenes' reports of about 290 B.C., but became best known through Ptolemy. In his *Geographia,* Ptolemy states that "Cory, a promontory of India is opposite the promontory of the Island of Taprobana, which formerly was called the Island of Symondi, now by the natives Salica." Pliny also wrote about *Taprobana*, noting that it "was long ago considered to be another world, with the name *Land of the Antichthones.*" Pliny — and those who copied him — reported that "the people of Taprobane do not take any observations of the stars while at sea; indeed there the Great Bear is not visible." This island of *Taprobana* is today almost universally accepted to have been Ceylon; however, most Renaissance commentators and mapmakers understood it to be Sumatra.[146]

The issue was not relevant to most medieval mapmakers, since Sumatra was not known *per se* and their Indian Ocean was too abbreviated to make the location of *Taprobana* meaningful. Although medieval *mappaemundi* typically placed *Taprobana* in the southeasternmost corner of the world, that position was virtually off the southeast coast of Arabia, straddling the Red Sea and Persian Gulf. Examples of this include the *Taphana* of the Hereford *mappamundi* (ca. 1290), and the *Tabrobana* of the world map in the *Rudimentum Novitiorum* of 1475 (fig.39).

Geographers' woes over the identities of *Taprobana* and Sumatra came with the widening of the Indian Ocean and the revelations of Marco Polo and other travelers in Asia. They were further fueled by maps which placed Ceylon off the continent's southeastern end, in the same position later correctly occupied by Sumatra (e.g., Mela, fig. 29, and the 'Peutinger' map, fig. 33). Abraham Cresques began the confusion in the *Catalan Atlas* of 1375 (fig. 40), where he depicts a large, east-west oriented *Trapobana* off the Southeast Asian coast, in addition to an island of '*Jana*' in the position of Ceylon. Fra Mauro reinforced the error in his map of 1459.

The question of the visibility of the Pole Star supported the identification of *Taprobana* as Sumatra. Polo, Odoric, and (copying them) Mandeville reported that the Pole Star is not visible from Sumatra. This would also have been true of southern Ceylon had it been as large as Ptolemy had dictated. But when the Ptolemaic conception of Ceylon was finally brushed aside with the Portuguese voyages into the Indian Ocean, Ceylon no longer extended over the equator, and Sumatra became the obvious choice for such an evidently equatorial island. The same consideration, we will see, helped support an exaggerated idea of the southern extent of Java.

Mandeville made it easy to envision *Taprobana* off the eastern bounds of the world, informing us that "toward the east part of Prester John's land is an isle good and great, that men clepe Taprobana, that is full noble and full fructuous."[147] The Arab pilot Ahmad ibn Majid, who may well have influenced Portuguese mariners on the question, furthered the confusion by transferring the term "Serendipity" *(Sirandib),* properly designating Ceylon, to Sumatra. "Opinion differs," he wrote in 1462, "as to the name Sirandib. Some say that it is the name of the island of Ceylon and some say of Sumatra [but] what is certain is that the equator is confused with the Valley of Sirandib."[148]

Ludovico di Varthema wrote in his *Itinerario* of 1510 that "in my opinion, which agrees also with what many say, I think that [Sumatra] is Taprobana." Maximilian of Transylvania, in his account of the Magellan expedition, noted that the Portuguese had sailed by way of Africa to Calicut and thence to *Taprobana*, "which is now called Zamatara [Sumatra], for where Ptolemy, Pliny, and other geographers placed Taprobana, there is no island which can possibly be identified with it." The error not only placed *Taprobana* in Southeast Asia, but it brought a large family of other islands to Southeast Asia with it, since Ptolemy related that "there are many islands around Taprobana, which are said to number one thousand three hundred and seventy-eight."

The belief that Sumatra was *Taprobana* received the most venerable Renaissance endorsement when the influential Gerard Mercator accepted it in the mid-sixteenth century. The error also led Mercator to disturb the placement of other features so that they would conform to the new position of *Taprobana.* The most egregious blunder to result from this was Mercator's dismissal of the correct, millennia-old placement of the Ganges River to the east of the Indian subcontinent, in favor of the Canton River in southern China, which roughly lies in the same position relative to Sumatra as Ptolemy had prescribed for the Ganges relative to Ceylon. As a result, on his globe of 1541 and world map of 1569, Mercator labels the Canton River as *Cantan flu olim Ganges* (Canton River, formerly known as the Ganges), and places the city of *Cantan olim Gange* by its estuary.

Confusion of Sumatra and *Taprobana* was imported into the English language in the middle of the sixteenth century with Richard Eden's translation of Münster's *A Treatyse of the Newe India.* In the latter half of the seventeenth century the English poet Milton, in his *Paradise Lost,* painted the same image of Sumatra, writing that "from India and the Golden Chersonese, And utmost Indian isle Taprobane . . .". The first person to correctly note that *Taprobana* is Ceylon (excepting *Taprobana's* early correct position in Ptolemaic maps) seems to have been João de Barros.[149]

The sequence in which Ptolemy's *Geography* is organized may also have helped sanction the idea that *Taprobana* was Sumatra. Ptolemy divided Asia into twelve maps, beginning with Asia Minor and moving progressively eastwards. India is map ten, and Southeast Asia/China is map eleven. His twelfth and final map of Asia is *Taprobana,* implying (contrary to his text) that *Taprobana* lies 'after' (to the east of) map eleven, the logical order if *Taprobana* were Sumatra. (One would have expected map eleven to be *Taprobana,* and map twelve to be Southeast Asia and southern China).

The confusion between Ceylon and Sumatra occurred in reverse as well. In 1538, Sebastian Münster, pioneering the use of the name 'Sumatra' on a published map, knew it was a modern term for the *Taprobana* of old, but transferred to it the wrong *Taprobana* — he mistakenly applied it to Ceylon (fig. 49). Authoritative precedent for the transposition of 'Sumatra' to Ceylon had already appeared on such manuscript maps as that of Zorzi/Columbus (fig. 51).

Java Major and Java Minor

Java Major and *Java Minor* were popularized by Marco Polo, and almost certainly refer to Java and Sumatra respectively. These two names proved to be even more slippery than *Taprobana,* being tossed about by explorers and mapmakers alike. Both were on occasion used for Borneo, and *Java Minor* was applied to the island of Sumbawa by Duarte Barbosa, a Portuguese official who reached India in about 1500. Later, the standardization of the modern names of Sumatra, Borneo, and Sumbawa had the effect of banishing the old nomenclature from Southeast Asia proper. *Java Major* and *Java Minor* passed the latter part of their lives by assuming the guises of new southwest Pacific lands: first as wayward islands far to the south, then as regions of the continent of *Terra Australis* and, in their final incarnation, as Australia.

The term 'Java' was used for both Sumatra and Java. Some Arab geographers, and the traveler Ibn Battuta, used *Jáwa* to denote Sumatra, and *Mul-Jáwa* for Java. Although in many instances the references to 'Java' are ambiguous, the more reliable Arab

geographers make fairly clear that their '*Jáwa*' is Sumatra. The thirteenth-century Ibn Sa'id, for example, describes a strait, two miles wide, in a place called Bintan, through which ships must pass to the east of *Jáwa* to reach China — obviously Sumatra and the Singapore Strait. The 'Java' of some later medieval *mappaemundi,* such as the *Catalan Atlas* (ca. 1375; fig. 40), is as likely to have been Sumatra as Java. Thus the term 'Java' of old might be roughly comparable to our modern sense of the word 'Indonesia'. Hence António Galvão, writing in the mid-sixteenth century, remarked that "the cosmographers called [the Indonesian islands] the *Jaoas,* though they now have different names."

But why would Sumatra, with a total area of about 473,600 square kilometers, come to be known as *Java 'Minor',* while Java, with only about 126,520 square kilometers, was *Java 'Major'*? Before medieval European adventurers first began to bring back data about Southeast Asia, Arabic and Asian pilots already had some knowledge of Sumatra's full compass, and had a fairly accurate idea of its circumference. As compared to the correct figure of about 3,700 kilometers, ninth century Arabic sources cite 800 *parasangs,* or about 4,500 kilometers, while in the thirteenth century Marco Polo reports a figure of "2,000 miles or more" (whatever the value of the mile he was quoting). Extreme figures came from Ludovico di Varthema, who gave its circumference as 4,500 miles, and Benedetto Bordone, who in his *Isolario* of 1528 pared Sumatra's girt down to 1,200 miles.

Java, by comparison, was virtually unknown along its southern coast. Arab pilots believed that no good ports existed to the south, and its austral coastline remained largely mysterious to Europeans until the seventeenth century. As a result, fanciful estimates of its size blossomed. Polo, merely repeating what he heard from Chinese, Malay, or Arab acquaintances, reported that "the experienced mariners of those islands who know the matter well, say that it is the greatest island in the world, and has a compass of more than 3000 miles." Two other important early travelers in Southeast Asia, Friar Odoric and Nicolò de' Conti, give the same estimate of Java's size. Conti reports that "There is in India far within, almost at the furthest end of the world, two Islands, and both of them are named *Lava* [Java], the one is two [thousand] miles in length, and the other of three [thousand], towards the Orient, and they are known in the name, for the one is called the great, and the other the less." Yet Conti also describes having visited the island of Sumatra, which he considered to be the *Taprobana* of old, confounding the identities of the two Javas, since geographers following his text needed to create another *Java Minor,* separate from Sumatra.

The Earliest Post-'Discovery' Charts of Southeast Asia

Even the finest printed maps lagged far behind what was known by cosmographers privy to the inner circles of Portuguese and Spanish intelligence. Extraordinary manuscript charts demonstrate that radically new data, discarding all traces of classical and archaic Southeast Asia, were available by the later 1520s. In about 1519 the chartmakers Lopo Homem, Pedro Reinel, and his son Jorge Reinel, assisted by the artist Gregorio Lopes, produced a voluptuous work known as the '*Miller Atlas*'. It is believed to have been commission-ed by Portugal's King Manuel for François I of France. Illustrating the transition between the old and new, its chart of Southeast Asia located the southern extent of Malaya correctly for the first time, with 'Singapore' transliterated recognizably *(Simgapura)* and placed at a latitude of 2° north, which is close to its true position. Malacca is shown twice under different spellings, probably following a text such as the *Suma de Geographia* of Martín Fernández de Enciso, which dates from about the same year and also duplicates Malacca.[150] Sumatra *(Taprobana Insula)* is now mapped from first-hand cartographic (rather than textual) data, finally taking on its correct rudimentary shape, position, and orientation.

But the chart's depiction of the numerous small islands to the south, and large islands to the east *(Java Major, Java Minor* and *Candin)* hark back to earlier days. The erroneous diagonal orientation of the islands east of Sumatra may be rooted in Arab navigational texts, which stated that the '*Javas*' (Sumatra and Java) and 'islands of Timor' (i.e., Lesser Sunda islands) were orientated in line with Sumatra, so that their coasts line up, with those to the east lying progressively farther south. The Homem-Reinel chart keeps the northwest-southeast orientation, but deviates from the Arab texts in that it keeps the islands at the same latitude. The islands are highly stylized, and many of the smaller ones, in particular, appear to be arbitrary. Indeed, most of the chart's island world is far more reminiscent of the *Catalan Atlas,* a century and a half earlier, than the maps which were soon to be produced by the best Iberian geographers a mere two decades later. For example, Diego Ribero,who was a Portuguese chartmaker in Spanish employ, compiled a world map in 1529 which depicts an astoundingly accurate Southeast Asian mainland, betraying not even a hint of Ptolemy's geography or the theoretical configurations of Martellus and Cantino which replaced it.

Johann Ruysch (1507)

The most individual of early printed maps to break away from an archaic depiction of Southeast Asia is a world map attributed to Johann Ruysch. The Ruysch map is on a fan-shaped projection which is 'split' through Indochina, in the position of the Mekong River (the two ends have been joined for the detail illustrated in figure 53). It was published in the 1508 Rome editions of Ptolemy's *Geographia,* and it is also known to have been included in a volume of the *Geographia* with the 1507 date still on the title page, and the book in its original binding with no evidence of having been made up at a later date. However, it is not called for in the index and could not have appeared until the very end of 1507 at earliest, since the map records Portuguese activity in Ceylon in that same year. This extraordinary map shows a wealth of new Portuguese data in the Indian Ocean and Southeast Asia, which it attempts to reconcile with data from Marco Polo and Ptolemy. For its virtues, its flaws, and its resourcefulness, its depiction of Southeast Asia is the most interesting to be found in this new harvest of printed maps. That Ruysch had access to superb Portuguese sources is evidenced most obviously by his depiction of India and Ceylon, which for the first time on a printed map have been given their correct relative proportions (rather than the subcontinent being absent, and Ceylon vastly oversized), and by his mention of Portuguese activity in Ceylon in 1507.

Ruysch's pattern for the Southeast Asian mainland is unique. A single peninsular subcontinent is indented on its southern end by a bay whose sides form, in effect, two smaller, secondary peninsulas partitioned by a (Ptolemaic) river. The bay formed by the indentation is unnamed on early strikes from the copperplate, such as that illustrated. But Ruysch (or his editors) later identified the bay as being the *Sinus Magnus,* that is, the troublesome 'Great Bay' of Ptolemy, adding the term to later states of the map; most surviving examples contain it.

The actual identity of Ptolemy's *Magnus Sinus* aside, Ruysch's map may represent the first instance of a true delineation for the Gulf of Siam on a Western printed map. The entire region lies too far to the south, extending past the Tropic of Capricorn — a vestige of the southerly latitude accorded to *Cattigara* and the Great Promontory in Ptolemaic geography. This southerly displacement may also reflect the influence of fifteenth century Arab navigators, who placed Singapore at 5 *isba'* as measured against the Pole Star, which again corresponds to a position well below the equator.

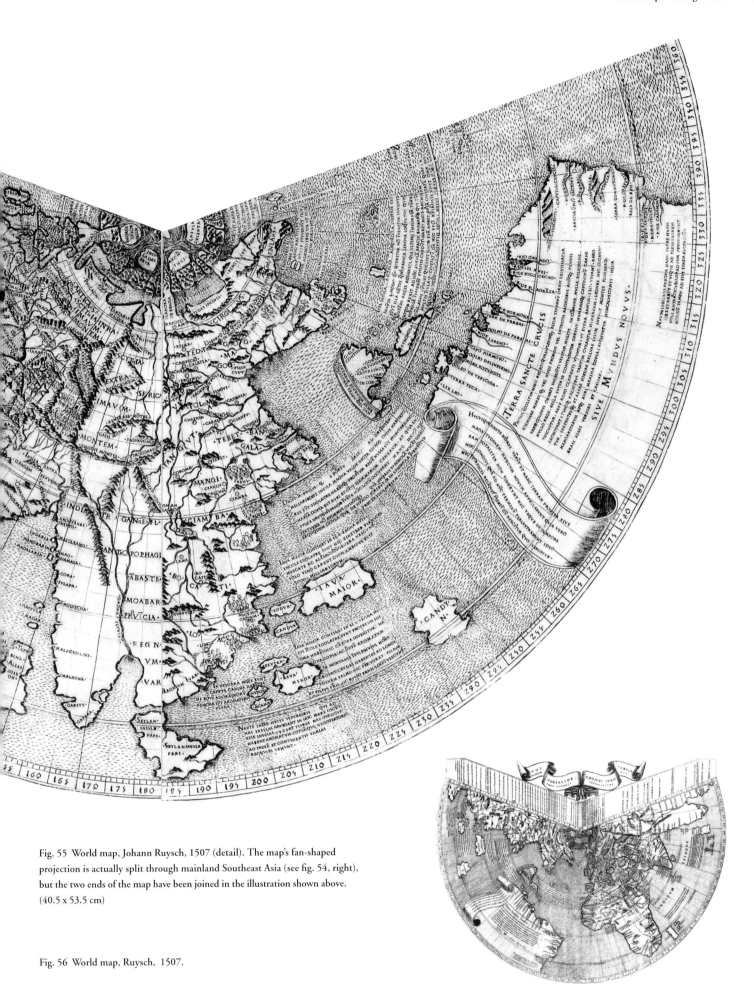

Fig. 55 World map, Johann Ruysch, 1507 (detail). The map's fan-shaped projection is actually split through mainland Southeast Asia (see fig. 54, right), but the two ends of the map have been joined in the illustration shown above. (40.5 x 53.5 cm)

Fig. 56 World map, Ruysch, 1507.

Malacca

On the Malay Peninsula itself we see *Malacha* (Malacca) four years before the Portuguese conquest. Early writers generally agreed on Malacca's virtues as a commercial center: it was situated at the point where the wind systems of the Indian Ocean and China Sea met; it provided a better and safer natural harbor than Singapore; and it was easier to defend. Tomé Pires (1515) summarized the entrepôt's commercial significance bluntly, stating that "whoever is lord of Malacca has his hand on the throat of Venice."

Just what was this nerve-center of Southeast Asian commerce like? Malacca was of such importance to Arab traders that it is one of the few places mentioned in Arab navigational texts where the commentary is not restricted solely to commercial interests. Shortly before the Portuguese arrived on its shores Ahmad ibn Majid, an Arab sea captain well-acquainted with the place wrote that the people of Malacca

> have no culture at all. The infidel marries Muslim women while the Muslim takes pagans to wife. You do not know whether they are Muslims or not . . . The Muslims eat dogs for meat for there are no food laws. They drink wine in the markets and do not treat divorce as a religious act.[151]

In the latter sixteenth century, the Dutch traveler Linschoten gave this view of the city under the Portuguese:

> [Malacca] is inhabited by the Portuguese, together with the natural born countrymen, which are called Malayos: there the Portuguese hold a fort, as they do in Mozambique and is (next to Mozambique and Ormuz) the best and most profitable [fort] for the Captain throughout all India. There is likewise a Bishop [who is under] the Archbishop of Goa, this is the staple for all India, China, the Islands of Moluccas, and other Islands thereabouts; it has great traffic [and deals with ships sailing] to and from China, the Moluccas, Banda, and the islands of Java, Sumatra [and] Siam, Pegu, Bengals, Choramandel, and the Indies: whereby a great number of ships . . . load, unload, sell, buy, and barter, and make a great traffic out of all the Oriental countries.

But living conditions were harsh for Europeans. Linschoten tells us that in Malacca there

> "dwell some Portuguese, with their wives and families, although but few, not above a hundred households . . . The cause why so few Portuguese dwell [therein] is because it is a very unwholesome country, and an evil air as well for the natural Countrymen, as for strangers . . ."

Historians differ on when Malacca was founded; the most widely accepted theory is sometime in the early fifteenth century. Its founders, possibly renegade corsairs from Sumatra or Java, may at first have been nominally subject to Siam, but soon aligned themselves with China.

Singapore

At the tip of the peninsula, in the position of Singapore, Ruysch plots *Gapara.* This is an abbreviated form of *Bargimgapara,* a place name which appears in the same relative position on the Southeast Asian subcontinents of the manuscript maps of Cantino (1502) and Caveri (ca. 1504-05), and the printed maps of Waldseemüller, 1513 and 1516 (1513, fig. 60). The Arab word '*bahr*' refers to land, hence the term *Bargimgapara* is simply a corruption of '*bahr Singapura*'. The Caveri map identifies

Bargimgapara as a place where one can find "perfumes, also sandalwood, benzoin, storax, aloes and lead."

The name *Singapur* is mentioned in Portuguese documents as early as a letter to Albuquerque in 1513,[152] and is first known by actual report in the writings of Tomé Pires. Pires noted that *Sijmgapura* "has a few *Celate* villages; it is nothing much"; the settlement appears on a chart attributed to Francisco Rodrigues of about the same date. But the term 'Singapore' itself would not be found until the Homem-Reinel chart of ca. 1519, and does not appear on printed maps until the Gastaldi world map of 1546 and his map of Southeast Asia from 1548.

Ruysch fitted Ptolemaic locations around the Portuguese data. The *Daona* River, which Ptolemy had flowing into his Great Gulf, here empties into Ruysch's primitive Gulf of Siam in the position of the Chao Phraya River, that is, at the apex of the gulf, just below modern-day Bangkok.

The two slight indentations on either side of the southern end of the western (i.e., Malayan, rather than Indochinese) peninsula probably reflect an early awareness of trade routes across the mountainous peninsula. Rather than harbors or inlets, these indentations are probably the estuaries of the Muar and Pahang Rivers in southern Malaya, and represent the earliest trace of an imaginary waterway which many later mapmakers believed cut right across the peninsula (see page 187, below).

Venturing eastwards around the Southeast Asian peninsula, the core of Ruysch's Portuguese data ends and one backtracks two hundred years, entering an entirely different world. In order to map the rest of Southeast Asia, Ruysch, like his immediate predecessors and successors, had to sit down and make sense of the section in Marco Polo's *Description of the World* where he records his return journey to the West in the 1290s. Although Ruysch did not always follow the book verbatim, it provided him with the data for a fertile Southeast Asian landscape. We will try to reconstruct the reasoning Ruysch used in mapping his Southeast Asia by following Polo's return voyage from China to the Indian Ocean. When worthwhile, we will interrupt Polo's voyage with short diversions which follow the travels of Friar Odoric, Ibn Battuta, Nicolò de' Conti, and the lore of John Mandeville.

Champa

On the upper right of the detail of the Ruysch map, there is a peninsula with a city named *Zaiton.* This port, from which the Polos began their odyssey through Southeast Asian waters, is Ch'üan-chou (Quanzhou), in the Amoy harbor region of China.[153] From *Zaiton,* they escorted a young woman named Kokachin, chosen to be the new queen of Persia, "1500 miles" across a gulf to *Ciamba* (Champa), the ancient coastal kingdom of what is now central Vietnam, known to the Chinese as *Lin-yi.* Polo visited the Chams toward the end of their civilization, which had flourished for over a thousand years, from about the second to fourteenth centuries. Polo considered Champa to be the easterly beginning of Southeast Asia, what he called 'Lesser India' and described as the region extending from Champa on the east through Motupalli (eastern India) on the west.

Ruysch places *Silva Aloe* in southern Champa. The name refers to a forest of the valuable aloe plant which Polo said the king of Champa offered as part of his annual tribute to the Kublai Khan. In the mountains above Champa, as well as further south along the Vietnamese coast, Ruysch indicates *silva ebani* (forests of ebony), the wood of which, Polo tells us, was used for making chess-men and pen-cases. Polo used the term *bonús* for ebony, from the Persian *abnús.*

According to Polo, no woman of Champa may marry until the king has seen her first, for "if the woman pleases him then he takes her to wife." Modern historians do not think that Polo was exaggerating when he reported that the king of Champa had 326 children (the actual number varies slightly) from his many wives. Odoric of Pordenone, who reached Champa from Java in about 1323, concurred. The king of Champa "had so many wives and concubines, that he had three hundred sons and daughters by them." Odoric was awed by the king's collection of elephants as well, noting that "this king hath 10,004 tame elephants, which are kept even as we keep droves of oxen, or flocks of sheep in pasture." The good Friar was of the opinion that Champa is "a most beautiful and rich country, and abounding with all kinds of victuals."

Once every year, according to Odoric, fish come swimming toward the coast of Champa "in such abundance, that, for a great distance into the sea, nothing can be seen but the backs of fishes." The fish cast themselves onto the shore where, for a period of three days, men "come to take as many of them as they please." The remaining fish then return to the sea, and another kind of fish offers itself in the same manner. The story of the fish in Champa inspired the fourteenth-century author John Mandeville to concoct an island called *Calonak* for his book of travel lore. Mandeville, with his usual flair, romanticizes Odoric's description of Champa, imaginatively elaborating upon it. The phenomenon of the multitude of fish casting themselves upon the shore of Champa, Mandeville explains, is said to be the result of the king's prodigious siring of children. "Because that he fullfilleth the commandment that God bade to Adam and Eve, when God said, *Crescite et multiplicamini et replete terram . . .* therefore God sendeth him so the fishes of diverse kinds of all that be in the sea, to take at his will for him and his people."

New first-hand reports about Champa came soon after Ruysch's map, with the *Suma Oriental* of Tomé Pires (1515). Pires related that Champa was an agricultural kingdom with no ports suitable for large junks, but with a few towns situated on its rivers. Its principal exports were dried fish, rice, textiles, pepper, and the highest quality calambac (aloe-wood), which mainly went to Siam.

Sumatra, Java, and the South China Sea

When Ruysch came to placing Java and Sumatra on his map he needed to align them properly relative to Champa. Polo had explained that if one were to sail 1,500 miles south-south-east from Champa, one would reach Java *(Java Major),* and Ruysch positions the island according to this instruction. Polo never insinuates that he himself ever visited Java; but Odoric of Pordenone did. Odoric tells us that when he left Sumatra, he "traveled further unto another island called Java, the compass whereof by sea is 3000 miles," thereby repeating the common overestimation of the island's size. According to Odoric, Java was "thoroughly inhabited, and is thought to be one of the principal islands of the whole world," with cloves, cubeb, nutmeg, and many other spices and victuals. He reported that there was a king of Java with seven vassals under him, who had many wars with the "khan of Cathay". The king has

> a most brave and sumptuous palace, the most loftily built, that ever I saw any, and it hath most high greeses and stairs up to the rooms therein contained, one stair being of silver, the other of gold, throughout the whole building.

Ibn Battuta, like Odoric, reached Java from Sumatra. He believed Java (his *Mul-Jáwa*) to be two-months' journey in length, and to contain the territories of *Qáqula* and *Qamára,* after which the excellent aloes *Qáqulí* and *Qamárí* are named. Reaching port in

Qáqula he found it to be "an infidel land", either because its people were not Muslim, or if they were, then at least not sufficiently orthodox from his way of thinking. Junks were docked in *Qáqula*, "ready for making piratical raids, and also for dealing with any junks that might attempt to resist their exactions, for they impose a tribute" on every arriving junk.

The Motionless Sea

A phenomenon noted by both Battuta and Odoric, which they encountered somewhere *en route* between Java and Indochina, was the motionless sea. Leaving Java, Battuta reached a "sluggish or motionless sea" tinged with a reddish color "which, they say, is due to soil from a country in the vicinity." It is clear from Ibn Battuta's account that traversing these difficult waters was not chance misfortune, but rather an expected and routine part of a well-established itinerary for which they came prepared: three smaller vessels accompanied the junk to tow it through the stillness, and in addition,

> every junk has about twenty oars as big as masts, each of which is manned by a muster of thirty men or so, who stand in ranks facing each other. Attached to the oars are two enormous ropes as thick as cables; one of the ranks pulls on the cable [at its side], then lets go, and the other rank pulls [on the cable at its side]. They chant in musical voices as they do this, most commonly saying *la'lá, la'lá.*

The crossing was also familiar enough that the sailors knew how long to expect it to last. Battuta and his crew

> passed thirty-seven days on this sea, and the sailors were surprised at the facility of our crossing, for they [usually] spend forty to fifty days on it, and forty days is the shortest time required under the most favorable circumstances.

Friar Odoric also described sailing through such a sea while *en route* from Java to Champa, but his account paints a more ominous image of it. He placed the sea near a country called *Panten* (Bintan?), or *Tathalamasin*, and he refers to it as a 'dead sea' *(mare mortuum)* which flows continually to the south, and into which "whosoever falleth is never seen again."

To where did this sea lead? Battuta reached a land he called *Tawálisí*, "it being their king who is called by that name." They put into port at a town called *Kaylúkarí,* which was said to be one of the land's finest and largest cities. Its people, who were often at war with the Chinese, have a reddish hue to their skin, with "handsome men [who] closely resemble the Turks in figure." They were probably in Indochina.[154] Battuta's company asked after the prince to offer him a present, but learned that King of *Tawálisí* had appointed his daughter, Urdujá, as governor of the city. She was celebrated as a skillful archer and warrior, and had been granted the city by her father after repelling their enemy in a valiant and fierce battle. Battuta at first declined to meet with her "because, being infidels, it is not lawful [for a Muslim] to eat their food." But the princess insisted, and Battuta accepted robes, rice, buffaloes, sheep, and other supplies for their voyage to China. Conversing in Turkish, she asked Battuta about "the pepper country" of India, noting that she wished to conquer it herself, "for the quantity of its riches and its troops attracts me." Leaving *Tawálisí*, Battuta sailed for seventeen days and reached China.

Battuta, like Polo, left China from *Zaiton.* He boarded a junk belonging to a sultan from Sumatra, and retraced his earlier route only in reverse, heading back to the country of *Tawálisí*. On this occasion, they sailed for ten days with fair winds, but as we approached the land of Tawálisí, the wind changed, the sky darkened, and it rained heavily. We passed ten days without seeing the sun, and then entered a sea which we did not know. The crew of the junk became alarmed and wished to return to China, but that was out of the question. We passed forty-two days not knowing in what sea we were.

Battuta and his company then witnessed what appears to have been a freak atmospheric or meteorological phenomenon, the stuff from which *Arabian Nights* epics were born:

On the forty-third day there was visible to us at early dawn a mountain, projecting from the sea at a distance of about twenty miles from us, and the wind was carrying us straight towards it. The sailors were puzzled and said "We are nowhere near land, and there is no record of a mountain in the sea. If the wind drives us on it we are lost."

All on board prayed and sought the mediation of Muhammad. But later, when the sun had risen,

we saw that the mountain had risen into the air, and that daylight was visible between it and the sea. We were amazed at this, and I saw the crew weeping, and taking farewell of one another.

The crew explained to Battuta that what had appeared at first to be a mountain was actually the *rukh*, a bird of such enormous size that it could easily devour the entire company. Marco Polo heard a similar story in the Indian Ocean — he was told about a bird on Madagascar which could carry off elephants. Pigafetta, sailing through western Indonesia, heard a similar tale in the South China Sea (*Sino Grand*) where there "is found a very large tree, where certain birds called *garuda* live, that are so large that they can carry a buffalo, or an elephant to the place where the tree is, which is called Busathaer." No ships can approach within three or four leagues of this tree without peril. Its fruit, which are larger than watermelons, are called *buapangganghi.*

Ibn Battuta and his shipmates were spared the clutches of the *rukh*. A favorable wind turned their vessel away, and two months later they arrived once again at *Jáwa* (Sumatra). The sultan there had just returned from a raid in which he had acquired a large number of captives; he lodged Battuta "in the usual manner" and gave him two girls and two boys. After a two months' stay on Sumatra, Battuta took a passage on a junk sailing west back to India, reaching *Kawlam* (Quilon) in time for Ramadán in January of the year 1347.

Marco Polo and the Question of *Lochac*
Marco Polo, whom we left at Champa, did not make the voyage between Java and the mainland, and did not refer to any motion-less sea. Leaving Champa and sailing seven hundred miles south-south-west, Polo passed two uninhabited islands, Ruysch's *Sodur* and *Candur* (Polo's text has him leaving 'Java' instead of 'Champa', but this is generally agreed to be an error).[155] These belong to the Pulo Condore group, a collection of small islands lying off the southeastern tip of Vietnam, to the southeast of the Mekong Delta (in Malay, *pulau* = island, *kundur* = gourd). Pulo Condore was an important landmark and stopping point for pilots guiding their vessels between Southeast Asia and China.[156] Although Polo reported that no one lived on Condore, the principal island of the group was later settled; William Dampier, for example, gives a description of its people when he visited Condore four centuries later.

Leaving Pulo Condore, we come to one of the great riddles of Polo's text. Polo claims that from the island pair they sailed 500 miles, "and then one finds a province which is on the *terra firma,* which is called *Lochac.*" He describes this kingdom as being very isolated, "in such a strange and out of the way place [that no one] can go upon their land to do them any harm." Said to be inaccessible even to the Kublai Khan and his thirst for tribute, *Lochac* boasted an abundance of brazilwood, gold, and elephants, and was the source of the cowrie shells which the entire region used as money, and had its own language. What was this kingdom, "such a savage place that few people ever go there"? Ruysch placed his kingdom and city of *Loac* to the west and southwest, on the Southeast Asian mainland, which agrees with the majority of modern scholars, who hold *Lochac* to be Thailand. If so, the name may refer to the ancient city of Lop Buri; in Chinese, Lop Buri would have been *Lo-Hui,* or *Lo-Hui Kuok* (*kuok* = country).[157]

The versions of Polo's text probably read by mapmakers in the early sixteenth century offered no direction for the voyage from Condore to *Lochac,* simply stating that he sailed five hundred miles "beyond" Condore. Some geographers, such as Ruysch, assumed that he was continuing in the same general direction, in this case south-southwest; others used the missing direction as an opportunity to adjust for other pieces of the puzzle. Another version of Polo's manuscript, however, states that he sailed from Condore "per scirocco", the term 'scirocco' meaning 'to the southeast' according to the Venetian compass. This rendering, published by Ramusio in Volume II of his *Navigationi* (1559), inclined Mercator in 1569 to identify *Lochac* with *Terra Australis* (see page 159 below), in 1688 swayed Coronelli to suspect that *Lochac* lay in Australia (fig. 83). Today, it supports the theory that *Lochac* was Borneo.[158] South-western Borneo, specifically the region known to sixteenth century Europeans as *Laue, does* indeed lie in a southeasterly direction from Condore. Polo certainly could have misconstrued Borneo as being continental, a 'terra firma' — his other descriptions of *Lochac* could apply equally well to Borneo or Thailand. More than two centuries later, Pires, who believed Borneo to be comprised of many islands, learned from his sources in Malacca about an 'island' called *Laue* whose people traded extensively with Java and Malacca. A few years later, Pigafetta, who actually reached Borneo, wrote of "a great town called *Laoe* [or *Laoc* in another manuscript] that lies at the end of the island towards Greater Java [i.e., somewhere in southern Borneo]." *Laue* as a Bornean place-name is found as early as the Penrose map of ca. 1545, and on such sixteenth century maps of Borneo such as the de Bry 1602 (fig. 102).

None of these solutions is without flaws, and none of Polo's descriptions of *Lochac's* people, kingdoms, resources, religion, or customs clearly favors a particular option. As with the other riddles associated with Marco Polo's account of his travels, we who scrutinize his words often err in treating his text as a gospel which must provide an answer. Polo, who scarcely could have imagined that scholars seven centuries in the future would devote their careers struggling to make sense of his story, may simply have made a mistake. Ruysch, of course, knew nothing of of Lop Buri or the Bornean place name of *Laue.* Modern scholars have argued over the identity of *Lochac* and other points based on etymological analysis. They have scrutinized Polo's description of the realm, and its position in the itinerary in relation to 'known' places, particularly Condore and Bintan. But these considerations exist only from our vantage point in history, not from Ruysch's. Polo's statement that *Lochac* was on a *terra firma* meant either that it lay on the Asian continent, or that there was another landmass of major proportions lying in the China Sea.

Chiang Mai: the Kingdom of Lan Na

Upcountry from *Loac,* we find a place called *Lama,* situated near two rivers, and to the north of it, a lake of the same name *(Lama La[cus]).* This is probably the kingdom of Lan Na, whose principal metropolis, Chiang Mai, was founded in 1296, round about the time when Polo arrived back in Venice.[159] Ruysch plotted the kingdom at 15° north latitude, as compared with the actual 18½° for the older city of Lamphun, and 19¾° for Chiang Mai itself. The lake north of *Lama* is Lake Chiang Mai, the legendary source of Southeast Asia's rivers, which we will visit later (page 152, below).

The kingdom of Lan Na is found on European maps as early as the Leardo map of 1448 (as *Llana*). It also appears on the Behaim globe of 1492 (*lanna,* fig. 48) and is recorded by Waldseemüller as a city (*Lamia civitas*) in 1507. Its namesake lake is mapped by Behaim (*lamacin see*) and twice by Waldseemüller, once simply as *lamia,* and again in a more mountainous situation as *lama lacus,* said to be where pearls are obtained.[160] Tomé Pires learned about Chiang Mai while in Malacca, and referred to it in his *Suma Oriental* (1512-15) by the name *Jangoma,* by which name, in several variations, it quickly became known to mapmakers. Pires correctly noted Chiang Mai's position relative to other kingdoms, learned that it was a source for musk, and that the king of Siam waged war against it.

There is textual evidence of possible early European knowledge of Chiang Mai. *The Travels of Sir John Mandeville,* though largely a collection of folklore, includes a description of a city called *Jamchay* which lies in "Ind the more" (i.e., Southeast Asia). '*Jamchay*' was a common variant of *Jangoma* — in the later sixteenth century Ralph Fitch, the first English visitor to the kingdom, referred to it as '*Jamahey*'. Mandeville wrote that "*Jamchay* is a noble city and a rich [one] and of great profit to the Lord, and thither go men to seek merchandise of all manner of thing," and that it was a vassal to the Khan. He believed *Jamchay* to lie five miles from the head of the river *Dalay,* perhaps the *Thalay* of Friar Odoric. In the decade and a half after the Ruysch map, the Portuguese learned a bit more about Chiang Mai when they helped King Rama T'ibodi II of Siam wage a successful war against Lan Na in 1515.

Below *Lama,* Ruysch has marked *Bocati,* which is probably Cambodia, in its approximately correct position. *Boja* is Pali for 'country', thus *Bocati* is a corruption of *(Cam)boca(ti).*[161] But the kingdom to the west of Cambodia, *Regnum Var,* is misplaced. There is, in fact, a people known as the Var, feared for being *anthropophagi,* eaters of human flesh, though the Var actually live in the inland mountains. We will visit these people again in 1522, on a map by Lorenz Fries.

The Philippines

In Ruysch's South China Sea, below *Iava Minor,* there is an inscription which refers to an archipelago of precisely 7,448 islands. Marco Polo related that "according to the testimony of experienced pilots and seamen that sail upon [the China Sea] and are well acquainted with the truth it contains 7,448 islands, most of them inhabited." There are minor variations in Polo's figure, such as 7,459 or 7,440. Other early mention of islands in the eastern sea are far more general and less detailed than Polo's. Friar Odoric made only a passing reference to the Khan's islands in the eastern sea "which are at the least 5000," and Mandeville repeats the same, speaking of "Ind and the isles beyond Ind, where be more than 5000 isles."

Many historians dismiss all these islands simply as figments of a seemingly universal island-mythology with no connection to a real archipelago. The belief that the ocean sea was dotted with islands was certainly common in both Western and Eastern traditions, and can be found in Buddhist cosmology, which Polo's Chinese hosts probably followed. "Sixty thousand inhabited islands are scattered in the spaces between" the continents, Gervaise noted, when describing the Buddhist cosmology of Siam four centuries later.

It is unlikely, however, that Polo's archipelago is simply a case of island mythology. In reality it probably represented the Philippines (comprised of approximately 7,100 islands) and neighboring Indonesian islands on the major trading routes. The report about the 7,448 islands contained specific details, and accompanied his news of *Zipangri* (Japan), which was of course factual. Chinese bureaucrats whom Polo would have met certainly did have knowledge of the Philippines. During the eleventh and twelfth centuries the Song Dynasty, China established regular trade with the Spice Islands through the Sulu Sea and set up trading posts in the Philippines, becoming familiar with parts of the archipelago, as well as with northern Borneo. So, given Chinese contact with the Philippines by the time of Polo's visit, it is eminently reasonable that the archipelago would have been mentioned in a discussion about islands in the ocean sea. Still, the accuracy of the figure of Polo's 7,448 islands was probably only coincidental; the alternative would raise the question of how early Chinese, Arab, or native pilots were able to inventory the vast archipelago with such precision.

While this was one of Polo's digressions about places which he himself had never visited, his description, though general, also suggests that these islands are indeed the Philippines and perhaps some of the Indonesian islands that are located along the same trade route. They are said by Polo to lie in the waters off China — in the "Eastern Sea of Chin". When "the ships of Zayton and Kinsay do voyage thither they make vast profits by their venture," though "they lie so far off from the main land that it is hard to get to them." The shuttle takes

> a whole year for the voyage, going in winter and returning in summer. For in the Sea there are but two winds that blow, the one that carries them outward and the other that brings them homeward; and the one of these winds blows all the winter, and the other all the summer.

Polo also states that these islands "are so far from India that it takes a long time also for the voyage thence." And finally, while Polo explains that the sea in which these islands lie is part of the earth's one ocean, in other words the encircling ocean sea of medieval and ancient thought, he is quite explicit about placing them in the waters off China, rather than in the Indian Ocean:

> Though that Sea is called the Sea of Chin, as I have told you, yet it is part of the Ocean Sea all the same. But just as in these parts people talk of the Sea of England and the Sea of Rochelle, so in those countries they speak of the Sea of Chin and the Sea of India, and so on, though they all are but parts of the Ocean.

According to Polo, Chinese mariners know the make-up of these islands so intimately because "their whole life is spent in navigating that sea." All the islands of this archipelago

> produce valuable and odorous woods like the lignaloe, aye and better too; and they produce also a great variety of spices. For example in those Islands grows pepper as white as snow, as well as the black in great quantities. In fact the riches of those islands is something wonderful, whether in gold or precious stones, or in all manner of spicery.

About a century and a quarter after Polo learned of these islands from his hosts in China, the second Ming emperor tried to establish a military presence in the Philippines, but failed.

Despite Polo's description of the islands as lying in the Sea of China, the term was sufficiently vague from Ruysch's perspective to allow him to place the archipelago in roughly the position of Indonesia, along with all his other data from Polo. This is approximately the same position accorded the islands on the *Catalan Atlas* of 1375, in which *Zaiton* is placed on the south Asian coast (since the map had no Southeast Asian peninsula), with the 7,448 islands lying to the south and west of it. Ruysch's 'China Sea' was already overflowing with the Caribbean islands he had learned about from the recent Spanish voyages to the new world, and in fact the overlap of Asian and Caribbean data had already convinced him that Japan and Hispaniola were one and the same island.

A more faithful rendering of the Polian Philippines is seen later, in the maps of Sebastian Münster, 1540 (fig. 72). Münster separated China from North America by a modest ocean and thus was able to place Polo's 7,448 islands further to the north, scattered indiscriminately across the China Sea from Vietnam northward to the latitude of Korea, with Japan to the east.

After Magellan made the first recorded European discovery of the Philippines, Peter Martyr identified them with the vast archipelagoes reported by earlier writers, though without specifying Polo's 7,448. He wrote: "In my opinion [the Philippines] are those concerning which many authors have written, but in different senses; for according to some they should number a thousand, and according to others three thousand or still more; they lie not far from the Indian [Asian] coast."

Multiple Ceylons and Sumatras

Now let us return to Marco Polo, whom we had left at *Lochac*. When Polo and company departed its shores, they sailed 500 miles and reached *Peutan*, "which is a very savage place" with forests of sweet-smelling hard woods. This is the island of Bintan, which was an important stop-over on the sea route through Southeast Asia, being mentioned in many of Arab texts; Polo was now in the waters near Singapore. About 100 miles southeast of Bintan, Polo continues, lies the island of *Java Minor* (Sumatra), the itinerary thus fitting nicely into place. Due west of Sumatra lie *Neucā* and *Agama*, the Nicobar and Andaman Islands, which are well-positioned relative to *Java Minor*, given the vagueness of Polo's text (though they should be northwest of the island), but placed on the wrong side of Malaya because of *Java Minor's* erroneous location.

Rounding Ruysch's peninsula to the west, we leave the world of Polo in the late thirteenth century and re-enter the realm of Portugal at the turn of the sixteenth century. Here, Ruysch confuses Sumatra and Ceylon as badly as any of his colleagues. Straddling the two ends of the map (but joined in figure 55) Ruysch shows a correctly placed Ceylon (*Prilam*) and the Indian subcontinent, freed, for the first time on a printed map, of the old Ptolemaic distortions, but he also creates an island off Malacca and designates it "*Taprobana alias Zoilon*" (*Taprobana*, also known as Ceylon). Alexander the Great, we learn from an inscription, visited this island, confirming that Ruysch was confusing it with Ceylon (though Alexander, of course, never reached Ceylon either). Finally, Ruysch places yet another island marked as *Seylan* south of the Indochinese coast, in approximately the correct relative position of Sumatra.

It was understandable that mapmakers, reconstructing these islands from existing textual accounts, would duplicate and confuse Ceylon, *Taprobana*, and Sumatra. Marco Polo went from Sumatra (*Java Minor*) to the Nicobars (*Necureran*), the Andamans (*Angamanain*) and Ceylon (*Seilon*). But the direction he traveled from Sumatra to Nicobar was imprecisely given as 'north' (rather than northwest), which some mapmakers then plotted with some

liberty as slightly northeast (see, for example, Fries, 1522; fig. 62). Adding to the ambiguity, Polo gave no directions at all for the sail from Nicobar to Andaman. Perhaps this was because he was simply continuing in roughly the same direction, but the omission let mapmakers assume any direction that 'worked' — many chose south. In the final leg of the Sumatra-Ceylon trail, Polo sailed "about a thousand miles in a direction a little south of west," and reached the "Island of Seilon", which Ruysch, the errors at each segment of the journey having compounded themselves, situated far away from India.

Finally, some versions of Polo's text may have been misunderstood as invoking the island twice. The first occurs after leaving *Angaman* (Andaman Islands), when "steering a course something to the southward of west, for a thousand miles, the island of Zeilon presents itself." A glance at Ruysch's map will confirm that Ruysch heeded this instruction with great care; his *Seylan* is in the position of Sumatra only because his *Agama* is misplaced. But the text's description of Ceylon is in some versions interrupted by a lengthy digression on India, after which Polo returns to the topic of the island to relate "certain particulars . . . which I learned when I visit-ed that country in my homeward voyage," Or in other versions, "let us turn to a delightful story I forgot to tell when we were dealing with Ceylon." In consequence, some mapmakers, in the course of extrapolating nomenclature from the text, may have transcribed the name twice. Further complicating any mapmaker's attempt to discriminate between these multiple Ceylons are the sometimes varying descriptions accorded them in travel lore. Mandeville speaks of an island called '*Silha*', which, as with Odoric's *Sylan*, we know to be Ceylon because within it lies the great mountain upon which "Adam and Eve wept" (Adam's Peak); yet Mandeville states that this island is 800 miles in circumference, far less than the size generally accorded Ceylon, which, combined with its variant spellings, would have suggested to mapmakers that it was a different island.

Friar Odoric and Nicolò de' Conti also contributed to the confusion. Traveling "further by the Ocean-sea towards the south, [passing] through many countries and islands," the good Friar reached the island of *Sylan*, the island with "the Mountain where Adam mourned for his son Abel." But the itinerary suggested an island further east toward the 'ocean sea', that is, toward the ocean encircling the continents, in this case the China Sea. Conti stated that he traveled to Sumatra and *later* went to what he called *Java Minor* and *Java Major*, implying that *Java Minor* was not Sumatra.

Some later mapmakers would also use the term *Ceilon* to refer to the Philippine island of Panoan or southern Leyte, but this came from *Selani*, an island mentioned in the accounts of the Magellan voyage by Pigafetta and Maximilian, and bears no relation to Ruysch's island.[162]

Gauenispola

Due north of *Taprobana* on the Ruysch map lies *Gauenispola* (*Gaspula*), small islands just north of Sumatra which had for centuries provided pilots with their first landmarks when approaching the Indonesian islands. Early Portuguese rutters refer to the entire archipelago of little islands off the north Sumatran coast by the name, though Marco Polo, as well as most printed maps, identify a specific island as *Gamispola*. Marco Polo, after describing Sumatra in detail, said that "I will tell you of a very small island that is called Gauenispola," but that is the last we hear of the island — he either forgot to do so, or the passage is missing from the extant versions of the text. Tomé Pires described *Gamispola* as "ten or fifteen islands three or four leagues round, [some of which] are inhabited by a few people. They have water and a great deal of fish and firewood. They all have quantities of sulphur, which supply Pase and Pedir."

During the late sixteenth century, the first part of the island's name was gradually mistaken for a Spanish or Portuguese surname, Gomez or Gomes. In the mid-century, Barros called it *Gomispola*, and later in the same century Linschoten referred to it *Gomespola*. Eventually mapmakers, aware that the second part of the name (*pola*) was 'island' in Arabic, made it *Pulo Gomez* as if it were named 'Gomez Island'. It continued to appear on maps through the eighteenth century, still being identified as *Pulo Gommes* on the charts of Mannevillette in 1775 (see also Gastaldi, page 142, below).

Candy

Perhaps the most enigmatic island in Ruysch's 'Pacific' is *Candy,* the most easterly of his Southeast Asian islands. This unknown island also appeared on the great 1507 world map of Waldseemüller, but is otherwise unknown to printed maps. It is found, however, on manuscript maps such as the 1457 Genoese world map, and the 1492 globe of Martin Behaim. *Candy* may have originated as an island reported by Odoric as *Dondin,* or with an island called *Sandji* that was mentioned by al-Idrisi. According to an annotation on the 1457 Genoese map, a "monster fish" was taken from the waters off an island of *Candia* in the Indian Ocean, and taken to Venice. Behaim noted his *Candyn* as a place which lies "foot against foot" (i.e., on the part of the earth opposite Europe), where "when it is day with us they have night, and when the sun sets with us they have their day."

There is no persuasive answer for its identity, though several Indonesian islands, and even the Philippines, have been proposed. Conceivably, *Candy* may be yet another duplication of Ceylon, born when the island was referred to by the name of the inland city of Kandy, the last of its ancient capitals. Although the island of *Candy* made but brief appearances on the stage of the printed map, *Candin insula* could be found through to the end of the sixteenth century in Gemma Frisius' inventory of islands lying in the eastern ocean.[163] Frisius' coordinates agree with Ruysch's placement: 250° east of the Canaries, at a latitude 24° south.

Amerasia

To the northeast of Ruysch's *Candy,* due east of *Zaiton,* lies a triangular-shaped island which, on an earlier state of the copperplate, had been a peninsula jutting out of the Chinese mainland. That 'Asian' peninsula was Cuba. Columbus had insisted that Cuba was the eastern tip of a continental Asian promontory, and on his second voyage (1493-94) he went so far as to force his officers to make sworn statements to that effect. Ruysch, at first, had accepted this premise. Although he changed his mind before the map was published in the *Geographia,* his original intention is known from vestiges of plate erasure which reveal an immense peninsula jutting out to the southeast from the Asian mainland with the name 'Cuba'. Even in its published state, the now-unlabeled Cuban island retained the name *Fundabril* ('end of April') on its southeastern tip, a condensed form of name accorded the mainland American peninsula on Cantino-based maps.[164] Since the name 'Cuba' was erased from the plate along with the peninsula, the identity of the land as being Cuba was, ironically, more definite in Ruysch's mind when he recorded it as a Southeast Asian peninsula than when he mapped it as a Caribbean island. Waldseemüller, on his *Carta Marina* of 1516, associated the full eastern continental bulk of Asia with Cuba, rather than just an Asian promontory, designating his Cantino-inspired North America as *Terra de Cuba Asie Partis* (Land of Cuba, part of Asia).

Ruysch, however, was by no means confident about his decision to make Cuba an island, however. He leaves the island undefined on its western coast, and veils the question by placing a 'fig-leaf' over the crucial area — a ribbon hides the open end of the island, with an inscription stating that "as far as this the ships of Ferdinand [i.e., Columbus and the Spanish expeditions] have come."

Although Ruysch correctly understood South America to be a previously unknown continent, he envisioned North America as the previously unknown northeastern shores of Asia. As we have seen, the Caribbean was thus synonymous with the South China Sea. The various West Indian islands were simply part of an extended family of Southeast Asian islands, and it was only later that the terms 'East' Indies and 'West' Indies arose to sort them out. Ruysch's own explanation of his confusion over Japan and Hispaniola is insightful:

> M Polo says that 1500 miles to the east of the port of Zaiton there is a very large island named Sipagu [Japan]. . . but as the islands discovered by the Spaniards occupy this spot, we do not dare to locate it here . . . being of the opinion that what the Spaniards call Spagnola is really Sipagu.

Both Hispaniola and Japan, in fact, had been judged candidates for the 'Southeast Asian' island of Ophir. Columbus believed Ophir to be the 'Asian' island of Hispaniola, while others believed it to be Japan, since it lay closer to the Asian mainland.[165]

Martin Waldseemüller and Vesconte Maggiolo, 1507-16

About the time Ruysch's map was published, Martin Waldseemüller was preparing a new set of maps for another edition of Ptolemy's *Geographia.* "I think you are aware," wrote Waldseemüller to a colleague in April of 1507, "that I am about to print in the town of St. Dié the *Cosmographia* of Ptolemy after revising it and adding some modern maps to it." But the atlas project, which may have been beset by financial problems, was abandoned. It was not until 1513, in the city of Strassburg, that the work was finally published, and without any acknowledgment of Waldseemüller whatsoever; the attribution of the atlas to Waldseemüller is generally accepted, but not proven. In a dedicatory letter, the two Strassburg lawyers state that they have salvaged the volume from "six years of neglect" by their "energy, expense, and direction". The six year time lapse cited by the lawyers in 1513 jives with the anticipated 1507 publishing date implied in Waldseemüller's letter, and the woodblocks for some of the maps (in particular the world map) can provisionally be dated ca.1507.

The 1513 atlas' map of Asia (fig. 57) was the first 'modern' map devoted to the continent. It is rooted in the Cantino model of 1502, and is constructed on the quadratic plane projection used by nautical charts. The term *Sinus Magnus* now designates the South China Sea, while 'Indian Ocean' *(Mare Indicum)* is spread over both sides of the Southeast Asian peninsula. On the main part of the continent, *India Intra Gangem* (India within the Ganges) is placed far to the east, and is a mistake for *India Extrum Gangem.*

One place name found on both the Waldseemüller and Cantino maps would soon become familiar to Europeans. This is *Maitbane,* which is Martaban, an important port at the mouth of the Salween River in Burma. We will return to Martaban in 1548, via a map of Gastaldi. Following Cantino, Waldseemüller also places *Mallaqua* in its 'correct' position relative to the 'wrong' Southeast Asian peninsula, which extends much too far south.

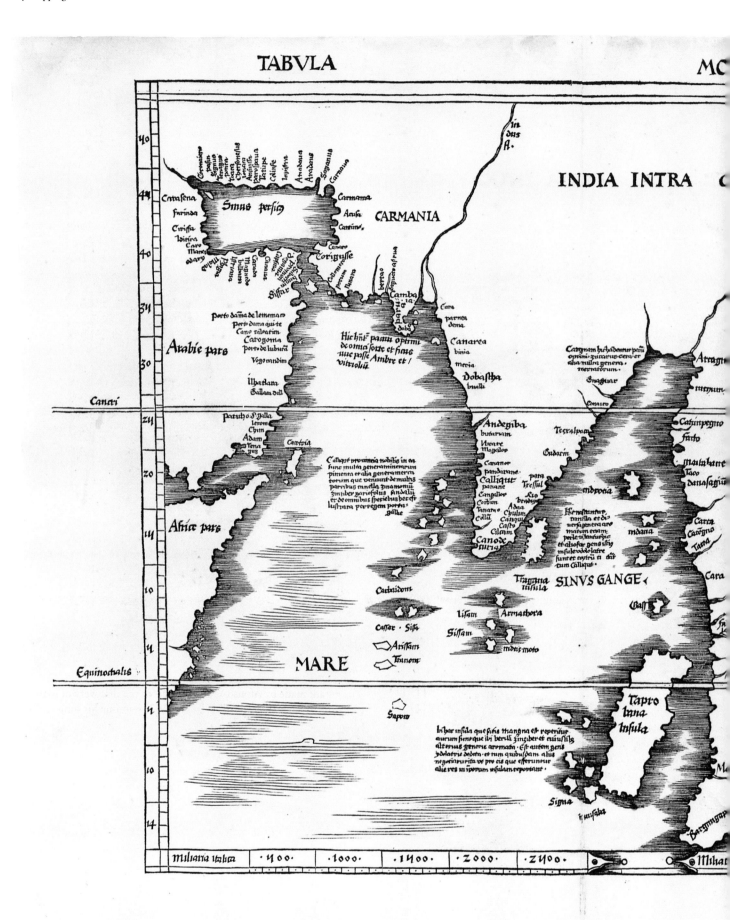

Fig, 57 Asia, Martin Waldseemüller, 1513. (40 x 51 cm)

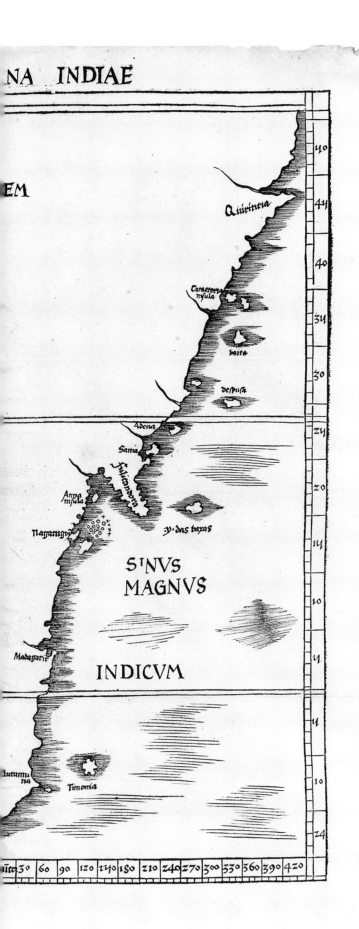

Nagaragoy is *Lugor* (Nakhon si Thammarat), in southern Thailand, to which the Caveri map of ca. 1504-05 attributes "rubies and other stones of great value," as well as other items.

Indochina

Indochina is rudimentarily depicted on the 1513 Waldseemüller and all Cantino-based maps as a small, sharp peninsula at the northeastern corner of the main peninsula, with the term *fulicandora* designating the peninsula or its cape (Waldseemüller's *fulucandoia*). The term *fulucandora* is probably derived from the *Condore* (Con Son) islands off the Mekong River Delta.[166] '*Fulu*' is a corruption of '*pulo*', the use of 'f' for 'p' being a common error due to the lack of the English consonants p, q, ch, and v in Arabic. The roots of *fulucandoia* can perhaps be better seen on the Ramusio map of 1554 (fig. 74), where a vestige of the Cantino Indochina has been carried over as an extra cape in Indochina marked *capo pulocanpola* — 'cape pulo Condore'.

Pulo Condore is also the origin of the islands that lie immediately off the 1513 map's Indochinese peninsula, called *y das Baixos* ('island of the baixos', *baixos* being a Portuguese word referring to shoals which appear at low tide). Waldseemüller's reference to these "islands of shoals" is extracted from an inscription accompanying the Condore Islands on the Cantino map; the inscription reads *ilha das baixos chamada fullu candora esta o norte em iij pulgadas,* 'island of the shoals called pulo condor is north three *pulgadas* [of the equator]', the Portuguese word '*pulgada*' *(polegada* = 'inch') used for the Arabic *isba'*.

The 'Seven Cities'

At that time, there were some who associated our Cantino/ Waldseemüller proto-Indochina with the fabled 'seven cities' of Christian legend. The story goes that seven bishops and their followers fled the Moorish invasion of the Iberian Peninsula in 711, and sailing west from Portugal reached a land in which each of the bishops founded a city. The exodus, Martin Behaim wrote on his globe of 1492, comprised one archbishop from Porto (Portugal), and "six other bishops, and other Christians, men and women, who had fled there from Spain, by ship, together with their cattle, belongings, and goods." Ferdinand Columbus, son of Christopher, wrote that their ships and rigging were subsequently destroyed so that the people would not think of returning to Spain.

Reference to these 'seven cities' is found as early as 1475, in a grant made by Afonso V of Portugal for the discovery of islands in the ocean sea; eleven years later, João II sent one his subjects "to discover a large island or mainland by the coast, which is supposed to be the Island of the Seven Cities." That mainland, of course, would have been envisioned as part of Asia, as indeed can be seen on a chart from a manuscript atlas of ca. 1510 attributed to Vesconte Maggiolo (fig. 58). At the tip of Maggiolo's Cantino-based Indochina peninsula lie the *zebez cita,* the 'seven cities' of the bishops.[167] As early as 1424, the island of Antilia appeared on Genoese portolan charts situated in the Atlantic, a location which came to be understood as the Caribbean and/or the South China Sea. In this respect, Maggiolo's decision to place the Seven Cities on Indochina was eminently logical.

The myth of the Seven Cities was transposed to the North American interior in 1536, when the Indians of the American Southwest told the Spanish about Cibola, a Zuni region, with seven large cities. Thus the legend of Antilia and the seven cities was transformed into the Seven Cities of Cibola, which were sought by Coronado on his march northward from Mexico into the American interior in 1540.

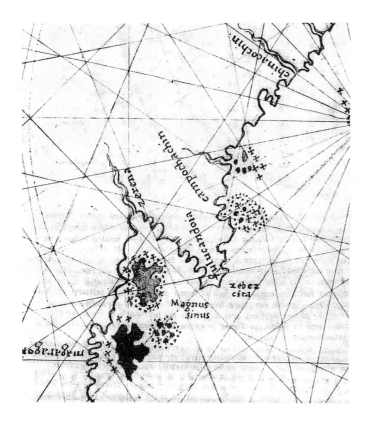

Fig. 58 The eastern coastline of the Malaya Peninsula, showing the mythical Seven Cities (*zebez cita*) in Indochina. A detail from an anonymous Portuguese manuscript atlas of ca. 1510, attributed to Vesconte Maggiolo. [The Hispanic Society of America, 1911, from a manuscript in the British Museum]

All ships bound from Batavia to Siam have instructions from the Company [V.O.C.] to put in, if possible, at Puli Timon for food and water, this island being very commodiously seated for this purpose, about half way from Batavia.

Kaempfer notes that the island "consists of scarce any thing but rocky precipices," though its inhabitants find patches of flat ground for their dwellings and gardens. It had an abundance of wood and water to supply visiting ships.

Pulau Tioman's importance continued through the eighteenth century. The standard English pilot guide to Southeast Asian seas of the eighteenth century, the third book of *The English Pilot* (fig. 59), gives detailed instructions for sailing from Tioman to Condore, and thence to Amoy (China). The pilot book describes Tioman as "the greatest of all the neighbouring islands, and so high that you seldom see the top by reason of the fogs and mists that lies upon it."

Waldseemüller's World Map from the 1513 *Geographia*
While Waldseemüller's map of Asia faithfully follows the Cantino/Caveri model, his world map from the same atlas (fig. 60) is a unique configuration which combines both the Martellus and Cantino subcontinents. Differences in nomenclature between the two maps suggest that the Asia map was intended to represent complete continent, with no additional subcontinent lying off the map to the east. The world map places Malacca and *Bargamgapara* (Singapore) on the 'Cantino' (western) peninsula, but transfers the remaining Southeast Asian nomenclature, as well as some Indian place names, to the Martellus (eastern) peninsula. Inconsistencies within the 1513 atlas can also be found between its world and America maps.

Pulau Tioman
Off the eastern shores of the peninsulas lies the island of *Timonia,* which is Tioman Island. Known to Arab sailors as *Tiyuma,* its prominence on the Waldseemüller and Cantino maps testifies to the island's importance in Southeast Asian commerce in the pre-European period. As with Malacca, Waldseemüller places Tioman too far south, but 'correctly' in relation to the map's misplaced Malaya and other associated features.

Why would the little island of Tioman figure so importantly on the map — and thus in the eyes of pilots whose sources were tapped? Now a paradisiacal tourist destination, Tioman was an important island in the history of Southeast Asia, being a key port-of-call on the route between the Indian Ocean and China. The island is mentioned in Arab geographies as early as ca. 850; they describe Tioman as having fresh water for anyone who desires it, as well as being the source of aloeswood of the sort called *Hindi,* and camphor.[168]

Pinto stopped at Tioman in 1555, where he was "exposed to many dangers from both the storms and the treacherous natives on shore." After three Portuguese *naos* arrived at Tioman, Pinto, followed the traditional route of generations of Arab and Asian seamen before him, and sailed from the island to Indochina. In the early seventeenth century, shortly after the creation of the Dutch East India Company, the Dutch sailor Willem Bontekoe had already indicated the importance of Tioman in the Company's affairs, noting that "from the island of Pole Timon to [Pulo Condore] the course is straight N.N.E. following the charts." Later in the same century, Tioman was described by William Dampier as "a place often touch'd at for wood, water, and other refreshments," though his own vessel passed it by without stopping. Engelbert Kaempfer, a German scientist who traveled to Southeast Asia in the employ of the Dutch East India Company in the latter seventeenth century provided an especially comprehensive account of Tioman. The island was a specified port-of-call for V.O.C. merchants bound for Ayuthaya:

Fig. 59 Pulo Tioman, from *The English Pilot*, 1703. (9.5 x 9 cm)

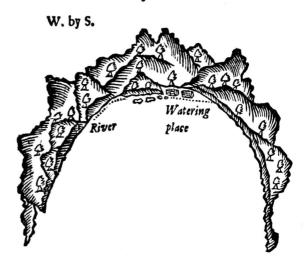

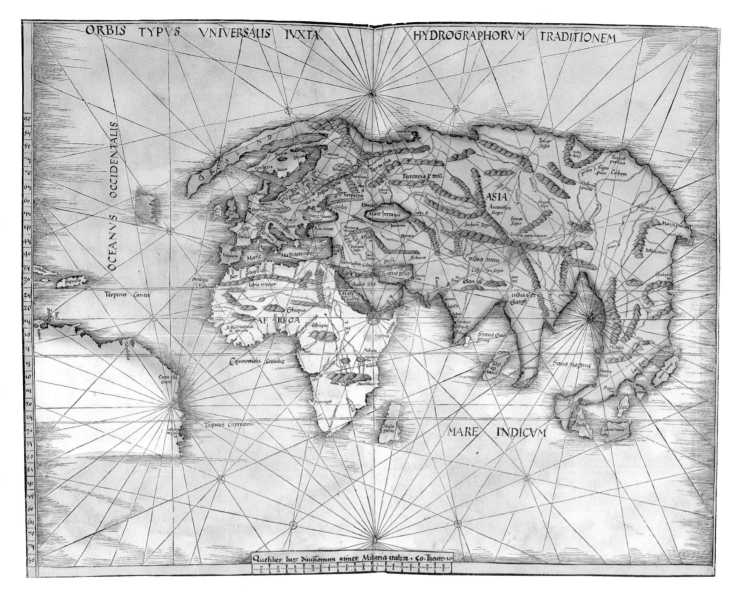

Fig.60 World map, Martin Waldseemüller, 1513. (44 x 57.5 cm)

The publishing history of the atlas suggests that the maps were made several years before the publication of the atlas in 1513, and there is empirical evidence for this in the case of the world map. A separate example of the map is extant which bears the name *America* inserted on the woodblock to denote the New World, apparently struck before Vespucci fell from Waldseemüller's grace shortly after 1507.[169]

Ludovico di Varthema and Waldseemüller's *Carta Marina*, 1516
In 1516, Waldseemüller constructed another variation of the Cantino model, a monumental map of the world on twelve sheets. This work added new islands to the Cantino prototype, some of which, such as the island of Timor placed just off the southeast coast of Malaya, appear to have been derived from unpublished sources. Tomes Pires described Timor in his *Suma Oriental* of 1515, and excerpts from it or a similar manuscript must have been available to Waldseemüller to enable him to map the island at this date. Other novel places on Waldseemüller's map appear to have been drawn from a newly-available text, the *Itinerario* of the Bolognese adventurer, Ludovico di Varthema. For example, just below the Southeast Asian subcontinent, there now lies the island of Banda (*Bandam*), which was described by Varthema as being "a

very flat and low country", about one hundred miles in circumference, and "very ugly and gloomy." He left a reasonable description of the island's nutmeg tree, which seems to have impressed him more than the people, of whom be spoke disapprovingly. Rather than cultivate the spice, the Bandanese "leave nature to do her own work," but once ripe "every man gathers as much as he can, for all are common."

Varthema then sailed twelve days to reach the Moluccas, which Waldseemüller mapped to the south of Banda as *Monocn* (Varthema's *Monoch*). The Moluccas, of course, lie to the north of Banda, but Varthema's account did not specify a direction, and the other references contained in his text logically pointed Waldseemüller southwards. The Bolognese traveler wrote that "the air is a little more cold" in the Moluccas than in Banda, suggesting that he had headed away from the equator into cooler climes. Varthema also noted that the North Star was not visible, which it should have been had he sailed to a more moderate latitude in the northern hemisphere. He described the islands' clove trees and observed that cloves were sold for twice the price of Banda's nutmeg, but were purchased by measure rather than weight. He thought even less of the Moluccans than he did of the Bandanese.

Continuing southwards, Waldseemüller maps a small island which he identifies as *Burney*. This is the same spelling that Tomé Pires used for the island of Borneo and appears here five years before the Magellan expedition actually reached the island in 1521. Borneo, in fact, lies due west of Banda and the Moluccas, and is vastly larger than either. But Varthema wrote that the vessel he traveled in from the Moluccas to *Bornei* sailed about two hundred miles "constantly to the southward," and indicated that it was a small island, "somewhat larger" than Molucca "and much lower". Hence Waldseemüller was following Varthema's words faithfully in his placement and sizing of of Borneo. Although this segment of Varthema's account is so flawed as to suggest that he had reached Buru rather than Borneo, he then states that he sailed five days southward from *Bornei* and reached Java, which is consistent with the identification of his *Bornei* as Borneo. The conflict is a major enigma surrounding his travels in Southeast Asia. Immediately south of *Burney* there is a large island corresponding to the relative position and approximate shape of Sumatra, but since Sumatra, following Cantino/Caverio, already lies due west of the peninsula in a vertical orientation, Waldseemüller has identified the new island as *Giava sev Iava Insula La Maxima* ('Java, or Java Major'), the italicized spellings perhaps following Varthema.

Lorenz Fries (1522)

The Cantino/Caverio and Waldseemüller models of Southeast Asia evolved even further in a new edition of Ptolemy's *Geographia* published by Lorenz Fries in 1522. "Lest we seem to claim the merits of others," Fries states in the preface to his atlas, "we declare that these maps were originally constructed by Martin Waldseemüller." Fries adds, however, three new maps not found in the 1513 atlas, one of the world (see page 96 and fig. 50), one of Southeast Asia, and one of China and Japan. With them, Fries offered his readers fully five disparate views of Southeast Asia under the same binding:

1) The traditional Ptolemaic rendering as found in the Ptolemaic world map and the regional Ptolemaic map.
2) A modern world map derived from the version found in the 1513 Waldseemüller *Geographia* (fig. 60).
3) An entirely new and modern world map, apparently of Fries' own invention (fig. 50).

Fig. 61 Detail of 'Lamai' from the map of Asia by Lorenz Fries, 1522. The inscription notes that silver and silks are transported from Lamai to Malacca.

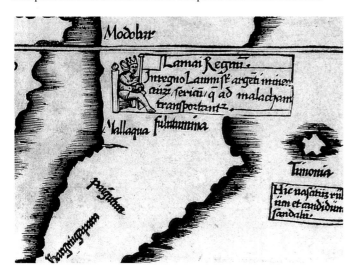

4) A modern map of Asia derived from the corresponding map in Wladseemüller's 1513 *Geographia* (fig.57).
5) A pair of maps, one devoted to Southeast Asia, the other to China and Japan (but including parts of Southeast Asia), based on the great cordiform world map composed by Waldseemüller in 1507, which was itself rooted in the Martellus model (figs. 62 & 63).

The Map of the Asian Continent
The modern map of Asia is geographically unchanged from the 1513 version as regards the delineation of the coasts, but is now embellished with annotations and vignettes. One of these, a generic image of a king lying above Malacca in the southern portion of Malaya, represents a kingdom not found on the 1513 prototype. It is said to be the kingdom of *Lamai* (fig. 61). This is a corruption of the *Lamia* of the 1507 Waldseemüller map (the last two letters being reversed), which we have identified as Lan Na (Chiang Mai). The only missing ingredient to clinch the identity of Fries' *Lamai* as Lan Na is the kingdom's namesake lake, which appeared on the Ruysch and Waldseemüller maps, but is lacking on the Fries version.

The Maps of Southeast Asia and China/Japan
Although the Fries map of the Asian continent does not have Lake Chiang Mai, two other maps in his atlas do. One of these is a separate map of Southeast Asia (fig. 62); the other is of China and Japan (fig. 63). Both show a 'Chiang Mai' (?) lake *(Lamia Lacq)* which feeds one of Southeast Asia's principal rivers; yet neither one shows a kingdom of that name, nor any of the other terms used to denote Lan Na.

The Fries map of Malaya, Indochina and the Indies is entitled *Tabula moderna Indiæ Orientalis* (*New Map of Eastern India*, fig. 17).[170] This is the first printed map of European origin devoted to Southeast Asia and its islands, and as such is a cartographical landmark, codifying the recognition of Southeast Asia as a distinct entity. Since its geography is taken from Waldseemüller's large 1507 cordiform world map (see fig. 45) — whose Southeast Asia, in turn, was descended from a common ancestor of the old Martellus model — the Southeast Asian subcontinent upon which Fries has mapped the various regions of Southeast Asia, and around which its various islands have been fitted, is simply the phantom vestige of the old Ptolemaic land bridge. The now-abandoned shores of Ptolemaic Malaya and Indochina are seen in the upper left corner.

Java and Bintan
The map's nomenclature is derived principally from the adventures of Marco Polo and, as in our previous discussion of the Ruysch 1507 world map, we will reconstruct Fries' reasoning by following Polo's return voyage from China to the Indian Ocean, making short diversions to include the travels of other early European visitors to Southeast Asia. The large, northeasternmost island, *Iava Maior,* is Java, mapped from textual references rather than any known cartographic prototype of the island. It is in roughly the position of Borneo, with which it sometimes became associated by later mapmakers. Fries, quoting Polo, states that its compass is 3,000 miles.

The dimensions of the small island of Bintan *(Peutam)* are more exaggerated than those of Java, but since Polo offered no estimate of the island's size, it was left to guesswork. Bintan's importance was probably growing around the time Polo passed by in the 1290s. It is recorded in the geographical treatise of the thirteen-century scholar Ibn Sa'id, composed about the time Polo set off for China, and a comparison of this segment of Polo's route with that described by Ibn Sa'id shows many similarities.

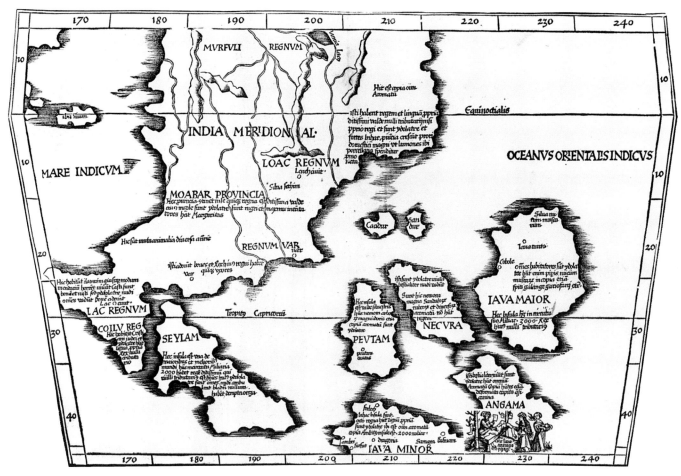

Fig. 62 Southeast Asia, Lorenz Fries, 1522. (27.6 x 42.6 cm) [Jonathan Potter]

After the Portuguese capture of Malacca in 1511, the son of the deposed sultan established a rebel base on the island of Bintan, from where he launched raids on shipping in the Malacca Straits. In 1526 the Portuguese seized Bintan, hoping to end these attacks and safeguard their hold on the peninsula. The sultan's supporters fled to Sumatra (Fries' *Iava Minor*), whose coastal regions were largely Moslem.

Like the sultan's followers, Polo went from Bintan to Sumatra, and so Fries, like Ruysch, correctly places Java Minor just below Bintan. "About 100 miles south-east of Bintan," Polo explains, "lies the island of Lesser Java. For all its name 'tis none so small but that it has a compass of two thousand miles or more."[171] Fries notes the 2,000 mile figure in an inscription on the island. Depending on the value of the mile Polo was quoting (which he himself doubtfully knew), this may be close to the true figure of about 2,300 miles circumference. Polo explains that Sumatra has eight kingdoms, each with its own king, each with its own language. He and Friar Odoric skirted Sumatra in opposite directions, since Polo was *en route* from China to Persia, while Odoric reached Sumatra on his outbound trip to China.

Sumatra

(Compare to the Waldseemüller prototype, fig 65). Fries was able to offer a more complete reconstruction of Polo's Sumatra than had Ruysch, because his scale was larger and allowed greater detail, despite the rougher woodblock medium, compared with Ruysch's copperplate world map. Following Polo's text as literally as possible, Fries places the Sumatran city of *Ferlec* opposite Bintan, since that is the kingdom Polo reached after his hundred mile voyage south

from Bintan. *Ferlec* would have been within a degree of the equator on the northern coast of Sumatra. "But let me premise one marvellous thing," the astute merchant informed his audience, "and that is the fact that this island lies so far to the south that the North Star, little or much, is never to be seen!" Polo distinguishes the townspeople of *Ferlec*, merchants who have converted to Islam, from the people dwelling in the mountains, who "live for all the world like beasts." These mountain people were reputedly cannibals, and were probably the Batak of north Sumatra. Friar Odoric mentions them as well.

From *Ferlec*, Polo and his company traveled northwest to *Basman (Basma),* which is "also an independent kingdom, and the people have a language of their own," *Basma* is believed to be the '*Pasei*' of the Malays. Unscrupulous merchants in *Basma,* Polo warns the reader, 'manufacture' human pygmies by fixing up a kind of monkey which is small and "has a face just like a man's," selling them to gullible customers. In *Basma,* Polo encountered 'unicorns', which were rhinocerii. For a medieval man raised with the lore of the pure and magical unicorn, the Sumatran rhino must have been quite a shocking encounter. "'Tis a passingly ugly beast to look upon," Marco says of the 'unicorn', "and is not in the least like that which our stories tell us of being caught in the lap of a virgin." Reflecting — one can assume facetiously — on the imagery of a rhino "in the lap of a virgin", Polo finishes by commenting that the beast, "in fact, 'tis altogether different from what we fancied."

Polo left *Basma* and next reached the kingdom of *Samara* (above the 'R' of Fries' *Iava Minor*), from which the name Sumatra was derived, and the same as Odoric's *Sumoltra* and Ibn Battuta's *Samudra* or *Sumutra*. Odoric placed the kingdom of Sumatra "towards the south" on the island, and Polo insinuates a southerly location as well by explaining that he could not see the pole star while there. Polo spent fully five months in *Samara*, "waiting for weather that would permit us to continue our voyage." He and his colleagues built a fortified camp there to protect themselves from "nasty and brutish folk who kill men for food," but was able to trade with the islanders "for victuals and the like; for there was a compact between us." They ate rice, since "the people have no wheat", enjoyed "the best fish in the world," and drank a wine made from a tree "like little date-palms" which was "a sovereign remedy for dropsy, consumption, and the spleen." Odoric remembers his *Simoltra* as a kingdom "where both men and women mark themselves with red-hot iron in twelve sundry spots of their faces."

Ibn Battuta's itinerary was more like Odoric's, than Polo's. He reached Sumatra from the northwest, either from Burma or the Andaman/Nicobar Islands. He described the island, which was visible when he was still half a day's journey from it, as verdant and fertile. The local people met his company in small boats, bringing coconuts, bananas, mangoes, and fish, for which each visiting merchant was by custom expected to pay according to his means. He referred to the port's village as *Sarhá*, which translated as 'houses' or official establishments. Tin and "unsmelted native Chinese gold" were used for exchange there.

Ibn Battuta believed the name of the actual island, *Jáwa,* to be the origin of the name of the incense called *jáwí.* He traveled by horse to "the sultan's capital, the town of Sumutra", which he describes as "a large and beautiful city encompassed by a wooden wall with wooden towers." When the sultan of *Sumutra* sent Battuta two slave girls and two men servants as gifts, the amír who presented them apologized: "the sultan says to tell you that this present is in proportion to his means, not to those of [the sultan of India]."

Now let us rejoin Marco Polo. Leaving the kingdom of Sumatra, Polo came next to *dragoria,* seen on Fries' map due west of *Samara.* In this kingdom, sick people whom sorcerers believe will not live are suffocated and their bodies are then cooked and consumed in total, lest their remains beget worms which then die for want of food and in turn visit "sin and torment" upon the dead person's soul — Ludovico di Varthema related a similar story about Java at the turn of sixteenth century. Reaching the southwest coast, Polo reached *fursut* (Fansur), which "produces the best camphor in the world." The tenth-century Arab geographer Mas'udi noted that *Fansur's* camphor "is only found there in large quantities in the years that have many storms and earthquakes."

We then come to the important kingdom of *Lambri*, the port at which Odoric reached Sumatra and from which Polo left Sumatra. Consistent with Fries' map, *Lambri* was indeed, in all likelihood, on the northern shores of Sumatra, in the region of Banda Aceh. Odoric, leaving *Moabar* (India), sailed "thence by the Ocean Sea fifty days' journey southward," where he "came unto a certain land named Lammori," which is Polo's *Lambri.*

Because of its location, *Lambri* seems to have been the traditional port of call for vessels traveling between Sumatra and all points west, and as a result, its name was at times used to denote the island itself. Arab geographies referred to Sumatra as the island of *Lamuri* at least as early as 'Aja'ib al-Hind (ca. 1000). During the Song Dynasty, Chinese vessels leaving *Zaiton* are known to have waited out the winter monsoon on Sumatra at *Lanli-poï,* which is probably what Polo's company did.

Polo tells us that *Lambri* "produces abundance of brazil, besides camphor, and other precious spices in profusion." He obtained seeds for brazil here and brought them back home with him, but

they yielded nothing "due to the cold climate" of Venice. When Odoric visited *Lambri*, he was concerned with the more friarly matters of the mores of the people (which he may have misunderstood). Odoric says that he was chastised by the local people for wearing clothes, which they considered a sacrilege, since "God made Adam and Eve naked."

John Mandeville also wrote about Sumatra by the name *Isle of Lamary.* He repackaged Odoric's and Polo's stories about its people, and dwelled on its southerly position and the existence of a southern pole star. Most importantly, Mandeville was excited about the proof it offered of the earth's sphericity. He explained that on the island of 'Lamary', as well as

> in many others beyond that, no man may see the Star Transmontane [north star], that is clept the Star of the Sea . . . But men see another star, the contrary to him, that is toward the south, that is clept Antarctic. And right as the ship-men take their advice here and govern them by the Lode-star, right so do ship-men beyond those parts by the star of the south, the which star appeareth not to us.

Nor, Mandeville notes, can the north star be seen by people in the south,

> for which cause men may well perceive, that the land and the sea be of round shape and form; for the part of the firmament sheweth in one country that sheweth not in another country.

The Nicobar and Andaman Islands

Once the seasonal monsoons had shifted, Polo and company departed Sumatra from *Lambri,* and continued their westward voyage to Persia. "When you leave the Island of Java (the less) and the kingdom of Lambri, you sail north about 150 miles, and then you come to two Islands, one of which is called *Necuveran.*" This is the Nicobar group, which lies just northwest of Sumatra's northern tip, but Fries lacked three bits of data: the accurate direction 'northwest' rather than Polo's 'north'; the diagonal orientation of Sumatra (NW-SE); and the fact that Malaya partitions the ocean above Sumatra. As a result, Fries maps his *Necura* directly above Sumatra, between Bintan and Java. Polo reported that the people of the Nicobar island he visited (probably Great Nicobar), "have no king nor chief, but live like beasts. And I tell you they all go naked, both men and women, and do not use the slightest covering of any kind." Despite this, however, the people "have very beautiful cloths or sashes some three fathoms in length, made of silk of every color, [which they buy] from passing traders and keep them hung over rails in their houses as a token of wealth and magnificence." The island was rich in valuable trees, such as the brazil, red sandal, and coconut.

Polo does not specify the direction he sailed between Nicobar and his next landfall, which was *Angamanain.* This is an island in the Andaman group, the *Angama* of Fries' map, and the *Agama* of Ruysch. As with the Nicobars, Polo referred to the Andamans as a single island rather than a group of islands. He probably skirted both archipelagoes along their east or west coasts, and from this perspective the Andamans could well have appeared to be a single island about the size of Hainan, rather than a series of very narrow, closely-spaced, islands. Conti, who also traveled to the Andamans, believed that the name meant 'Island of Gold'. Polo reported that Andaman was "a very large island, not governed by a king, [whose people are] a most brutish and savage race." Ninth century Arabic annals agree with Polo's unflattering description of the islanders as cruel and cannibalistic.

The 'Dog-headed' People of Southeast Asia

Many European and Arab authors described a race of people in Southeast Asia, either in Burma or the Andaman/Nicobar Islands, whose heads resemble those of dogs. Polo states that "all the men of [Andaman] have heads like dogs." Friar Odoric also reported such a land. Leaving Champa, Odoric "traveled on further by the Ocean-sea towards the south, and passed through many countries and islands." One of these islands, called *Moumoran,* is 2,000 miles in circumference and is inhabited by people with the faces of dogs. Similarly, Ibn Battuta reported "the country of the Barahnakár . . . [whose] men are shaped like ourselves, except that their mouths are shaped like those of dogs; this is not the case of their womenfolk, however, who are endowed with surpassing beauty." Ibn Battuta's *Barahnakár* was probably in the Cape Negrais region of Burma, or the Andaman-Nicobars.[172] He found Muslims from Bengal and Sumatra living there, and noted that the king exacted from all visiting ships a tribute of cloth (with which they cover their elephants on feast days), a slave girl, and a white slave. Ibn Battuta and his companions gave all of these, as well as presents of spices and cured fish, for "if anyone withholds this tribute, they put a spell on him which raises a storm on sea, so that he perishes or all but perishes." John Mandeville, as one would expect, also described such people. He places them on "an island that is clept Nacumera," the Nicobars, though he believed it to be more than a thousand miles in circumference. "And all the men and women of that isle," Mandeville states, "have hounds' heads, and they be clept Cynocephales." Both Odoric and Mandeville (probably copying Odoric) write that the dog-headed people are said to worship the ox; Ibn Battuta notes that they follow neither the religion of the Hindus, nor any other.

Some versions of Polo's text describe how Andaman "lies in a sea so turbulent and so deep that ships cannot anchor there or sail away from it, because it sweeps them into a gulf from which they can never escape," which is reminiscent of the stories associated with Ptolemy's ten *Maniolae* islands. Some mapmakers of the later sixteenth century, mapping the Andamans as such and no longer finding a rationale for Polo's *Agama* in the Indies, transposed them to the New World where they were identified as a region of the northwest coast of North America.

From the Andamans, Polo and the delegation sailed "about a thousand miles in a direction a little south of west" to *Seylam* (Ceylon). Since Fries envisioned Andaman to the *south* (rather than north) of Nicobar, he assumed that Polo reached and departed Andaman from its north coast. Polo would have then sailed between Bintan and Sumatra (since Fries mapped the Nicobars and Andamans to the east, rather than west), reaching Ceylon on its southern coast. All this confusion left Ceylon directly off the southeast coast of Malaya, in the same relative position which Ceylon incorrectly shares with India. As a result, Malaya 'became' India, with Fries planting various Indian nomenclature upon its soil.

Var

Going ashore on the southern peninsula, we find ourselves in the kingdoms of *Var, Lac* and *Coilu.* What are these realms that Fries envisioned occupying the southern end of Malaya? As with the mystery of *Lamai/Lamia* versus *Lama,* trying to identify them illustrates the confusion that can result when similar sounding place names are transliterated. The Vac (or Wa) are a people belonging to the Lawa family who were much feared as cannibals, eating human flesh to assist in the fertility of the land.[173] If indeed the Var were the Lawa, Fries, like Ruysch before him, misplaces them, for they inhabited inland mountains, not the coastal region.

Coilu

Occupying the southern tip of the peninsula is the realm of *Coilu.* Several old Chinese chronicles describe a place called *Ko-lo* which was known as early as the Han Dynasty and which likely lay in Malaya, as assigned by Fries. The *Hsin Tang Shu* (*Old Annals of the Tang Dynasty,* A.D. 618-906), written by Liu-Hsü in the first half of the tenth century, contains a passage about a place called *Ko-lo* whose "walls are of piled stone; the towers, the palace, and the houses are thatched with straw." The Tang annals inform us that the soldiers of *Ko-lo* fight with

> bows and arrows, lances, and spears," and "decorate their banners with peacock feathers. In war a hundred elephants constitute a company, with a hundred men to each elephant . . . [each elephant has] a saddle like a cage, four men inside."[174]

An itinerary compiled by one Chia-Tan in about 800 A.D. contains a section about the voyage from *Kuang* (Canton) to the "Barbarians of the Sea," by which he referred to peoples of Southeast Asia. It records that there is a strait called *Chih* by the 'barbarians', the southern shore of which is *Fo-shih* (Sumatra), and that "some four or five days' journey over the water to the eastward of *Fo-shih* is the kingdom of *Ho-ling* (Java) . . . [and] on the northern shore of the strait is the kingdom of *Ko-lo*." Thus *Ko-lo* was somewhere in Malaya.[175]

But despite the perfect way in which *Ko-lo* fits the placement of our *Coilu,* and despite how seductive it would be to ascribe ancient Chinese sources to Fries' southern Malaya, our *Coilu* is more likely to be a kingdom mentioned in Polo's text, called *Coylū* or *Coilum,* which is not part of Southeast Asia at all, but is in India, probably Quilon, on the Malabar Coast. As knowledgeable an authority as Peter Martyr, writing in his fifth *Decade,* confused the Indian kingdoms of Cochin, Cananor, and Calicut with the '*Chersonesus*' (Malaya), writing that Magellan had "passed seven years at Cochin, Cananor, at Calicut in the Chersonesus, otherwise called Malacca." As we shall see, *Coilu* is not the only region of India that Fries erroneously placed in Southeast Asia.

Lac

The next region, *Lac,* is also found in Polo's text. Polo located the province of *Lac* as bordering on Russia, lying near the Black Sea to the north-northwest. He described it as a cold region inhabited by tall, fair people with wonderful fur garments. But Fries' inscription about *Lac* describes the realm's inhabitants as being naked — completely contrary to Polo's description of his Province of *Lac.* Evidently, Fries did not intend to record Polo's *Lac.* His *Lac,* rather, appears to be a corruption of Polo's *Lar,* which is "a Province lying towards the west when you quit the place where the Body of St. Thomas lies," that is, west of the province of Maabar (Madras). *Lar* is also said to be the place from which Brahmins come. Fries' corruption of the name *Lar* as *Lac,* and its placement in a Southeast Asian peninsula, rather than in India, had a good pedigree; both errors are found on the maps of such illustrious geographers as Martellus and Waldseemüller. *Lac* was always judged to be a mainland kingdom except on a map supposedly prepared by Robert Thorne in 1527, which copied its Southeast Asia from a world map in the 1515 *Margarita Philosophica* of Gregor Reisch, but made southern Malaya, which was *Lac,* into an island.[176]

Many mapmakers probably associated the kingdom *Lac* with the substance lac, from which lacquer was made, even if taking the name from Polo's text. There is specific mention of lacquerware as a product of *Teng-Liu-Mei,* a kingdom lying to the west of Cambodia

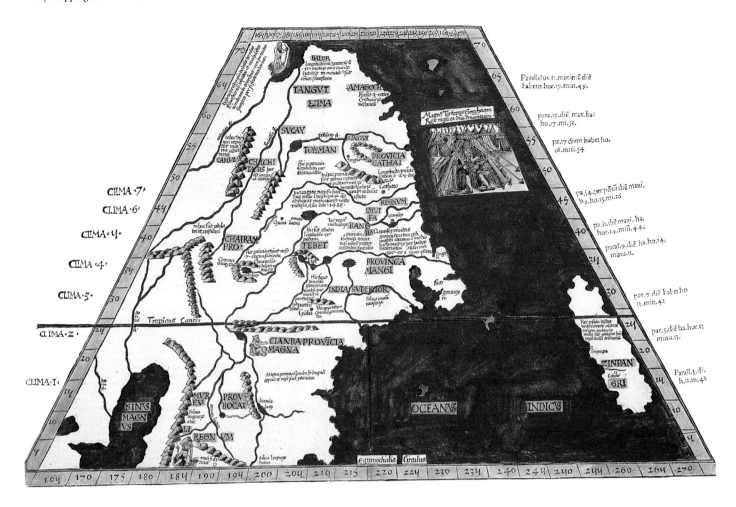

Fig 63 China and Japan, Lorenz Fries, 1522. (28.8 x 45.2 cm) [Roderick M. Barron]

(probably the northern part of the Malay Peninsula), in a Chinese chronicle written in 1226, about half a century before Polo's trip.[177] Burmese lac is mentioned by Ludovico di Varthema in 1505, in early sixteenth century Portuguese reports, as well as by the English chronicler Hakluyt in the latter part of the century. The Burmese lac was a coloring matter exuded by an insect and found as a resinous incrustation on tree branches. It was an important trade item, along with gems and silver, which was purchased by merchants from Gujarat at the entrepôts of Martaban and Rangoon. Indochinese lac, however, was the resin of the lacquer tree, or *son tru*. William Dampier spent about half a year in Tongkin (which coincidentally has a 'Lac Dragon Lord' in its history) in 1688-89 and commented on Tongkinese lacquerware at length. According to Dampier, the white "gummy juice" from trees was brought to Hanoi (*Cachao*) in vats every day by people living in the country. Though not unpleasant smelling, those who worked with the substance suffered, Dampier reported, from the poisonous qualities of its fumes.

Maabar, Mutfili and Lochac

Continuing north on Fries' map, up the peninsula from the province of *Lac,* we reach the province of *Moabar* which, lest his readers assume otherwise, Polo tells us is "not an island, but a part of the continent of the Greater India." This is Maabar, which is a region of the Coromandel Coast of India (it is not 'Malabar', as it might seem). Along with transposing some of India's kingdoms to Malaya, this error shifted some of India's spiritual and material allure to Southeast Asia as well, for it is in *Moabar,* the Venetian tells us, that the body of St. Thomas lies, and it is from *Moabar* that

pearls of monumental size are obtained. Further north on our confused Southeast Asian peninsula, crossing north of the equator, Fries has marked the kingdom of *Murfuli* (or *Mutfili*), yet another part of Polo's India mistakenly placed in 'Malaya'. *Murfuli* is, however, correctly placed in relation to Maabar; Polo explained that one reaches this kingdom "when you leave Maabar and go about 1,000 miles in a northerly direction."[178] *Murfuli* added yet more glamour to the Southeast Asian landscape, since from this region, we learn from Polo, huge diamonds are found in the rivers that are washed down from the mountains during torrential rains. Northeast of Maabar is Polo's *Loac Regnum,* which is probably Thailand, but may be Borneo.

The Fries Map of China and Japan

The other new map of eastern Asia which Fries added to Waldseemüller's atlas is entitled *Tabula Superioris Indiæ et Tartariæ Majoris* (*Map of Upper India and Greater Tartary*). Like the previous work, it is based either on the 1507 Waldseemüller cordiform map of the world, or on related regional maps by Waldseemüller which have not survived

On this occasion, Fries depicts all of China, Tartary, Japan, and Southeast Asia as lying north of the equator. In truth, all of Southeast Asia lies above the equator except for most of the Indonesian Archipelago, but the canvas upon which Fries was working — one that he had inherited from Martellus and vestiges of the Great Promontory model — pushed all the lands of Southeast Asia reported by Marco Polo, except for Champa (Vietnam), into to the southern hemisphere.

'The Great Province of Champa' *(Cianba Provicia Magna)* is represented by the seated figure of its king, marveled at by Europeans for his many wives, children, and elephants, and his forests' wealth of ebony and aloe. Below Champa is the province of *Bocat,* which, as we saw on the Ruysch map, is Cambodia *(boja* is 'country' in Pali, *Bocat = Cam*bocat*i*).[179]

Just east of Cambodia is Lake Chiang Mai *(Lamia Lacq),* the fabled source of Southeast Asia's major rivers and auspicious site for the founding of Chiang Mai. To the west lies some of Polo's Indian nomenclature, and the Ptolemaic 'Great Gulf'. In the lower left-hand corner is the original, 'correct' Malaya/Indochina of Ptolemy, abandoned as a home for Southeast Asian nomenclature by map-makers in favor of the new peninsula concocted from the vestige of Ptolemy's Africa-Asia land bridge.

This map shows the island of Japan *(Zinpangri)* occupying the position of the Philippines and waters north, and follows the 1507 Waldseemüller map by placing the cities of *Sinpaugen* and *Cobebe* in Japan. The first of these cities, *Sinpaugen,* is an arbitrary capital city for the island (a variation of *Zinpangri),* just as Fries placed a capital namesake city on the island in Bintan *(peutam civitas),* though no such metropolis is mentioned by Polo. The second city in Japan, *Cobebe,* appears to be a duplication of the kingdom of *Cobole* on Fries' *Java Major.*[180] Some cosmographies, such as that of Peter Apianus (1524), do note a *Cobale civitas* lying within the coordinates assigned Java, and a *Coloba civitatas* within those of Japan.

Fries' are the first printed atlas maps in which we find 'quaint' vignettes fancifully depicting Southeast Asian native curiosities. Such images were largely generic figures carried over from earlier *mappaemundi,* and Fries' immediate inspiration was probably Waldseemüller's monumental *Carta Marina* of 1516. But correspondence with his publisher, Grüninger, reveals that such naive images were hardly without controversy in their day. Albrecht Dürer, we are told, saw such illustrations and laughed in ridicule.[181] Some complained that this sort of material was for children and the uneducated, and likened the illustrations to playing-card pictures. Grüninger, protesting the association with playing-cards, insisted that he knew his audience's tastes. Many manuscript charts also indulged in such images, eliciting criticism from the Portuguese cosmographer Pedro Nunes. Nunes complained that such charts, though lacking all-important coordinates, "have much gold and many flags, elephants and camels on them".[182] Yet these flourishes could offer legitimate information that supplemented the purely geographical data; usually they illustrated ethnological details even if erroneous or misunderstood, gleaned from some of the same reports tapped for geographic data.

Fries' atlas was restruck, without modification to the maps themselves, in 1525, 1535, and 1541. The second issue was printed by Grüninger for the publisher Hans Koberger, and the latter two editions were published by Servetus, who was later executed by the Inquisition for 'heresy'.

Benedetto Bordone's Island Maps (1528)

The first European cartographer to publish separate maps of the Southeast Asian islands was Benedetto Bordone. Bordone, an illuminator, miniaturist and wood-engraver from Padua who had established himself in Venice by 1494, devoted himself to geography after becoming disillusioned with his earlier profession, astrology. His *Isolario,* published in 1528, was inspired by the manuscript *isolarios* used by medieval sailors in Mediterranean waters, and followed in a tradition that was well-established in Italy. Bordone, however, expanded his coverage to islands of newly discovered parts of the earth, including five maps of Southeast Asian islands. The orientation of these *Isolario* maps varies; each contains an arrow to represent north, and a cross to indicate east.

Bordone's vision for his *Isolario* could well date from as early as 1508, when he is first known to have published maps. A manuscript copy of the *Isolario* is extant which is thought to date from before

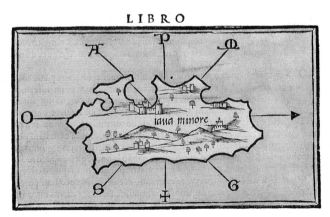

Fig. 64 *Java Minor* (Sumatra), Benedetto Bordone, 1528. (8.4 x 14.6 cm)

Fig. 65 *(left)* Sumatra, from the world map of Martin Waldseemüller, 1507.

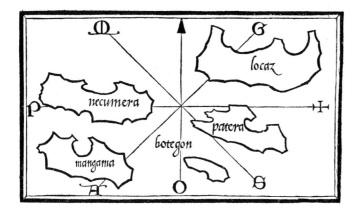

Fig: 66 Various Southeast Asian islands, including *locaz*, Benedetto Bordone, 1528. With this map Bordone helped plant the notion that *locaz* was an island, rather than a part of the Southeast Asian mainland. (8.4 x 14.5 cm)

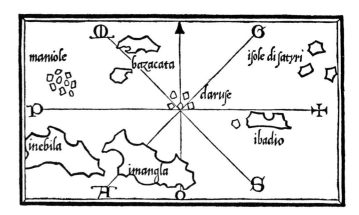

Fig. 67 Benedetto Bordone, 1528. The Southeast Asian 'male' and 'female' islands of *inebila* and *imangla*, along with islands from Ptolemy. (8.6 x 14.8 cm)

1524, and indeed the *Isolario's* maps of Southeast Asia offer nothing to reflect the discoveries of the previous two decades. Even the world map, which employs an oval projection, is reminiscent of an apocryphal map "in the round form of a ball" which Bordone apparently published in 1508.[183] But the permission to publish the book dates from 1526, and a few maps in the work (particularly one of Mexico City which was partly based on the second Cortés letter of 1524) clearly date the finished work close to time of publication.

Bordone's small maps are of very limited cartographic value, and he can be faulted for various errors (such as the duplication of Madagascar on the world map and the reversing of the Indus and Ganges Rivers). Nonetheless, his geography shows a degree of individuality within his Martellus framework, and the very fact that he devoted separate maps to the islands represents a symbolic step in Southeast Asian cartography.

Java, Pulo Condore, and Sumatra

The first of Bordone's maps of Southeast Asia depicts the island of Java *(iava maggiore)*, as well as two islands described by Polo as being uninhabited, *sondur* and *condur* (Pulo Condore). Correlation with Bordone's world map, which identifies places by key, shows that his *iava maggiore,* rather than playing its proper role as Java, is far off to the east and north relative to his Malaya, suggesting that he, like some of his contemporaries, either believed it to be Borneo or that it was not duplicated by the nomenclature of later discoveries.

Since Bordone's Malaya is patterned after Martellus, extending south to just past the Tropic of Capricorn, all his Southeast Asia islands are correspondingly offset. Thus when we note that 'Java' is placed to the north in the approximate position of Borneo, that is a relative statement; it is in reality still a long way south of the correct position of either island.

The following map is devoted solely to *iava minore* (Sumatra) (fig. 61), which Bordone orients on a north-south axis, probably following the Cantino model (reflected in the map of Waldseemüller, figs. 57 & 65).

Lochac, the 'Island' of Thailand

The subjects of his next map (fig. 66) are the islands of *necumera, mangama, botegon, patera,* and *locaz,* the Polian *Lochac* (Thailand or Borneo). Whereas Bordone's predecessors (such as Ruysch and Fries) heeded Marco Polo's statement that *Lochac* lay on the *terra firma* (mainland) and simply took a liberty when deciding what direction

'beyond' Champa and Condore Polo traveled to reach it, Bordone rendered it as an island (the *Isolario* makes no mention of *Lochac* being a *terra firma),* and positioned it to the south (as can be seen from his world map). By doing so, he initiated *Lochac's* cartographic journey to an austral location in the coming years at the hands of Gerard Mercator.

Bordone may have been influenced by a textual precedent for an insular *Lochac.* This can be found nine years before the *isolario* was published, in a Spanish cosmographical text, the *Suma de Geographia* of Martín Fernández de Enciso. The latter describes a Southeast Asian land named *jocat,* which the author states is an island, and which he claimed to be the Biblical Ophir. His *jocat* is probably a cross between Polo's *Lochac* and *Bocat* (Cambodia); the association of *jocat* with gold and elephants probably comes from Polo's *Lochac,* but the spelling and the identification with Ophir is more typical of Cambodia.

Dondin

Bordone's next map is of *scilam* (i.e., Ceylon, but here Sumatra) and *dondina,* which lies to the south. *Dondin* island was reported by Friar Odoric, and may be related to Ruysch's Southeast Asian island of *Candy* (see page 109, above). Bordone's orientation of the islands, however, suggests that he was following Mandeville rather than Odoric. From Ceylon, "going by sea toward the south," Mandeville wrote in his *Travels,* "is another great isle that is clept Dondun." Mandeville's description of *Dondun,* where the dead are consumed so that worms will not eat their bodies and then torment the deceased's soul, is reminiscent of Polo's description of the island of *Dagroian.*

The Male and Female Islands

The final map of Southeast Asia (fig. 67) covers the familiar Ptolemaic islands of *maniole, bazacata,* and *isole di satyri,* as well as *imangla* and *inebila,* two curious islands new to the repertoire of Southeast Asian lands found on printed maps. Bordone, repeating popular lore, explains in the *Isolario* that the island of *imangla* is inhabited exclusively by women, while *inebila* is an island on which only men live. If a woman of *imangla* gives birth to a girl, they keep the child, but if a boy is born, he is sent to *inebila* at the age of three.

The name of the island of women, *imangla,* is believed to derive from the Sanskrit word *mangala,* meaning 'fortunate'; that of men, *inebila,* from the Arabic *nabílah,* meaning 'beautiful'. The roots of these islands can be traced to the folklore of various peoples. As we have seen (page 55, above), early Arab travel narratives describe such a place in Southeast Asia. The Chinese official Chou Ch'ü-fei recorded them in 1178, perhaps repeating what Arab merchants trading at Kuang-hsi had told him. The islands are found in early Malay, Thai and Cambodian legend, as well as classical European lore about Asia. The meeting of a group of Scythian men and Amazonian women is described by Herodotus in the fifth century B.C., and in the first century A.D., Pomponius Mela repeated a story of an Isle of Women who live entirely without men. Other traditions cite an isolated island of women in the seas to the south of China; Arab and other traditions relate that these women conceive a child by the wind (a story which was repeated by Pigafetta), or by eating a particular fruit, or by plunging into the sea, or by looking at the image of their own face. This last version, a Chinese tradition which had been told to Magellan, may have inspired the narcissistic mermaids in the 1570 map of Southeast Asia by Ortelius (fig. 86). Malay tradition identifies the island of Enggano, south of Sumatra, as the island of women (see, e.g., the Linschoten map of 1595, fig. 92).

Various mapmakers placed imangla and inebila in different parts of Southeast Asia. In 1320, Marino Sanudo placed the island of *nebile* (i.e., *nabílah*) off the southeastern coast of Asia (see detail, fig. 44); Fra Mauro, on his world map of 1459, placed them (as *Mangla* and *Nebila*) south of Zanzibar. Bordone, like Sanudo, deemed them to be Southeast Asian isles; he explains in his text that the pair lies 120 miles northwest of *scilam* and *dondina,* which lie just southeast of Bordone's exaggerated Malaya, thus placing *imangla* and *inebila* very roughly in the relative position of Singapore.

Both Polo and Conti mention two neighboring islands, one peopled by men, the other by women. According to Polo, the men set out for the island of women every March, staying there for three months. Conti believed that sometimes the men visit the women, and sometimes the women visit the men, but that if either stays on the others' island for more than six months, the visitor dies.

Columbus believed he had discovered an island of women in the 'China Sea', and this is found on early maps as *Matinina* (presumably Martinique); Bordone includes a map of that island in the *Isolario* as well. At the end of the seventeenth century, the Venetian cartographer Coronelli was still grappling with the identity and whereabouts of *imangla* and *inebila,* and an Amazonian island can be found in European maps of the Philippines as late as the mid-eighteenth century.[186]

Shortly after the publication of Bordone's *Isolario*, the poet Camões created an Asian island from these traditions. In his *Lusiads,* the goddess Venus, as a respite for the heroism and sufferings of valiant Portuguese sailors, prepared an island "in the midst of the waves" of the Indian Ocean. On this island, "amidst an endless plenty of food and drink . . . loving, love-struck nymphs attend these mariners, in a mood to make them free of all their eyes may covet."

Sebastian Münster (1532 and 1538)

Sebastian Münster was born the year that Bartolomeu Dias reached the Cape of Good Hope (1488), and was in his teens when the adventurous Bolognese, Ludovico di Varthema, traveled through Southeast Asia in 1505-06. Thus he grew up in a world already acclimated to a European presence in Southeast Asia and to the expectation that newer and better knowledge of the world would be continually at hand. His maps are, in general, poorly executed, and his texts represent an uncritical acceptance of a wide range of sources, new and old, good and bad. Nonetheless, some of his maps record revolutionary new data not previously available to the European public.

The 'Holbein' Map (1532)
The earliest map associated with Münster which shows Southeast Asia, is the world map from the Basle edition of Huttich's *Novus orbis* (1532 and subsequent; detail, fig.68).[187] Hans Holbein is believed to have been responsible for the elaborate pictorial border — the ancestor of the *cartes à figures.* The map's alignment of place names varies slightly between examples of the map, because its nomenclature was printed with metal type inserted into a second block, which was printed as a separate strike, the register of which was not precisely controlled.[188]

The map's depiction of Southeast Asia is geographically little-inspired, combining the old Martellus peninsular subcontinent with the islands of Sumatra (as *Zeila,* or Ceylon), and Java, the latter offset far to the east. The Southeast Asian subcontinent is the site of 'India' on the north, Champa on the northeast, and *Sinar* (China) to the south of the equator. *Murfuli* (part of Polo's India) lies below *Sinaru,* and Malacca occupies the bottom on the peninsula, below the Tropic of Capricorn.

Pegu (Burma)
The Münster-Holbein map does, however, boast one new Southeast Asian land: *Regnum Pego,* the Pegu kingdom of what is now Burma, mapped just above Malacca on the subcontinent. Pegu was a Mon kingdom, probably founded in the early ninth century, pivoted along the delta of the Sittoung River. Conti visited Pegu in the early fifteenth century, and the Genoese trader Girolamo da Santo Stefano reached there in 1496. The actual Burmese term for their kingdom was *Bagó;* the European 'Pegu' came from the Malay term *Paigu.*[189]

Use of 'Pegu' on a printed map dates back as early as the 1516 *Carta Marina* of Waldseemüller, which denoted the central bulk of his Cantino-based Malaya as '*Pego Regnum*'; the term is also seen in the Homem-Reinels map from the Miller atlas of ca. 1519.

The poet Camões associated Pegu with the medieval myths of a dog-headed race, writing that Pegu "was once peopled with the monstrous breed of its solitary first inhabitants, a woman and a dog." Pinto, writing to the fathers of Portugal in 1554, offers the opinion that Pegu is equivalent to the Rome of its area. In the mid-1580s Ralph Fitch, an English merchant, related that Pegu "is a city very great, strong and fair, with walls of stone, and great ditches round about it." The streets are "the fairest that ever I saw, straight as a line from one gate to another, and so broad that ten or twelve men may ride affront through them." Fitch describes both an old and a new town of Pegu. Native and foreign merchants stayed in the old town, where all business was transacted. The suburbs of Pegu had houses made "of canes which they call bamboos," and are covered with straw. Goods were stored in a brick warehouse within the house to safeguard them in the event of fire.

The lower right quarter of the border depicts Ludovico di Varthema, who traveled to the Middle East, India, Burma, Malaya, Siam, and Indonesia between the years 1502-1508. The classical columns in the corner of the map hint at the belief, founded in the *Romance of Alexander,* that Alexander the Great reached (and claimed) the southeastern corner of Asia two millennia earlier.

The Münster-Solinus Map of Asia (1538)
Münster was the author of two maps specifically devoted to the Asian continent, both unrelated to the core of the classical texts they accompanied. The first (fig. 49) appeared in a 1538 edition of the *Polyhistory* of Solinus, a popular travel lore of the third century A.D. which helped perpetuate myths about the East; the second appeared in Münster's rendering of Ptolemy's *Geographia,* and in his *Cosmographia* (see page 127 & fig. 72, below).

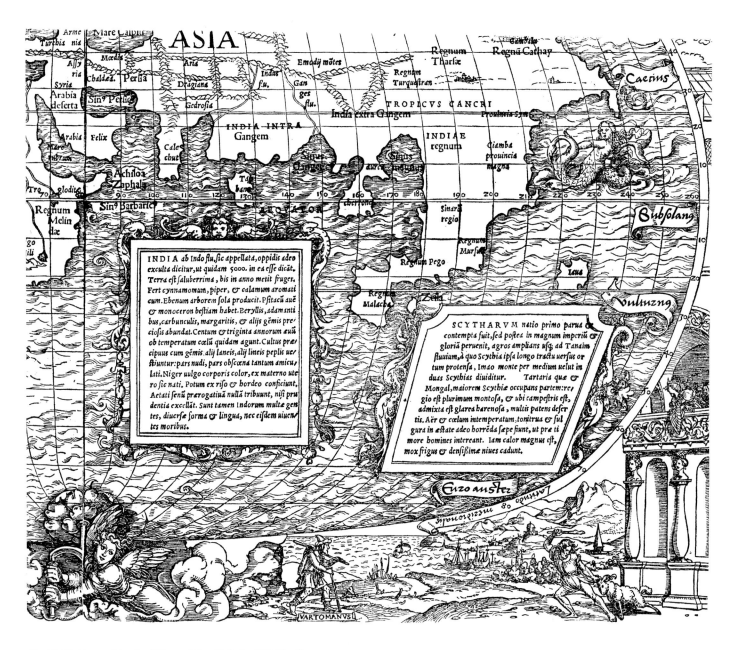

Fig. 68 World map, Sebastian Münster / Hans Holbein, 1532 (detail).
The figure depicted walking along the bottom of the map is the Bolognese traveler Ludovico di Varthema. (35.5. x 54.5 cm)

The Modern Name 'Sumatra'

On the 1538 map from the *Polyhistory,* we find the 'modern' name Sumatra used for the first time on a published map — unfortunately, however, it is mistakenly applied to Ceylon. This awkward christening of the name was an ironic product of the confusion between *Taprobana* and Sumatra, and the error had already appeared on the manuscript map by Bartolomeo Columbus and Alessandro Zorzi (1506-22, fig.51). The name is derived from the Sumatran kingdom known to Polo as *Samara,* to Friar Odoric as *Sumoltra,* and to Ibn Battuta as *Samudra.* It was used to designate the island in Arab navigational texts of the fifteenth century (where it is recorded as *Shumutra*). The precedent to use the name 'Sumatra' to denote the entire island seems to have been introduced to Europe by Nicolò de' Conti, whose report of his travels was written down by Poggio Bracciolini in about 1447. Girolamo da Santo, a Genoese merchant who crossed the strait from southern Malaya to Sumatra in 1497, also used the term 'Sumatra'.

The *Polyhistory* map is similar to the Asia section of the Münster-Holbein world map of six years previous (fig. 68), but lacks even that map's rudimentary representation of Java and Sumatra (*Zeila*). Its Ptolemaic (western) peninsula is identified as the fabled golden peninsula, *Aurea Chersonesus,* while the eastern promontory, which is a Martellus subcontinent, is the site of Malacca.

Mergui (Tenasserim)

This vestigial remnant of the old Ptolemaic land bridge does, however, contain a new and important location on its fat trunk: *Regnum Mursuli,* or Mergui (Tenasserim). Mergui was an ancient region in peninsular Burma on the Bay of Bengal, which Conti had visited in the 1420s. Conti's narrative relates that, after departing "the Island of *Taprobana,* and sailing fifteen days, he arrived by tempest of weather, entering of a river called *Tenaserin,* and in this region there be many Elephants, and there groweth much Brazil."

A far more detailed report about *Mergui* came from Varthema in the first decade of the sixteenth century. Varthema spoke of its silk making, its cockfights, its brazil-wood and benzoin,[190] and gave a good description of it as approached from the sea:

> The city of Tarnassari is situated near to the sea: it is a level place and well watered, and has a good port, that is, a river on the side towards the north . . . the houses of this city are well surrounded by walls.

One episode described by Varthema may be a clue to earlier knowledge of the region, discernable amidst the fourteenth-century travel lore of John Mandeville. Varthema relates how a local merchant, after learning that Varthema and his companion were strangers who had only been in Tenasserim for four days, requested that one of them 'deflower' his fifteen-year-old bride. Though the culture-shock made the Christian Varthema "quite ashamed at hearing such a thing," the merchant was quite insistent on the favor. He lodged the foreigners in his home for fifteen days until the young woman arrived and Varthema's companion "did for the merchant all that he had asked of him." This, indeed, was a tradition which persisted in the region of the Bay of Bengal for quite some time.[191] Mandeville had written of an Asian country with this custom, perhaps repeating the word-of-mouth reports of earlier visitors to the region. Describing "diverse Isles that be about in the Lordship of Prester John" (a mythical Christian king who ruled over the 'three Indias'), Mandeville wrote of a land

> where the custom is such, that the first night that they be married, they make another man to lie by their wives for to have their maidenhead: and therefore they take great hire and great thank. And there be certain men in every town that serve of none other thing . . . for they of the country hold it so . . . perilous for to have the maidenhead of a woman [that it can cause death].

Soon after Varthema's visit, Mergui became an important point for European traders. From his perch in Goa, Linschoten referred to Mergui by its modern name, Tanassarien, noted that the Portuguese "have great traffic unto this town," and that it receives a great deal of merchandise from Pegu and Siam. The kingdom lay on a stretch of coast which "runneth inwards like a bow", between two islands. He spoke of the wine which came from Tenasserim, which was shipped to Goa and throughout India in "Martaban pots". The Indians considered the Tenasserim wine superior to their own, and Indian women "are very desirous thereof", making good cheer and gossip with it when out of view from men.

Well-organized commerce was described by Jacques de Bourges, who reached Mergui in 1662:

> the customs officers established there to oversee the interests of the King of Siam came about our vessel, and listed the persons and goods on board, after which they sent an extract of all they had found to the governor and officials of Tenasserim.

Shortly afterwards, Gervaise described *Mergui* as "one of the most beautiful and safest [ports] anywhere in the Indies."

Barros' 'Hand Map' of Southeast Asia

The Portuguese authorities were even more reluctant to allow maps and descriptions of the Indies to be published than their Spanish counterparts. Ramusio, for example, noted in the introduction to his *I Navigazioni et Viaggi* that the account of Tomé Pires was not allowed in Portugal. But even Spain, which tolerated the publication of maps of America in 1511 and 1545, did not publish a significant map of Asia in the sixteenth century.[192]

Just as indigenous mapping in Southeast Asia stretched common European notions of just what constituted a map, so did the constraints of governments and the need for expediency to expand the cartographic medium in Europe. João de Barros (ca. 1496-1570), a Lisbon official privy to the spice trade and colonial affairs, wrote a history of his country's Asian endeavors, the *Décadas da Ásia*, which appeared in five volumes beginning in 1552. Although no maps were included until an expanded, posthumous version was published in Madrid in 1615, Barros offered his original 1552 Portuguese audience a map which was neither politically offensive nor required engraving, the skills for which were uncommon in Portugal.

Barros' answer was to form the human hand into a map of Southeast Asia. He instructed his readers to point their left hand toward their body, palm down, index finger straightened and separated from the thumb, with the remaining three fingers also separated from the index finger, these three curled so that they extend out only to the knuckle. The thumb represented India, the index finger Malaya, and the flesh between them Burma. The three curled fingers were Indochina, with their natural shape approximating that land's true southwest-northeast. contour. Barros' 'hand map', which even used features of the hand's landscape to locate interior kingdoms and topographical features, may well have been the most accurate representation of Southeast Asia available to the literate Portuguese public. It was, in fact, more accurate than some printed maps still in circulation. In his text, Barros refers to the reader's left hand to locate Siam, Chiang Mai, three *Laotian* kingdoms, Champa, Cambodia Malacca, the Burmese kingdoms of Arakan, Ava, and Pegu, the Mekong and Chao Phraya Rivers, and various mountain ranges.

The reader may find pointing the hand toward the body with the palm down to be awkward becuase it requires a sharp clockwise bend of the elbow and writst. In this case the hand cand be oriented with south to the top. Keep the index finger and thumb straight, but bend the three fingers on left (east). In this instance, the thumb is India with the wide space to its left being the Gulf of Bengal and the flesh of the hand making up the various kingdoms of Burma. The index finger is the Malay Peninsula, with singerpore at its tip. the space between the index finger and the three curled fingers to the left is the Gulf of Siam, and the crevasse formed where these fingers meet the hand is the Chao Phraya River. The knuckle of the next (middle finger) is Cambodia, the line formed between it and the next finger to the left is the Mekong River. The curving coastline of the remaining two fingers represents Vietnam. far to the north, at the joint where th hand meets the wrist, lies Lake Chinag Mai, the largelae which Barros believed to be the source of six of Southeast Asia's major rivers.

Chapter 10

First Maps from the Spanish Voyages 1525 – 1540

Inasmuch as I have been assured by those who have actually been there, that the Molucca Islands are rich in spices — the chief article sought by the said fleet — I order you, the said Fernando de Magellaes, to pursue a direct course to the above-mentioned islands, exactly as I have commanded you.

— letter from the King of Spain to Magellan, dated Barcelona, April 19th, 1519.[193]

The new crop of maps we have examined thus far placed some new nomenclature on archaic geography. New lands and new kingdoms were marked on existing delineations of Southeast Asia as best as they could be accommodated. But other maps were beginning to appear which were based on first-hand surveys and coastal charts, analyzing new reports to actually redraw the coasts themselves. Manuscript maps took the lead with this more difficult task to an even greater degree than they did with nomenclature.

The Magellan Voyage

The maiden circumnavigation of the world by the Magellan expedition was one of the most incredible of human achievements. Although only thirty-five people survived from the original crew of 280, it is astounding that the *Victoria* completed the voyage at all. Unlike most 'firsts' in the history of exploration, this is not simply the first *recorded* circumnavigation; the chance of there having been an earlier trip around the globe is virtually nil. While the honor of the first circumnavigation of the earth might best be shared by many members of Magellan's fleet, the question of exactly who was the first person to circle the earth is open to debate. If, as some historians speculate, Magellan had previously reached the Moluccas or Philippines from Malacca in 1512, he would have effectively circled the earth upon his landfall in the Philippines in 1522. If not, then Enrique, his Malay slave, would have been the first person to circumnavigate the globe, though only if he himself was a native of the Moluccas, which lie to the east of the Philippine island of Mactan, where we last hear of him. If, however, Enrique came from Sumatra, then technically the honor would fall to one or more of the seamen, depending on how far east each had sailed on previous voyages.

Antonio Pigafetta

Fortunately for us, one of the survivors was Antonio Pigafetta, who accompanied the expedition as a participating observer and who provided a fine chronicle of the voyage. Pigafetta was a 'gentleman' of high education, the son of a noble family of Vicenza in northern Italy, who was talented at languages and sought a career as a diplomat. While touring the major capitals of Europe with a Vatican mission in 1518, he learned of the expedition being planned by

Ferdinand Magellan, and requested permission to join it. King Charles I agreed, making him Magellan's private secretary. Pigafetta's journal of the three-year odyssey described kingdoms, people, and geography. He was assisted in his understanding of Southeast Asian lands by an interpreter, a slave from Sumatra or the Moluccas (sources differ on this) known as Enrique, whom Magellan had earlier purchased in Malacca. Since Magellan's own log was confiscated by Portuguese authorities in the Indies and is not extant, Pigafetta's journal is the only surviving comprehensive first-hand account of the voyage. Four manuscript copies of the book, made ca. 1525, are extant, three of which are in French, one in Italian. They include simple cartographic images, what might be described as bird's-eye view maps, of many of the Southeast Asian islands they visited (figs. 69 & 70). The first published rendering was a French edition of 1525, without maps. The presence of corruptions to the text suggests that Pigafetta had little or no involvement with that published version. One particular error in the printed text had an especially profound effect on the cartographic record of the voyage and, indeed, on an influential image of world itself (see Oronce Fine, page 125). Abridgements of Pigafetta's account (also without his maps) were published by such compilers as G. B. Ramusio and, loosely copying him, Richard Eden.

Maximilian of Transylvania and Peter Martyr

The first printed record of the voyage, however, was that of Maximilian of Transylvania, first published in Cologne in January of 1523 under the title *De Moluccis Insulis.* It was compiled from interviews with the captain of the *Victoria,* Juan Sebastián del Cano, as well as from "individual sailors who have returned with him" and whose testimony was able "to contradict and refute the fabulous statements made by ancient authors." Another chronicler of the voyage was Maximilian's mentor, the great chronicler Peter Martyr. Martyr was a member of the Spanish Crown's Casa de las Indias, as well as priest and confident to Queen Isabella, and tutor to Ferdinand and Isabella's children. He interviewed some of the survivors upon their return, and published a brief record of the voyage in 1530 as the seventh chapter of the fifth 'decade' of his *De Orbe Novo.* His record of the circumnavigation had some influence on printed maps, being occasionally discernable in the spelling of nomenclature, as well as in some of the geographic quirks.

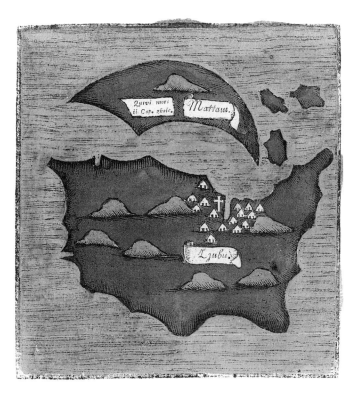

Fig. 69 Cebu and Mactan, Pigafetta, ca. 1525. A woodcut rendering in *Primo Viaggio Intorno al Globo Terracqueo,* based on the Italian manuscript, Milan, 1800. (15.3 x 14.6 cm) [Richard B. Arkway, Inc.]

Fig. 70 Molucca Islands, Pigafetta, ca. 1525. A woodcut rendering in *Primo Viaggio Intorno al Globo Terracqueo,* based on the Italian manuscript, Milan, 1800. The inscription in the upper left-hand corner, repeating an error which was prevalent through the first half of the sixteenth century, states that all these islands lie south of the equator. The tree depicted in the lower right-hand corner is a clove tree. (10 x 13.5 cm) [Richard B. Arkway, Inc.]

Early Evidence of the Voyage on Maps

The reports of the expedition's survivors had an immediate and profound impact on the manuscript charts of Spain's Casa de Contratación. The Philippines are found on a Spanish chart by Garcia de Torreno of 1522, the very year Pigafetta and the other survivors of the voyage returned, and the *Islas s. Lazaro* are found on a chart done about the same year which is attributed to Pedro Reinel. Juan Sebastián del Cano, who assumed Magellan's command after his death, is believed to have provided additional data found in the planisphere of 1527 attributed to Diego Ribero.

Printed maps incorporated elementary knowledge of the voyage by 1527, in which year a rudimentary woodcut map by Franciscus Monachus recorded the new-found strait below America, not yet named for Magellan. Southeast Asian islands from the Magellan expedition were placed on printed maps by Oronce Fine on a double-cordiform map of 1531 (fig. 53) and a true cordiform map of 1534-36 (fig. 71).

Diego Ribero

The charts by the first cosmographer of the Casa de Contratación, Diego Ribero, are among the truly extraordinary cartographic works of the early sixteenth century. Ribero's charts, of which a handful are extant, preserve the sophisticated image of the world which could be constructed at this early date from the confidential data collected by the Iberian crowns, and reflect early recognition of the vastness of the ocean Magellan had traversed, long before printed maps commonly conveyed this to the general public. In Castille as well, charts such as Ribero's were highly secret, and although confidentiality could not be maintained indefinitely, they were guarded well enough to keep printed maps two to three decades behind them.

On Ribero's chart, Southeast Asia has shed every stitch of classical encumbrances, with a well-formed western Malaya, Indochina, and Sumatra, and a wealth of islands, including many of the Philippines, as far east as *Gilolo* (Halmahera). The mainland has the single major designation of *Regno de Ansian* (Siam). Typical of the portolan chart genre, coasts for which no data was available were simply left blank (printed maps, in contrast, generally completed all coasts, guesswork and supposition sufficing where data were not available). On the mainland, Ribero leaves most of the Gulf of Siam and part of the Gulf of Tongkin undelineated; most of the islands, except for Sumatra, have coasts left open.

Oronce Fine (1531)

The map of the world on a double-cordiform projection by Oronce Fine (figs. 52 & 53) was the first printed map to plot any of the lands reported by the survivors of the first circumnavigation. It was probably sold as a loose-sheet in 1531, but has survived because it was included in the Paris editions of Huttich's *Novis Orbis,* 1532 and subsequent.

A Corruption of Pigafetta's Journal Misplaces the Indies and Shrinks the Earth

Fine, as we have seen, came to the conclusion that America was a subcontinent of Southeast Asia, neatly reconciling long-established geographic concepts with new discoveries. The final piece in his puzzle, the piece which made everything else 'work', was his severe abbreviation of the size of the Pacific Ocean. Yet such portolan charts as those by Diego Ribero, which directly tapped the reports given to the Crown by Pigafetta and the other survivors of the Magellan voyage, considerably broadened the Pacific and even Maximilian, based on his interviews with the crew of the *Victoria*

Fig. 71 (Detail) Fine, 1534 (Cimerlino, 1566). Islands of Southeast Asia, including Timor, Halmahera, Borneo, and several of the Philippines, lying off the western shores of the New World. Note Mexico (*Messigo*) marked on the mainland coast. (52 x 58 cm)

declared that the expedition had shown the ocean to be "more vast than mind of man can conceive." What, then, led the brilliant and well-informed Fine to do the opposite?

We can propose that the key to Fine's error lies in a corruption introduced into the printed rendering of Magellan's journal. Fine was long-settled in Paris when he composed the map sometime prior to 1531. His source for data from the Magellan voyage would therefore have been the original French edition of Pigafetta's journal, published in 1525, and sold in the shop of the official bookseller of the University of Paris.[194]

The printed rendering introduced occasional errors into the text. This author believes that one such error, which is persistent throughout the book, explains Fine's distortion of the Pacific and the ease with which he perceived America as a limb of Southeast Asia. Pigafetta determined his longitudes based on the distance traveled westwards from the all-important *línea,* the Line of Demarcation, running roughly down the western limits of Brazil. But our French editor or translator consistently refers to the figures as representing longitude *from the line of departure,* that is, west from Seville. There is no reason why citing distance from their "line of their departure" would have raised any suspicions in Fine's mind, hence he calculated Pigafetta's longitudes from the point of his embarkation in Spain and dramatically abbreviated the Pacific. Others made the same error; Richard Eden's *Decades,* for example, speaks of their position "in longitude from the place from whence they had departed."

To illustrate how this contorted the image of the world, we will look at the island of Palawan. Pigafetta's figure for Palawan was 179° 20' west from the line of demarcation.[195] Fine places the island at

about 175° west of the Canary Islands, his prime meridian (a convention established by Ptolemy). The difference in longitude between western Iberia (which Fine believed to be Pigafetta's zero-degree point) and the Canaries (Fine's prime) is small, only about 5° on Fine's map. If we add this 5° to his placement of Palawan at 175° west, we get 180°, 'precisely' matching Pigafetta's published figure of 179° 20' west. The printed journal's corruption of 'demarcation' to 'departure' cost the Pacific Ocean dearly, a breadth roughly the size of the entire Atlantic Ocean.

The Printed Journal's "Antarctic" for "Arctic"

Fine's misplacement of the southernmost part of Malaya also reflects his use of the printed French edition of Pigafetta's journal. That volume reported that "the Cape of Malacca [i.e., Singapore] lies in one degree and a half to the Antarctic" (south), and indeed, Fine meticulously followed that figure. But "Antarctic" was an error for "Arctic" — Pigafetta had meant $1^1/2$° north, not south, as we learn from later editions of the work. But the editors of the printed edition cannot be blamed for this far less serious error, since it also appears in at least one of the manuscripts of the journal. The transposition of "Antarctic" for "Arctic" was probably just a slip of the pen for Pigafetta, and the error was soon caught. Later printings of his journal, including those published by Ramusio in Italian and Richard Eden in English, correctly place the cape of Malaya at $1^1/2$° north latitude, a near-match for the actual figure of 1°15' north. Even the wrong figure used by Fine largely corrected the error, nearly universally accepted since the Martellus and Cantino models were embraced at the turn of the century, that Malaya extended well into southern latitudes.

Southeast Asian Islands

Squeezed into the gulf formed by the New World and Malaya, Fine has for the first time on a printed map recorded some of the Southeast Asian lands reported by the survivors of the Magellan voyage. Although the map is an exquisite example of woodblock cutting, the limitations of the medium make the Southeast Asian islands difficult to discern on his world map. If it is difficult for us today to decipher the nomenclature, already knowing what it 'should' be, it could have only been more difficult for those laboring over the map in 1531. Additionally, the unusual projection splits the earth at the equator which, though a brilliant move in many respects, meant that the islands of Southeast Asia are divided between the two hemispheres.

If we begin from the middle of Fine's three Asian subcontinents (Malaya) and move counter-clockwise along the equator (the perimeter of the hemisphere), the first island we come is *burney,* (Brunei or Borneo), making its printed post-Magellan debut (the Waldseemüller *Carta Marina* of 1516 charts a small island of *burney* below Malacca, based on Varthema; see page 113). Just to the left (northeast) of Borneo, we find the first appearance of any Philippine islands on a printed map (aside from Polo's 7,448 islands). Two are definitely identifiable: to the lower left of the 's' of *sur,* is *pluan,* the island of Palawan; and left (northeast) of Palawan is *Cobu,* the island of Cebu. East of Cebu lie *Nāgana* (*Nangana,* perhaps *Angama,* the Andaman Islands of Polo, but more likely *Inuagana,* Magellan's first landfall in the Philippines according to Maximilian's account) and *Siulan* (Suluan, Philippines). The island due south (to the upper right) of Cebu is indecipherable, but comparison with Fine's 1534 map shows that it is probably *Baibay.* North of it is *Moluc,* the Moluccas, and Southeast (above) the Moluccas is *Gilolo* (Halmahera). South of the equator, on the other half of the map, lies the southern portion of Halmahera, as well as *Ambuon* (Ambon), *Neto* (Ternate), Java, and *Minor* (i.e., *Java Minor*).

Fine's larger map of 1534 on a single 'true' cordiform projection (fig. 71) contains more detail. Timor, and the Philippine 'island' of *Baibay* (actually part of western Leyte), are now visible. Borneo is duplicated as a result of Fine's having combined Portuguese and Spanish sources, being recorded as *Bunæi* and *Porne.* The copper-engraved rendering illustrated, by Cimerlino, corrupts Fine's *pluan* (Palawan) to *phian.*

Sebastian Münster (1540)

Sebastian Münster created a new map of Asia as a supplementary 'modern' map for his rendering of Ptolemy's *Geography* (1540-1552). In it he attempts to integrate new data from Portuguese voyages into the Indian Ocean and from the voyage of Magellan across the Pacific. The map, which was one of a quartet that formed the first set of continent maps ever published, was also used in the numerous editions of Münster's *Cosmographia* (1544-1578). Despite its shortcomings and rough execution, Münster's contributions are of far more consequence than his errors.

In the illustration (fig. 72), Münster's map of Asia has been mated with that of America, making a single continuous map showing the Far East, Pacific, and New World. We need in fact to look to American shores to find two missing pieces of the Southeast Asian puzzle. *Cattigara,* the Southeast Asian emporium described by Ptolemy and sought by various European explorers, here lies on the coast of Peru because Münster believed Ptolemy's 'Great Bay' to be the Pacific Ocean. The pair of maps gives a typical mid-century idea of the extent of the Pacific Ocean itself — undersized, though not nearly as badly as many previous depictions.

The Asia map's title, *India Extrema* ('further' or 'outermost' India), reflects how the term 'India' continued to denote all of Asia beyond the Holy Land and Middle East. The title varied with different editions, as it was separately set and was not an integral part of the woodblock. Academics who now recognized the Americas as a separate continent from Asia (regardless of whether or not the two were connected along the north Pacific), were already aware that the name 'India' had taken on unintended meanings. Maximilian of Transylvania, chronicler of Magellan's voyage, summarized it by noting that "the natives of all unknown countries are commonly called Indians." Even at the late date of 1595, the map of America prepared for the Mercator *Atlas* was entitled *America sive India Nova* ('America or New India').

The course Magellan's fleet followed across the Pacific missed the lush Polynesian archipelagoes which could have provided periodic breaks and restocking of provisions. Consequently, only one archipelago was mapped by Pigafetta — and in turn by Münster — between the South American mainland and the western Pacific: this was the 'unfortunate islands', lying at a latitude of about 15° south, where the crew saw "nothing but birds and trees." Münster illustrates the galleon *Victoria* sailing past the pair of islands.

The Western Pacific

The most prominent land in Münster's Pacific is *Zipangri* (Japan). Since Marco Polo, whose text was well known to Magellan and Pigafetta, reported the location of Japan based on reports he heard in China, Magellan and his crew 'knew' they were passing the island while *en route* across the Pacific Ocean. Indeed, Pigafetta 'confirmed' Polo's report by stating that they passed close to *Cipanghu* while crossing the Pacific. But contrary to the typical and more accurate pre-'discovery' notion, he located the island at a latitude of 20° south. Richard Eden, corrupting that report, placed the island at "xx degrees from the pole Antarctic", insinuating 70° south latitude. Münster, however, followed the roughly correct latitude implied in Polo's text. The ocean between Japan and the Asian mainland is cluttered with an 'archipelago of 7,448 islands', the Philippines as reported by Polo (see page 107, above).

Münster sloppily duplicates a pair of islands at disparate latitudes. They are the *ins. pdonum,* which lie to the north of the equator on the America sheet, yet are found to be south of the equator on the Asia sheet, where they are marked as the *Insulae pdonū.* This is a corruption of *ladroni,* Magellan's 'thieves islands' of Guam and Rota; a similar corruption of spelling can be discerned in Richard Eden's *Decades,* where the same pair of islands is called the *Latronum.*

Due south of Guam and Rota on the America sheet one comes to the island of *Calensuan.* Given the island's neighbors, the sound of its name and the context in which it appears on the map, it might seem that *Calensuan* came from the Magellan voyage, perhaps as a corruption of *Calaghan,* or *Suluan,* or *Cagayan.* In fact our *Calensuan* has nothing to do with Magellan's expedition and can be found on maps compiled well before the voyage. For example, it appears as *Callenzuam* on the large 1507 world map of Waldseemüller in the approximate position of Sumatra, and again as *Calensuan,* just to the southwest of Sumatra, on a world map by Juan Vespucci (1524). Its origin is unclear, but it may be a variation of 'Ceylon'.

Fig. 72 Asia and America, Sebastian Münster, 1540 (each map 25.5 x 34 cm including title) [Mappæ Japoniæ].

Sumatra

Münster added his own touch to the usual confusion of the old Ptolemaic and Polian nomenclature. Sumatra is yet again identified as *Taprobana,* since as Münster himself related in his *Treatyse of the Newe India,* "the Island of *Taprobana* [is] now called *Sumetra, Zamata* or *Samotra.*" While the 1538 Münster/Holbein map was the first published map to use the term 'Sumatra', his 1540 map is the first to use it *for* Sumatra (it denoted Ceylon in 1538). But the correct usage of Sumatra meant that *Java Minor,* the Sumatra of old, no longer had a home. Münster, influenced by earlier maps which used *Java Minor* as Borneo, retains *Java Minor,* placing the island north of Java *(Iava Maior).*

Borneo

Münster also maps 'true' Borneo, his triangular-shaped island of *Porne,* leaving the old *Iava Minor* as an orphaned isle. Our intrepid Friar Odoric is believed to have touched on parts of Borneo in the fourteenth century, while Ludovico di Varthema may have made a cursory examination of the southwest coast of the island in about 1506 — his report, which was published in 1510, was subsequently tapped by Waldseemüller for his *Carta Marina* of 1516. Borneo was also described in some detail by Tomé Pires in his *Suma Oriental* of about 1515, but the first known European mapping of the island, based on actual observation, came in about 1525, when Pigafetta drew a simple cartographic representation of the island of *Burne* in his manuscript report of the Magellan expedition. It was the crew of the *Victoria* who provided the first substantive report of Borneo and it was upon this information that Münster based his own representation of the island.

Münster followed the mainstream view of the chroniclers of his day in correctly depicting Borneo as a single, large island. The only dissenting voice on this issue was Tomé Pires, who believed it to be "made up of many islands, large and small." Pires, gathering his data in Malacca, may have been influenced by Arab sailors. Sulaiman's navigational treatise, for example, described Halmahera as the largest island in the region, suggesting that he interpreted various landfalls on Borneo as being different islands. Some early chartmakers erred by placing Borneo too close to China, possibly because they were both sources of camphor, though Münster himself keeps his *Porne* a good distance from the Asian mainland. Camphor was an important commodity; Borneo's camphor was deemed the best, and was obtained, Linschoten wrote later in the century, from "trees as great as Nutte trees, and is the gumme which is within the middle of the tree."

The Philippines

Although Fine had represented the Philippines on his map of 1531, it was Münster's rough but fearless map that gave the literate public of Europe its first clear cartographic glimpse of any member of the archipelago. *Puloan,* located due north of *Porne,* is what Martyr described as the "marvelous fruitful island" of Palawan.

Moluccas, Timor and Halmahera

Münster depicts the Moluccas, Timor *(Timos),* and Halmahera far more clearly than had Fine in 1531. Halmahera *(Gilolo Vel Siloly)* is shown as a large, rectangular island, on a north-south axis, a configuration which was common before the island began to assume its characteristic spider-like shape. Pigafetta noted that *Gilolo* was controlled by 'Moors' along the coastal regions and inhabited by 'heathens' inland (Islam had first arrived in the Moluccas only about

VLAE, XVII·NOVA TABVLA·

Malaya Freed of the Great Promontory.

Münster improves upon Fine by depicting Malaya as a peninsula, rather than an India-shaped subcontinent. But since the map lacks latitude markings, it is not clear whether the southernmost part of Malaya is placed at a latitude of $1^{1}/_{2}°$ south, as *per* the error which Fine followed, or the same latitude north, as Pigafetta intended.[196] Münster depicts the Gulf of Martaban with reasonable accuracy and indicates a river at its apex (unnamed, but probably the Irrawaddy or Salween). *Pego* (Burma) is now in its correct position, rather than on the Martellus or Cantino subcontinents.

These breakthroughs did not prevent Münster from reminiscing over the ancient lands of gold and silver, *Argyre* and *Chryse*, which he identified by the terms *Regio argentea* and *Regio aurea*, locating them in Burma and Malaya, respectively. Nor has he completely rid himself of the confusion between Indian and Southeast Asian nomenclature which plagued earlier mapmakers: in the northeast of Malaya lies *Sindi,* the region of Sind in India from which the legendary Sindbad came.

The large gulf to the east of Malaya is the combined Gulf of Siam and Gulf of Tongkin. Since the peninsular shape of Indochina is lacking, *Cyamba* (Champa, Vietnam) is pushed up into southern China. That this was still largely mysterious territory is reflected in the words of the Portuguese chronicler Barros, who wrote that the easternmost section of Southeast Asia lay from the apex of the Gulf of Siam "to a famous cape which is at the easternmost of the firm land which we now know about."

Typography

The Münster maps of Asia and America are shown here in their first state, first issue. Although the original woodblocks served for thirty-eight years without geographic change, seemingly purposeless changes in nomenclature can be found in the various editions. This resulted from Münster having printed the place names on his maps by inserting metal type into the woodblock rather than by actually cutting the letters out of the block itself, a technique used at least as early as the great 1507 world map of Waldseemüller. Though not elegant typography, this method had distinct advantages — it was less time-consuming to produce the woodblock, while the letters were sharp and consistent, and could be changed or modified without damaging the block. The system did, however, cause some problems; the metal type appears to have occasionally fallen off, to be replaced only once discovered (if then), and sometimes replaced indiscriminately. As a result, many inadvertent states of the map exist, sometimes lacking a letter or a word, sometimes with the original type sloppily replaced, and other times with new type to replace a lost or broken piece.

The *Cosmographia* was an extremely popular book. Münster's important map of Asia, however, has in our day too often been glossed over as the quaint relic of a naive mapmaker. This neglect has been caused in part by the map's rough execution, but principally because a new crop of vastly improved maps began to appear by 1548, quickly making Münster's work obsolete. To make matters worse, the map was reprinted without improvement through to 1578, so that for thirty of its thirty-eight year publishing life, it was an anachronism.

a half-century before the Iberians). Other European expeditions had reached this important spice-producing island by the time of Münster's map, including those of Garcia de Loaisa in 1525 and Saavedra two years later.

Dwarfed by *Gilolo* are the Moluccas *(Moluca),* known to history affectionately as the Spice Islands. Varthema reached the Moluccas in 1505, disembarking on the "island of Monoch," which was probably Ternate. His published account of 1510 correctly notes that the island is a source of cloves:

> Here the cloves grow, and in many other neighboring
> islands, but they are small and uninhabited . . . The
> country is very low, and the north star is not seen from it.

Organized Portuguese exploitation of the Moluccas began with the expedition of António de Abreu and Francisco Serrão, which left Malacca in December of 1511 — immediately after its conquest. Skirting the northern shores of the Lesser Sundas to Flores, they turned north, finding the Banda and Ambon Islands, and Seram. De Abreu returned to Malacca, while Serrão made his way to Ternate, the only island of the Moluccas which Münster specifically names (*Taranata*). If Münster had to settle for a nearly arbitrary mapping of the Moluccas, he has nonetheless followed Pigafetta meticulously in locating precisely five members of the archipelago, and in presenting Ternate as the most important. As related by Richard Eden,

> the Islands of Molucca are five in number, & are thus
> named: *Tarenate, Tidore, Mutir, Maccbian,* and *Bacchian.*
> Of these, *Tarenate* is the Chiefest.

Maximilian also reports five Moluccan islands, "situated partly to the north, partly to the south, and partly on the equator."

Chapter 11

Giacomo Gastaldi's Three Models 1548 – 1565

Italy, though not the sponsor of ocean voyages to Southeast Asia, was a major cartographic 'think-tank' for digesting and sorting out the data that those expeditions brought back. During the late fifteenth and much of the sixteenth century, Italy was a pioneer in printed maps, both loose-sheet and in books. Italy's theoreticians secured geographic data from primary sources and molded it into a coherent whole.

The name of Giacomo Gastaldi dominates the cartography of Southeast Asia on printed maps throughout the middle decades of the sixteenth century. A native of Piedmont, Gastaldi was a brilliant cosmographer and engineer who was active in Venetian affairs, and was largely responsible for the flourishing of geographic disciplines in Venice during this era. He composed three fine maps of Southeast Asia, each of which provided the best and most inspired published rendering of the region in its day. The first was published as part of his edition of Ptolemy's *Geographia* in 1548, the second in 1554 as part of a collection of voyages by Ramusio, and the third was a separately published map of Asia of 1561, to which a supplemental sheet was added in 1565, thus extending the map's reach to south of the equator.

The Three Types and Their Sources

1548 *India Tercera Nova Tabula*

Although the maps in Gastaldi's issue of Ptolemy's *Geographia* are relatively small, they were handsomely engraved on copper, allowing more detail and clarity than their woodcut predecessors. The medium of copperplate had been virtually abandoned by mapmakers for four decades, since the 1507-08 Rome edition of Ptolemy's *Geographia*.[197] Gastaldi retains Ptolemy's full complement of maps, including the Ptolemaic rendering of Southeast Asia, which by this time was of no merit except as a testimonial to the reverence in which classical writings were still held. He adds, however, a fine new map of the Southeast Asian mainland and islands, entitled *India Tercera Nova Tabula*, as it was the third of his atlas' 'new' maps (fig. 73). It was extracted from a world map Gastaldi had made in 1546, and was a major contribution to the mapping of the region. Although this map was published in only

one issue in its original form, it was copied by other mapmakers and via such copies, it outlived its relevancy even longer than the map by Münster. In 1561, the map was re-engraved, roughly twice the size, for a new edition of Ptolemy by Girolamo Ruscelli, in which form it was re-published several times up until 1599. It was also re-engraved in a smaller format for the *isolario* of Tomaso Porcacchi, which was published in several editions from 1572 through to 1686.

1554 Ramusio *Terza Tavola*

Gastaldi's next landmark in Southeast Asian cartography appeared six years later, in 1554, in the second edition of Volume 1 of G. B. Ramusio's *Delle navigationi et viaggi*. The map is untitled except for being designated the 'third map' *(Terza Tavola)* in the book. Geographically, it is utterly unrelated to the *India Tercera Nova Tabula* of 1548.

The Ramusio map was first printed from a woodblock but that block, along with those for all the other maps from the *Navigationi* except for one of the Nile, was destroyed by a fire in the printing establishment of Tomaso Guinti in November of 1557. Curiously, while the woodblocks from the third volume of the *Navigationi* (covering America, plus one of Sumatra) were replaced by new, slightly rougher woodblocks, those for the first volume (including the present map) were replaced by more costly copperplate versions (fig. 74), which were first used in the third issue of volume 1 (1563). This volume was last reprinted in 1613.

Separately-published Maps and Made-to-order Atlases

Unlike the 1548 Gastaldi map, the 1554 Ramusio map was not copied by other atlas makers. However, a loose-sheet knock-off of it was published by Bertelli in 1564. Loose-sheet maps (sold as separate, unbound sheets) rarely survived the centuries unless the purchaser opted to have a selection of them bound as a made-to-order atlas. This practice anticipated the formal 'atlas' genre (which began with the 1570 *Theatrum* of Abraham Ortelius) and vastly increased the chance of a loose-sheet map's survival to our day. Such made-up atlases, often called 'Lafreri' atlases after Antonio Lafreri, their greatest proponent, were of 'modern' maps, which no longer relied on Ptolemaic geography or used classical texts as a vehicle.

They differ from the modern connotation of 'atlas' in that these early compilations were neither uniformly sized nor their contents methodically conceived. The next map in the Gastaldi trio was one such separately-published work; it never appeared in a book, but has survived precisely because some map sellers' clients included it among the maps they asked to have bound into a made-up atlas.

The Separately Published 1561 'Lafreri' Map and its Southern Supplement

The publishing history of this last member of Gastaldi's trio is the most complex of the three. In 1559, Gastaldi began issuing a map of Asia in three parts, though only the westernmost section appeared in that year. Parts two and three, of which the latter covered eastern Asia, came in 1561. In its original form, the map extended only to the equator, so that most of the Indonesian islands were not included. To remedy this, in about 1565, two narrow sheets were made by the great Italian engraver Paolo Forlani to supplement the main body of Gastaldi's map. These new sheets, which covered the region to about $17^{1}/2°$ south latitude, were probably added to shop copies of the map, and occasionally purchased as a supplement by people who already owned the main Gastaldi map but wished to have the full repertoire of Southeast Asian islands. This lower addition bears an inscription in the lower left corner which reads *si vende alla libraria di San Marco in Venetia* (this is sold in the book shop of San Marco in Venice), indicating the location of the shop of the publisher Bertelli. The Gastaldi/Forlani map was also re-engraved by Girolamo Olgiato, in which form all of Southeast Asia falls on a single sheet (fig. 75).

Gastaldi's Sources

All three maps tap the Magellan voyage for their mappings of the Philippines and other Southeast Asian islands, and reflect the marked variations in the printed accounts of the voyage, with differences in nomenclature, and even in itinerary. The second (Ramusio, 1554) and third (Gastaldi/Forlani, 1561/1565) maps are geographically related to each other, while the first (Gastaldi, 1548) stands alone.

The 1548 map from the *Geografia* relies far more heavily on data from the Magellan voyage than either the 1554 or 1561 maps, particularly in relation to the Philippines. But it also records certain features — notably, a remarkably accurate depiction of Palawan and a 'Gunung Api' (a small volcanic island) — which cannot be attributed to Magellan's discoveries and suggest that Gastaldi had access to advanced Portuguese sources.

The 1554 map from Ramusio's *Navigationi* benefited from several new sources of data which were not available in 1548. On the mainland, this results in an enormous advance over the 1548 map, while Ramusio's mapping of the island world is far more comprehensive, though not always better, than the 1548 map. Ramusio states in the preface to the *Navigationi* that the coasts of the 1554 map "are drawn according to the marine charts of the Portuguese, and the inland parts are added according to the descriptions contained in the first volume of this book."

Ramusio cites the history of Asia being compiled by João de Barros in Lisbon, the *Décadas da Ásia.* The second volume of this work, detailing Portuguese maritime adventures in the East from 1505-1515, had been published in 1553, just before the appearance of Ramusio's map in the *Navigationi* (volume one, published in 1552, covered the period from Prince Henry to 1505, and would have had little or no effect on Ramusio's map). With Barros' data, Ramusio believed, "a part of modern geography will be clearly illustrated, and it will no longer be necessary to struggle with the geography of Ptolemy."

Comparison of nomenclature and geographic descriptions in the *Décadas* with the map suggests that Ramusio used the work, but did not rely on it. River names, and the configuration of Lake Chiang Mai (page 222, below), for example, correspond generally, but not in detail.

António de Abreu and Francisco Serrão, who pulled anchor at Malacca in December of 1511 and sailed toward the Spiceries under Malay pilots, should have provided the foundation of first-hand European mapping of Indonesia, coming a scant few years after Varthema's informal spree through the islands. But their jaunt had no discernable impact on maps because news of the voyage was suppressed by the Portuguese authorities, and by the time word of it reached geographers, more current information had already rendered their data irrelevant. However, Serrão was certainly influential on the course of events. He established himself in Ternate as a renegade facilitator of trade between the Spiceries and Malacca, and wrote letters to his close friend Magellan, firing the latter's determination to reach the Moluccas with idyllic descriptions of the islands and the inference that they lay so far east as to be more easily accessible by a voyage west, around the New World.

Among the other voyages from which the Gastaldi maps draw some data are those of Loaisa and Salazar (1526), and Saavedra (1527-29). Some of the data on the Ramusio and 1561 Gastaldi maps appears to have been gleaned from the important chronicle of the Indies by Gonzalo Fernandez de Oviedo y Valdés, the *Historia general y natural de las Indias, islas y tierra-firme del mar océano.* Although first published in Seville in 1535, the relevant parts didn't appear until Part II, published in 1548. Finally, the 1554 Ramusio and 1561 Gastaldi maps reflect knowledge of the account of a voyage from Mexico to the Philippines by Juan Gaytan (often known by the Italianicized *Gaetano,* as coined by Ramusio) and Ruy Lopez de Villalobos, which Ramusio himself was the first to publish, in the first volume of his *Navigationi* (1550).

'Taqwim Albudan'

There is an interesting myth regarding the sources of the 1561 map. According to the Flemish mapmaker Abraham Ortelius, a younger contemporary of Gastaldi, the latter had based his map on information he obtained from an Arab geographer named Taqwim Albudan. Ortelius would seem to have been in a position to know, since he corresponded with the French geographer and mathematician Guillaume Postel, who had purportedly brought this 'Taqwim Albudan' or at least his work to Europe. Ortelius reveals this information in a legend on his own separately-published map of Asia, 1567, which (as he acknowledges) was largely based on the Gastaldi map.

In fact, 'Taqwim Albudan' is an historical folly. It was not the name of an individual at all, but rather the title of a geographic lexicon, the *Taqwim Albudan,* which was written by an Arab geographer named Abu'l-Fida, who lived 1273-1331. This work, itself an undistinguished compilation of earlier material, would have been quite worthless to Gastaldi. Gastaldi's map is derived directly from Spanish and Portuguese exploration, and is vastly superior to Arab knowledge of the region, prior to the advent of a European presence in Southeast Asia.

Comparing the Gastaldi Maps

For clarity, henceforth the 1548 map which appeared in the *Geografia* will be referred to as the 'Gastaldi 1548' (fig. 71); the 1554 map from the *Navigationi* as the 'Ramusio' (fig. 72); and the separately-published map of the Asian continent as 'Gastaldi 1561' (fig. 73). When we sail south of the equator, we will actually be

Fig. 73 Southeast Asia, Giacomo Gastaldi, 1548. (13 x 17 cm)

venturing onto the narrow southern supplementary sheet of ca. 1565, and so will refer to it as the 'Forlani', although the Olgiato version illustrated here makes the geography easier to follow by combining the main and supplemental maps into one. We will first look at these three maps' record of insular Southeast Asia together, and then peruse their depiction of the Southeast Asian mainland individually.[198]

Trying to reconstruct how Gastaldi compiled his data for Micronesia and the Philippines for this first crop of modern Southeast Asia maps is frustrating.[199] There are many ambiguities and discrepancies in the various published renderings of Pigafetta's text, and also in the account by Maximilian, so that even the most well-intentioned mapmaker attempting to extrapolate from accounts of the voyage must have often had to settle matters by simply following his nose. Further confounding the mapmaker's job (and our job as well) is the fact that the names of the islands represented, as transcribed by Pigafetta, Maximilian, and subsequent chroniclers, were transliterated and corrupted in various ways by copyists. Furthermore, many of the names, perhaps already similar sounding to European ears, were often confused with one another, or presumed to be others, or were otherwise transposed so that they blurred into each other. First-hand observers sometimes mistook part of an island to be a separate island altogether, while the mapmakers who plotted their data sometimes misconstrued distinct islands as different regions of a single island, or visa-versa. Finally, explorers who searched for specific islands discovered and named by their predecessors often reached different islands but erroneously deemed them to be the original landfall, and then mapped the new island under the original name.

To view the three Gastaldi maps, one must first erase from one's mind any image whatsoever of what the Philippine or Micronesian islands 'look' like; only then can one sympathize with the chaos of these Pacific incunabula. Whereas Sumatra and Java were already known for many years and provided some humble point of reference for east-bound pilots charting the Indonesian islands in their proximity, there was not as yet any such point of departure for mapping Micronesia or the Philippines.

Micronesia

Micronesia, the eastern periphery of Southeast Asia, is an important part of the Ramusio and 1561 Gastaldi maps, but does not appear at all on the Gastaldi of 1548. Gastaldi at that time did not yet have access to the important Micronesia data from post-Magellan voyages, and the 1548 work does not extend far enough east to include even the initial Magellan landfalls in Guam and Rota. Beginning with Ramusio, cartographers had the advantage of three new sources for Micronesia: Gonzalo Gómez de Espinosa, Alvaro de Saavedra, and Ruy López de Villalobos. Many of these early landfalls in Micronesia fell in the proper latitude band to be prominently recorded on de Jode's semi-cosmological cross-section of the earth (fig. 16).

Li Ladroni (the Marianas)
In 1521, after their torturous crossing of the Pacific, Magellan's fleet reached *Li Ladroni* (left border of the Ramusio), the 'thieves' islands, the appellation *ladroni* given by Magellan because of the islanders' supposed propensity for stealing. Pigafetta placed the islands at a latitude of 12° north. He described the Mariana islanders as an isolated people who live in freedom, accountable to no outsider (this in great contrast to Southeast Asia proper, where elaborate tribute systems existed). They traveled their own waters

Fig. 74 Southeast Asia, G. B. Ramusio. Originally published in woodblock in 1554, this is the copper-engraved rendering of 1563. (27.4 x 37.3 cm)

in black and red boats which reminded Pigafetta of the gondolas which shuttle between Venice and Fusine. Pigafetta speculated that the people of the *Ladroni,* prior to the arrival of the Spanish, believed themselves to be the only people in the world — if true, an interesting comment on the cosmologies of the Southeast Asian periphery. Maximilian's account of the voyage omits Magellan's landfall in the Marianas altogether.[200]

Magellan did not penetrate the Marianas further than Guam and Rota, yet both Ramusio and Gastaldi (1561) chart a full north-south chain of islands. Whose voyage, then, provided the basis for the Marianas on these maps? When the crew of the Magellan expedition set sail for home from the Moluccas in 1522, there were two remaining vessels in the fleet; the flagship, *Victoria,* continued on its maiden circumnavigation, while the other ship, the *Trinidad,* under Gonzalo Gómez de Espinosa, was assigned to search for a return route east, across the Pacific to Panama. The discovery of such a return route would be of immense benefit; but the problem was to locate a latitude in which favorable westerly winds could be harnessed. Sailing north in search of such winds, the expedition "discovered fourteen islands which were full of an infinite number of naked people, the said being the same color as the people of India . . . they are from 12 degrees as far as 20 degrees north of the equinoctial line." They were back in the Marianas. A Chamorro guide they brought with them was able to identify all fourteen of the islands in the archipelago by name. Espinosa's charts were later confiscated by the Portuguese in Ternate, and thus it came about that the various islands of the Mariana chain appear on early Portuguese charts.

Oviedo's *Historia* provides another possible source for improved Mariana data in the Ramusio and 1561 Gastaldi maps. Gonçalo de Vigo, a Galician who deserted the Magellan fleet, spent five years in the Marianas until he was picked up by Loaisa's expedition in 1526. De Vigo knew the Chamorro language, as well as some Malay, the *lingua franca* of Southeast Asia, and was thus quite an asset for the Spanish. He passed on his information about the islands to the authorities in New Spain (Mexico), who in turn were informants for Oviedo. De Vigo explained that the Marianas group comprises thirteen islands lying due north-south up to a latitude of 21° north.[201] Ramusio approximates the north-south oriented chain, and indeed depicts precisely thirteen members, as Oviedo specified; like most cartographers of the era, he does not extend the Marianas chain far enough north.

In a move that probably influenced mapmakers, Peter Martyr, in his fifth *Decade,* indiscriminately groups the Marianas and Philippines together as "a multitude of islands," which are "separated from one another by narrow channels and extended throughout a distance of five hundred leagues." He believed that Magellan had given the name 'archipelago' to these islands "because of their resemblance to the Cyclades of the Ionian Sea", though Pigafetta does not mention this in his journal. Martyr jumbled up Micronesia, the Philippines, and other islands still further by remarking that "the largest of their islands is called Borneo", when he appeared to be describing the Marianas.

Isole del Rey ('Islands of the King')

The *Isole del Rey,* shown as a large, diamond-shaped archipelago on the Ramusio, and as small 'x's on the 1561 Gastaldi, was derived from the expedition of Alvaro de Saavedra (1527-29). Saavedra crossed the Pacific from the vicinity of Acapulco, Mexico, in 1527, reaching Guam, Mindanao, Tidore, and Halmahera. Saavedra, like Espinosa before him, tried to discover a latitude with westerly winds for the return voyage to Mexico. He made two attempts, first sailing to a latitude of 14° north and then to 31° north. While meandering

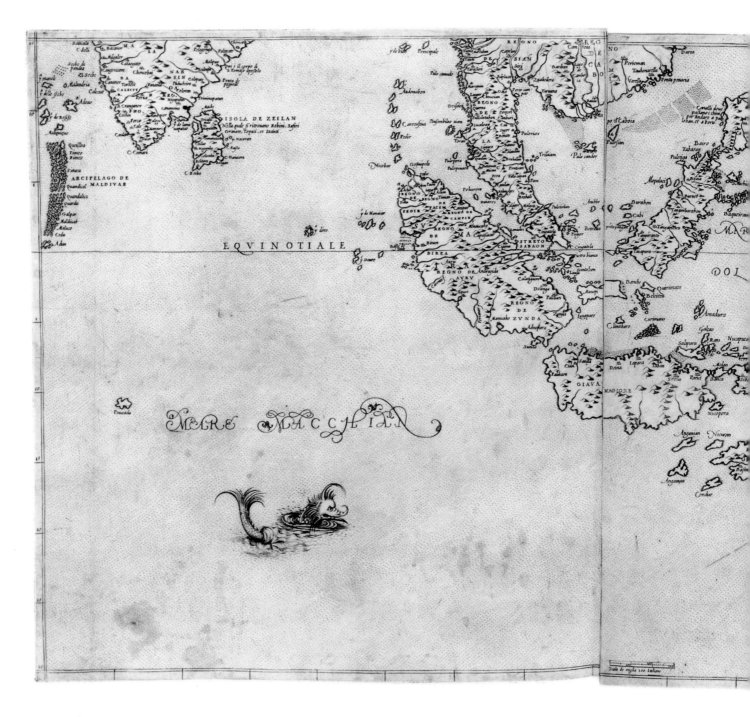

Fig. 75   Southeast Asia, Girolamo Olgiato, ca. 1570. This is a re-engraving of the Gastaldi Asia of 1561
and the southern sheet added to it by Forlani ca. 1565. [Richard B. Arkway]

about in the western Pacific searching for these winds, Saavedra
encountered new islands in Micronesia, among them the *Isole del Rey.*
One of the sailors in Saavedra's crew recorded that early one morning
in 1528,

> with the sun already up, we saw land when we were about
> one league from it. This land was baptized by Captain
> Alvaro de Saavedra the islands of Los Reyes [the kings],
> because on the day we saw them it was the feast of the
> Kings [the Epiphany].[202]

These are undetermined members of the Caroline or Marshall
Islands. By the end of 1529, Saavedra had died at sea, and the
surviving crew returned to Halmahera. Thirteen years later, on the
evening of Christmas day in 1542, one of Saavedra's crew, Antonio

Corso, was back in the Pacific again as a pilot under Ruy López de
Villalobos. Villalobos led a fleet of six ships from Navidad on the
Pacific coast of New Spain in 1542. He discovered various Pacific
archipelagoes such as *Santo Tóme, La Nublada, Roca partida,* and
Placer de siete brazas, before reaching the western Pacific at the end of
that year. On Christmas day the crew sighted an archipelago which
they believed (probably wrongly) to be the *Reyes* of Saavedra, and
described it as being round, which probably influenced the shape of
the islands on the Ramusio map. It was surrounded by deep water,
and they could not find a place for a secure anchorage.

Corali Isole ('Coral' Islands)

The next day, the Villalobos fleet arrived at a group of islands which
they first named after San Esteban, whose feast day it was. Later they

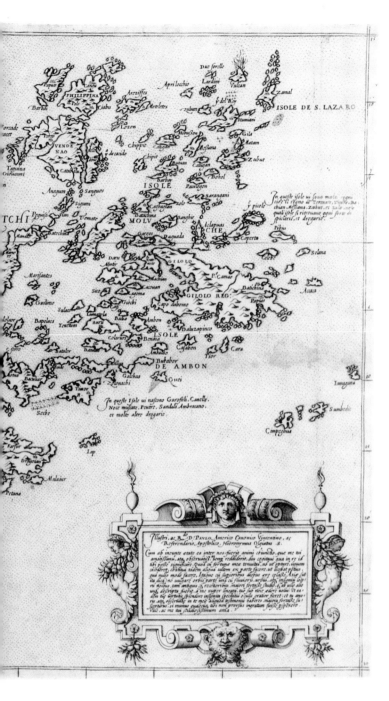

volume of the *Navigationi* (1550). Leaving the *Corali isole,* the fleet sailed 35 leagues west (Gaytan recorded 50 leagues) and on January 8th of 1543 sighted a group of ten or more islands. "As they seemed green and beautiful," Gaytan wrote, "we gave them the name of *li Giardini.*" The crew did not cast anchor there.[204] The 1561 Gastaldi marks them as *Due sorelle Lardini* ('two sister gardens'), and later in the century (e.g., the Plancius map of 1594, fig. 90), the archipelago was put down as the garden isles 'of no merit' (*al la Desaprovechada*).

Matelotes

On January 21st of 1543, a couple of weeks after Villalobos visited the 'garden' islands,

> having made 100 more leagues [being now 200 leagues from the Giardini islands], at a latitude of 10°, we passed by a small island, well populated, which appeared very beautiful. We did not anchor here. Rather, Indians came out with proas making the sign of the Cross with their hands and they were heard to say in Spanish: 'Buenos dias, matalotes', reason for which we baptized it Matalotes.

(The term *matalotes* means 'friends' in Spanish or 'sailors' in Portuguese). Augustinian priests aboard noted that the island was small but high, with black rocks, fresh-looking and with beautiful palm trees. They wished to rename the island after St. Idlefonso. The crew was not able to learn how the people had come to know a greeting (supposedly) in their tongue, since the island was isolated and lay outside of trade routes with other islands.[205] The island is recorded by both Ramusio and Gastaldi (1561).

Vulcan

Another island reported by the Villalobos expedition is *Vulcan,* which is mapped by Gastaldi (1561) as an erupting volcano just west of the Marianas. Hakluyt reported that in September of 1543 Villalobos, having resupplied in the Philippines, "sailed for certain days with a fair wind" until the winds slacked just below the Tropic of Cancer. Among the islands they encountered were a group of four which they christened *los Volcanes.* Afterwards they discovered an island called *Farfana,* "beyond which there standeth an high pointed rock, which casteth out fire at 5 places." Gastaldi (1561) appears to have combined *Volcanes* with the active 'high pointed rock' lying beyond *Farfana.* We will meet up with *Farfana* island on the Ortelius *Indiae Orientalis* of 1570 (page 169 & fig. 86, below).

Apri l'Occhio

Both Ramusio and Gastaldi (1561) map an 'open the eyes' island, *Apri l'occhio,* due west of the Marianas, in the northern Philippines. The origin of this island is not known, as it has not been found on any early Iberian source documents. It later appeared in a corrupted spelling on Vaz Dourado's charts of about 1575, as *abeio.* In fact, *Apri l'occhio* may have been a standard cautionary phrase, and may not have been intended to depict a specific island. The same feature is found on other early Portuguese charts in unrelated waters — for example, on a portolan chart of the New World from an atlas attributed to Vesconte Maggiolo.[206] On that chart there is a seemingly arbitrary spattering of dots in the seas to the north of Hispaniola which are similarly accompanied by the remark *apri lochio.*

In any event, the 'open the eyes' island of Ramusio and Gastaldi (1561) and the same map's sputtering volcano, mark the beginning of a long-lived cartographic curiosity: henceforth, up until the early eighteenth century, charts of these waters commonly depicted navigational hazards to the west of Guam, even though ships routinely crossed those waters without mishap.[207]

christened them the *Archipiélago del Coral,* because of the beautiful coral found there. According to one account, a fine specimen came up with the ship's anchor. Most of the island's people, who were described by a missionary accompanying the fleet as "well-proportioned and pretty, but badly-clothed and ill-mannered," fled by canoe to neighboring islands. Ramusio maps the *Corali isole* at 10° north, just north (below) of the *Isola del Rey.*[203]

Li Giardini ('Garden' Islands)

Like the *Corali isole,* the Garden Islands are believed to be members of the Marshall Islands, and were also mapped from the reports of the Villalobos expedition. Ramusio marks them in the western corner between the *Rey* and *Corali* islands. An account by Juan Gaytan, a sailor on the expedition, was included by Ramusio in the first

The Philippines and Borneo

As a result of the bustling commerce between Malaya and other parts of the Indies, the existence of the Philippines was known to the Portuguese in Malacca — and thus to Magellan — several years before Magellan's actual discovery of the islands. In fact, the Portuguese did not even have to 'discover' the Philippines, for the Filipinos had already discovered them. In about 1515 Tomé Pires, resident in Malacca, wrote that there were some five hundred *Luções*, "some of them important men and good merchants." living in Malaya, in the district of *Mjmjam* (Dinding) — a reference to the Filipinos six years before Magellan's discovery of their island homeland. In his *Suma Oriental*, Pires reported that

> the Luções ['Luzones', i.e., Philippines] are about ten days' sail beyond Borneo. They are nearly all heathen; they have no king, but are ruled by groups of elders. They are a robust people, little thought of in Malacca. They have two or three junks at the most. They take the merchandise to Borneo and from there they come to Malacca.

> The Borneans go to the lands of the Luções to buy gold, and foodstuffs as well, and the gold which they bring to Malacca is from the Luções and from the surrounding islands which are countless; and they all have more or less trade with one another.

The term *Luções,* in fact, became an alternate term for the Philippines. António Galvão, an early governor of the Moluccas, claimed that in 1545 a Portuguese sailor named Pero Fidalgo was blown by contrary winds north from Borneo and reached *dos Luções* islands, named for their inhabitants. The landfall appeared on some Portuguese charts and influenced Ortelius on his *Maris Pacifici* of 1589.

Just as the Portuguese had learned of the Filipinos before 'discovering' their islands, so word of Portugal's presence in Malacca reached the Moluccas and the Philippines long before de Abreu and Serrão arrived in Banda in 1512, and Magellan got to Cebu in 1521. When Magellan stopped off in Cebu, a trading vessel from *Ciama* — either Siam or Champa — had just called at its port, and a merchant who remained to gather merchandise was very familiar with Portuguese exploits in Malacca.

Some historians have speculated that Magellan may well have reached the Moluccas, or even the Philippines, nearly a decade before his Pacific crossing in 1521. He had been one of the men under Albuquerque's command responsible for safeguarding Malacca and the region of Singapore following their victory in Malacca in 1511, and conceivably could have accompanied the expedition of de Abreu and Serrão as far as Banda in 1512, or been sent on a secret mission into the South China Sea. But if, as records indicate, Magellan signed for salary in Lisbon on June 12 of 1512, then he could not have participated in that voyage.

The expedition of de Abreu and Serrão was, nonetheless, a principal factor in Magellan's determination to reach the Spiceries. Serrão was a close friend of Magellan he wrote to him from the Moluccas extolling the virtues of life on the islands, and indicating that they lay very far to the east — hence the rational of reaching them by circumnavigating the New World to the west. There is also a tantalizing statement left for us by António Galvão, the early governor of the Moluccas, which claims that Serrão, after his shipwreck, "went back as far as the island of *Midanao*". It is very doubtful, however, that Serrão reached the southern Philippine island of Mindanao.

First Sighting of the Philippines

In any event, Magellan made the first recorded European discovery of the Philippines soon after leaving the Marianas. Sailing west, the expedition sighted the mountains of the Philippine island of Samar piercing the ocean surface, and made landfall on the small island of Homonhón, which lies just to the south of Samar and was, at that time, uninhabited. The islands were named in honor of St. Lazarus, on whose feast day they touched shore.

Pigafetta explains that Samar itself was avoided because "the Captain wanted to land on an uninhabited island . . . in order to be safer, to take on water." Each of the three Gastaldi maps refers to Homonhón by a different name. Gastaldi 1548 maps it as *aguada* (latitude 8½° north on the right side of the map), from the Spanish for 'watering place'. Ramusio records it as *Humunu* (4° north latitude on the left side of the map), the spelling by which the island's name was recorded by Pigafetta. And Gastaldi (1561), following Maximilian rather than Pigafetta, labels the island *Accaca* (3½° south latitude at the right end of map), this term being Maximilian's corruption of *aguada*.

The Accuracy of Pigafetta's Longitude

Pigafetta placed this maiden Philippine landfall at 161° west of the Line of Demarcation. How one interprets the accuracy of this figure depends on what longitude one assigns the Line. He may have envisioned it as did the Spanish geographer López de Velasco (ca. 1575), who placed the Line 30° west of Cape Verde. That cape lies about 17½° west of Greenwich, making Pigafetta's longitude 208½° west of Greenwich as compared to Homonhón's actual location at 234¼° west of Greenwich. Pigafetta's longitude, then, would be 25¾° too far east, an error favoring his sponsors, the Spanish Crown. Pigafetta, we will see, was far more successful in determining the longitude of their goal, the Moluccas.

Landfall in the Philippines

Homonhón was also known as *Buena Señal* (good sign), "because we found two very clear springs, and gold, and white coral in large quantities," and fruit. This name was used for the island by Gerard Mercator on his large world map of 1569 (fig. 76), and, copying that work, by Abraham Ortelius on his *Indiae Orientalis* of 1570 (fig. 86).

One of the most inept accounts of Magellan's voyage was that of Richard Eden. Eden has Magellan landing on the 'uninhabited' island of *Zamal,* confusing Homonhón with Samar proper. Such inaccurate accounts are very important to the study of early maps for the very reason that they, rather than original source documents, were normally those which educated the general public, and were often relied upon by the makers of published maps as well. On his 1561 map Gastaldi, who was similarly confused, places *Zamal* (Samar) as the southernmost of the Mariana chain; in fact, Gastaldi appears to have engraved Samar there as an afterthought, superimposed on the southern *Ladrones* after they were already in place (see fig. 87). This error tricked such cartographers as Abraham Ortelius, who left *Zamal* stranded in the approximate position of Guam (fig. 86).

While Magellan and crew were replenishing their store of fresh water, nine people from another island came by. Gifts and food were exchanged, after which the islanders "made signs with their hands indicating that within four days they would [return with] rice, coconuts and many other things"; the promise was kept. Magellan's crew, based on Homonhón for a week, learned what they could about the neighboring islands from the natives and from their own excursions. The islanders' own home was *Zuluan,* by which spelling it is recorded by Ramusio. This is Suluan, a tiny island just east of Homonhón off the southeast of Samar.

They set sail again from Homonhón on the 25th March and "took a west-southwest course among three [four] islands, that is to say *Canalo, Huinanghan, Hibusson,* and *Abarien.*" These lie just east of the point where southern Leyte and northern Mindanao nearly meet. Ramusio places these islands together to the northwest of Homonhón *(Humunu),* naming the first and fourth: *Zenalo* and *Abarien.* The expedition anchored off the small island of Limasawa just south of Leyte, which Pigafetta called *Mazava.* This appears on the Ramusio map as *Messana,* and is placed in approximately the correct position in relation to the Marianas, but wrong in relation to his misplaced Homonhón.

Maximillian's account may have contributed to the confusion. He recorded that the uninhabited island was *Inuagana,* and that visiting Filipinos spoke of an island to the west named *Selani* (probably southern peninsular Leyte just north of Panoan) which was inhabited, and where "an abundance of everything necessary for life was to be found."[208] According to Maximilian, the fleet sailed toward *Selani* but never reached it because a storm drove them to *Massana. Selani* is mapped by Gastaldi 1548 as *Caylon,* and in 1561 as the little island of *Selana* (lying just below the equator on the far right), but is not mapped by Ramusio. These were based on nothing more than Maximilian's citation of native advice, or on the account of Pigafetta, which mentions *Selani* later on.

The island of *Messana* — a small island, despite the size accorded it by Ramusio — was not continuously inhabited, but rather served as a retreat and hunting ground for two kings, who were brothers.[209] Pigafetta refers to one brother, Kolambu, as 'the king of Mazana'; the other was named Siagu (or Siani/Siain).[210] They were from the 'islands' of *Buthuan* and *Calaghan,* which Gastaldi (1561) records as *Butuan* and *Calugan.* These two locales are in fact Butuan and Caraga, which are regions of north-eastern Mindanao, not separate islands. In addition, Gastaldi (1561) maps an island by the name of one of the brothers — the island of *Sian.*

It was on the little 'retreat' island of Limasawa, Ramusio's *Messana,* that the Spanish first had extensive contact with the Filipinos, wining and dining with the islanders and learning what they could of the region. Here, also, was the first recorded Filipino encounter with a European map: Magellan, after having shown Kolambu the Spaniards' swords, shields, and armor, "led him to the deck of the ship, and up to the poop deck, and he [Magellan] had his sea chart brought to him and the compass." Magellan probably had several charts to chose; for according to the expedition's expense records, twenty-four charts and one globe were acquired in three different transactions, of which the globe and one chart were given to the king, leaving twenty-three charts for the five vessels.[211] With chart and compass displayed, Magellan had his interpreter describe the strait through which they had reached the ocean and its islands, and the length of their voyage.

Cebu

It was now the end of March 1521. Easter fell on the last day of the month, and Magellan seized the opportunity to indoctrinate the Filipinos in the Christian faith. After ceremonies, a cross was planted on Limasawa's highest mountain. Magellan and Kolambu made a blood compact of their friendship, and discussed to which isle the visitors should sail next. The decision appears to have been an easy one: Cebu. When Magellan asked Kolambu and the other Filipinos "which was the best port for revictualing," they replied that "there were three of them. Ceylon [Panoan or southern Leyte], Zubu [Cebu], and Calaghan [Caraga, part of Mindanao], but that

Zubu was the largest and had the most trade" The chief offered to guide them to Cebu, and Magellan accepted the invitation.

Taking a northwest course from Limawasa, "we passed among five islands . . . Ceilon, Bohol, Canghu, Baibai, and Catighan." Bohol is still known by Pigafetta's spelling, and is recorded as such by Gastaldi in 1548 and 1561, though it is not identified by Ramusio. *Canghu* is Canigao, a small island lying to the southwest of Leyte; *Baibai* is part of western Leyte, which Gastaldi (1561), following Pigafetta, maps as a separate island *(Baybay),* placing it below *Papuas* (Panglao or Negros). *Catighan* has not been identified with any certainty. Gastaldi mapped it on the 1561 map, and it is probably the *cāningar* of his 1548 map and the *Candigan* of Ramusio, yet in 1545, the Spanish chartmaker Santa Cruz places the island, as *Candigar,* south of Mindanao in the position of Sarangani, which is altogether in the wrong place for *Catighan.*

Magellan and crew sailed toward Cebu. Pigafetta wrote that since Kolambu "could not keep up with us, we waited for him near three islands, that is to say Polo, Ticobon, and Pozon." These three are islands of the Camotes, sandwiched between Leyte, Bohol, and Cebu. *Polo,* which is Poro Island, is not shown by either Gastaldi (1548) or Ramusio, but recorded as as part of *Philippina* Island by Gastaldi in 1561. *Ticoban* has been identified as the modern Pasijan, and *Pozon* as Ponsón. Ramusio maps *Pozon* as a small island just south of *Messana.*

But here we experience a case of the queer 'blurring together' of the complex Philippine nomenclature. Ramusio's island of *Pozon,* so close in spelling and sound (at least to the European mapmaker) to '*Lozon*', may have inspired Gastaldi in mapping Luzon on the 1561 map. If Gastaldi believed (reasonably) one to be a duplication of the other, Ramusio's *Pozon* helped seed the first appearance of the island of Luzon on a printed map. This will fit into place in a later segment of our voyage.

"On the seventh of April at midday," Pigafetta relates, "we entered the port of *Zubu* [Cebu], passing by many villages, seeing many houses on tree trunks, and we approached the city." Pigafetta's description of Philippine dwellings would still sound familiar to many modern visitors to the rural parts of the islands:

> The houses are of wood, built of planks and bamboo on large pilings raised above the ground, which they have to enter by means of ladders, and they have rooms like ours. Under the houses they keep their pigs, goats, chickens.

It was on Cebu that the Spaniards formally began their task of establishing trading links with the Indies. The ruler of Cebu, whose name was Humabon, was well-experienced at trade, and figured importantly into Magellan's dealings and experiences — the most northeasterly of the 1548 Gastaldi's Philippines, an island named *Huban,* may actually be a confusion with the name 'Humabon'. After firing the fleet's guns, Magellan sent an ambassador ashore to Humabon, assuring him of their friendly intentions and desire to trade. The king explained the tariffs he required for trade in Cebu. Magellan, in response, set out his own terms, demanding Spanish exemption from tariffs, as well as the islanders' conversion to Christianity. Humabon explained that a 'Junk of Ciama' (*Ciama* being either Siam or Champa) had just passed through, trading gold and slaves and he summoned one of that vessel's agents — described in the printed account as a 'Moor' — who had remained behind to gather merchandise, to confirm that they indeed had themselves paid the tariff. The foreign merchant, believing the Europeans to be Portuguese, warned the king that these were the very people who had subjugated Calicut and Malacca, and that perhaps they should be appeased. The king said he would confer with his people about his policy toward the foreigners. Pigafetta, observing the Cebuans' procedures for trade, noted that they measured commodities with "wooden balances, with a cord in the middle to hold it by, on one side is the weight, and they are very similar to ours."

The king "had a meal of many foods brought, all on porcelain plates, with several vessels of wine"; this and subsequent references to porcelain testify to Ming trade with the Philippines. The 'wine' was the *tuba* of the Filipinos, which Pigafetta called *arrack*. Richard Eden's version related that four of these "vessels made of the fine earth called Porcellana" were filled with wine made from date trees (the 'date' tree was a coconut; Eden, however, had taken his account from Martyr's Latin, who in turn had no word for 'coconut').[212] The Filipino liquor was perhaps a similar intoxicant to that enjoyed by Marco Polo on Sumatra two-and-a-quarter centuries earlier.

In 1548 Gastaldi marked Cebu as *cubu*, due east of Palawan (*polagua*). Ramusio, however, appears to have combined different sources which refer to the island by different spellings, for he charts Cebu twice: once as *Cyābu* (immediately to the west of *Messana*), and a second time, to the southwest, as *Zubut*. Pigafetta reported that the island lies at a latitude of 10° north and Gastaldi (1548) following Pigafetta, placed the southern coast of the island at that latitude. Ramusio, on the other hand, located both his Cebus slightly to the south, while Gastaldi (1561) similarly duplicates Cebu, once as *Ciabu*, judged to be a region of the island of *Philippina* rather than a separate island, and again as the island of *Zubut*, in the eastern end of the archipelago.

Humabon's nephew, who was a prince (he was married to the king's eldest daughter), along with Kolambu, the trader from Ciama, and other functionaries, met with Magellan to exchange friendly words and intentions of trade. The prince led some of the Spaniards, Pigafetta among them,

> to his house and showed us four girls who were playing on very strange and soft instruments, and their way of playing [was] somewhat melodious. One was playing a little drum our way, but it was placed on the ground. Another was beating a stick, with its head wrapped in palm cloth, upon the bottom of two instruments made like long drums. Another [girl played on] another but larger [instrument] in the same manner. The last one [played on] two similar ones, one in each hand, beating them in rhythm and making a very soft sound. These girls were rather pretty, almost white [in complexion] and as tall as ours. They were naked, except that from the waist down to the knees they wore a cloth made of palm to cover their nature. Some of them were all naked, having long black hair and a little veil around the head, and they always go barefoot. The prince made us dance with three of the naked ones. Then we ate a light meal and later returned to the ships. [By the way], these types of drums are made of metal, they are made in the land of the Great Sine, which is China. There they use bells like ours which they call *Agon*.[203]

Pigafetta also mentions stringed instruments on Cebu, relating that "this people plays a viol with copper strings."

Mactan and the Death of Magellan

It was during the company's stay in Cebu that Magellan's unyielding determination to extract commercial and spiritual concessions from the Filipinos led to disaster. Although the commander was able to use threats of force to evade the tribute paid by other trading vessels, as well as to 'convert' reluctant local chiefs to Catholicism, his burning of entire villages whose people resisted conversion, and his attempt to subordinate the entire

archipelago under a chief acting as his puppet, proved foolhardy. Cilapulapu, or Lapu-Lapu, chief of part of a small neighboring island which Pigafetta called *Mattam* or *Matan* (Mactan), defied submission to this strange faraway place called Spain even though another chief on the island, named Zula, had already yielded to Magellan. On April 27 of 1521, Magellan and several of his crew were killed in an attempt to subdue Lapu-Lapu. As a symbolic tragedy for historical romanticists, this is rivaled only by the death of James Cook two and a half centuries later. "Thus," Peter Martyr observed, perhaps wryly, "did this brave Portuguese, Magellan, satisfy his craving for spices."

The 1548 Gastaldi places Mactan off the southeast of Cebu and spells it *matā* (= *matan*). This being the island where Magellan was killed, it is curious to note the coincidence of sounds between Gastaldi's *matā* and the word for 'kill' in Spanish (*mata*) and some dialects of Italian, such as Gastaldi's own Venetian. Many readers of the 1548 *Geographia,* sipping wine and burning midnight oil, must have had a macabre joke about it. Mactan appears as *Matam* on the 1561 map, but is not identifiable on the Ramusio.

From Cebu to Mindanao

After leaving Cebu, the remaining crew sailed to Bohol, which is marked by Gastaldi 1548 and 1561 as being to the southeast of *matā*. On the northwest coast of Bohol, the men set fire to their vessel, the *Concepcion,* since they no longer had sufficient hands to man her. From Bohol, the fleet — now consisting of only two ships — sailed southwest, and skirted the island of *Pauiloghon,* on which "there are men as black as Ethiopians." True to Pigafetta's description, Gastaldi (1548) has mapped this island as *Negros;* Ramusio and Gastaldi 1561 both mark it *Papuas.* This is either the large island known today as Negros, or the smaller island of Panglao, which lies just off the coast of Bohol.[214] Continuing due south, the two vessels reached a large island. They had arrived at Mindanao.

Mindanao

Gastaldi (1548) records Mindanao as *mendana;* Ramusio and the 1561 Gastaldi, following a spelling used by Oviedo, Peter Martyr, and the Spanish cartographer Santa Cruz, use *Vendanao* (Ramusio) and *Vendenao* (1561). The Ramusio version is remarkably fine, unmistakably recording the principal features of the southern coast: the Zamboanga Peninsula, the Moro Gulf, as well as the Davao Gulf and the peninsula forming its eastern bounds. As regards the shape of Mindanao, the 1561 map was a small step backwards.

As admirable as the shape of Ramusio's Mindanao is, both he and Gastaldi (1561) missed the clues that were already available to be able to fix the island's latitude more accurately. The pilot Juan Gaytan, who was among the survivors of the disastrous 1542 expedition under Villalobos, made an official report on the expedition in 1547 or 1548, from which Ramusio composed an account for the first volume of his *Navigationi* (1550).[215] Gaytan reported that Mindanao extended from 5° or 6° to 11¹/₂° north latitude (actually 5° 33′ to 9°49′), quite an accurate figure, given the large island's complex shape. Gaytan's approximation of its circumference was even more impressive — 380 leagues, which is about 2,445 kilometers, against the actual figure of about 2,700 kilometers.

Although Gastaldi did not have the benefit of Gaytan's data when making the 1548 map, he did have excellent figures from Pigafetta, whom we left on his arrival in Mindanao. By this time, the Magellan expedition no longer had their Sumatran interpreter — either he had been killed along with Magellan, or he was in

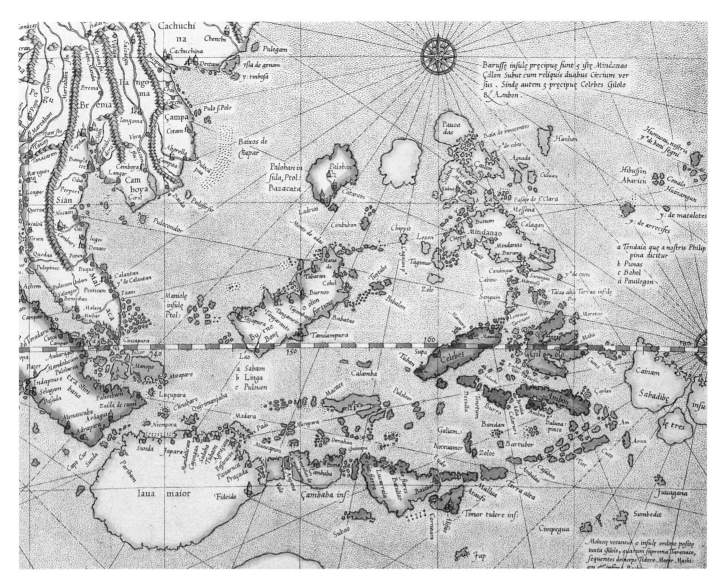

Fig. 76 World map (detail), Gerard Mercator, 1569. (134 x 212 cm) [Maritiem Museum 'Prins Hendrik', Rotterdam]

collusion with the dissenters on Mactam — and so it was left to Pigafetta, who was an adept student of languages, to replace Magellan's former slave in that role as best he could. Pigafetta

> went ashore alone with the king to see this island. When they entered a river, many fishermen presented fish to the king. Then the king and his chieftains took off their clothes, and, singing, they began to row, passing by many dwellings, on the river, and he arrived at his house at two o'clock at night and it was two leagues from the mouth of the river to the house of the king.

Pigafetta was brought into the king's house, where "they came upon many torches made of bamboo and palm leaves, for supper was approaching." He stayed overnight in the house, and the next morning set out exploring the neighborhood. As on Cebu, Pigafetta describes various aspects of Mindanao life in detail. He visited the queen, where

> several porcelain vessels were hanging in the house, and four bronze cymbals [gongs], one larger than the others, and two smaller ones to ring with.

Pigafetta seems to have realized by now that *Butuan* and *Caraga,* the realms of their old friends, the kings Kolambu and

Siagu (Siani/Siain), were in fact part of Mindanao, noting that "this part of the island is of a piece with Buthuan, and Calaghan," and that it "overlooks Bohol, and shares a boundary with Mazana [Limawasa]." Gastaldi (1561), however, defers to the earlier insular inference, mapping both *Butaun* and *Sian* as separate islands. Pigafetta's hint, however, was correctly deciphered by Mercator in 1569 (fig. 76).

Pigafetta did not as yet introduce the name 'Mindanao'. For the moment he identified the island simply as *Chippit* (Maximilian has it as *Gibeth*), and indeed they were at Quipit, a good harbor on the northwest side of the island. Pigafetta assigns *Chippet* a latitude of 8° north, which was an extremely accurate figure (the modern city lies at 8° 4'). Gastaldi (1561) again gets himself into trouble here by attempting to put in more detail than he had previously recorded in the 1548 and Ramusio maps. Not realizing that Quipit is a region of Mindanao, he maps *Chipit* as a small island about 4° east of Mindanao. In fact, he maps the 'island' twice, once as *Chipit*, and again as *Chippit*. He may have been fooled by the variant spellings, believing them to represent two different islands, and because Pigafetta mentions Quipit again when they returned to Mindanao later on. Neither is truly placed at the 8° latitude measured by Pigafetta, although the westerly one is close.

First Mention of Luzon

We now get the first glimpse of the principal island in the Philippine archipelago. Pigafetta was told that "two days from [*Chippit*] to the northwest is a large island called Lozon." This was Luzon, but given the weakened state of the Magellan enterprise and the determination of the survivors to reach the Moluccas (which lay in the opposite direction), the recorded European discovery of Luzon would have to wait. Magellan, had been specifically instructed by the king of Spain "that first and foremost, before sailing elsewhere, you proceed without fail to the said Moluccas."[216] Thus, Gastaldi, in 1561, with nothing more than Pigafetta's second-hand mention of the island to go on, introduced the first definite reference to Luzon on a Western printed map. Just as Japan was arbitrarily mapped by European cartographers using second-hand reports, well before its 'discovery' (see, for example, Münster's *Zipangri,* fig. 72), so too was Luzon placed on Gastaldi's 1561 map on the basis of a textual reference.[217] While it is true that Portuguese sailors may have reached Luzon in 1545, and that word of Luzon could also have come from Luzon merchants trading in Malaya, or from Japanese sailors (who knew it as *Ruson* or *Roson),* the spelling and placement of the 1561 Gastaldi map's little island indicate that it was born from nothing more than Pigafetta's allusion to it. Gastaldi simply judged a two-day sail to the northwest of one of his *Quipits,* engraved an island and labeled it *Lozon.* He probably fussed over Ramusio's island of *Pozon,* and out of concern that Pigafetta might have unknowingly recorded the same island twice, kept Ramusio's *Ponsón* but left it unnamed, adding his *Lozon* just north of it. In fact, his duplication of Quipit may have been caused by his mapping of Luzon, since his northern Quipit lay in exactly the same position relative to his Luzon as the southern Quipit does in relation to the island, now unnamed, which had previously been Ramusio's *Pozon.*

Pigafetta himself actually met, and spoke with, a native of Luzon. Leaving Borneo on 29 July 1521, the Spanish, believing they were being ambushed, attacked some junks and captured "the son of the king of Lozon," who was the "captain general" of the king of Brunei. Pigafetta explained that the man from Luzon had come with the junks "from a great city called Lao" which was at the end of Borneo facing Java (in other words, the southwest coast), which they claimed to have just sacked. This may have been the kingdom of *Lochac* visited by Marco Polo in the 1290s, if the latter was not Siam; later in the present (sixteenth) century, we will see more of this kingdom of *Lao,* alternatively identified as *Laue* (page 178 below). The Magellan crew also met traders from Luzon when they put into Timor. In 1589, Abraham Ortelius was influenced by reports of Luzon when composing his *Maris Pacifici* — although he did not map Luzon as such, he divided the archipelago into the *Philippinas* in the south, and the '*Lucoes*', the *Islas de Lucois,* in the north.

From Mindanao to Palawan

From Quipit, the pair of Spanish ships sailed southwest until they reached *Caghaian* (Cagayan Sulu), which Pigafetta accurately placed at 7½° north latitude (the small island is centered at 7°). Yet again, Gastaldi (1548) charts his *Caguyan* as prescribed, while the 1561 map places its *caghagian* to the south in the region of Molucca. Pigafetta reported that the island was sparsely inhabited by exiles from Borneo, and that it barely provided enough food for its people. Unable to supply the Spanish with the commodities they needed, the expedition left the island, and continued west-northwest until reaching an island called *Pulaoan* by Pigafetta. This was, of course, the island of Palawan.

Palawan

Palawan is recorded by Gastaldi (1548) as *polaguã* and by Ramusio as *Palobã.* Since Palawan is a very long, very narrow, diagonally oriented island which the Spanish crew did not circumnavigate or even explore extensively, no cartographer could yet have mapped it confidently based on the expedition's reports. Gastaldi, however, succeeded sufficiently well in rudimentarily portraying the island's shape and orientation on the 1548 map so as to suggest that the Magellan voyage could not have been his only source for mapping it. Information from Portuguese voyages to China from Malacca may have touched on the island and provided Gastaldi with some hints of the island's shape, just as they may have led to the mapping of the *Costa de Incõis* on some Portuguese charts and the Ortelius map of China (fig. 88). Neither the Ramusio nor 1561 map's depiction of Palawan rivals that of the 1548 map, either in shape or placement.

Pigafetta placed the island at a latitude of 9° 2′ north; if this measurement was accurate, this means they reached Palawan roughly midway along its coastline, on the south end of Island Bay. But since the stop at Palawan was simply a diversion for resupplying and nothing more, they would have gone no further north than necessary. Whereas in 1548 Gastaldi (probably correctly) understood their landfall to be toward the south of the island, Ramusio envisioned it along the island's northern coastline. Given that Pigafetta's latitudes tended to be slightly north of true, and that they approached from the south, we can speculate instead that they probably landed toward the southern part of the island's eastern coast.

Curiously, Palawan is found on some of the separately-published Italian world maps of the mid-century as *Pola Guan,* which seems to be *Gauenispola,* the little island off the northwest coast of Sumatra first known to the West from Marco Polo, in reverse. When these mapmakers, who were largely inspired by Gastaldi, learned that *pulo* meant 'island', they probably suspected that *Gaunispola* was a corruption of *Pulo Guanis,* and associated it with the *Pulaoan* of the Magellan expedition, thus identifying Palawan as *Pola Guan* (see also page 108, Ruysch, above). *Gauenispola* itself is recorded on all three maps in its proper position.

Borneo

In contrast to *Cagayan de Sulu,* the crew found ample supplies on Palawan, such that "it could be called the Promised Land. For if they had not found it, they would have gone very hungry. The king made peace with them," and they made a blood compact, a rite with which the Spanish were by now familiar. Now sailing to the southwest of Palawan, after ten leagues another island was sighted: they had reached Borneo. Here we get some of Pigafetta's finest imagery: as they approached the island, it "seemed to be moving slightly upward," perhaps because of the distortion of the air caused by "the fires known as the corposant" [although Pigafetta does not make the connection]. They traveled fifty leagues "from the beginning of this island" to a harbor.

> On the following day, July ninth, the king of this island sent us a proa, which is like a galley, very beautiful with the bow and stern worked in gold; on the bow it had a white and blue banner, and topped with peacock feathers. Some people were playing on stringed instruments and drums. With this proa came two almadias, which are like fishing boats. And eight old men among the chieftains, came aboard their ship, and sat on a rug in the stern and presented to them a painted wooden vessel full of betel leaves and areca nuts, which is the fruit that they always chew with jasmine and orange blossoms, covered with a

cloth of yellow silk, two cages of chickens, a pair of goats, three vessels full of rice wine distilled by an alembic, and some bundles of sugar cane, and they also made gifts to the other ship, and, after embracing them, took leave.

Like Münster in 1540, both Gastaldi (1548) and Ramusio use the term *Java Minor* for Borneo, rather than Sumatra. In 1561, Gastaldi abandoned the term *Java Minor* for Borneo since, as we will see later, he decided on a new location and identity for it as an antipodean island. All three Gastaldi maps record the kingdom of Brunei, where the Spanish landed.

Pigafetta placed Borneo (that is, the sultanate of Brunei) at latitude 5° 15′ north, which indeed corresponds to the correct figure for the northern part of the large bay on whose southern end the modern city of Bandar Seri Begawan lies. Once again, the 1548 Gastaldi follows Pigafetta's latitudes more carefully than Ramusio or the Gastaldi of 1561. *Burneo* lies on the eastern shores of the Ramusio map, whereas on the Gastaldi (1548) it lay on the west coast; by 1569 Mercator correctly located it midway along the island's northern shores (fig. 76). As for determining the size of Borneo, Pigafetta gathered an idea in itinerary-language from his hosts: according to people he conversed with on Borneo, the island is of a size that it would take three months to circumnavigate by boat.

Following various adventures around Brunei, the Spanish expedition explored the Sulu Sea and its islands, searching for the elusive Moluccas, the fabled Spiceries. After passing reefs and bountiful seaweed, they came to the islands of *Zolo* and *Taghima* (Jolo and Basilan, just southwest of Mindanao), "near which pearls are found." Gastaldi (1561) places his *Solor* and *Taguina* correctly in relation to Mindanao, as does Ramusio, though a bit more sloppily, with his *Zolo* and *Tagliman*. The 1548 Gastaldi charts Pigafetta's *Colo* and *Taguima* the most inaccurately of the three maps. Food and water were difficult to come by during this episode in their voyage, and they eventually returned to Mindanao, this time on the south side of the island, for supplies. They reached the Moro Gulf in southern Mindanao, around the Zamboanga Peninsula.

The Modern Name 'Mindanao'

"Taking a northeast course," wrote Pigafetta, "we went to a large city named Mangdando," which he correctly realizes is "on the island of Batuan and Calaghan," in other words, on Mindanao, though Gastaldi (1561) did not heed this detail, making them three separate islands. This is the first mention of Mindanao's modern name.

The crew captured a group of local rulers, one of whom advised the Spanish that the 'spice' islands they were seeking lay to the southeast. The Spanish, as a result, changed their vessels' course. Sailing to the southeast brought them to the island of Sarangani, which lies just off the southernmost cape of Mindanao. Ramusio charts Sarangani in the correct position relative to Mindanao on east-west axis, though too far to the south from it. Near by, one sees *Candigar,* which is Sarangani or its sister island, Balut. The mapmaker Santa Cruz used *Candigar* for Sarangani in 1545, drawing from the voyages of Loaisa (1525) and Saavedra (1529). Below *Cādingar* on the 1548 Gastaldi lies an island named *Taguima* (probably Basilan, just south of Mindanao at Zamboanga), which was mentioned by Santa Cruz, who grouped it with other islands to which he assigns a latitude of between 6° and 9° north latitude.

Other early European explorers reached Sarangani as well. The luckless Villalobos crew passed nearly one and a half years on *Sarangā* and *Candigar,* coming upon a pirates' lair in the latter. Unable to secure their necessities from these islands, Villalobos sent a vessel to the north to find food. After unsuccessfully attempting to find provisions on Mindanao, Villalobos' scouts continued to an island which was called *Tendaia* by its people. This was probably the island of Samar, though possibly Leyte.

The Modern Name 'Philippines'

In contrast to Villalobos' fruitless experience on the coasts of Mindanao (according to some later accounts, the Portuguese had convinced the islanders not to give any assistance to Spaniards), the citizens of *Tendaia* were extremely kind and generous. They enabled the crew to put together the food and fresh water they so desperately needed. Inspired by the benevolent reception, the Spaniards dubbed the island *Filipina* after the then Crown Prince of Spain.

Thus, the modern name of the archipelago was given by Villalobos in 1543. Ramusio christens the printed cartographic use of the term 'Philippines' with his *Filipina* island, a thin isle which he places to the east of Mindanao. But whether Villalobos' island was Samar or Leyte, both lie to the north of Mindanao, not east. In 1561 Gastaldi caught this error and created a new, large island of *Philippina,* which he properly locates just north of Mindanao. On it he places kingdoms which are actually island names: *Ciabu* (Cebu), *Polo* (Poro), and *Cangu* (Canigao).

Ramusio's same narrow, vertical *Filipina* island is retained by Gastaldi in 1561, but has assumed a curious new guise — it is now *y delle dóne,* the 'island of women'. The original source of this transformation is a mystery, but we can make a plausible theory about it. In his narrative, Pigafetta recalled that an "old pilot of the Moluccas" told him about an "island where there are only women, who conceive by the wind, and if there is a male, they kill him." Although Pigafetta considered this story nonsense, it is possible that Gastaldi, having inherited from the Ramusio map an island whose feminine name, *Filipina,* he had already used for another island, judged the 'island of women' to be the closest replacement for what had been an island with a woman's name. Another possible explanation is that the notion of an 'island of women' may have been imported from Arab sources. However, although an 'island of women' is indeed found in early Arab travel narratives of Southeast Asia, claims made by some of Gastaldi's contemporaries that he tapped Arab sources are dubious (see page 131, above).

San Lazaro

Magellan dubbed the Philippines the *Archipelago of San Lazaro* at the very beginning of the expedition's wanderings through the archipelago, applying the term to Homonhón, Suluan, Samar, and the little isles in their vicinity. It is not clear how many subsequent landfalls were still deemed to be part of the San Lazaros. Pigafetta explained that "thereabouts [Homonhón] are many neighboring islands. Hence we called them the St. Lazarus Archipelago because we stayed there on the day and feast of Saint Lazarus. This region and archipelago is 10 degrees latitude north and 161 degrees longitude from the Demarcation Line." Ramusio and Gastaldi (1561) use the term *Arcipelago di San Lázaro* in its limited sense, denoting the sea and islands in the vicinity of Panaon.

But because Ramusio and other mapmakers had understated the distance between the Marianas and the Philippines, the distinction between the two was lost; and the *Ladrones* appeared to be part of the *San Lazaros.* Beginning with Ramusio, we see a tendency to gently push the old term over to the east, indiscriminately blurring into various early landfalls in Micronesia what would later be regarded as the 'New Philippines' (see figs. 13 & 20). In 1561, Gastaldi took this transference one step further when he replaced the term *Ladroni,* for the Mariana chain, with *S. Lazaro* and, as we saw earlier, identified the Philippine island of Samar *(Zamal)* as its southernmost point.

The Spiceries and Indonesia

Departing Mindanao for the second time, the Magellan crew "set their course east by south, in order to find the Molucca islands." Pigafetta gives a cursory mention of the many islands they passed during this part of their journey. He gives an impeccable reading of $3^{1}/2°$ north for one called *Sanghir,* today known as Pulau Sangihe, a small island group lying roughly midway between Mindanao and the Celebes. One of the men captured at Saranghani to guide them to the Moluccas escaped by swimming to Sangihe at night. Yet again, Gastaldi (1548) obeys Pigafetta, placing *sangil* as prescribed, while Ramusio's *Sangil* is pushed a bit to the south.

Passing several more islands, they finally reached their goal and "discovered four islands rising high in the east, [which] the pilot whom they had captured, said . . . were the Moluccas." Having thus sighted four of the five small islands which at that time were the only source for the cherished clove, the determination of the coordinates of these islands was a foremost task of the expedition.

Francisco Serrão and the Longitude of the Moluccas

When the Spaniards reached Ternate, they rejoined the tracks of other Europeans who had arrived in Indonesia from the west, and in so doing metaphorically tied the earth together. They were now in 'known' territory, the farthest eastern European outpost, where Magellan would have been reunited with his close friend from his days of service to the Portuguese Crown in Goa and Malacca, Francisco Serrão. But Serrão was also dead. "Seven months had not yet passed," Pigafetta learned, since Serrão had been "poisoned with betel leaves" by the king of Tidore. As it was now the second week of November 1521, that would place Serrão's death in mid to later April, ironically coinciding with the death of Magellan himself. Pigafetta met his widow, "a woman he had taken in Greater Java" (though other sources suggest she was the daughter of the sultan of Tidore), and their son and daughter. The Spanish also met another Portuguese, Pedro Alfonso, who sought a passage for himself and his wife to Spain.

Francisco Serrão provided the earliest first-hand Portuguese estimate of the position of the Moluccas, and was the source of Magellan's vision of the islands when he was preparing for his expedition. Serrão had placed the islands considerably farther to the east that they actually lay, an error which no doubt contributed to Magellan's conviction that a Pacific crossing from the east would be an advantageous alternative to the circum-Africa route. Serrão's misplacement of the Moluccas probably appeared all the more credible since it erred in Spain's favor, not that of his own Portugal; if Serrão were correct, then the islands lay on the Spanish side of the Line of Demarcation, and he would have effectively been evicting himself from his occupancy of them.

But although Serrão served Lisbon, he likely had a personal interest in exaggerating the islands' distance from Malacca. Serrão had sailed to the Spiceries with de António de Abreu at the end of 1511 with a fleet of three ships. The seas were new to the Portuguese, but they had Malay seamen to guide them. During the outward voyage, Serrão's ship, an Indian vessel taken at Goa, became damaged to the point of unseaworthiness, and a Chinese junk was purchased in Banda to replace it. When the three vessels set about their return to Malacca, Serrão's junk was separated in a storm and shipwrecked on an uninhabited island. He purportedly survived by tricking some pirates into preying upon them, and then ambushing the pirates as they landed and commandeering their ship. In any event, Serrão eventually reached Ternate, where he seized the opportunity for a colorful, symbiotic relationship with the island's sultan. Serrão,

accustomed to the hard and uncertain life of a Portuguese seaman, was suddenly a respected and pampered member of the elite of the very island group Europeans had for centuries longed to reach. Serrão had fallen, so it must have seemed to him, into an earthly paradise. From the sultan's point of view, Serrão promised to be the ambassador of lucrative trade with Portugal. That relationship would, in turn, provide the military strength to defeat his rival, the ruler of Tidore, the island immediately to the southeast, separated from Ternate only by a narrow strait.

Serrão, however, could maintain his lofty role only if both the Ternatean sultan and the Portuguese authorities believed that his intercession was beneficial. By offsetting the Moluccas to the east, Serrão accomplished two things. First, he made a routine shuttle between Malacca and the Moluccas seem less practical, and thus the role of a permanent ambassador more important; secondly, since it placed the islands within Spain's jurisdiction geographically, Portugal would only have reasons of pre-eminence and precedent — a role which Serrão was personally in a unique position to fulfil — to claim them as their own.

Pigafetta and the Longitude of the Moluccas

Pigafetta's estimate of the longitude of the Moluccas was markedly superior to the longitude he had assigned the Philippines — an illustration of the fact that although the *direction* of longitude error was almost certainly politically influenced, the fact that the figure was inaccurate was not itself necessarily deliberate. What we, in hindsight. might see as politically motivated deceptions of longitude, may in fact have been simply the temptation to opt for the most advantageous figure within the range of possible truths. Had Pigafetta been deliberately distorting longitude in the favor of his patron, he would have pushed the Moluccas at least as far to the east as he had the Philippine island of Homonhón, not less so.

Whereas Pigafetta had placed their landfall in Homonhón at 161° west of the Line of Demarcation, he put Tidore at 171°, although the island actually lies to the east of Homonhón. Analyzing his figure the same way we did with Homonhón, we find that his longitude for Tidore, a far more important prize, is more accurate. The island lies about $232^{1}/2°$ west of Greenwich, compared to Pigafetta's adjusted figure of $218^{1}/2°$ (his figure plus the distance from Greenwich to the Line), leaving an error of 14°, roughly half the error of the Homonhón figure. Since his longitude was determined by dead-reckoning, and thus errors were cumulative, this more accurate figure was simply the result of some errors canceling out others. The expedition's extensive east-west meanderings from Mindanao to Palawan and Borneo, then back to Mindanao, must take credit for these corrective 'errors'.

The Portuguese and the Latitude of the Moluccas

Whereas longitude was determined by dead-reckoning, with each calculation being based on the previous estimated figure, latitude was based on absolute astronomical measurements and could be determined with reasonable precision, independent of previous readings. Even so, portolan charts from earlier in the sixteenth century tended to place the Moluccas too far south. Although these discrepancies may seem small for such an early date, the reason for this inaccuracy is unclear. As with the longitude question, it has commonly been attributed to deliberate Portuguese misinformation, based on the premise that Lisbon wished to keep others off the scent of their precious Spiceries for as long as possible, and so mis-stated their true location. Soon after the present trio of Gastaldi maps were compiled, mapmakers like Abraham Ortelius became aware of the inaccuracy of the figures given by João de Barros, who presented the official Portuguese opinion on the matter. In 1567, after the last of the three Gastaldi maps was published, the geographers Guillaume Postel and Ortelius are known to have criticized Barros' coordinates for the Moluccas.

In fact, several sources of uncertainty in the methodology for determining latitude could introduce small errors. The position of the Pole Star itself, that great anchor in the northern sky, slowly changes. It was about $3^{1}/2°$ off true celestial north at the turn of the sixteenth century, as opposed to just over one degree today. At the same time, the calibration of astrolabes and other instruments was imperfect. Solar declination tables, or 'regiments of the sun', were used to correct latitude figures obtained by measuring the sun's elevation, but some seamen complained that faulty tables in their almanacs introduced error into the readings from their astrolabes. A pilot on the voyage of Miguel Lopez de Legazpi (1565), determined that variations in almanacs were causing discrepancies in readings between ships, since one of the vessels "had been using almanacs made in Spain but we in the flagship used almanacs made in Mexico which had been derived from those of Spain, adjusted for the time it takes the sun to arrive at Mexico."[218] Taking a reliable reading on the deck of a sea-borne vessel was also a major problem. In the mid-sixteenth century a contraption was proposed which involved a suspended chair that would remain level despite the pitching of rolling of the ship, though this idea is not known to have been used.

The equatorial region itself presented new challenges for determining latitude. The Pole Star is too low to be used for such observations — English sailors complained of "losing the pole" several degrees further north. The star's apparent position could also be distorted by the refraction of its oblique path through the atmosphere near the equator, so that the star might be 'seen' even though it were actually below the horizon. Since the Moluccas straddle the equator, the path of the sun directly overhead should also have marked whether an island was to the north or south of it, but the fact that the sun itself has an apparent span of part of a degree again left room for small errors and uncertainty.

Although there is no star in the southern hemisphere truly comparable to the Pole Star, Indonesian and Arab sailors had long ago learned to use southern stars for navigation. Ludovico di Varthema, sailing in equatorial Southeast Asian waters only seven years before Serrão, inquired about these stars. En route from Borneo (or possibly Buru) to Java, he asked the captain of the ship how he steered with the north star no longer visible, and the captain

> showed us four or five stars, among which there was one which he said was opposite to our north star, and that he sailed by the north because the magnet was adjusted and subjected to our north. He also told us that on the other side of [Java], towards the south, there are [people] who navigate by the said four or five stars opposite to ours.

Interestingly, Barros, the official Portuguese chronicler, reached the conclusion that the Moluccas themselves were a relatively recent arrival in the Pacific. Since the Portuguese reported finding seashells when digging in the earth and at the roots of trees, and since the islanders themselves seemed to lack any oral tradition of a long history on the islands, he reasoned that the Moluccas were submerged until fairly recent times.

Pigafetta and the Latitude of the Moluccas

The Magellan fleet's expense records provide us with an idea of the navigational and mapping instruments used on the voyage. These included twenty-one quadrants (which indicated the altitude of celestial objects observed through a sight hole), of which fifteen were 'bronzed', one wooden astrolabe, six metal astrolabes "with their ruled lines", six hour glasses (for measuring elapsed time, and thus estimated distance sailed), six pairs of compasses, and a total of thirty-five or thirty-six magnetic needles.[219] The cross-staff, another instrument used for determining latitude, could not be used between 20° north and 20° south at any time of the year because the sun's altitude was too high even in mid-winter, and thus was useless for charting Southeast Asia.

The expedition had lost some of its mapmaking expertise to the Mactan violence. Both Magellan, whom Pigafetta had spoken of as an "expert in making sea charts", and Andres de San Martin, a professional geographer, were among those killed.[220] Despite the handicaps, Pigafetta and the other surviving members of the crew succeeded remarkably well in this foremost task of mapping the Moluccas.

The latitudes assigned by Pigafetta to the Moluccas are generally very sound. He located them only slightly too far south, except for the southernmost one, Bacan. Tidore, the first of the group they visited, was measured by Pigafetta as lying 27′ north and all three Gastaldi maps place it roughly according to this figure, which is hardly distinguishable from the island's actual latitude of 37′ to 45′ north.

Pigafetta had even better success with Ternate, which he placed at 40′ north, a virtual match with its true latitude spanning 45′ to 51′ north. But Ternate mysteriously floats about on our trio of maps. Gastaldi (1548), who otherwise follows Pigafetta meticulously, places his *Tereneta* on the west coast of Halmahera at about 2° north, while Ramusio and the 1561 map, probably influenced by Portuguese models, shove their *Terenate* down to a latitude of between $2^{1}/2°$ - $1^{1}/2°$ *south*.

Now we reach the island of Motir, and with it the additional sheet engraved by Forlani to extend the 1561 map into southern latitudes. "Mutir," prescribes Pigafetta, "lies exactly under the equincotial line" (this island, just above Makian, actually lies 26′ to 29′ north of the equator). Of our three maps, the 1548 Gastaldi's *motil* is just above the equator, Ramusio's straddles it, and on the Gastaldi/Forlani sheet of ca. 1565 it spans about 2° to $3^{1}/2°$ south.

Pigafetta placed Makian about half a degree too far south, believing it to lie 15′ below the equator, when in fact it spans 17′ to 23′ north. Here all three maps roughly follow Pigafetta, Gastaldi (1548) the most faithfully.

Bacan is the final member of the Moluccan quintet. Pigafetta places Bacan at 1° south latitude (the island spans 19′ to 53′ south). Since Pigafetta states (correctly) that it is the largest of the Moluccas, Forlani (1565) depicts it vastly oversized, but correctly places its southern coast at about 1° south, while Ramusio places his *Bachian* $1^{1}/2°$ north. The Gastaldi of 1548 is the most correct.

Peter Martyr admitted to complete bewilderment regarding the location of the Moluccas. He wrote that "the Spanish think [the Moluccas] are five thousand leagues from Hayti, that is to say twenty thousand miles according to the Italian measure; but I think they are mistaken." Having been told that "the greatly desired Moluccas . . . lie seventy-five leagues from the equator," and that according to survivors of the Magellan voyage, "they are within ten degrees," Martyr then tried to work out a set of consistent distances based on the conflicting values accorded to leagues and degrees by "ancient philosophers" and Spanish sailors. He concludes, however, by saying: "let any one who can, understand this; for myself, I give it up."

If the Portuguese indeed used misinformation to thwart the access of other European nations to the Moluccas, it extended beyond the question of the islands' location; Pigafetta's journal suggests that the Portuguese had attempted to falsely portray the

Spiceries as dangerous to navigate. "The Portuguese used to say," Pigafetta recalled when they reached the islands, "that they could not navigate it because of the extensive shoals, and the dark sky." Contrary to this, however, Pigafetta found that "around all these islands as far as the Moluccas, the shallowest bottom that we found, was one hundred and two hundred arms lengths."[221]

When the Magellan expedition reached the Moluccas, Ternate exerted central rule over the other islands, and while Tidore and Bachan had local sultans, Motir and Makian, according to Pigafetta, "have no king, but are governed by the people." The sailors busied themselves gathering their cargo of cloves from the five islands, filling both remaining ships of the expedition. Pigafetta noted that the fetish for the spice was not indigenous to the islands: Although 'Moors' had inhabited the islands for fifty years, the 'pagans', some of whom still lived in the mountains, "did not previously know how to appreciate cloves."

From the Moluccas Onward

Halmahera *(Gilolo)* is prominent on all three Gastaldi maps. Even before mapmakers knew anything of the island's shape, Pigafetta learned from indigenous sources that it was large, "so big that it takes four months to go around it with a proa." Halmahera was entirely misshapen by Gastaldi in 1548. On the Ramusio and Forlani maps, the island rudimentarily boasts two of its four 'arms', but they are facing west, rather than east — an error found after the Ramusio map in Barros' *Década III* of 1563. Barros stated that Halmahera (which he called *Batochino do Moro*) faces islands to its west and enfolds them in three arms of land. Possibly Barros, or his sources, may have consulted a separate chart of the island on which orientation was not marked, with north and south thus being mistakenly reversed.

The island of Sulawesi (the Celebes), which lies to the west and south of Halmahera, appears in a crude fashion on the Ramusio and Forlani maps, but is altogether lacking on the 1548 map. Magellan, on whose voyage the 1548 Gastaldi map was so dependent, brought no word of the island; news of Celebes came shortly afterwards, from such sources as Oviedo. The wild shape of the island, which is remarkably similar to Halmahera but much larger, eluded the makers of published maps into the seventeenth century — the vague contour shown in Ramusio's maiden depiction of the island was little improved upon even on the maps published by Valentijn in the early seventeenth century.

While loading provisions and cargo in the Moluccas, one of the fleet's two remaining vessels, the *Trinidad,* developed a serious leak. Despite the efforts of local divers to find its source, she was judged unfit to attempt the voyage back to Spain. On the 21st of December, the *Victoria* began its solo journey home, guided by two Malay pilots through the complex of islands that lay to their south and west. Pigafetta continued to record his observations as they wove their way through the isles. They passed Buru, which lies just west of Seram, and which Pigafetta accurately placed at 3¹/₂° south (the island spans 3°-3¹/₂° south latitude). Ramusio's Buru meticulously follows Pigafetta, while Forlani errs to the south, and the 1548 map to the north. The island of Ambon is shown twice by Gastaldi (1548), once correctly as a small island at about the correct latitude, but then repeated as a large island in the position of Flores.

Pigafetta was a little less successful with Banda, the original source of nutmeg and mace, plotting it at 6° south, compared with its actual position of 4¹/₂°. The island is found on the Ramusio map, its southern coast at Pigafetta's figure. Ramusio also places Pigafetta's *Mallua* at the prescribed position of 8¹/₂° south, while

Gastaldi (1548) puts it at about 6° (Molu is today a small island just north of Kepulauan Tanimbar, but the name may have referred to the main island, which spans about 7° to 8° south).

Soon they reached Timor, which Pigafetta records as being inhabited by "pagans" (that is, not by "Moors"). He reports that ginger and diverse kinds of fruit come from Timor, and that "people from Java, the Moluccas, Lozon and from all these parts come here to trade for sandalwood," *Lozon* being another pre-'discovery' reference to the northern Philippine island of Luzon. Richard Hakluyt, the late sixteenth century promoter of English expansion, explained that white sandal "is wood very sweet and in great request among the Indians; for they grind it with a little water and anoint their bodies therewith." The Spanish also indulged in a bit of trading here themselves, exchanging "red cloth, vinegar, iron, and nails" for precious woods.

Pigafetta placed Timor at 10° south, which is the southern limit of the island on the 1548 Gastaldi and Ramusio maps. Gastaldi (1548) correctly aligns the island's southern shores at that latitude, while Ramusio has the island straddle the coordinate. Forlani, conversely, decided that the 10° south figure marked the island's northern coast. Timor actually lies between 8° 20′ and 10° 22′ south.

Gunung Api

To the northwest of Timor, Gastaldi (1548) graphically depicts an erupting volcano named *Ocape*. This is *Gunung Api*, which is the Malay for 'fire mountain', a descriptive name bestowed upon small, active volcanic islands. The best-known of these is the isolated *Gunung Api* which pierces the Banda Sea roughly 200 kilometers north of Timor and probably served as a natural 'lighthouse' that gave Southeast Asian pilots their bearings when they made their crossing from the southern Indonesian islands to the Spiceries; it is this 'fire mountain' that Gastaldi would appear to have mapped.

In 1512, António de Abreu and crew sailed eastward as far as Flores and then (as recorded by António Galvão) "set their course toward the north of a small island called *Gumuapè,* because from its highest point streams of fire run continuously into the sea, which is a wonderful thing to behold." This was the first of many such reports about the virile volcano. William Dampier, for example, described it at the end of the seventeenth century (see page 240, below). Ramusio places his *Guinape* on the island just east of Java (Lombok/Sumbawa), suggesting that his sources had passed the volcanic island which lies off the northeast tip of Sumbawa and is now known as Gunung Api or Sangeang Island. Vessels sailing east or west along the northern coasts of Java and the various Sunda islands routinely passed Sangeang. A third 'fire mountain' of consequence to European sailors was the *Gunung Api* of the Banda group. Since this volcano is separated from the main Banda islands only by a narrow strait, its eruptions could be a nuisance to pilots entering the islands (see Valentijn, page 235 and fig. 142).

Mare Lantchiodol

Directly above Timor on the Gastaldi (1548) and Ramusio maps is *Malva* (present-day Pulau Alor or Ombai), where Magellan's crew stayed for two weeks to make repairs to their ship despite, Pigafetta tells us, the island being inhabited by cannibals.

They were now in open seas, which Ramusio and Gastaldi (1561) identify as being the *Mare Lantchiodol.* 'Lantchiodol' is an ancient term which was used by Pigafetta, who related that on February 11th, 1522, they "left the island of Timor, swallowed up in the great sea called Lantchidol, and set their route west southwest . . . for fear of the Portuguese." That is to say, they sailed to the south of Sumatra through the southern seas of *Lantchidol* to avoid a confrontation with the Portuguese. In ancient times, *L'Antichthones* was used to denote southern Asian realms. It was employed by Pliny in the first century to refer to *Taprobana* (Ceylon), but translocated into Southeast Asian waters by Pigafetta and Gastaldi because they believed *Taprobana* to be Sumatra.

Pigafetta now tries to describe to his reader the long chain of islands which ultimately leads to the Southeast Asian mainland. "There are many islands, one after the other, all the way to Greater Java, and the Cape of Malacca, and they are, Eude, Zanabutum, Cile, Araranan, Moin, Zumbona, Lomboch, Chorum, and Greater Java, Maiaoa." Notably missing is Bali, which is actually mentioned by Pigafetta, but mistakenly (though understandably) as a town on Java. Some of the many small islands that fill the Indonesian seas on the 1561 map come from this inventory.

Borneo, Sumatra, Java

Whereas the three Gastaldi maps relied extensively on Pigafetta for the Philippines and eastern Indonesia, their mapping of Java and the western Indonesian islands was based on Portuguese sources. Like Münster in 1540, Gastaldi (1548) and Ramusio put a 'modern' Sumatra in its approximately correct position, in addition to the old *Iava menor,* placed in the position of Borneo. But while Münster's *Iava minor* was a duplication of true Borneo (which he mapped as *Porne*), the *Iava menor* of the 1548 Gastaldi and Ramusio maps is actually Borneo itself. Gastaldi (1548) has more detail, recording Mount Kinabalu (*S. Pedro*), which is the highest peak in Southeast Asia outside of western New Guinea.

A curious fate befalls *Java Minor* on the 1565 Forlani map. Breaking with Gastaldi (1548) and Ramusio, Forlani exorcised the ghost of *Java Minor* from Sumatra and Borneo, and fabricated a new, non-existent island of *Giava Minore* to the southeast of Java. Not only has the question of *Java Minor* been thus shoved out of sight, but, in addition, various other places from Marco Polo not otherwise accounted for — *Felech, Dragoian, Lambri* and *Basma* — have been placed as regions of that island (as indeed they were regions of Sumatra). Surrounding this new '*java minor*' are various other Polian islands, including Sondur, Condore, the Nicobars, and the Andamans. Of these, *Malaiur,* found just off the southeast corner of *Giava Minore,* but believed by modern scholars to have been either part of southern Malaya or neighboring Sumatra, offers a premonition of confusion yet to come. In 1569 Gerard Mercator, who was probably influenced by the southerly location assigned to it by Gastaldi, identified this southern 'island' of *Malaiur* with a southern continent (see chapter 12).

On *Camatra* (Sumatra), Gastaldi (1548) records the kingdoms of Pedir and *Pazer* (Pasai), both of which promptly received trade and diplomatic missions from Albuquerque after the capture of Malacca. Pedir had long been a major center of the pepper trade; just six years before the Portuguese took Malacca, Ludovico di Varthema crossed the strait and reached *Pider* (Pedir), which he believed to be "the best port of the whole island."

The 1548 Gastaldi map also shows *Campar* (Kampar) and *Ardagui* (Indragiri). These states had been vassals of Malacca and following the Portuguese conquest they had sent emissaries to Albuquerque to concede their loyalty to Portugal. The island of *Funda* between Sumatra and *Iava mazor* (Java) is an early appearance of the term *Sunda,* the name of a region of western Java, whose people speak a different language to the rest of the island and after which the strait separating Java and Sumatra is named. On the

Ramusio map, Sumatra's kingdom of Aceh *(Atjeh),* visited by members of da Cunha's expedition in 1506, is mistakenly placed in the southeast, rather than the northwest, part of the island; this is corrected by Gastaldi in 1561 (as well as by Ramusio himself on his 1556 map of Sumatra, figure 35). Of the three maps, Ramusio's alone actually labels Sumatra as *Taprobana,* breaking with Barros' minority opinion that *Taprobana* was Ceylon. All three maps record the island of Bangka off the southeastern coast of Sumatra; although in the mid seventeenth century Wouter Schouten found Bangka to be uninhabited, despite being "large and fruitful", Abbé de Choisy found it well-populated in 1685, and any solitude the island might have enjoyed ended when vast deposits of tin were discovered there in 1709.

Gastaldi's depiction of Java is also a dramatic improvement over previously published maps. Even the earliest map of the trio, the 1548 Gastaldi, has 'modern' nomenclature on Java, notably the city of *iapara,* correctly plotted near where the Dutch settlement of Batavia would be established (today's modern city of Jakarta). Japara was a major link in the trading route between the Moluccas and Malacca. In 1513, nervous about the Portuguese conquest of Malacca, Japara sent a fleet in a bid to take the all-important emporium, but failed.

Since the Magellan crew, now in Portuguese-held seas, sped home without further diversion, Pigafetta allowed second-hand sources and folklore to enter his text, in the same manner that Marco Polo had used similar material to fill his narrative's empty spaces — and with similar results. By the end of his journey through Southeast Asia, Pigafetta introduced diversions reminiscent of Polo, and even Mandeville, complete with the story of birds which carry away elephants, which was the *rukh* we heard about from Ibn Battuta.

Gulf of Tongkin, Hainan, Paracels, and the 'China Sea'

The Gulf of Tongkin appears in a recognizable fashion for the first time on a printed map on the 1548 Gastaldi map, with its northern shores accurately placed at about 21° north, and the Leithou Peninsula (unnamed) correctly forming its northeast bounds. There is not, as yet any trace of the island of Hainan. Ramusio christens that island six years later, referring to it as *Aliosar,* though the modern name *Aynam* had already been cited by Pires in 1515. Mercator introduced the island's correct name on his world map of 1569 (fig. 76), using the same spelling as Pires.

Another island group christened at the hands of the Ramusio map are the Paracels, indicated as a bank with the inscription *Canali donde Vengono gli liquij.* Ramusio's inscription on the Paracels is better explained by Gastaldi in 1561, who identified them as "*Canalli dove passapo i chini per andare a palohan, et a boru*" (the channel through which the Chinese pass *en route* to Palawan and Borneo).

The Gastaldi of 1548, whose insular region is more compressed than the Ramusio or Gastaldi of 1561, gives the western Pacific the general term *Provincia di Maluco,* the 'Province of Molucca', in honor of the Spiceries. For Ramusio and Gastaldi (1561), the sea in which these islands lie is now called the China Sea (*Mare de la China*). The European idea of a 'Sea of China' dates back at least to Marco Polo over two centuries earlier. Pigafetta, near the end of his chronicle, when he is for the most part repeating second-hand reports, mentions that "beyond Greater Java, toward the north, there is a gulf of China that is called Sino Grand." The term *sino* is pivotal here. It could be from the Latin *sinus* or Portuguese *seno,* that is, gulf, making Pigafetta's *Sino Grand* a reference to the old 'Great Gulf' of Ptolemy (as indeed Gastaldi had labeled the sea in 1548 with his Italianicized *Golpho Grande*). Alternatively *sino* could refer to 'China'. While Gastaldi (1548) understood *sino* as 'gulf' *(sinus),* Ramusio understood it as 'China', neatly making the transition from the old Great Gulf to the modern China Sea by way of a linguistic coincidence.

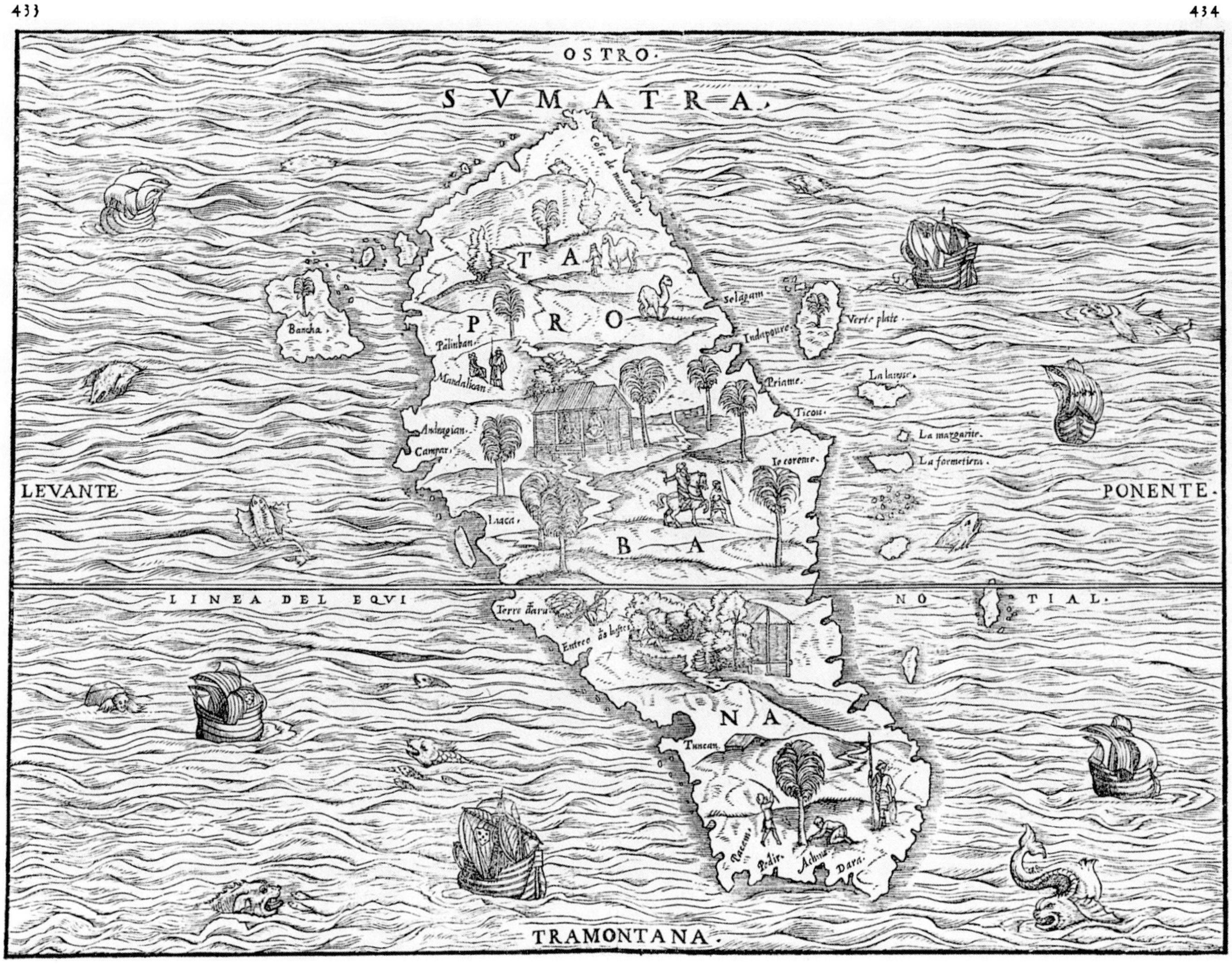

Fig. 77 Sumatra, G. B. Ramusio, 1556 (1565) (27 x 36.6 cm).

The Mainland on the 1548 Gastaldi Map

Malaya

Gastaldi's depiction of the Malay Peninsula and Indochina, though a 'modern' rendering, still bears distant roots in Ptolemy's Golden Chersonese. Like a sculptor pushing and squeezing his clay before it dries with the passing of time, Gastaldi tried to mold the geographical knowledge of his day into Ptolemy's 'Golden Peninsula' rather than abandon it altogether for an entirely new sculpture just yet. He took Ptolemy's Southeast Asian mainland (compare his Malaya to that of Ptolemy, fig. 27), and corrected the latitude of Malaya to extend southward to $1^1/2°$ north, as correctly reported in Pigafetta's journal, rather than the $1^1/2°$ south found in the manuscript and first printed version of his journal, and used by Fine in 1531, or the 3° south, as *per* Ptolemy. The upside-down leaf shape of Ptolemy's southern Malaya (which may have been Sumatra) now leans correctly to the east, and the cape which on Ptolemaic maps probably represented southern Indochina (the emporium of *Zaba,* cited by Ptolemy at 4° 45′ north latitude) now lies just above 10° (the correct figure is just above 8°). Comparing these two peninsulas with their Ptolemaic ancestors shows that Gastaldi equated Ptolemy's *Sinus Perimulicus* with the Gulf of Siam — a name which, indeed, he retains on the map as *Golfo Permuda.*

Singapore

One of the important names christened by Gastaldi is Cape Singapore *(c. cinca pula),* forming the southernmost point of Malaya, the relative position held by *Sabana* on Ptolemaic maps. The origin of the name *Singapore* is not definitely known. Some accounts suggest that it was named *Singapura,* or 'Lion City', after an animal encountered at the river mouth, though lions do not inhabit the region. Another explanation is that the name derived from *singgah* and *pura,* meaning 'the city where one breaks one's journey.' A related interpretation was expressed early on by Albuquerque and Barros, who believed the name meant 'treacherous delay.'[222] Finally, *Singapura* may indeed refer to 'Lion City', but not in the literal sense. Followers of Bhairava-Buddhism at the court of the early Majapahit used the term 'lion' to refer to their prowess and to wild meetings, and some cities — *Singarajya* in Bali, and *Singasari* in Java — were named after that theme.[223] Our *Singapore* may be another such 'city of prowess'.

Singapore, which had probably been intermittently used as a trading and fishing center since ancient times, flourished during the fourteenth century, and effectively controlled the Keppel harbor passage between Pulau Karimun and Pulau Batu Puteh. Both Siam and the Majapahit kingdom of Java were envious of control of the passage, which had probably been under Javanese rule during the latter part of the fourteenth century. By Gastaldi's time, Singapore had likewise become vital to the Portuguese for its strategic position on the eastern entrance to the Malacca Strait; 'foreign' control of it could usurp the Portuguese hold on Malacca. On the 1548 map of Southeast Asia and the 1546 world map upon which it was based, Gastaldi brought the name 'Singapore' to a wide audience.

Cambodia and Burma

In our newly-formed Indochina, the modern name of Cambodia *(Camboja)* appears here recognizably on a printed map for the first time (though Cambodia may have been the *Bocati* of Ruysch and Fries). Such an early use of the name 'Cambodia' is not surprising, since the name had already been used in a letter printed in 1513, from King Manuel to Pope Leo X, recording contact between Cambodian envoys and Albuquerque in Malacca. Tomé Pires also mentions Cambodia in ca. 1515, describing it as a kingdom extending far into the hinterland whose king "does not obey anyone" and is frequently at war with Burma and Siam. It is said to be the source of various products and foodstuffs. The Cambodian city Gastaldi has indicated in vignette is a merely arbitrary marking of a capital city — that metropolis would at this time have been Lovek, as the Cambodian rulers had deserted their capital of Angkor in 1431 due to attacks by Ayuthaya, establishing their new capital further east. In about 1570 the Cambodian king, Satha, relocated his court near the old site, though this apparently proved to be less than ideal, as by 1593 the capital was again moved to Lovek. Gastaldi also introduces the modern name 'Burma' *(Berma),* and marks the Burmese kingdoms of Pegu and Martaban, which we have already seen on printed maps, as well as Ava and Arakan.

Martaban was an important port, though Pires considered it to be treacherous on account of its rushing tides. In 1519, Antonio Corrêa closed a commercial treaty with Burma at a large temple in Martaban. According to Maffei's account, the Portuguese priests and officials who signed the treaty intended to do so in mockery, believing they had no obligation to honor any agreement made with pagans. However, the story continues that when Corrêa deceitfully began to swear his honor to a book of songs, rather than a Bible, he happened to open to an adage advising against making light of one's vows. Believing (the story goes) that the event had been a message from God, he then swore to the terms of the treaty in sincerity. Some European traders used Martaban as a port for smaller vessels wishing to sell wares to the Burmese.

From the port of Martaban, officials oversaw the transport of goods to the customs house in Pegu, while the traders were given a permit to travel to Pegu by themselves. The customs officials who scrutinized the cargo were particularly worried about the smuggling of diamonds, pearls, and expensive textiles. Once their goods had cleared customs, a trader usually rented a house for half a year to do his selling, of which 2 per cent went to a broker. According to the account in Samuel Purchas' massive compilation of voyages, *His Pilgrimes* (1624-26), neither gold nor silver was allowed as exchange; rather, the currency was made of copper and lead, and was minted by each individual. Mendes Pinto, writing to the fathers of Portugal in 1554, related that in Martaban there is a wonderful reclining idol lying among forty-eight stone 'pillows', which the Burmese call the 'god of sleep.' Subsequent visitors, such as Alexander Hamilton described these 'reclining Buddhas', which are still found in Southeast Asia today.

Ava was another kingdom that would later become part of what we know today as Burma. Gastaldi correctly indicates Ava along the (unnamed) Irrawaddy Valley, although the river erroneously veers far to the northeast. Founded in about 1365, the name of the kingdom comes from *In-wa*, meaning 'entrance to the lake'. Just to

the east of Ava lies *Capelan R[egio],* a mountainous region first mentioned by Varthema from which the Burmese obtained rubies, sapphires, and spinels. "The sole merchandise" of Pegu, according to Varthema,

> is jewels, that is, rubies, which come from another city called Capelan, which is distant from this thirty days' journey; not that I have seen it, but by what I have heard from merchants.

According to the English traveler Ralph Fitch, who went through Burma and on to Chiang Mai in the mid-1580s, Ava and *Capelan* were a six days' journey apart. Ava was said to have once had exclusive control of the *Capelan* mines.

Far to the north lies the kingdom of *Erancangui,* which is Arakan, wildly misplaced inland. This error stems from Portuguese sources which mistakenly reported that Arakan had no port, probably because its port was far upriver; Gastaldi interpreted the faulty report to mean that the kingdom lay upcountry. Ramusio also places his *Araquam R[egio]* far inland, but now adds the *city* of *Arocam* at the coast, beginning the kingdom's return to its proper place on the 1561 Gastaldi map.

It was at Arakan that Nicolò de' Conti first set foot on Southeast Asian soil in the early fifteenth century and traveled overland to Ava. In 1518 the Portuguese João de Silveira entered Arakan, but Portuguese interest in the kingdom was limited because better trading was to be had both with the neighboring kingdom of Chittagong (marked for example, on a map by Langenes, fig. 107, as *Chatigam*), and Pegu. Arakan soon became a pawn for European interests, and as a consequence, for Pegu as well (see page 189, below).

Siam and Other Mainland Kingdoms on the 1548 Map

Davsian, a kingdom north of Ava, is *Sian* (Siam), corrupted from the *d'Ansian* (i.e., 'kingdom of Siam') found in such charts as the Ribero of 1529. *Nagogor,* on the east coast of Malaya, is Lugor (Nakhon si Thammarat), a city in southern Thailand, the *Nagaragoy* of such earlier maps as that of Waldseemüller, 1513 (fig. 57). In their battle for control of Malacca, the Portuguese quickly acquired a fascination with Siam, both because of its purported wealth and power, and because of its king's rivalry with the Malaccan rulers whom Albuquerque was fighting. Thus Albuquerque sent an envoy, Duarte Fernandes, to Ayuthaya even before he had secured Malacca. The Portuguese were comforted to learn that the Siamese ruler, Rama T'ibodi II, was preoccupied with war with Chiang Mai, and not interested in threatening the balance between those vying for Malacca. Another Portuguese mission went to Ayuthaya in 1518, this time establishing a treaty with Rama T'ibodi II. The Portuguese were allowed to settle in Siam and to trade at Ayuthaya, Lugor, Patani, and Mergui, while the Siamese were allowed to settle in Malacca, and were promised guns for their war against Chiang Mai.

Comche China (Cochin-China), a term coined by the Portuguese and first used to denote any part of Vietnam other than Champa or the Mekong Delta region, is marked beside Indochina. This term may have come from the Malay word *Kuchi*, which itself probably came from the Chinese '*Chiao-Chih*', or Tongking. Tomé Pires explained that 'Cochin China' is the term used in Malacca, "on account of *Cauchy Coulam*". The Portuguese may have added 'China' to 'Cochin' to distinguish it from the Indian region of 'Cochin'. Barros considered Cochin-China to be the least known of the Indochinese kingdoms (after Cambodia and Champa). He deemed Western ignorance of Cochin-China to be the result of the treacherous and stormy seas off its coast, and the fact that the

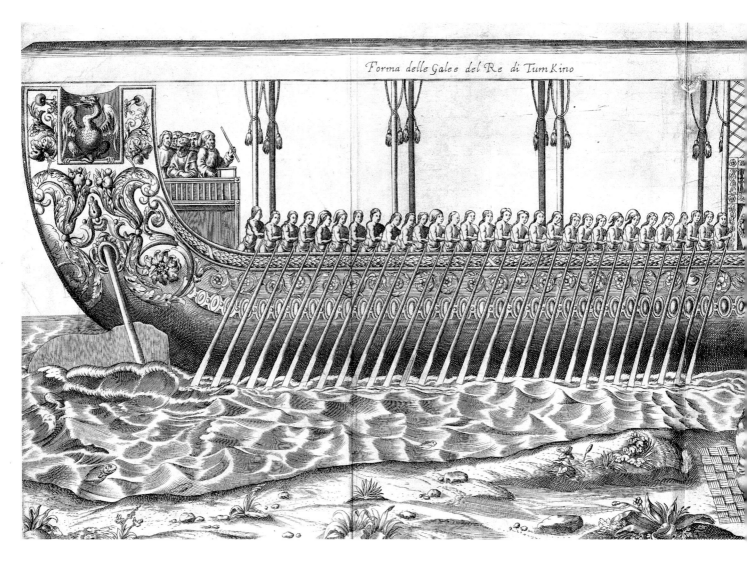

Forma delle Galee del Re di Tum Kino

Cochin-Chinese seldom engaged in maritime ventures. A century later, the Jesuit Marini complained that geographers wrongly used the term 'Cochin-China' to denote all of Indochina because they were ignorant of the region and its history.

Pires reported that this kingdom was, like Champa, more of an agricultural than a seafaring nation, though he believed it to be blessed with large, navigable rivers, and knew that its merchants routinely traveled to China, and occasionally to Malacca. Cochin-China was closely tied to China by culture, commerce, and even by royal marriage. Familiarity with Tongking — evidenced by the appearance of its name on European maps — increased when the Portuguese established themselves on Macao, which provided a practical base for reaching out to Vietnam. Once Tongking was placed to the north of Indochina, the name Cochin-China was used to refer to Annam, the part of Vietnam centered around the city of Hué.

The Mainland on the 1554 Ramusio and 1561 Gastaldi Maps

In general, what can be said about the mainland of the 1554 Ramusio map applies equally to the 1561 Gastaldi map. We will refer to the Ramusio only, when the Gastaldi differs in a meaningful way.

While the shape of Malaya on the 1548 Gastaldi map retained a hint of the Golden Chersonese of Ptolemy, Ramusio's has the vague memory of the old Cantino peninsula (see Waldseemüller Asia, fig. 57), likewise corrected for latitude, with the orientation of its southern tip also properly turned around. Vastly superior to the Malaya of the 1548 map, this is the foundation of modern Malaya, Indochina, and indeed of the Southeast Asian mainland. A couple of printed world maps of about the same time as the Ramusio also show improvements on the mainland, notably the rather crude Lopez de Gómara map of 1552-53, and, more especially, the Michele Tramezzino map of 1554.

Vietnam

The Red River makes its debut, here, and is shown with the city of *Cochinchina* (Hanoi) marked on its banks. Continuing south, we come to a peninsula jutting out due east which is labeled *Capo pulocanpola*. This name is a corruption of 'Pulo Condore', the island off the Mekong Delta, and is a variation of the *fulucandoia* of the proto-Indochinese peninsula of the 1502 Cantino map and the several subsequent maps in the same family (see page 109, above). Ramusio's is closer to the truth, with Cantino's *fulu* now more properly spelled as *pulo* (island).

The Mekong River

After this anomaly we pass through Champa *(Campa)*, and get our first glimpse at the mighty Mekong and its extensive delta, marked by its modern name *(Mēcon)*, which according to Camões meant 'Prince of Waters'. Pinto's *Travels* refers to the river as *Pulo Cambin,* though this appears to be a corruption of the text since *pulo*, as Pinto must have known, means 'island'. Many later mapmakers simply referred to the Mekong as the 'River of Cambodia'.

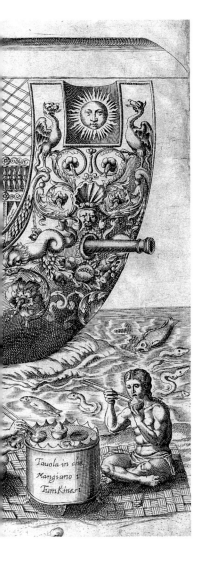

Fig.78 The galley of the king of Tongking, and (lower right) the manner of their dining with chopsticks. From Marini's *Delle Missionide' Padri della Compgania di Giesu, Rome*, 1663. Abbé de Choisy relates a vivid description of these vessels which he heard during his 1685 voyage to Siam, and which closely matches Marini's illustration. "Each galley [of Cochin-China] has thirty oarsmen on each side," (which is precisely the number shown}) "and there is only one man to an oar. The poop and prow are free and the place for the officers. There is nothing so well-ordered as these. The oarsmen must always watch their captain, who by the action of his baton has them carry out all his orders. Everything is in such unison that a conductor does not do better in co-ordinating the rhythm of his musicians than a captain of a Cochin-Chinese galley is obeyed with the movement of his baton." The 'conductor' stands on the left 'podium', baton in hand. "The outside of the galley is varnished black and the inside has a red lacquer in which can see one's reflection. All the oars are gilded." [Courtesy of Martayan-Lan, New York]

about 2600 to 25,000 square kilometers, and thus explorers seeing the lake at different points in its cycle would report quite differing scenarios. Ramusio simply designates the lake as *Lago*, while Gastaldi uses *Lago de camboia* (Lake of Cambodia). 'Tônle Sap' is a Cambodian term meaning 'great lake', just as the Chao Phraya River of Thailand was long called by the Thai word for river, '*menam*'.

Siam and the Chao Phraya River

Continuing west through Cambodia to the apex of the Gulf of Siam, Ramusio also christens the Chao Phraya River with the name *Menam*.[224] Barros referred to the river as *Sião* ('river of Siam'), and noted that the Siamese called it *Menão*. The term 'Siam' itself is now recognizably marked by Ramusio as *Syam*, and the kingdom's capital, Ayuthaya (*Odia*), correctly appears on the banks of the Chao Phraya.

The Evolution of a Trans-peninsular Waterway

Turning south and traveling down the Malay Peninsula, we pass through the important trading state of Patani (*Patan*), before arriving at Singapore (*C. de cimcapula*) at the southeastern tip. Here Ramusio introduces the most curious quirk in the early mapping of Malaya: a trans-peninsular waterway. Forty years before Gastaldi's first map of Southeast Asia, in 1507-08 (fig. 55), we saw how the Southeast Asian peninsula of Ruysch had opposing east-west indentations near its southern end. On the 1548 Gastaldi map, these indentations have become very pronounced, nearly cutting the peninsula in two. Finally, on the 1554 map, Ramusio completes the transition: what had initially been indentations now completely pierce Malaya as a trans-peninsular waterway, the entire southern segment of the peninsula south of Malacca severed from the mainland by a strait. The Gastaldi of 1561 goes a step further, accentuating the inlet just south of the waterway on the west, as though a new waterway were forming. We will explore this strange, non-existent strait on a 1598 map of Lodewijcksz (see page 187, below).

Western Shores of Malaya

North of Malacca, Ramusio attempts to depict the myriad steep, spectacular, tiny islands which dot the waters of the western coastline of the Malaya Peninsula, principally between Phuket and Mergui, in what is now Thailand and Burma. These islands gained a reputation as pirates' lairs, dampening the usefulness of the harbors along the mainland coast, and interfering with trade with Phuket. Ramusio is the first to record Phuket (*Iuca loam),* which, along with Selangor (*Solongor),* supplied Malacca with tin for export. Further north, another source for tin is *Tavay,* which Ralph Fitch passed en route from Pegu to Malacca in early 1587. From "the Island of Tavoy," he wrote, "cometh great store of tin which serveth all India, the Islands of Tenasserim, Junkseylon [Phuket], and so many others."

Burma

Entering Burma, the Irrawaddy and Tenasserim Rivers (both unnamed) now appear, and on their banks lay the first record of Rangoon on a printed map. This is the metropolis of *Dogon,* by which name Rangoon was then known. Cape Negrais, which was an important landmark for early European mariners, juts to the west (rather than the south), and is marked by its modern name *(C. nagraes)*. As described by Ralph Fitch, Negrais "is a brave bar and hath 4 fathoms water where it has least." Pinto remarked that Portuguese mariners, when rounding Cape Negrais, knew to look for a gilded pagoda shining in the sun on its shores; this landmark

The course of the Mekong was guesswork, the river being portrayed as a nearly straight, unencumbered seaway to the far interior of China. Thus with this naive view of the Mekong's course, we see the inauguration of two immensely important fallacies. The first was the dream that the river might provide an easy commercial highway into the heart of the Celestial Kingdom, a folly which taunted European fortune-seekers well into the late nineteenth century. Secondly, since Europeans perceived the Mekong as defining Siam's eastern boundary, the arbitrary course assigned the mysterious river — lacking its dramatic eastward swing north of Phnom Penh — greatly diminished the West's perception of Siam's size. This error, as we will see later, helped Siam avoid foreign domination in the nineteenth century.

Tônle Sap

Both Ramusio and Gastaldi 1561 rudimentarily depict the Tônle Sap, the lake fed by a tributary of the lower Mekong whose northwest shores border the ancient metropolis of Angkor. But the size of the lake differs markedly between the two maps, Ramusio showing it as a relatively small lake confined to the immediate delta region, while Gastaldi's spans nearly 5° latitude. Why would there be such a discrepancy between the lake on two maps whose mainlands are otherwise closely related? The breadth of the Tônle Sap does in fact change radically, expanding and contracting as the lower Mekong actually reverses direction twice a year. Like a living lung of the Mekong basin, the fish-abundant lake expands and contracts with the season, varying in size nearly ten-fold from is the Hmawdin temple, which still survives. It is also mentioned at the turn of the eighteenth century in the standard pilot book for English vessels in the Indies, *The English Pilot*: "Upon the South Point of Cape Negrais . . . turning in is on a small Hill, a little Pagood, and 4 Teddy Trees, standing singly to be seen very remarkable." In the 1750s, the island of Negrais was considered as a trading post for the English East India Company. Since Negrais is misaligned on the map, Burma's rivers empty eastward. Along the closely-spaced, roughly parallel rivers of the Irrawaddy, Tenasserim, Chao Phraya, and Mekong, are the situated kingdoms of Pegu, Ava, Arakan *(Araquam)*, and Burma *(Berma)*, as well as Capelan, identified for its mines of precious stones (*M. Capelangā*), rather than as a kingdom in its own right.

Chiang Mai and Lake Chiang Mai

South of Ramusio's Burma and the Capelan mines, in the mountains east of the Mekong, lies *Iamgoma*. Though misplaced far to the south and east, this is the first 'modern' reference to Lan Na, the civilization encompassing Lamphun, Chiang Mai, Chiang Rai, and other cities in what is now northern Thailand (Lan Na was probably the *Lama* of the 1507 Ruysch and Waldseemüller maps).

Though now part of Thailand, Lan Na was an independent kingdom at this time. Ramusio's *Iamgoma* is one of several variations of the name by which Chiang Mai was known to Europeans. The Chiang Mai language, which is still widely spoken in the north of Thailand, is similar to Laotian, as are many of the peoples' customs. The English merchant Ralph Fitch, who visited Chiang Mai (which he called *Jamahey*) at the end of 1587, noted that Chiang Mai "is a very fair and great town with fair houses of stone, well peopled," with very large streets. The men of Chiang Mai are "very well set and strong," and the women "much fairer than those of Pegu." It took the Siamese about fifteen days to reach here from the northern frontier of their kingdom, according to the later French traveler, La Loubère, "for they are Journeys by water, and against the Stream."

Pinto described some towns of Lan Na in his usual colorful fashion. He followed a large river, called the *Angegumá*, and arrived at

> a small, strongly walled city called *Gumbim,* in the kingdom of *Jangomá,* which was surrounded on the inland side for a distance of five or six leagues by forests of benzoin and fields of lac, products that are carried to the market in the city of Martaban and transshipped from there on many *naos* to different parts of India, the Straits of Mecca, El Quseir, and Jiddah. There is also a large supply of musk in this city, which is much better than the Chinese kind, which they also ship to Martaban and Pegu where we Portuguese buy it and carry it for resale to *Narsinga*, Orissa, and Marsulipatam.The women in these parts are generally very fair and beautiful; they go dressed in clothes of silk and cotton and wear gold and silver bangles on their feet and thick chain-link collars around their necks. There is a great abundance in the land of wheat, rice, and meat, and especially honey, sugar, and beeswax, of which they have an enormous supply. This city and its environs, which extend for ten leagues around, provide the king of *Jangomá* with an income of sixty thousand gold *alcás,* or the equivalent of 720,000 *cruzados* in our money.

Pivoted 0n the crossroads between several rival kingdoms, Chiang Mai is blessed with a mixture of rich cultural legacies. But La Loubère, sent on a mission to Siam by Louis XIV, correctly noted the great tragedy of such a location, "adjoining to several Kingdoms," is that Chiang Mai was "more subject than another to be ruined by War." At the time the Ramusio map was published Chiang Mai was, in fact, beginning to enter a long period of decline, precipitated by the ruling élite's lavish spending habits, a succession crisis, and other destabilizing events in the early sixteenth century. Chiang Mai fell to Burma in 1558 with barely a struggle, and soon all of Lan Na was under Burmese control. By the time of its rebirth in the late eighteenth century, its civilization had been reduced to a shambles.

Lake Chiang Mai

Beginning with Ramusio, European maps commonly depicted a large Lake Chiang Mai that served as a Himalayan source for Southeast Asian rivers, usually the Chao Phraya and rivers to the west. The Mekong seldom arose from the lake, and the Red River rarely, if ever.[225] A particularly explicit portrayal of this lake can be found on a seventeenth century portolan chart by Antonio Sanches (fig. 80). The existence of such a lake was reported by various European observers, notably by João de Barros in his *Décadas da Asia*:

> This lake lies 200 leagues north [about 950 kilometers] in the inland, and six rivers rise from it. Three of these join to form the river that flows through Sião [the Chao Phraya]. The other three flow into the Gulf of Bengal. One of these three gets its name from the Caor Kingdom . . . the second is the river of Pegu . . . and the third flows through Martaban.

On his hand map (see page 123, above), Barros located the lake at the point where the wrist and hand meet — quite far north indeed. Ramusio includes an Italian translation of extracts of Barros' work, though his depiction of Lake Chiang Mai and its rivers does not precisely follow Barros' text. *Lago de chiamay*, which was the *Lama La[cus]* of Ruysch, dominates the interior of Southeast Asia and forms the common source for the Chao Phraya, Irrawaddy, Tenasserim (or Salween), and *Caor* (Brahmatutra?) Rivers. The original woodcut version of the Ramusio map (fig 79), also places a city of *Chiamay* on the southern shores of a 'Lake Chiang Mai' (*Lago de chiamay*). Curiously, no 'Chiang Mai' enjoys the lake on the copperplate version of the map.

Perhaps the most elaborate early European report about Chiang Mai and Lake Chiang Mai came from Pinto, who described a 'forbidding' region of the Southeast Asian interior

> with large mountain ranges, inhabited by many [animals and wild herds] roaming around in such huge numbers, that it is impossible for a man to grow anything there to feed himself . . . and in the middle of this country or kingdom, which is what it was formerly, there is a big lake the natives call *Cunebeté,* though others call it the Chiang Mai . . . The writers who have described it state that the lake measures sixty *jaus* in circumference with three leagues to a jau, and that all around it there are many mines of silver, copper, tin, and lead, which are in constant production and yield huge quantities of these metals which are then carried by merchants in elephant and yak caravans to the kingdoms of the *Sornau*, or Siam, [and other kingdoms].

Elsewhere, Pinto paints another picture of Lake Chiang Mai. Traveling through central Asia and bound for Cochin,[226]

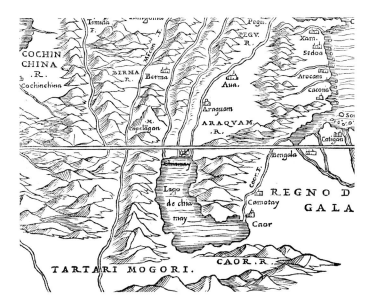

Fig 79 The original (woodcut) 1554 version of Ramusio's map placed the city of Chiang Mai on Lake Chiang Mai. In this example, both the name of the city and the icon representing it have been crossed out, apparently because the map's owner dutifully 'corrected' it upon seeing that the newer copperplate rendering (1563) did not show it.

> we came out on the lake of *Singapamor,* called *Cunebeté* by the local people, which, according to the information they gave us, measured thirty-six leagues in circuit. There we saw such an enormous variety of birds of all different species, that I would not dare attempt to describe it. In this lake of *Singapamor,* which is carved into the heart of the country by some admirable work of nature, four deep, wide rivers have their source.

Pinto's report that the lake is called *Cunebeté* by the local people found life in some cartographers' creations. Vincenzo Coronelli, for example, refers to the lake by that name on his large globe of 1688. Pinto laments that perhaps Portugal might derive greater benefit from this upland country and its lake, "at less cost in blood and all that goes with it," than it has from its exploits in India — a comment he makes about other places as well, such as Siam.

Pinto claimed that the king of Siam (which would have been King P'rajai, who reigned 1538-46) actually reached the lake when he led an expedition against Chiang Mai. "After six days of marching through enemy territory . . . he arrived at Lake *Singuapamor,* which most people call Lake Chiang Mai," in the vicinity of which the king captured "twelve very noble and wealthy towns that were fortified with walls, moats, and ramparts."

By the time of Ramusio's map, the lake had already become such an accepted feature that is was even accorded a place in Western poetry; Camões, toward the end of the *Lusiads,* speaks of "the River Menam [Chao Phraya], that rises in the great Lake Chiamai." From whence did this fictitious lake come?

One theory is that it was inherited from Hindu/Buddhist iconography and cosmology in which the center of the universe was the sacred mountain called Sumeru. From Sumeru, the four continents were pivoted along the four cardinal directions, with the inhabited world (Sanskrit *Jambu-Dvipa*) lying to the south. This concept was sometimes poetically compared to a lotus flower. A river flowed from the top of Sumeru and emptied into a lake *Anavatapta* (or *Anotatta*), from which four great rivers flowed, one in each of the four cardinal directions. The *Traiphum,* for example, describes

Anotatta Lake and its rivers in detail. There were probably many variations of this tradition, and the basic idea of a spiritual mountain in the Himalayas was widespread in Southeast Asia.

Another theory cites an old Thai chronicle that refers to an auspicious lake that influenced the selection of Chiang Mai as the site for the capital of Lan Na. Thus, our *Lago de chiamay* may be that fortuitous lake, having taken on a cartographic life of its own. Until modern times there was in fact a small lake in Chiang Mai, called *Nong Bua,* which according to tradition influenced the site chosen for the founding of the city in 1296. That lake, however, has now been filled in and is lost to the overbuilding of Chiang Mai. All that remains of *Nong Bua* is a Thai song from the 1950s in which a girl extols the lake's beauty in the tree-clad moonlight and sings of her sadness to leave it. Although the Lake Chiang Mai of Ramusio and other mapmakers did not remotely reflect the modest size of *Nong Bua,* Europeans in Ayuthaya being told of the lake may have been misled by the Thai word for lake — *satnam* — for it can mean anything from a pond to a major body of water.

There is also a small lake lying to the north of Chiang Mai, which could appear to feed the Meping River, which runs through Chiang Mai and ultimately flows into the Chao Phraya River. This lake lies in Chiang Dao and is situated beside a cave, which could certainly have attracted attention.[227] But even if *Nong Bua* or another historical lake was associated with Lake Chiang Mai, the role of such a 'real' lake must be seen as part of a broader Buddhist cosmological concept in which *Nong Bua* is simply another player. Lakes were auspicious topographical features in Southeast Asian cosmography, and *Nong Bua* was an embodiment of that tradition, not its origin.

Parallels in Europe
Regardless of the roots of our enigmatic lake, it is interesting to note the legend's similarity to the Judeo-Christian belief that four great rivers flowed from Paradise (see Schedel, fig. 38). Medieval European tradition generally located Paradise and the four rivers' source on a mountain at the eastern end of the world — in approximately the same place as Sumeru. Some medieval maps (see, for example, fig. 36) do show a configuration remarkably suggestive of our *Lago de chiamoy.* The influential Pierre d'Ailly, in his *Imago Mundi* (ca. 1400) painted an image of Paradise that was particualrly analogous to the Oriental Sumeru: a mountain, which reached near the moon with an immense lake which fed the rivers of Paradise. There were other early writers who envisioned Paradise near the circuit of the moon, both for its poetic value as well as to explain why Paradise was exempt from the ravages of the Deluge.

Other deeply-rooted Western European precedents may have also helped to prime the European psyche to be receptive to any indigenous suggestion of such a serendipitous lake in the Himalayas. Pliny wrote that, "some authorities say that the Ganges rises from unknown sources, like the Nile . . . but others state that its source is in the mountains of Scythia [central Asia]." In one version of the legend of Prester John we learn that the four rivers of Paradise all arise in a spring in the mountains of 'India'. Nor did the distance of the Himalayas from the actual rivers traditionally associated with Paradise contradict these beliefs, because early Christendom had long ago learned to explain such logistical glitches by the existence of subterranean water passages through which the rivers flowed for a great distance. The cosmographical cross-section of the earth by de Jode (fig. 41), for example, shows Lake Chiang Mai as the source for four rivers, and depicts the lake's waters as connecting to a large abyss deep inside the earth.

A S I A

Catalao.

Ciuno

Mogoll

INDIA

Benĝala

Canbaia

Canboiete

Concan

Orixa

Golfo de Bengala

Cuchin China

Pegu

decan

Canboĵa

Piao

Samala

Sei
lao

Sa
ma
tra

Borneo

IAVA

Por Antonio Sanches em lixb
oa. em oanno de 1641

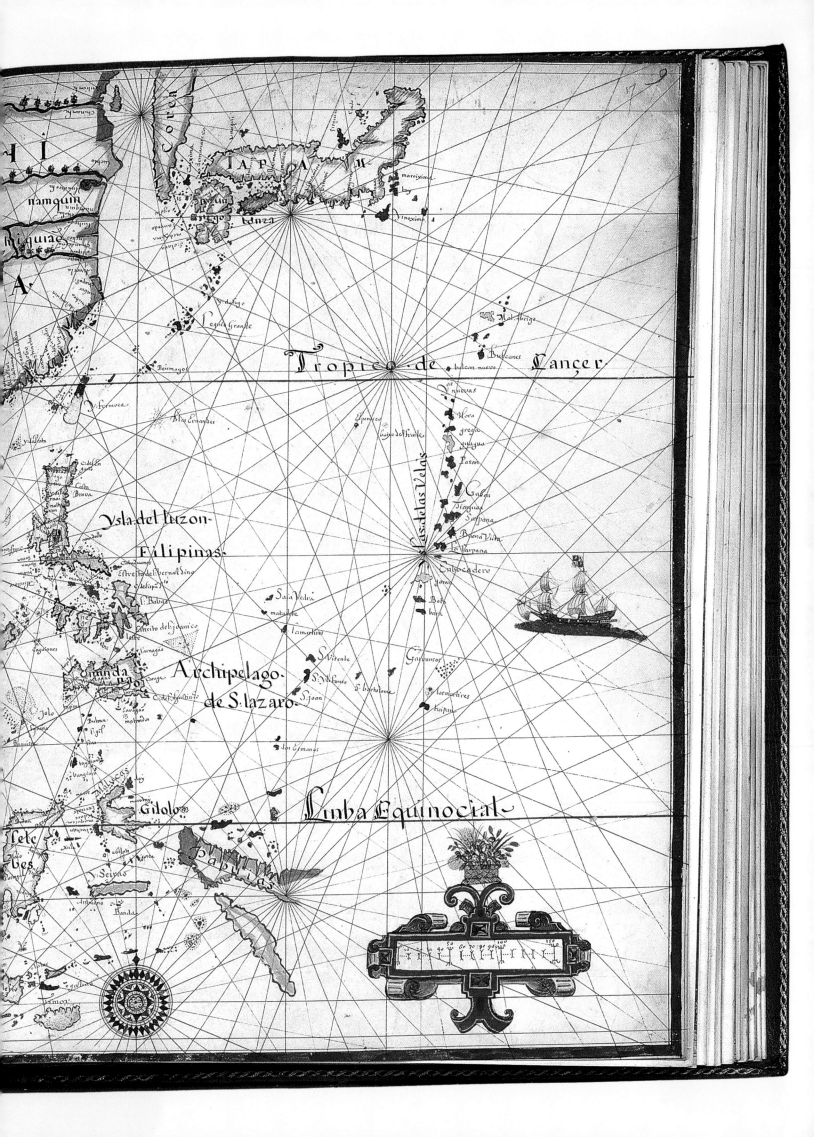

159

namquin

hi quiao

A.

Couen

IAPPAH

mareixima

kay

Vinexima

Mol. Abrigo

Tropico de bulcan nuevo Cancer

Bulcanes

as nuevas

Uora

gregta

dyagua

Posan

Galan

Tiagnua

Sarpana

Buena Vista

La Sarpana

Embocadero

Yoran

Boti

beni

Ysla del luzon

Filipinas

Yslas de las Velas

Safa Vedra

matalotc

losmartires

SVitente

SNicolano

S.barholome

Garranpos

Sjoan

losmartires

Tuipan

Archipelago

de S lazaro

dos Ermanas

Chanda

Jolo

Gilolo

Linha Equinocial

Papuas

Banda

Fig 80 Portolan chart of Southeast Asia, Antonio Sanches, 1641. From an atlas entitled *Idrographisiae Nova Descriptio.* Manuscript on vellum (45.5 x 66 cm). The chart is a brilliant illustration of the mythical 'Chiang Mai Lake' from which major rivers of Southeast Asia were believed to originate.

Sanches has gilded certain places with gold to indicate where there is a strong Portuguese presence — for example, Timor and Macao. Gold has also been used to illuminate places where there is no Portuguese influence but which nonetheless figure importantly in Portugal's history and the legacy of its exploration — Taiwan, for example. [Koninklijke Bibliotheek, The Hague, 129 A 25]

European Evolution of the Lake

Nicolas Gervaise, a French priest sent to Siam in 1683, passed nearly four years there and learned the language of the country. He reported that the Siamese believe the Chao Phraya River originated in "a great lake which was discovered some years ago in Laos." He accepted this theory, although some of his colleagues suspected that the river was a branch of the Indus. M. Le Clerc, a Frenchman who explored the Chao Phraya and sought local opinions about its source, offered a different view. As related by Abbé de Choisy in 1685, Le Clerc traveled up the Mekong "to the Laotian frontier, found it very narrow, and the inhabitants there assured him that three days' travel further up it was only a tiny stream which came out from the mountains." Shortly afterwards, La Loubère wrote that many geographers believed the lake to have been named after Chiang Mai, though he himself was skeptical of the existence of the lake altogether. In 1687 he further questioned Siamese who had participated in the sacking of Chiang Mai some thirty years earlier, and found that they "do not know of that famous Lake, from whence our Geographers made the River *Menam* arise and to which, according to them, this City gives its Names: which makes me to think either that it is more distant than our Geographers have conceived, or that there is no such Lake."

Some later mapmakers retained Lake Chiang Mai but virtually disassociated it from Southeast Asia. At the turn of the eighteenth century, for example, Guillaume de L'Isle depicted his *Lac de Chaamay* as the source of two relatively minor rivers, one which joins the Ganges at its gulf, the other which forms a distant tributary of the Irrawaddy.

In 1690 Engelbert Kaempfer, resident in Ayuthaya, noted that the Siamese view of the source of the Chao Phraya River "differs from what is represented in our maps; for they say, that it takes its rise like the Ganges of Bengale in the high mountains of Imaas [Himalayas]" and that various arms of the river penetrate Cambodia, Pegu, and Siam, and join with the Ganges. As a result some maps, such as those by the mid-eighteenth century French geographer Philippe Buache, show the Ganges and the major rivers of Southeast Asia all originating from the *Imaas Mountains.*

Lake Chiang Mai, Lake Tsinghai, Lake 'Kiang'

Gastaldi depicted the lake differently in 1561 than on the Ramusio. With better knowledge of Central Asia came word of a large lake in west central China, whose name he understood as being *cayaroay.* This is Lake Ch'ing-hai, or Tsinghai (Koko Nor), which lies in the province of the same name, in whose mountains the Yellow River also has its source. Being situated in the northeastern part of the Tibetan plateau, it was perfectly positioned to play the role of the remote, mountain-clad source of Southeast Asia's rivers. Tibetan folklore, in fact, relates that Ch'ing-hai was once a subterranean lake below what is now Lhasa.

Pinto may have been hearing stories of Ch'ing-hai when, in Nanking (China), he was told of a lake in the interior of Asia, called *Fāostir.*

> Branching out from this lake, which is twenty-eight leagues long, twelve wide, and tremendously deep, are five of the most torrential rivers that have been discovered to date.

Thus the 1561 map replaces Lake Chiang Mai with *cayaroay Lago,* and even keeps the city of Chiang Mai *(chiamai)* on its southern shores. Gastaldi positions his lake just north of the Tropic of Cancer, the identical position accorded Lake Chiang Mai by Ramusio, and retains a hint of its shape. But this idea had little influence. Linschoten (fig. 92), hands back the job of river sourcing to Lake Chiang Mai, and placed Lake Tsinghai (for Linschoten, *Lacus Cincui Hay*) further north, impressively close to its correct latitude (Linschoten's is 36$^{1}/2$° to 40$^{1}/2$°, actual is about 36$^{1}/2$° to 37$^{1}/4$°). These two theories may have been combined by Ortelius in 1570, on his map of the Asian continent (see page 168, below).

Marco Polo

A map of China published in 1655 by Joan Blaeu and based on the travels of the Jesuit Martino Martini also placed two northern lakes as the sources for Southeast Asia's rivers. *Kia[ng],* the larger and more northern lake lies in the region of Lhasa, and is reminiscent of earlier conceptions of Ch'ing-hai and Lake Chiang Mai. Located much too far west to be derived from contemporary knowledge of the Chinese province of Kiangsi, even with longitude uncertainty, our *Kiang* Lake comes from Marco Polo's *Kian* (or *Kian-suy*). Around the kingdom of Kiang, Blaeu has placed the letters S I: F A N, probably the province of Sindafu, where Polo encountered the Kiang River. Polo placed his *Kian-suy* as a river in the province of *Sindafu* (Szechwan?)[228] The map's inclusion of the kingdom and city of *Mien* (Burma) is another throw-back to Marco Polo.

In this Martini/Blaeu concept, Lake *Kiang* was the source for the Ganges, the Red River, and three Burmese rivers, two of which joined each other at *Mien* (Irawaddy and Salween?). The second river source is a small unnamed lake further south; it is the source for the *Lukiang* River, which splits at *Brema* to form the Chao Phraya and Mekong Rivers.

Thus, our fortuitous lake quickly took on an adventurous life of its own, since the truth of Lake Chiang Mai could not easily be settled by practical experience. If the seventeenth-century French ambassador Gervaise is to be believed, even Thai sailors sent from Ayuthaya by the king to discover the source of the Chao Phraya found the terrain so confusing that after a while they "were greatly astonished to find themselves back again at almost the same spot as that from where they had started." The question remained unsolved even longer than such celebrated riddles as *Terra Australis* and the source of the Nile. Not until the early twentieth century were the sources of the rivers clearly determined and the lake proven to be mythical.

In summary, the idea of a great lake in the Asian interior from which the major rivers of the Southeast Asian mainland flowed, seems to have been an Asian tradition, perhaps re-interpreted, but not invented, by Europeans. Europe's own history and traditions, however, made the adoption of the symbolic lake all the more natural and appealing. Actual lakes which affected the selection of a site for settlement, such as the old Nong Bua of Chiang Mai, may be part of that tradition. Even if so, Nong Bua and Lake Chiang Mai are simply part of a larger, more ancient inheritance rooted in indigenous and Hindu-Buddhist thought.

Other Works by Gastaldi and the Italian School

Ramusio's Sumatra, 1556

Although it was the Portuguese and Spanish who brought Europe most of the revelations about Southeast Asia during this period, mariners under French auspices also tried their luck at overseas exploration during the 1520s. Giovanni da Verrazano attempted and failed to find a nortwest passage via the top of North America to Cathay in 1524. Vessels from Dieppe reached the island of Diu in the northern Indian Ocean in 1527. Two years later, a pair of ships commanded by two brothers, Jean and Raoul Parmentier, reached the Maldive Islands by running the Portuguese blockade, and then continued east to the west Sumatran port of Ticon. France, for a brief moment, was a new contestant in Southeast Asia.

But their luck ran out in Sumatra. Both Parmentier brothers died of fever, and trade with Sumatra was difficult. Only one ship returned to Dieppe, with far too little in bounty to encourage any further sacrifice of lives or resources for French voyages to Southeast Asia.

Two years after Ramusio's map of Southeast Asia, in 1556, the third volume of the *Navigationi* was published. While primarily concerned with the Americas, the work also related news of the Parmentier voyage and included a woodcut map of Sumatra (fig. 77), which was the first separate map of any Southeast Asian island based on actual observation, to be published (in contrast to the Bordone maps of three decades earlier, which were conjectural mappings based primarily on Marco Polo). The general contour of the island is remarkably good for its day, and the various islands lying off its western coast are shown in detail, as is Banca on the eastern coast. The port where the Parmentier brothers anchored, Ticon, is recorded, and the manner in which the island's coast forms a promontory just to the northwest of Ticon is accurately mapped. Ramusio has also corrected the erroneous placement of Aceh from his general map of Southeast Asia of two years earlier.

The truly peculiar aspect of the map is its complete omission of Malaya, and its omission of the west coast of Java. Even the exaggerated width of the Malacca Strait shown by Ramusio on his general map of two years earlier does not account for the absence of Malaya. Were it not that the equator is clearly marked on the map, it would appear that Ramusio was influenced by the notion, at this time gaining favor, that *Java Minor* (which he might have 'correctly' transposed to Sumatra) was not an island due south of Malaya as previously believed, but rather an island lying in the western Pacific much further south.

Wall Maps and Globes

Maps created for public or semi-public display date back to ancient times in Italy. As we saw in Chapter 4, Augustus had one made for the citizens of Rome at the turn of the Christian era. By later medieval times, large, beautiful maps were clearly in vogue for the wealthy. The manufacture of globes of wood or metal is described by Giovanni Campano, who flourished ca. 1261-64. A large map of the world, in the form of a wall fresco, was painted in a special room in the Venetian Ducal Palace in about 1400, offering its image of the world to a few generations of privileged folk until it was destroyed in a fire in 1483.

In sixteenth-century Italy, with a heightened interest in the distant parts of the world and the willingness of the wealthy to commission lavish works, art mingled with cartography more than ever before. Not surprisingly, the talents of Gastaldi were among those tapped for grand map projects. With the memory of the old palace map probably still alive, the Council of Ten in Venice commissioned Gastaldi to make maps to be painted on the walls in the Ducal Palace. Gastaldi, whom they referred to as "Master Jacopo of Piedmont," had already done a map of Africa for the 'Sala dello Scudo' in 1549, and in August of 1553, a wall in the church of San Nicolò was set aside for Gastaldi to make a full-size draft, in charcoal, of a map depicting various discoveries in America and Asia. From this draft, the maps were painted on a wall in the Ducal Palace, in the room now known as the 'Sala delle Mappe'.

According to Gastaldi's commission, the Portuguese historian João de Barros was to be consulted for the geography of 'China', and Marco Polo for that of 'Cathay', although these could not have been more than minor sources for the complete project. These maps do not survive in their original form; they were restored and re-worked in the eighteenth century. Geographically, the original map of the Far East probably resembled that prepared for Ramusio (fig. 74), since both were done about the same time, and Ramusio's name is given as the author of one of the maps.

In Florence, Duke Cosimo de' Medici was keenly interested in fresco maps, and specifically in the geography of Southeast Asia. After establishing his own room for wall maps, he acquired the 1554 planisphere of Lopo Homem, a chart which is noteworthy for its extensive new data on the Far East, the East Indies, and the Liu-ch'iu islands, including Taiwan. Cosimo then secured a book from Portugal written by a Portuguese pilot who claimed to have made the voyage from Lisbon to the Moluccas fully fourteen times. Cosimo gave the book, which contained a great deal of geographic information, to the artist Ignazio Danti to assist him in making a series of mural maps painted on the wooden panels of the cupboards. Fifty-three such mural maps were made during the period of 1563-75, including two of Southeast Asia.[229] These maps are still to be found in the room now known as the Sala della Guardaroba (in the Palazzo Vecchio). The maps of Southeast Asia hang one above another; that on the bottom covers the mainland and its nearest islands (1573), and the one above it is one devoted to the islands (1563). Geographically, they are typical of the Italian school of the period, exhibiting, for example, the characteristically sharp Indochina and angular Malaya of Ramusio, and the elongated Borneo of Gastaldi.

Danti created other maps for the Vatican to which time has been less kind. In 1580 he was brought to Rome by Pope Gregory XIII, whose mandate of 1582 put into effect the calendar now used by most of the world, and who allied with Philip II of Spain to fight the spread of Protestantism. Gregory commissioned Danti to create, with the help of Italian artists, maps of Italy for the new Galleria del Belvedere. Once the parochial contribution was completed, Danti then supervised the painting of many other maps in the cosmographic loggia, which was part of the Raphael rooms in the Vatican Palace. A world map and several regional maps were executed, among them one of Malaya. None of these maps, however, has survived.

Chapter 12

Tangling with Terra Australis and Snared by the Linea

In addition to the manner in which geographers interpreted new reports, two other forces were juggling Southeast Asian lands on the map during the sixteenth century. One was the confluence of the facts that medieval travelers had given no coordinates for their landfalls, that the names they had ascribed were being displaced from their traditional homes by more current data, and that a vast southern continent had been 'discovered'. The combination of these three factors resulted in the belief that some Southeast Asian realms were perhaps regions of the southern continent. The other new force affecting the map was the Line of Demarcation which separated Spanish and Portuguese claims to the Pacific. The fact that no reliable method existed for determining longitude, exacerbated by the inclination to err in one's own kingdom's favor, meant that islands were placed at widely disparate longitudes.

Terra Australis

Since the canvas inherited by mapmakers of this period was already filled with data from older sources, mapmakers' attempts to integrate new reports often required a tricky dance. One casualty of this dance was that some parts of Southeast Asia became entangled with *Terra Australis,* the great mythical southern continent. In fact, misunderstood Southeast Asian nomenclature actually helped create the Renaissance incarnation of *Terra Australis.*

The Historical Basis for Terra Australis
The concept of a great southern continent dates back to antiquity. Theoreticians in ancient Greece as early as Herodotus postulated that a large antipodean continent existed which balanced the known continents of the northern hemisphere, and the concept is found on some medieval *mappaemundi,* such as those illustrated in figures 30 & 36. Portuguese pilots seeking a southern route around the New World during the first decade of the sixteenth century probably believed they had found the austral continent when, upon reaching the gaping mouth of the Plata River (separating what are now Uruguay and Argentina), the river's southern banks were so distant that they first appeared to be a separate landmass. Finally, the continent was truly 'discovered' when cosmographers judged that the land forming the southern shores of the strait discovered by

Magellan (in reality the islands of Tierra del Fuego) in December of 1520 was indeed *Terra Australis,* a concept formalized by Oronce Fine on his world map of 1531 (fig. 52).

Soon, land-bound theoreticians were considering the possibility that some of the Southeast Asian lands reported by Marco Polo and other medieval travelers were actually regions of *Terra Australis* or its neighboring islands. Such realms as *Lochac, Java Major,* and the various kingdoms which Polo described on *Java Minor,* had been orphaned from the known regions of Southeast Asia by recent exploration and their modern nomenclature.

More discoveries were associated with the continent after a route east across the Pacific, from the Philippines to New Spain, was discovered. In 1565, Andrés de Urdaneta determined the long-sought latitude at which the proper winds could be found for such a voyage, after studying the charts and reports from the unsuccessful attempts by Villalobos two decades earlier. Now armed with the ability to sail both west and east across the Pacific, the Spanish dispatched exploratory voyages from Peru to search for other regions of the southern continent. Various landfalls, such as Mendaña's discovery of the Solomon Islands in 1567, led some geographers to expand the reach of the austral continent on their maps.

Java-la-Grande
Between about 1540 and 1566, some cartographers of the Dieppe school made maps depicting *Java-la-Grande,* a large austral land comprising Sumatra and Java.[230] Prominent among these map-makers were Jean Rotz, Pierre Desceliers, and Guillaume le Testu. Many *Java-la-Grande* charts, such as that in the Vallard atlas of 1547 (fig.3), were extravagantly illuminated. On the Vallard chart, reasonably accurate ethnographic vignettes, probably derived from the 1529 expedition to Sumatra by the Parmentier brothers (whose procession would be that depicted on the map), blend into increasingly fantastic imagery on the top and sides. Typically, such maps depicted an austral continent comprising *Grande Jave* and *Petite Jave,* the two regions being separated by a narrow strait or river.

This incarnation of an austral Southeast Asia was probably the result of the patching together of various Portuguese sectional charts of Indonesian islands.[231] Unidentified, or wrongly identified, charts were fused onto the southern shores of Java and Sumbawa, often at different scales and orientations. This was assisted by the fact that the southern coasts of these islands were not known, and portolan charts usually left such coasts open, allowing mapmakers to stitch several charts together without appearing to contradict them.

The idea was also popularized in geographic texts. Fonteneau's *Cosmographie* of 1544, for example, stated that Java was part of *Terra Australis*. Although the idea of a large, austral Java never caught on in mainstream published maps, as a minority view it lingered until both Francis Drake and Cornelis de Houtman reported sailing through open ocean south of Java at the end of the century. Even Linschoten, in his *Itinerario* of 1596, wrote that the southern part of Java "is not yet discovered, nor by the Inhabitants themselves [well] known. Some think it to be firm land, [and parcel] of the country called Terra Incognita [i.e., *Terra Australis*]."

Not everyone accepted the premise of a southern continent, or associated Southeast Asian lands with it. "The region of Molucca," the Jesuit Francis Xavier wrote from Ambon in 1546, "is all islands, and, until now, no one has discovered a continent." Nor did Sebastian Münster interpret Magellan's *Tierra del Fuego* to be continental at all, believing that "the land which [Magellan] had on his . . . left hand [south], he supposed to be Islands."

Mercator, Terra Australis, and Southeast Asia

The most influential mapmaker to be led astray by the rush to *Terra Australis* was Gerard Mercator. Mercator was born in a village near Antwerp in 1512, the year de Abreu and Serrão made the maiden Portuguese excursion into Indonesian waters, and was ten years old when the survivors of Magellan's circumnavigation returned home. By the time Mercator reached adulthood, Tierra del Fuego and various other sightings of islands in the southern oceans had been

interpreted as being parts of the same southern land. Mercator, influenced by Fine's world map of 1531, accepted this premise, and when he composed his great world map of 1569, he added Southeast Asian nomenclature from Marco Polo to the austral brew.

A few maps and texts of the preceding half century had set the stage for Mercator's analysis. First, the 1519 *Suma de Geographia* of de Enciso and the 1528 *Isolario* of Bordone identified *Lochac* as an island, rather than on the Southeast Asian mainland. Secondly, Ramusio's published version of Polo's text (1559), introduced the direction 'southeast' that Polo sailed to reach *Lochac* from Indochina, rather than leaving the direction unspecified. Lastly, the 1565 Forlani extension of the 1561 Gastaldi map placed *Java Minor* and *Malaiur* (Malaya or Sumatra) in the southwest Pacific. Mercator explains his considerations via annotations on the 1569 world map. Marco Polo "saw several provinces and islands facing this continent, and noted their distances from Lesser Java." Mercator then explains that he has deduced from Polo's statements that *Java Minor* is neither (as some of his contemporaries believed) the island of Borneo, nor any island to the east of *Java Major*. This is because Polo stated that *Java Minor* "bends so far to the south that neither the Arctic Pole or . . . Ursa Minor . . . can be seen," and that in the island's kingdom of *Samara* neither bear is visible. It is easy to sympathize with Mercator, since Polo's itinerary corroborated the impression of an extreme southerly latitude.

Mercator's interpretation of the data was amplified by the reports of Varthema, who was still alive in Mercator's youth. The great traveler had reached Java from *Bornei* by sailing five days "toward the south" — a reasonable itinerary if one assumes that his *Bornei* is indeed Borneo. But Varthema also claimed that he had

Fig. 81 *Terra Australis*, Cornelis Wytfliet, 1597. (23 x 29 cm) [Jonathan Potter, London]

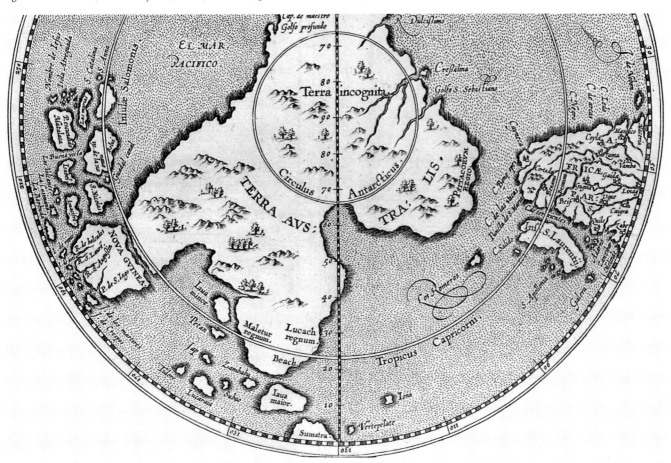

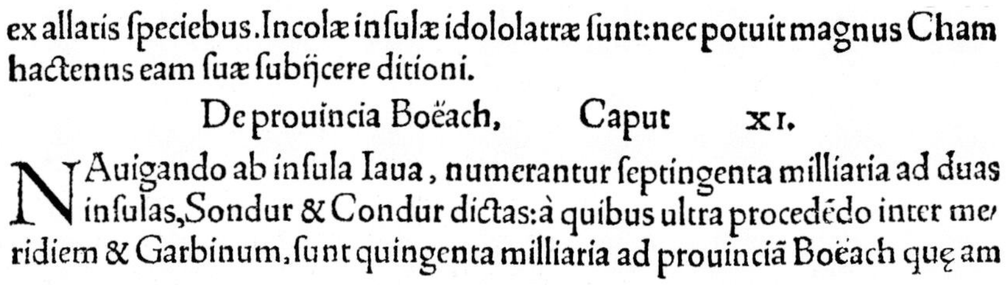

ex allatis fpeciebus.Incolæ infulæ idololatræ funt:nec potuit magnus Cham
hactenns eam fuæ fubijcere ditioni.

De prouincia Boëach,   Caput   x i.

NAuigando ab infula Iaua, numerantur feptingenta milliaria ad duas
infulas,Sondur & Condur dictas:à quibus ultra procededo inter me-
ridiem & Garbinum,funt quingenta milliaria ad prouinciã Boëach quę am

Fig. 82 This extract from the 1532 *Novus Orbis Regionum* of Huttich of the account of Marco Polo is the origin of the bogus kingdom of *Beach*. The text's *Boëach* is a corruption of Polo's *Lochac*.

previously reached *Bornei* from the Moluccas by sailing "constantly to the southward" about two hundred miles, though in fact Borneo is due west of the Moluccas, requiring a much longer voyage, either to the north and then west, or else to the south, then west, and then north. If a geographer plotted Varthema's stated route, Java would lie considerably further south than its true location.

Even more influential were Varthema's reports about Java's climate and the duration of its day. He reports that his ship's captain described the southern part of Java as having a day of only four hours, and that it is "colder than in any other part of the world." In fact, Varthema cited Java's "extreme cold" as one of his reasons for leaving the island, suggesting to a geographer that it lay quite far south indeed.

"Ludovico di Varthema," Mercator explained, "tells us that on the southern side of Greater Java toward the south there are certain peoples that navigate by constellations directly opposite to our Septentriones, and this to such a degree that they find a day of four hours, that is, in the 63rd degree of latitude; all this he repeats from the lips of an Indian skipper." Having calculated the latitudes at which this phenomenon would occur, Mercator considered the island's circumference assigned by Polo, and naively invoked the name of Mandeville, the medieval lord of travel mythology, to corroborate the data.

Extrapolating from all this, Mercator reached the conclusion that Polo's *Lucach* and *Maletur* were, like Magellan's *Tierra del Fuego*, regions of *Terra Australis*, and his 1569 world map, which reflected this conclusion, was widely copied for many years. Mercator's concepts are illuminated on a map of *Terra Australis* by Cornelis Wytfliet (fig. 81); *Lucach Regnum* is Thailand (or Borneo), and *Maletur Regnum* is Malaya (or Sumatra), and *Beach* is a corrupted duplication of *Luchac*. Off the austral coast lie the islands of *Java Minor*, *Java Major*, whose erroneously southern location has brought *Zambaba* (Sumbawa) along with them.

Beach
Between *Lucach* and *Maletur*, there lies a newcomer to the Polian neighborhood, a kingdom called *Beach*. From whence did this new, non-existent 'Southeast Asian' coastal kingdom of *Beach* come? The error can be traced to the 1532 *Novus Orbis* of Huttich, which was published in Paris with the map of Fine, and in Basle with the map of Münster/Holbein. This volume contained the text of Marco Polo's journey, but it mistakenly corrupts Polo's *Lochac* to *Boëach* (fig. 82), which in turn was shortened to *Beach*. Thus Mercator duplicated *Lochac* with *Beach*, giving cartographic life to the Huttich volume's error, and starting a precedent which continued well into the following century.

The Enigma of an Antipodean 'Brazilian' Waterway to Malacca
"On the twelfth day of the month of October, 1514," a reporter wrote from Madeira in that year, "a ship from the Land of Brazil arrived here to restock its provisions." His report about the voyage was quickly circulated in a German pamphlet entitled *Copia der Newen Zeytung ausz Presillg Landt*. It recorded that the vessel had

sailed south of the "Cape of Good Hope" and had found Brazil, which had a point extending into the sea. After they had navigated for nearly sixty leagues to round the Cape, they again sighted the continent on the other side, and steered toward the north-west. Driven away by the Tramontane, or north wind, they retraced their course, and returned to the country of Brazil.[232]

Following this itinerary, some maps from the early sixteenth century — notably the Schöner globe of 1515 and the 1531 world map of Fine — placed a land called '*Brasilie Regio*' *(The Kingdom of Brazil)* in *Terra Australis,* in line with Africa or the Indian Ocean.

The report's 'Cape of Good Hope' was probably referring to a point in the New World which was analogous to the southern African cape — in other words, a passage around the southern tip of South America. The overwhelmingly likely candidate for such a cape would have been the land forming the northern shores of the entrance of the Rio de la Plata, which lies at nearly the same latitude as southernmost Africa, and which at first encounter was believed to be (like the Cape of Good Hope) a cape leading into the Indian Ocean. Two hundred kilometers wide at its mouth, and maintaining its sea-like proportions for more than three hundred kilometers upriver, the southern shores of the river (the report's "continent on the other side") would not have been visible until the vessel had sailed far into it. Since the New World was commonly envisioned as a final, monumental Southeast Asian subcontinent, it was entirely sensible that from "the land of Brazil it would not be more than six hundred miles to go to Malacca," and that, as stated in the report, the people of this land "have a memory of St. Thomas." The grave of St. Thomas was traditionally placed in southern Asia; such sources as Marco Polo had specified the Indian region of Maabar, and Maabar in turn had been placed by some geographers of the day (e.g., Fries, fig 62) in Southeast Asia. As the German newsletter pointed out, "that they have a memory of St. Thomas is highly credible, for it is known that he lies buried behind Malacca on the Siramath coast on the Gulf of Ceylon [that is, in India]."

St. Thomas was also transferred to the 'American' region of Southeast Asia. In the early sixteenth century, the Spanish historian Las Casas wrote that Portuguese missionaries in Brazil reported that the Indians there preserved the memory of St. Thomas, and that the Indians even showed them his footprint on the bank of a river.[233]

Confusion over the entrance to the Rio de la Plata as a passage-way to the Indian Ocean began as early as 1508, when Vicente Yáñez Pinzon and Juan Dias de Solís reached its mischievously confusing mouth; reports of voyages such as theirs may be the origin of heresay about a strait which fueled the dreams of Magellan as a youth. Whereas the *Terra Australis* of Fine was born from the Tierra del Fuego, Schöner's earlier version depicts the southern bank of the Rio de la Plata, as per the voyage in search of Southeast Asia described in the *Newe Zeytung auss Presillg Landt*.

The newsletter's voyage is probably an account of Don Nuño Manuel and Cristóbal de Haro, who in 1514 explored the estuary of the Plata for sixty leagues and who would have believed, logically,

Fig. 83 An unmounted gore from a globe by Coronelli, 1688. Coronelli notes the possibility that the newly-discovered land of New Holland was the Lochac of Marco Polo. (45 x 28 x 18 cm)

that Malacca was only 600 leagues to the northwest. In their eyes, one need only have completed the passage through the Plata 'strait' to enter the Indian Ocean, where the Spice Islands lay, and beyond them, Malacca.

Southeast Asia and Australia
The story of Southeast Asia on an austral continent did not end with Dutch voyages around the south of Java, since the recorded discovery of Australia opened new shores on which to relegate confusing Southeast Asian nomenclature. The Australian phase of *Lochac's* austral journey is best seen on works of the Italian geographer, Vincenzo Coronelli, whose ideas were espoused on several globes and maps using the same geography. Coronelli, on his 110 cm globe of 1688, explained that some people believed newly-discovered New Holland to be Marco Polo's *Lochac*, while others believed that New Holland was *Java Minor*. Although the 1688 globe (fig. 83) explains these theories without suggesting a conclusion, the association of Australia with *Lochac* led Coronelli to place an elephant there.

Just as *Terra Australis* had been the logical solution for Mercator, Australia now provided an answer for Coronelli to the riddle of *Lochac* because Polo reported that it was on *terra firma* (i.e. continental), it was isolated, and because his text, particularly Ramusio's edition, implied a southerly location.

Literary Allegory and an Antipodean Southeast Asia
An antipodean *Beach* and *Maleteur* also figured in the whimsical geography of an early seventeenth-century literary fiction. In London, in 1605, Joseph Hall, a bishop in Exeter, published a satire told by way of an imaginary voyage to *Terra Australis*. "Over against *Moleture* and *Beach*," we are told, lies "the newly discovered *Womandecoia*," a joke on the name *Wingandekoa* of Virginia Indian origin used by such mapmakers as Ortelius.[234] "The onely garrison" of this realm of women is *Shrewes-bourg*. A reversal of stereotyped gender roles is found here, such that "if any woman use her husband somewhat gentler then ordinary (as some of them be tender hearted) she is presently informed against, cited to appear before the Court Parliament of *Shrewes-bourg*, and there indicted of high treason against the state."

In the neighborhood of *Womandecoia, Moleture* and *Beach* there is also "an Island called *Ile Hermaphrodite*, or more properly, *Double-sex*," which lies "not far from *Guaon*, the last Isle of the *Moluccaes*" (*Guoan* can be found, far to the southeast of Timor on the Linschoten map of 1595). On the island of *Hermaphrodite*, anyone so unfortunate as to be of a single sex, that is, who "have not shown themselves perfect both in begetting, and bringing forth," are made slaves to the rest and treated as deformed monsters. Halls' *Ile Hermaphrodite* may have in part been inspired by 'serious' passages in Mandeville's *Travels*, which describes a place in Southeast Asia that is strikingly similar, lacking only the facetiousness.

The Line of Demarcation and Southeast Asia

During the fifteenth century, both Spain and Portugal tested their prowess on the seas to their south and west. When the 'right' to various Atlantic islands and African shores was contested by the two Catholic kingdoms, treaties were drafted, and the intervention of the pope, being the only worldly authority they deemed to be higher than themselves, was enlisted.

Papal Bulls and Treaties
The 1479 Treaty of Alcáçovas gave the Guinea coast, the Azores, the Madeiras, and Cape Verde Islands to Portugal, and reserved the Canaries for Castile.[235] This treaty sowed the seeds of future strife by obligating Spain to "never disturb the King and Prince of Portugal and their heirs in their possession or quasi-possession . . . [in any] . . . islands, coasts, or lands, discovered or to be discovered, found or to be found . . .". But "discovered or to be discovered", in the minds of the treaty's authors, essentially meant Africa and its islands; the dramatic landfalls of subsequent decades lent unanticipated meaning to it.

In 1493 Columbus, sailing westwards under the Spanish flag, announced that he had reached the southeastern outskirts of Asia. Meanwhile Portugal, who had earlier pursued the westward option undertaken by Columbus, had more recently devoted her resources to finding a route around the southern part of Africa, and had already reached the Indian Ocean. It would be only a short matter of time before the two countries' paths would cross again in the Far East. Thus, Spain and Portugal fought over Southeast Asia even before they had 'discovered' it. Spain secured four papal bulls in 1493 which effectively divided up the non-Christian world between the two Catholic countries, her territory lying to the west of a designated demarcation line *(Línea de Demarcación),* Portugal's to the east. This effectively confined Portugal to Africa and her Atlantic islands, denying her sovereignty over any part of the New World, and leaving Asia ambiguous. Although the pope's failure to address the 'other' side of the *línea* has sometimes been interpreted to mean that he believed the earth to be flat, and that he therefore considered the Atlantic line to have resolved the entire issue, it is far more likely that the pope simply deemed the eastern side of the *línea* to be a future source of messy controversy, which he could, for the moment, spare himself. Since longitude could not be determined with any degree of accuracy, the eastern *línea* was a Pandora's Box best left undisturbed until Europe's explorers forced it open.

Recognizing the superiority of Portuguese sea power, and perhaps aware that the very fickleness of the papacy, which at the moment pleased them, would likely succumb to other incentives in another day, Spain agreed to negotiate once again. In June of 1494, Spain and Portugal worked out a new agreement known as the Treaty of Tordesillas, which moved the *línea* westwards, 370 leagues (approximately 1770 kilometers) west of the Cape Verde Islands, placing it roughly between the bulk of South America and the easterly bulge which is Brazil. It was because of this line, of course, that today Brazil is Portuguese-speaking, while the rest of South and Central America is Spanish-speaking. Spain and Portugal "covenanted and agreed that a boundary or straight line be determined and drawn north and south, from pole to pole, on the said Ocean Sea, from the Arctic pole to the Antarctic pole," which was to be "calculated by degrees [of longitude], or by any other manner, as may be considered best and readiest."[236] Spain would claim "all lands, both islands and mainlands, found and discovered already, or to be found and discovered hereafter," lying to the west of the line, and Portugal all those lying to the east. Pope Julius II sanctioned this agreement in 1506.

But what of the other side of the world? The vast majority of educated people assumed the earth to be spherical, and knew that if Southeast Asia proved to be a desirable asset, someone would eventually have to draw the continuation of the *línea* around to the other side of the sphere, 180° opposed from the demarcation line which defined the western boundary of Brazil. But, since science would not provide a reliable answer to the longitude question for nearly three centuries, more conflicts were inevitable. The *línea* could be pushed and shoved according to each power's respective interests. The question became inescapable when Portugal took Malacca in 1511 and began serious exploration of the Southeast Asian islands the following year.

In 1514 Pope Leo X, responding to an embassy sent by King Manuel of Portugal, issued the bull *Praecelsae devotionis,* which formally extended the *línea* around the other side of the earth, through the Pacific. Leo X was avidly interested in cosmographical matters, and is said to have had Peter Martyr's letters read to him at dinner; however rudimentary their appraisal of the southwest Pacific, Martyr's letters probably suggested to Leo X that his 1514 bull would not settle the matter. The Spiceries would not be easily won. Eight years later Magellan's *Victoria* returned to Seville, and Portugal's King John III formally protested Castile's perceived violation of Portuguese sovereignty to Charles I. According to the Spanish historian Las Casas, the fact that the strait Magellan anticipated finding would lie within Spanish domain was the reason he switched allegiance from Portugal to Spain.

The Psychology of the Línea de Demarcación

The question of the *línea* produced strange episodes of intrigue and mystery; beyond the actual question of longitude, it was, to an

extent, a state of mind. By sailing to the west from Spain, Magellan was approaching Southeast Asia via Spain's side of the *línea,* making it easier — psychologically — for Spanish geographers to bend the line in her favor. The length and difficulty of Magellan's route, however, kept the search for a new passage very much alive. The ideal passage would have been through Central America, which lay at an advantageous latitude, and was already known to be an isthmus. Since the line of demarcation was not an absolute gauge, the sheer accessibility of the Indies was seen as an important advantage in establishing ownership. Peter Martyr makes no question about this: "The explorers are devoured by such a passion to discover this strait," he wrote, "that they risk a thousand dangers," and that "if indeed a passage between the South and North Sea is discovered, the route to the islands producing spices and precious stones will be shortened, and the dispute begun with Portugal . . . will be eliminated."

Pigafetta claimed, according to information obtained from a Portuguese sailor he encountered in the Moluccas, that "the king of Portagalo had enjoyed Malucho already for ten years secretly, so that the king of Spagna might not learn of it." Martyr accepted Portuguese primacy in the Spiceries, but not its sovereignty over them: "our people admit," he wrote, "that the Portuguese name is known in those islands, and that Portuguese subjects have been there, [but] assert there was only a single man [presumably Francisco Serrão], who was a fugitive fleeing from judgement for his crimes [defying orders to return to Malacca], and that, moreover, no other proof of commercial relations can be produced."

Complicating the formidable scientific problems involved in determining longitude was the fact that cartographic 'evidence' and favorable 'authoritative' judgements could be bought and otherwise influenced. Jorge Reinel, the younger of two highly respected father-son Portuguese cartographers, fled from his native country to Spain just at the time when Spanish authorities were making preparations for the new voyage to be commanded by Magellan. He may have been the source for the mapping of the Moluccas on the Spanish *padrón* (the official chart given by the Spanish crown to its pilots). Jorge's father, Pedro, subsequently brought his son back to Portugal. When a conference held in Badjoz-Elvas in 1524 to determine ownership of the Moluccas ended in deadlock, with neither side able to agree on where to place the Indies on their maps, the intrigue continued. Lopo Homem, a Portuguese cartographer who acted as an expert witness for his country's representatives at the conference, is believed to have offered to sell his services to the Spanish.

Nor was the ownership question settled when Spain conditionally relinquished the Moluccas to Portugal in a confusing treaty signed at Zaragoza in 1529. Both sides agreed that a line needed to be drawn in the Pacific, from pole to pole, to separate Castilian and Portuguese spheres of influence in the Pacific. But, since no one could accurately place such a line, the treaty provided for Portugal's monopoly of the maritime spice trade until science provided a method of doing so. Portugal paid King Charles 350,000 *ducats* for him to relinquish any claims to the Moluccas, which would be returned if the ownership question was eventually settled in Castille's favor. In the interim, Portugal agreed not to fortify the islands further.

Thus, for many years, cartographers and chroniclers fought a tug-of-war with the *línea,* dragging and pulling it so that the rich markets of Southeast Asia and the Spice Islands would fall within their domain. The Portuguese chronicler, Barros, located the Moluccas three hundred leagues (1950 kilometers) east of Malacca, when in fact they are about 530 leagues (3400 kilometers) east. Far better estimates existed; the prolific proselytizer Francis Xavier, writing from Goa in 1542, stated that it was 1000 leagues from Goa to the Moluccas, and 500 leagues from Goa to Malacca, or in other words

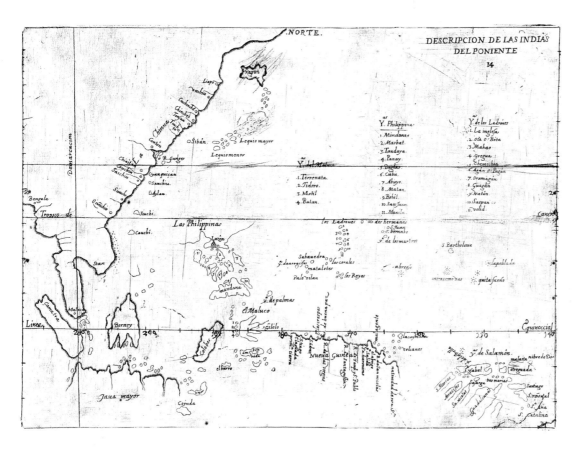

Fig. 84 Southeast Asia, Herrera, 1601. At the start of the seventeenth century Spain broke its silence on published maps of the Indies, and in Madrid, in 1601, Antonio de Herrera y Tordesillas published his *Historia general de los hechos de los Castellanos en las Islas i Tierra Firme del Mar Océano*, which included a map of Southeast Asia based on the politically bent charts of López de Velasco (ca. 1575-80). Like its manuscript prototype, the engraved map, *Descripcion de las Indias del Poniente*, fixes the line of demarcation to Spanish advantage. (20.5 x 28.3) cm)

that it was 500 leagues (3200 kilometers) between Malacca and the Moluccas. This was four years before Xavier himself reached the Moluccas, so his figures were based on information already available in Goa. Spain's mapmakers worked under a corresponding political pressure, pushing the *línea* to run along the west coast of the Malay Peninsula, granting them not only the Moluccas, but also most of the Southeast Asian mainland, virtually all of the Indies, as well as Japan, Korea, and most of China. The mapmakers and the sovereigns they served did not know themselves how far they were stretching the truth. For example, when Villalobos sailed to the Philippines from Mexico in 1542, the strong North Equatorial Current may have caused him to underestimate his dead-reckoning of the distance traveled, resulting in an 'honest' appraisal that the islands lay on the Spanish side of the line.

López de Velasco

In a manuscript map of ca. 1575-80, the Spanish cartographer López de Velasco, official historian of Castile and the Indies to Philip II, mapped the Moluccas fully 40° too far to the east. The López de Velasco chart was disseminated in printed form in a book by Antonio de Herrera y Tordesillas. Published in Madrid in 1601 (fig. 84), it was re-engraved for a Dutch issue of the same work, which appeared in 1622. This work marked the *línea* running through Malaya, with *Sian* (Siam) bulldozed east into Indochina, thus keeping Thailand within Spanish dominion (Siam did, in fact, claim suzerainty over most of continental Southeast Asia, and even the Portuguese chronicler, Barros, included a large part of Indochina in his description of Siam). Nonetheless, as Peter Martyr observed, "had the Portuguese accepted the decision of the plenipotentiaries of Castile, they would have been compelled to recognize that, not only the Moluccas and the islands touching upon China, the great gulf, and the promontory of Sartres and Gilolo, but also even Malacca had all along been usurped by them."

The Macao and Manila Dioceses

The Iberian powers' spheres of influence were finally solidified with Pope Gregory XIII's creation of a Portuguese diocese of Macao and adjacent islands in 1576, and three years later of the Spanish diocese of Manila. Spain still tried to place her imprint on the map of the Southeast Asian mainland, though this was but a brief and tragic episode. In the later 1500s, the king of Cambodia had entreated the assistance of the Spanish in Manila to help his country defend itself from Ayuthaya. Thus when Lovek finally fell to Siam in 1594, there was already a Spanish foothold in Cambodia. Manila and Madrid considered seizing the opportunity to establish a protectorate over Cambodia with the hope of eventually creating a Spanish empire in Indochina, controlling the Mekong Delta, ultimately conquering Siam, and perhaps even China. But hostility from the Muslims in Cambodia, and the taking of Phnom Penh at the turn of the century, effectively ended Spanish dreams of expanding to the west.

In place of those dreams, however, some found visions of expanding to the east. Luzon was an ideal perch from which to venture to Micronesia, which lay unquestionably within the Spanish side of the *línea*. Moreover, England and Holland had become worrisome new rivals in the competition for the traditional goals in Southeast Asia, making the various islands of the Marianas, Carolines, and Marshalls more realistic aspirations for the Spanish. Formal Spanish expansion to the islands did not begin for another century, however, since neither Manila nor Madrid seemed anxious to expend resources on a new goal in the Indies which promised neither a source of riches nor even a stepping-stone to one.

Chapter 13

1570 – ca.1600: Diversity in a Transition to Standardization

The end of the sixteenth and the beginning of the seventeenth centuries marked an interesting epoch in Southeast Asian mapping. Portuguese supremacy in the East was starting to look vulnerable, but the Spanish were well settled in the Philippines and showed occasional signs of an interest in Taiwan and Indochina. Meanwhile, Dutch, English, and French entrepreneurs were anxious for the chance to flex their muscles in the East. The emergence of major map publisher houses, together with the advent of their countries' 'East India Companies', was soon to lead to a wide standardization of maps, both as regards content and presentation. One event, in particular, can be seen as marking the point when this homogeneity, began to take hold: in 1570, Abraham Ortelius introduced his monstrously influential atlas, the *Theatrum Orbis Terrarum.*

In the interlude between the old and the new, however, there was a brief flurry of marvelous and individual compositions which reflected the fruits of a century of exploration, but just escaped the relative rigidity and conformities of the coming era.

Ortelius and Related Maps

Why was the *Theatrum* such a huge commercial success? It was methodically conceived, uniformly sized, and was based on current knowledge, rather than classical sources. The atlas was handsomely engraved, lavishly produced, and in the course of its 42 year history was published in translation in several major Western European languages, including English. In-house colorists were employed to produce illuminated examples for an extra charge. The atlas appealed to many different segments of the public, from those interested in geography to those who simply liked beautiful books or wanted to impress their neighbors. The integration of eye-tickling motifs into the cartographic canvas — in the case of the map of Southeast Asia, these include angry whales, tossing ships, and vain mermaids — foreshadows the highest Dutch mapmaking tradition in which art and cartography merged in new ways. Further distinguishing himself, Ortelius usually acknowledged the cartographer upon whose work a particular map was based. The *Theatrum's* maps remained the most influential published models through to the end of the century, and derivatives were disseminated even into the early eighteenth century, when the last miniature Italian knock-off was engraved.

The Ortelius Maps of Southeast Asia and of the Asian Continent

Ortelius' *Theatrum* contained a map of Southeast Asia, *Indiae Orientalis* (fig. 86), as well as a continental map depicting Asia as a whole, *Asiae Nova Descriptio* (fig. 85). Although the two maps propose very different geographies for Southeast Asia, both were published without change in the *Theatrum* from 1570 to the last edition of the atlas in 1612.[237]

Ortelius' map of Southeast Asia largely follows the great 1569 world map of his friend and colleague, Mercator, while his map of the continent is a reduction of his own separately-published map of Asia of 1567, which is in some respects superior to Mercator's work. In a legend on the 1567 map, Ortelius explains that he based his own map on the 1561 Gastaldi map of Southeast Asia and states that Gastaldi had derived the latter, without acknowledgement, from the Muslim 'geographer' Taqwim Albudan. However, as we saw earlier (page 131, above), *Taqwim Albudan* was simply the title of an archaic Arab geographic dictionary which was irrelevant as a source of information for Gastaldi's map. Ortelius is more credible when he claims that his map improves upon Gastaldi's model:

> However, we have surpassed it [the 1561 Gastaldi map of Asia], not without gain as is commonly said, so that it is absolutely not a barefaced imitation. For, firstly, the things that were missing in the Gastaldi version for a complete and definitive map, we have added in the right places; secondly, we have by no means excluded the borders and whatever else it seems possible to supply for geography and a detailed knowledge of these countries.[238]

Taiwan

The *Asiae Nova Descriptio* and *Indiae Orientalis* are the first published works to definitely map Formosa, and to identify it by that name *(Fermosa).* For the the Ryukyu chain itself, Ortelius uses the terms *Lequiho* and *Lequio*, forms of Liu-ch'iu. This was one of Ortelius' own contributions — neither the Italian models, from which the *Asiae Nova Descriptio* descends, nor the 1569 world map of Mercator, from which Ortelius adapted the *Indiae Orientalis,* names Formosa (though Mercator depicts the chain of Liu-ch'ius trailing Japan). But what, precisely, is Ortelius' *Fermosa* island? A look at the Liu-ch'iu chain on the *Asiae Nova Descriptio*

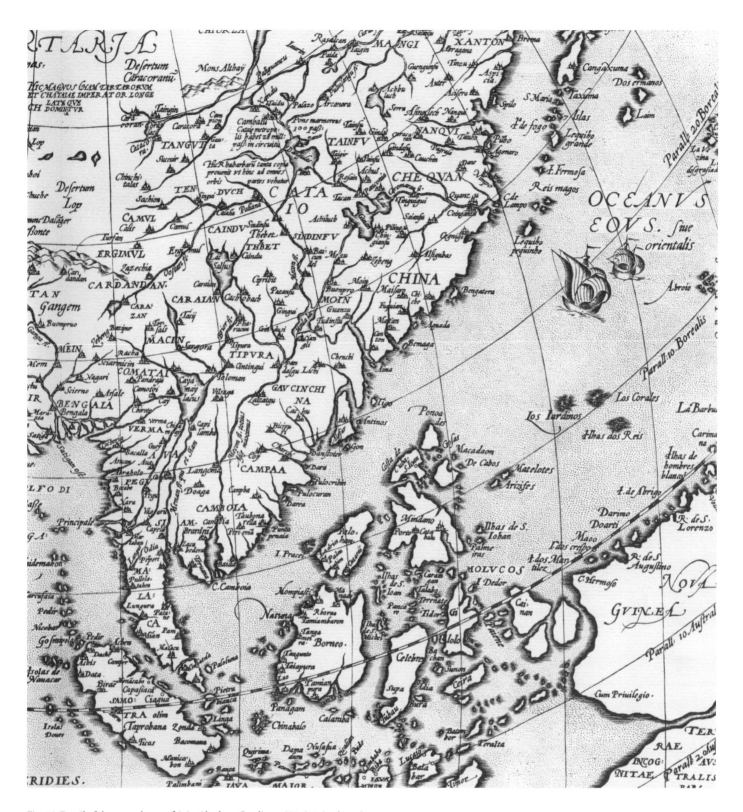

Fig. 85 Detail of the general map of Asia, Abraham Ortelius, 1570. (37.5 x 49 cm)

suggests that Ortelius did not mate island and name properly, in that his term *I. Fermosa* would appear to have been misapplied to one of the smaller Liu-ch'iu islands to the north. True Taiwan would seem to be his island of *Lequiho pequinho* ('small Liu-ch'iu'), which is at the correct latitude, being centered at about 25° north (Taiwan spans 21° 55′ to 25° 15′). In fact, however, all these islands, from *Lequiho pequinho* to *I. Fermosa,* are probably all parts of Taiwan. This becomes clearer when looking at more advanced maps — a good example is the Plancius map of ca.

1594 (fig. 90) — in which an enlarged 'Formosa', 'small Liu-ch'iu', and an unnamed island in between them, together form a rudimentary depiction of Taiwan. One of the little isles could be *Tayouan* Island which later became joined to the main island by silting, and from which the modern name of Taiwan is derived.

The Arab pilot Ahmad ibn Majid had described Taiwan (which

Following page: Fig: 86 Southeast Asia, Ortelius, 1570. (35 x 49.8 cm)

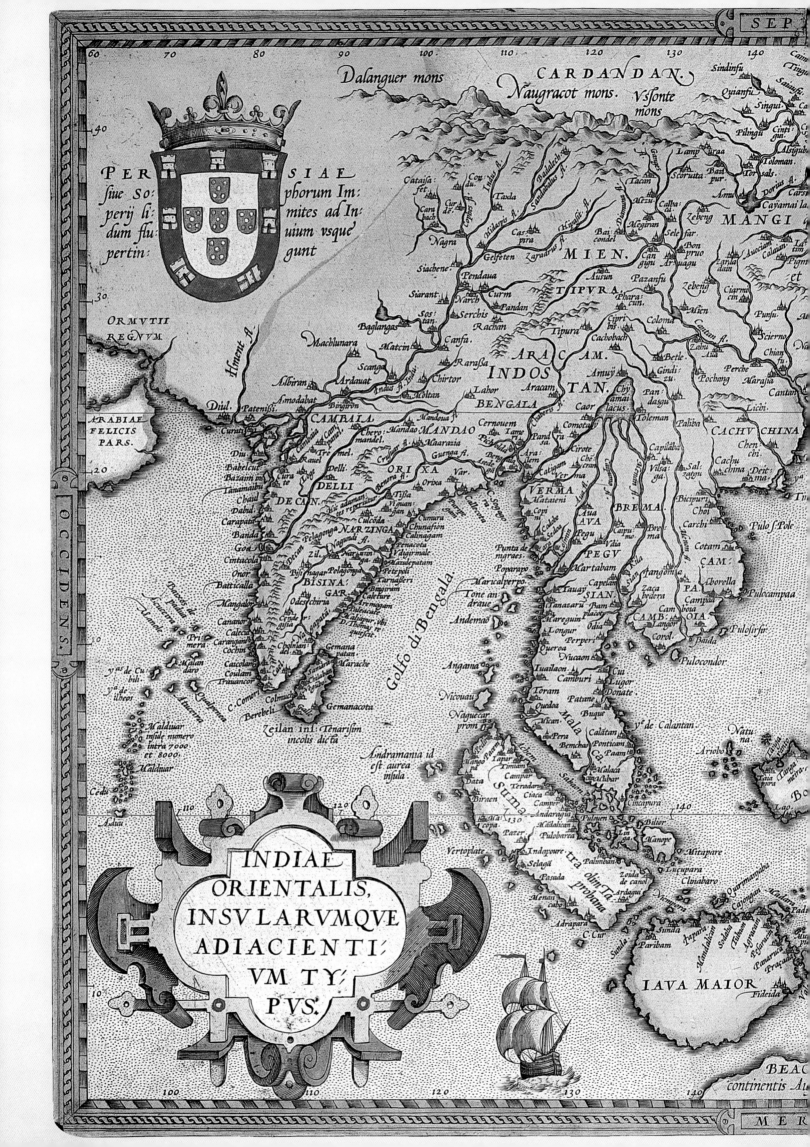

100   170   180   190   200   210   220   230   240

AMERICAE, siue Indiæ Occidentalis pars.

40

Quiuira

Cicuie

Tiguex

Mare Cin   Inf. de Miaco   Satyrorum inf.

Campu

Torza   Saendeber

Tam

Tartalho   Chela

Zangia ins   *Pilbo*   *Miaco academia*   *Bandu*

Dinlai

CIN   *Fuſſum*   *Homi*   *Amaguco*

Gieza   Done   IAPAN.   Hanc inſulam M. Paul:

Scabana   *Menlai*   *Negru*   Venet Zipangri vocat.

NA.   *Malao*   *Fraſon*

Liampo opp.   *Hormar*

Baraquis   *Cangoxina*   OCEANVS   30

Chequeam   et prom.

Mazacar

Lequio maior

vᵃ Fermoſa   Los dos hermanos

Reix magos   Los Bolcanes

Bergatera insula

Lequio minor

Tropicus Cancri   Malabrigo

de benjaga

ORIENTALIS   La farſana   20

Bata de innocentes

Pauao das   *Hanhan*   Humunu vel yᵃ   *Restinga de ladrones*

Cataram   *Callon*   *Aguada*   di buoni ſegni

Cenalo   *Zamal*

Culuan   *Hibuſſon*   ARCHIPELA:

Philippina   *Abarien   Huinangan*   GO DI S.   Inf. de los corales   10

Bohol   *Pasaie de S. Clara*   yᵃ de mata   Los uardines

Mesſana   *Calagan*   lotes

Paulogon   *Butuan*   yᵃ de arrecifes   LAZARO

Cimbubon   *Chippit   Mindanao.*   Inſulæ Moluccæ celebres ob maxima

Lozon   *Buran*   aromatum copiam, quam per

Candingar   *Buran*   totum terrarū orbem trans

Tagimar   yᵃ de cocos   ferunt, 5. ſunt, iuxta Gilolo

Zolo   *Cahnio*   Ma Talao, alijs Tarrao inſulæ   nempe, Tarenate, Tidore,

Sanguin   Doy   Motir, Machiã et Bachiã

Bibalon   Morotay

Babatao   Circulus æquinoctialis.   yᵃ de creſpos   190

Tantampura   160   *Mamoro*   180

Calamba   Supa   *Gilolo*   yᵃ de los martires   Hermoſo.

Tubar   Celebes   *Guebe*

Macace   *Varson*   Cainam   NOVA GVINEA

Pulola or   *Aylon*   quam Andreas Corſalus Ter-

Galiam   *Bandan*   *Am*   ram Piccinaculi appellare vi-

Noce Amanor   *Daluza pinca*   *Aruu*   detur. An inſula ſit, an pars

Zolot   *Cuto*   continentis Auſtralis incer-   10

Eude   *Tior*   tum eſt.

Timor Tidore inf.   *Jupagna*

Aliſao   Cum Privilegio.

Subao   *Cimpegua*   Sumbedit

Jap   ORIENS

160   170   180   190   200

was known to the Arabs as *al-Ghur* or, following the Chinese, *Likiwa*) in a navigational text of 1462. The archipelago, including Formosa, was also reported by Pires in his *Suma Oriental* of 1515, and appeared in some *rutters* as early as about 1550. Europeans had landed — and were shipwrecked — on Taiwan's shores by the mid-century, as the island's northern coast is depicted on charts of the Portuguese chartmaker Diego Homem dating from 1558. A narrative written in Lisbon in 1548 records that in 1542 "two Portuguese . . . while going in a junk to trade on the coast of China, were carried by a storm to the island of the Lequios where they were well-treated by the king of those islands, through the intervention of friends with whom they had traded in Siam."[239] Although the island quickly developed a reputation as a magnet for shipwrecks, the strategic value of its location was also recognized. It was a stepping stone to China and Japan which was not yet under the clear control of any major power, and it lay directly in the path between Manila and China. Any power seizing control of the island could disrupt the Philippine-Chinese trading route, and it could serve as a base for launching direct attacks against the Spanish strong-holds in the Philippines. In order to pre-empt this the Spanish attempted to establish a base in Taiwan in 1592, but were forced to turn back because of storms. The Dutch founded the Castle Zeelandia on the southwestern coast of *Tayouan Island* (modern Taiwan) in 1624 and the Spanish succeeded in placing a settlement at Keelung (northern tip) in 1626, but were driven out again in 1642.

Lake Chiang Mai

Both the Ortelius maps add their own twist to Lake Chiang Mai, mythical source of Southeast Asian rivers and auspicious site for the founding of the kingdom of Lan Na. The *Indiae Orientalis* places a new kingdom along the lake's shores — *Toleman* — which is apparently derived from the kingdom of that name which was reported by Marco Polo. Polo stated that there was a large gulf which "extends for a two-month's sail toward the north, washing the shores of Manzi on the south-east and of Aniu and Toloman besides many other provinces on the other side." Extrapolating from the map and Polo's text, it seems that Ortelius — who did not copy this feature from Mercator, but rather interpreted it himself — believed (correctly) that *Manzi* (Mangi) was in southern China, and (incorrectly) that Lake Chiang Mai was the 'gulf' which washed its shores, which led him to place Polo's *Toloman* along the lake. Other mapmakers, notably Gastaldi, had interpreted Polo's *Toloman* to lie in the American Northwest. Interestingly, the *T'ung Tien,* a Chinese encyclopedia dating from the late eighth century, records a place called *To-lo-mo,* but this has been identified as lying in western Java.

On the *Asiae Nova Descriptio,* Ortelius appears to be mixing Lake Chiang Mai with the Lake Tsinghai of the 1561 Gastaldi map (see page 156, above). He refers to Lake Chiang Mai as *Cajamaÿ lacus,* and connects it with a 'salt lake' *(lac salsus)* far to the north, in the mountains of China, by way of a river.

Borneo, Moluccas, and Halmahera

The *Indiae Orientalis* follows Mercator in erroneously depicting Borneo after the elongated model of the Italian maps, and in its use of *Monte de adas* ('mountain of spirits') for Borneo's Mount Kinabalu, rather than the more usual *San Pedro.* 'Spirit mountain' may have expressed a Spanish visitor's own impression of the mountain, or was translated by him from local Bornean traditions.

While Ortelius deferred to Mercator regarding such recent data as *Monte de adas,* he broke with his elder colleague's passion for making modern data 'work' without abandoning the Ptolemaic model altogether — for example, identifying Borneo as having

been the *Bone fortune* of Ptolemy. Ptolemy placed the island of *Bone fortune* on the equator to the east of *Taprobana,* but since Mercator believed *Taprobana* to be Sumatra (rather than Ceylon), Borneo was a logical choice for the 'island of good fortune', since it lay in approximately the same position relative to Sumatra as the Ptolemaic *Bone fortune* did to Ceylon. Ortelius omits that opinion from his *Indiae Orientalis.*

More seriously, however, Mercator's determination to make geographic features square with his belief that Ptolemy's *Taprobana* was Sumatra led him to conclude that the Canton River was in fact the Ganges River; Ortelius disagreed, labeling the Canton River correctly, but followed Mercator in assigning the name *Chaberis* to the Ganges itself.

Both of the Ortelius maps are an improvement upon those of Gastaldi, Ramusio, and Forlani as regards the positioning of the all-important Moluccas. In 1567, three years before the *Theatrum* was published, Ortelius received a letter from the French astronomer and mapmaker Guillaume Postel in which he complained about the faulty coordinates of the Moluccas assigned by Barros. Whatever the reason for the errors on the Italian maps and in the Portuguese chronicles, a more correct rendition had now entered mainstream publishing.

Halmahera (*Gilolo*) is also an improvement over earlier maps, particularly on the *Indiae Orientalis.* We saw earlier how Gastaldi, perhaps misguided by Barros, inverted Halmahera's limbs (page 146, above). But both Ortelius maps now correctly depict the island as having four arms that unfold to the east.

The First Traces of Irian Jaya

Both of the Ortelius maps also depict a few islands off the western coast of New Guinea called *Cainan,* which represent an early rendering of Irian Jaya. In 1526 a Portuguese pilot, Jorge de Menzes, was blown past Halmahera and sighted New Guinea, making a landfall somewhere in the northwest. Saavedra reached it two years later. In Urdaneta's account of the latter's attempt to return to New Spain from the Moluccas, he records that

> In the course of the voyage [they] were anchored because of contrary winds at some islands with negroes they call Papuas, these islands being located about 200 leagues east of the Moluccas.

Villalobos also reached the island in 1545, and was responsible for naming it New Guinea. Richard Hakluyt recorded the rediscovery of New Guinea by Villalobos, and lamented the loss of the original name *Papuas.* According to Hakluyt, Villalobos

> sent from the Island of Tidore [a] ship towards New Spain by the south side of the line [i.e., below the equator] . . . they sailed to the coast of Os Papuas, and ranged all along the same, and because they knew not that Saavedra had been there before, they challenged the honor and fame of that discovery. And because the people there were black and had frisled hair, they named it Nueva Guinea. For the memory of Saavedra as then was almost lost, as all things else do fall into oblivion, which are not recorded, and illustrated by writing.

In the latter part of the seventeenth century, Irian Jaya would often be confused with the island of Seram, or else judged to be a separate island from the mainland of New Guinea. On a chart of Asia dating from about 1689, the English hydrographer John Thornton compared its "high land on the sea sides" as being "like the island Formosa".

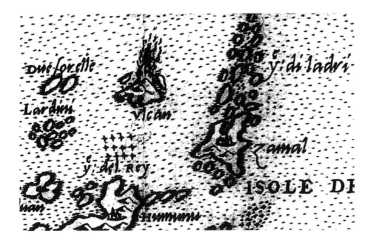

Fig. 87 (detail) Marianas with Samar superimposed over Guam, from Gastaldi's map of Southeast Asia, 1561. [Clive A. Burden]

Micronesia and the Philippines

The term *Archipelago de San Lazaro*, which Magellan originally applied to the Philippines while he was in the region of southern Samar, became mixed up with Micronesia because accounts of the voyage appeared to group members of the two archipelagoes together. Samar *(Zamal)* became superimposed on Guam, the southernmost island of the Marianas, on the plate for Gastaldi's 1561 map (fig. 87). and as a result was left behind in the same position when the rest of the Philippines was disentangled from Micronesia, first by Mercator in 1569 and then in turn by Ortelius on the *Indiae Orientalis*. The latter leaves it stranded out in the open Pacific below the right-hand mermaid. Since Samar was the first Philippine island to be mentioned by Pigafetta, the term *San Lazaro* stuck with it, now erroneously designating the Micronesian archipelagoes of the Garden and Coral Isles. The Marianas are represented only by *Restinga de ladrones* (barrier reef of the Ladrones), and lack the correct north-south orientation found on the Ramusio map; Ortelius' continental map omits the 'thieves islands' altogether. And just as Micronesia has now been designated the *San Lazaro* islands, in the eighteenth century the Carolines inherited the name 'New Philippines' (fig. 13).

The *Indiae Orientalis'* depiction of two mermaids with mirrors may hark back to the old lore regarding an 'island of women' who conceived by beholding their own image — one missionary report actually located an amazonian island to the south of the Marianas, near Ortelius' two figures.[240] Directly below the left mermaid lies an island labeled *Humunu vel ja di buoni segni* ('Humunu, actually known as the island of good signals'). This is the uninhabited Philippine island where Magellan first landed on March 17th of 1521, offering the crew a 'good signal' because there they found spring water and indications of gold.

Near the mermaid and whale lie the islands of *Malabrigo, Los dos Hermanos, Los Bolcanes,* and *La farfana.* The data for these four archipelagoes comes from Bernardo de la Torre's 1543 attempt to cross the Pacific from west to east, which was part of Villalobos' expedition of 1542-46. As related by Richard Hakluyt,

> the 25 September [of 1543] they had sight of certain islands, which they named Malabrigos (that is to say, The Evil Roads). Beyond them they discovered Las dos Hermanas (that is, The Two Sisters). And beyond them also they saw 4 islands more, which they called Los Volcanes. The second of October they had sight of Farfana, beyond which there standeth an high pointed rock, which casteth out fire at 5 places.

Ortelius has corrupted the gender of *las* and *hermanas* to the masculine, and *volcanes* to *bolcanes*, while the place name *Farfana* comes from the Spanish *huerfana,* meaning 'orphan'.

Two more islands from the Villalobos expedition lie due east of Mindanao. *Yᵃ de matalotes* is the 'buenos dias, matalotes' island (see page 136, above), with *Yᵃ de arrecifes* ('reef island') lying about 30 leagues further to the west (Ortelius reverses their position). Juan Gaytan noted that on the latter there were "many human dwellings and many palm groves." Villalobos placed it 35 leagues west of Matalotes, and stated that "on account of the reefs that came out of it, we could not anchor at it." Indians came out to meet then in canoes and they baptized it *Islas de Arrecifes.*

The island of Luzon appears on the *Indiae Orientalis* as the small island of *Lozon* due west of Mindanao; like the earlier, more correctly positioned appearance of the island at the hands of Gastaldi in 1561, Ortelius' depiction is conjectural. He places *Lozon* as a close neighbor to *Chippit*, which is the Quipit region of Mindanao; Pigafetta had reported that Luzon lies a two-days' sail to the northwest from Quipit, but Ortelius, following Mercator, reverses them.

The *Asiae Nova Descriptio* includes some fresh data from the voyage of Legazpi, who reached the Philippines from Mexico in 1565, determined to discover winds which would permit a return east across the ocean. One island discovered by the expedition is *La Barbuda,* one of the Marshall Islands, which can be found on the right extreme of the map, just north of New Guinea. The pilot Estéban Rodriguez explained the reason for the name *Barbuda* in his logbook: "There were about 100 Indians at this island, comely with beards; for this reason, we named this island Los Barbudos [the Bearded Ones]. It is in 10 degrees and ¹/₄."[242]

Ortelius accepted Mercator's belief in an imposing southern continent and in the theory that Marco Polo reached its shores. The *Indiae Orientalis* extends far enough south to depict the northernmost promontory of *Terra Australis* protruding below Java, with the Polian kingdom of *Beach* (a corruption of *Lucach*) occupying its shores. This northern extension of *Terra Australis* was drawn by incorporating the 'extra' *Java Minor*, shown by Forlani (fig. 75), into the continent.

Following Italian models, the *Asiae Nova Descriptio* depicts the southern coast of Indochina more accurately than the flattened shores of the *Indiae Orientalis,* but lacks the easterly bulge of Indochina and the Gulf of Tongkin, which the *Indiae Orientalis* competently represents. The *Asiae Nova Descriptio* also records the Paracels as distinct islands, labeled by their modern name *(I. Pracel).* These had been alluded to by Ramusio in 1554 as 'channels' *(canalli).* On the *Indiae Orientalis,* the old 'satyr' island of Ptolemy's Indian Ocean has been shoved to the north of Japan, in waters labeled *Mare Cin.* After the word *Cin* is empty space, suggesting that the engraver meant to incise *Mare Cinpagu* (Japan Sea); correlation to the 1569 Mercator map, however, confirms that *Cin*, for 'Sea of China', is correct.

The text on the verso of the *Indiae Orientalis* comments that

> There are near unto this country many goodly islands, which here and there lie scattering in the main Ocean, so that it may justly be termed the World of Islands . . . the Moluccas [are] famous for the abundance of spices which they yearly yield and send into all quarters of the world . . . From Achen in Samotra, and Bantam in Iava Maior our Merchants, this other day brought letters unto his Highness, so fairly and curiously written in that character and language, as no man will scarcely believe but that he hath seen them, especially from so barbarous and rude a nation.

Fig: 88 China, Abraham Ortelius, 1584. (37 x 48 cm)

The Ortelius Map of China, 1584

In 1584, Ortelius composed a map of China, *Chinae, olim Sinarum regionis, nova descriptio,* based on data compiled by Luis Jorge de Barbuda, a Portuguese in the service of Philip II of Spain (fig. 88). Ortelius shows Mindanao, and some islands to the north, but only *Cubo* (Cebu) is marked. Formosa is better formed than on his 1570 map of Asia, but the most curious part of Southeast Asia on this map is the long, diagonal island just north of Borneo. This is, of course, the position and orientation of Palawan, yet the island depicted here is clearly not based on information from Spanish expeditions approaching from the east, as those shores have been left hypothetical. In fact, if we look back at some earlier Portuguese charts, we find a series of closely-spaced islands forming a coastline similar to Ortelius'. The Lisbon chartmaker Reinaldo de Carvalho, in a map of 1563, depicts this archipelago, labeling it as *Costa de Incōis,* and similar configurations are found on the sea charts of Sebastião Lopes and Fernão Vaz Dourado; they are commonly accompanied by comments about a Chinese presence in these islands' waters. Whereas Magellan and subsequent Spanish explorers had approached Palawan from the east, our *Costa de Incōis,* like the interesting depiction of Palawan found on the 1548 map by Gastaldi (fig. 73), probably represents the occasional sightings of Palawan's western shores by Portuguese vessels heading round the Southeast Asian mainland, bound for China.

On the early example of the Ortelius *Chinae* illustrated here, this landmass is not identified. Later impressions, however, bear the name *Las Philippinas* — the first time on a printed map that the name is applied to the archipelago rather than one island. Five years later, on his *Maris Pacifici,* Ortelius divided the Philippines between the *Philippinas* to the south, and the *Lucoes* to the north (though Luzon itself is not shown). The transference of the term to include the whole archipelago had already occurred in manuscript maps.

Mazza: Early Appearance of Luzon (ca. 1590)

Ortelius flourished just a bit too early to have mapped the 'true' island of Luzon. That omission was subsequently corrected by the Italian geographer Giovanni B. Mazza on a separately published adaptation of Ortelius' map of the Asian continent. Probably dating from about 1590, Mazza adds *Luconia* (Luzon) in a roughly recognizable shape.

Sebastian Petri's Map of Sumatra (1588)

The *Theatrum* was also used as a model for a revised edition of Sebastian Münster's *Cosmographia* in 1588. The industrious Münster had died of the plague in 1552, but his original 1540 woodblocks were republished without improvement through 1578, by which time they were pitifully obsolete. Whereas the 1540 maps offered a wealth of new ideas to their audience in their heyday, the new 1588 maps of the continents contributed nothing. Probably prepared by Sebastian Petri, and still cut in woodblock, they were simply mediocre copies of the maps in the 1570 *Theatrum* of Ortelius. But the new *Cosmographia* did contain an interesting map of Sumatra and the southern part of the Malay Peninsula. Though essentially an enlarged extract of the corresponding section of Ortelius' *Indiae Orientalis,* the map adds some nomenclature and topographical details. It includes an inset of an elephant which had appeared on Münster's Ptolemaic map of Ceylon in 1540. Since Sumatra had become the preferred candidate for *Taprobana,* the elephant was now placed on the map of Sumatra. Although the updated *Cosmographia* never achieved the popularity of the original, several editions were published until 1628.

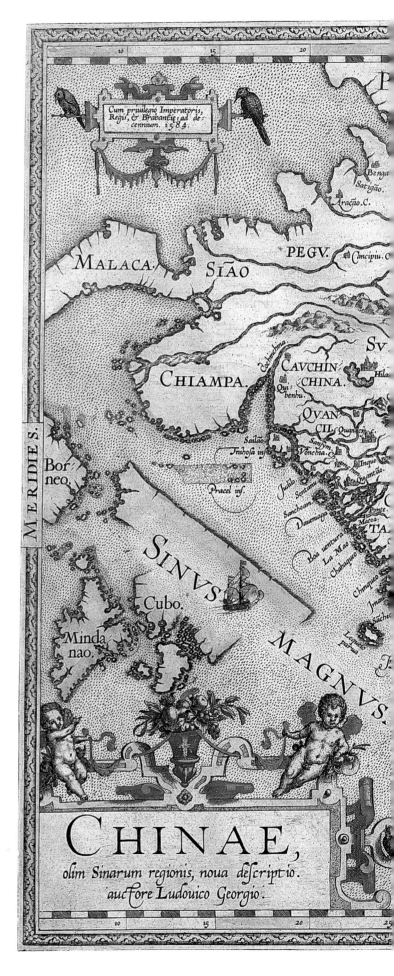

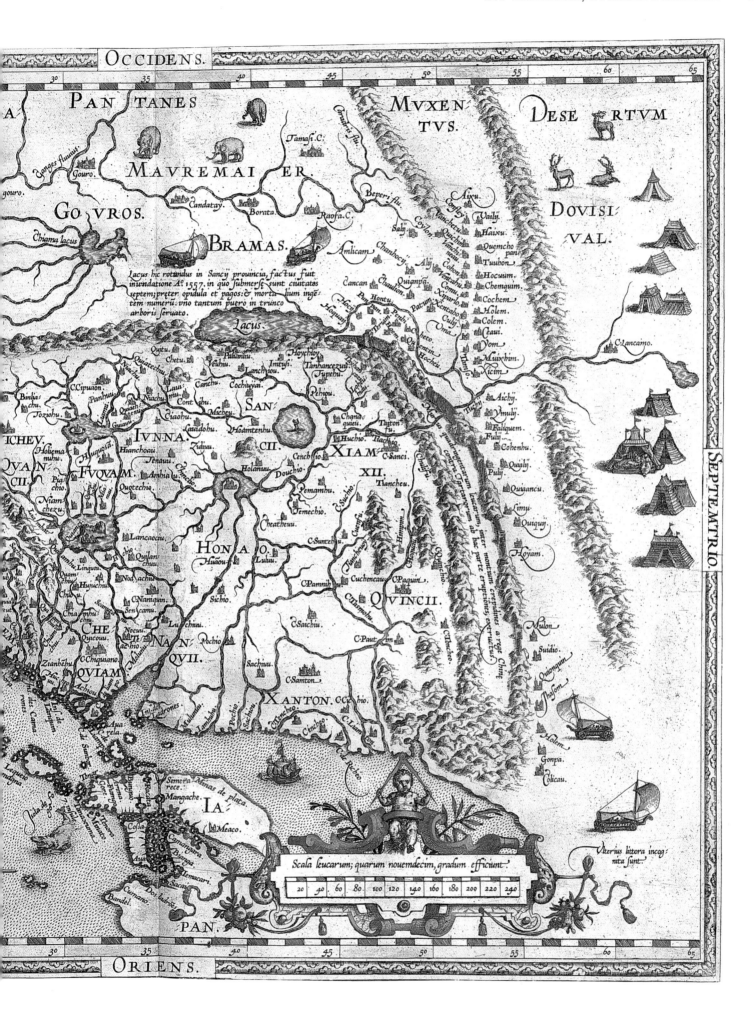

PAN TANES

MVXEN
TVS.

DESE RTVM

MAVREMAI ER.

GO VROS.

DOVISI
VAL.

Ganges fluuius.
Gouro.

Chiama lacus

Candatay.   Borata.

Tamaſi.C.

Raofa.C.

Beperi flu.

Aixu.

Vaily.

Haixu.

BRAMAS.

Amlicam   Chanbocy.

Saly

Ceylon.

Quemcho
pani

Tuuhon.

Lacus hic rotundus in Sancij prouincia, factus fuit
inundatione Aº. 1557. in quo ſubmerſæ ſunt ciuitates
septem; prǣter opidula et pagos: & morta hium ingē-
tem numeriū. vno tantum puero in trūnco
arboris seruato.

Cancan

Hocuum.

Chemquim.

Cochem.

Holem.

Colem.

Lacus.

Oytu.

Chetu.

Palamhu.

Veihhu.

Imtuſi.

Hoychloy

Tanhancezuij
Jupehu.

Ceaui.

Dom.

C. Iancamo.

Muixhim.

Xim

Onatechnu.

SAN

Lanchiou.

Pehiou.

Aichij.

Ymulij.

CCipuaon.

Panhui

Laua.
Cont.

Canchu.

Cochiuya.

IVNNA.

CII.

Micheu.

Hoamtenhu.

Chana
guieu.

Huchio.

Tlach io.

Taiton
fu.

C. Sanci.

Faliquem.

Fulij

Cohenhu.

FVQVAM.

Amhia u.

Zidhu.

Holanuu.

Douchio.

XIAM

XII.

Tiancheu.

Pulij

Quialy.

Quiaancu.

Quotechiu.

Pemamhu.

S.Suichio.

Limu.

Quiquij.

Cheatheuu.

Hoyam.

HON A O.

Huãou-fi

Luhuu.

C-Suntzehu.

Himpun.

Lancaociu.

Quian
chiu.

QVINCII.

Mulon.

Suidio.

CHE

NAN
QVII.

Pochio.

Lu chnni.

C. Pamnih

Cuchencau

CPaquin.

C.Saichiu.

C-Paut in.

Outonquin.

Thiſom.

QVIAM

Sachiuu.

C. Samton.

XANTON.

CC hio.

Holem.

IA

Meaco.

Gonpa.

Colicau.

PAN.

Scala leucarum; quarum nouemdecim, gradum efficiunt.

| 20 | 40 | 60 | 80 | 100 | 120 | 140 | 160 | 180 | 200 | 220 | 240 |

Vlterius littora incog-
nita ſunt.

SEPTEMTRIO.

The Curious Case of the Island of St. John

An enigmatic island, new to printed maps, appears on Ortelius' 1570 map of the continent. This is the 'Island of St. John'. Ortelius did not invent the island, but effected its peregrination to Southeast Asia from the Micronesian region of earlier manuscript maps. To retrace the story of this island, we need to backtrack to Portuguese charts of a few decades earlier.

San Juan was probably already common to official Portuguese charts of the 1530s. A portolan chart attributed to Gaspar Viegas, dating from ca. 1537, depicts the island of *San: Joā,* lying to the east of Mindana, just below a latitude of 6° north, in a configuration that is typical of charts of the middle decades of the century. These Iberian maps show the island at too great a distance from Mindanao to be considered one of the Philippines — all the more so since the tendency at this time was to understate east-west distances in the western Pacific.

But it was as a close neighbor of Mindanao, not as a Micronesian island, that St. John passed most of its life. Ortelius lays the foundation for this cartographic folly on the 1570 *Asiae Nova Descriptio* with a group of two islands immediately to the east of Mindanao, called *Ilhas de S. Iohan,* the direct descendants of the true *San Juan* of Portuguese charts. Ortelius also identifies *S. Ioan* as a town on Cebu, immediately north of Mindanao, and there is an archipelago marked *Ilhas de S. Ioan* scattered between Borneo and Mindanao. Shortly after the Ortelius maps, in about 1575, *San Juan* was re-born as a single Philippine island lying off the northeast coast of Mindanao at the hands of the Spanish mapmaker Juan López de Velasco.

This non-existent island became a standard landmark in Philippine cartography throughout the seventeenth and much of the eighteenth centuries, and was recorded in earnest as late as 1850.[243] Thus *St. John* island raises two essential questions: where did the island come from; and why did it migrate from Micronesia to the Philippines?

The answer to the first question lies in discoveries made in Micronesia by members of Magellan's crew attempting to sail from Southeast Asia eastwards, back across the Pacific to America. After the *Concepcion* was set afire on the island of Bohol following the loss of Magellan and other members of the crew in the skirmish on Mactan, the expedition had two ships remaining: the *Victoria* and the *Trinidad.* Both these vessels reached the Moluccas and loaded their hulls with spices, their departure for home expedited by fears of Portuguese retaliation. But while the *Victoria* continued westwards to complete the first successful circumnavigation of the globe, the *Trinidad,* under the command of Gonzalo Gómez de Espinosa, attempted — unsuccessfully — to return east across the Pacific and reach Panama. While searching for westerly winds, Espinosa charted the Mariana chain in detail (see page 135, above), and discovered an archipelago in the Carolines which they named *San Juan.* According to one pilot's account, on 6th May, 1522, they "made two small islands, which could be in about 5 degrees more or less, to which they gave the name of the islands of San Juan."[244] Another account explains that "on the way to New Spain, because the winds were scarce, they steered to the northeast. In 6 degrees on the north side, [they saw] two islands, to which they gave the name of San Juan."[245] Unable to locate the necessary winds for the return voyage across the Pacific, Espinosa, in desperation, returned to the Moluccas, where he was greeted by 300 Portuguese who were building a fort on Ternate. Espinosa and his men were arrested, and were held by the Portuguese for several years.

News of their discoveries quickly passed to their Portuguese captors. Espinosa was shipped from Ternate to Banda and from there to Malacca, before being finally shipped to Cochin (India). After ten months in Cochin, Espinosa wrote a letter to the king of Spain, informing his sovereign that the Portuguese authorities "took from me all the nautical charts, the logbooks, the astrolabes, the quadrants, the nautical books with all the equipment of the pilots and more" (dated January 12th, 1525). In this way the charts and records of Spanish discoveries in Micronesia — including our *St. John* island — were added to Portugal's repertoire of geographical data, and as a result the Spanish discovery of *San: Joā* island makes an early appearance on Portuguese charts.

Which of the Carolines was *St. John*? Of the many archipelagoes in the western Carolines which lie in the possible path of a vessel sailing northeast from the Moluccas, the two most likely candidates are the Sonsorol Islands and Palau, with Sonsorol fitting the evidence rather more closely. Accounts by Antonio Galvão and a Genoese pilot who was also on the voyage, both state that *St. John* comprised two small islands, as indeed does Sonsorol. As for location, *St. John* was said to lie at a latitude of about 5-6° north.[246] The Sonsorol pair again conforms to the description better, lying at 5° 19′ and 5° 20′ north, while the principal islands in the Palau group span between 7° 8′ and 7° 44′.

Palau, however, conforms more closely to the latitude of the *San: Joā* of early charts, and is more perfectly positioned for the island's migration to the northeast coast of Mindanao. The problems with Palau as a candidate are that it should have constituted a major discovery, hardly befitting the relatively minor attention paid to the 'smaller' islands the Spanish discovered. Furthermore, it consists of one major and several smaller islands, not the two islands mentioned by eye-witnesses. Arguably, Espinosa could have sighted two of the lesser Palauan islands, or did not necessarily have an accurate idea of the size of *St. John*, and even less so his Portuguese captors who confiscated his charts. We have ample evidence from the Philippine islands how different landfalls on a large island could be mistaken for several individual, small islands (for example, Quipit on Mindanao). Nor would such an error have been eventually caught, since once *St. John* migrated to northeastern Mindanao, the expectation of a large island in the true position of Palau vanished.

Juan López de Velasco and Antonio de Herrera y Tordesillas

That brings us to the second question in the story of *St. John*. How did the island — whether Sonsorol, Palau, or some other island — end up a stone's throw from Mindanao? There were two reasons. One was an attempt by mapmakers in the later sixteenth century to correct previous errors regarding the relative east-west placement of the various Philippine and Micronesian islands. As we have seen, the Marianas and the Philippines were bunched together by earlier cartographers, and they were sometimes associated with the wrong group when later cartographers subsequently attempted to sort them out — for example, Gastaldi's hitching Samar to the Marianas in 1561 (see page 138, above). The other reason was that the name *San: Joā* was suspiciously similar to that of an island which did indeed lie off the northeast coast of Mindanao, just east of Cebu — the island of Siargao.

Thus, it appeared to some cartographers — specifically, to the Spanish chartmaker Juan López de Velasco when composing a set of maps in about 1575 — that *San: Joā* was in fact Siargao, misplaced to the east on the older charts because of the longitude problem. The fact that he opted to make lists to organize the islands by archipelago shows that he was trying to sort out the prevailing chaos. He concluded that the two islands were in fact

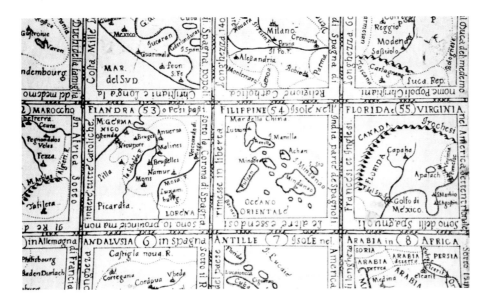

Fig: 89 Detail from a cartographic board game, Antonio Lucini, ca. 1665. Players began in the lower left of this game, moving counter-clockwise around the board in diminishing circles until reaching a panorama of Venice in the center. The game square with the map of the Phillippines is extraordinary for its S Sio island to the northeast of Mindanao, the island finally given its correct identity as Siargao Island rather than the bogus 'Philippine' island of *San Juan*, which was a misplaced Micronesian archipelago. Each map 5 x 5.2 cm. [Courtesy Clive A. Burden, Ltd]

duplications, and that the one cast adrift in the Pacific was the erroneous one. López de Velasco's charts, with their 'Philippine' island of *San Juan*, were widely disseminated in engraved form in the *Historia general de los hechos de los Castellanos en las Islas i Tierra Firme del Mar Océano* of Antonio de Herrera y Tordesillas, first published in Madrid in 1601 (fig. 84). The maps were more widely disseminated in a Dutch copy of the book, which was published in Amsterdam in 1622 with re-engraved copperplates. They depict a single island of *San Juan* off the northeast of Mindanao at about 8° north latitude, as though it were a corner of Mindanao which had been severed from the main by a strait. *San Juan* is numbered '10' on the map's list of Philippine islands.

In what would seem to be a poetic quirk of fate, the puzzle of *San Juan* was solved by a board game (fig. 89). The game consisted of 153 squares, each containing one map, and was engraved by Antonio Lucini, who made the plates for Robert Dudley's *Arcano del Mare* (see page 205, below). Players sliding their token onto the game square containing the Philippines found themselves in a rough representation of the archipelago in which *San Juan* has been replaced with *S Sio* (*San Siargao*), correctly identifying the island which indeed lies immediately northeast of Mindanao, and whose name was likely part of the confusion over *San Juan*. But not only was this board game an unlikely place to find anything of geographic significance, it was also an aberration, not the beginning of a new trend. The imaginary *San Juan* still had a century of life left to it.

The phantom island could be found on maps for nearly two centuries. In the early eighteenth century, François Valentijn, a long-time servant of the Dutch East India Company, prominently mapped the island in his important *Oud en Nieuw Oost-Indien* of 1724-26. In 1797, an atlas from the voyage of La Pérouse contained two charts of the Pacific between the Philippines and California. One, which the title identifies as being from a Spanish chart taken by George Anson in 1743, has no *St. John* island. The other, said to be from a Spanish chart acquired by La Pérouse, depicts Mindanao in a markedly different shape, with *Sn. Juan* hovering in its usual place off Mindanao's northeast coast.[247] Thus improvement was inconsistent — La Pérouse acquired his chart with *St. John* in about 1787, more than four decades after Anson had obtained his chart without *St. John*. Finally, early Jesuit maps based on Caroline Islander information coincidentally placed both Palau and Sonsorol (our two likely identities of the original *St. John*) off the northeast corner of Mindanao, in the place where the fake island had been (fig. 13).

Other factors may have contributed to making a new island from the northeast coast of Mindanao. In 1525, a fleet under García Jofre de Loaisa was sent from La Coruña to exploit Magellan's discovery of the Philippines. Herrera y Tordesillas, in his text, records that Loaisa visited various provinces of Mindanao; anchored on the east coast of Mindanao, he dubbed the region *vizaya* (also *bisaia* or *bicaia*). Another expedition was sent to the Philippines in 1527, this time from Mexico, under Alvaro de Saavedra. But here, Herrera y Tordesillas notes that Saavedra "went to Mindanoa and Vizaya and other islands lying at eight degrees," insinuating that *Vizaya* was itself an island. Although the name *Vizaya* suggests that the island was named for the Biscaya region of Spain, some deserters from Saavedra's crew referred to *Vizaya* as having been the indigenous term for the land, and indeed the central Philippines and its people are still known by this name (*Visayan*). Villalobos, however, was clearly honoring his native Spanish province when he gave the name *Málaga* to the north of Mindanao (found, for example, on the Linschoten map of 1595, fig. 92). López de Velasco made this region a separate island and transferred the name *St. Juan* to it.

Other Islands and Island Books

The Velasco/Herrera map contains three lists of islands, which provide a numerical key to four members of the Moluccas, eleven of the Marianas *(Ladrones),* and eleven of the Philippines. Its heading for the Philippine list finally established the modern name 'Philippines' for the entire archipelago, rather than an individual island or region. However, Mindanao long maintained effective autonomy from Manila and was not always considered a Philippine island by Spain's rivals. The British captain Thomas Forrest, who spent some time in Mindanao in 1775 as a guest of the sultan, claimed that the island was not part of the Philippine archipelago, and that the only reason the Spanish "sometimes call it a Philippine [was to] enlarge their own dominions." He cited French and Spanish sources to demonstrate that Mindanao was actually *adjacent* to the Philippines, and noted that "the Spaniards, though they have subdued the north coast of the island, never conquered the whole."

Following page: Fig. 90 Indies, Petrus Plancius, 1594 (this example published by J. Visscher, 1617). (39.5 x 55.5 cm) [Paulus Swaen Old Maps Internet Auction]

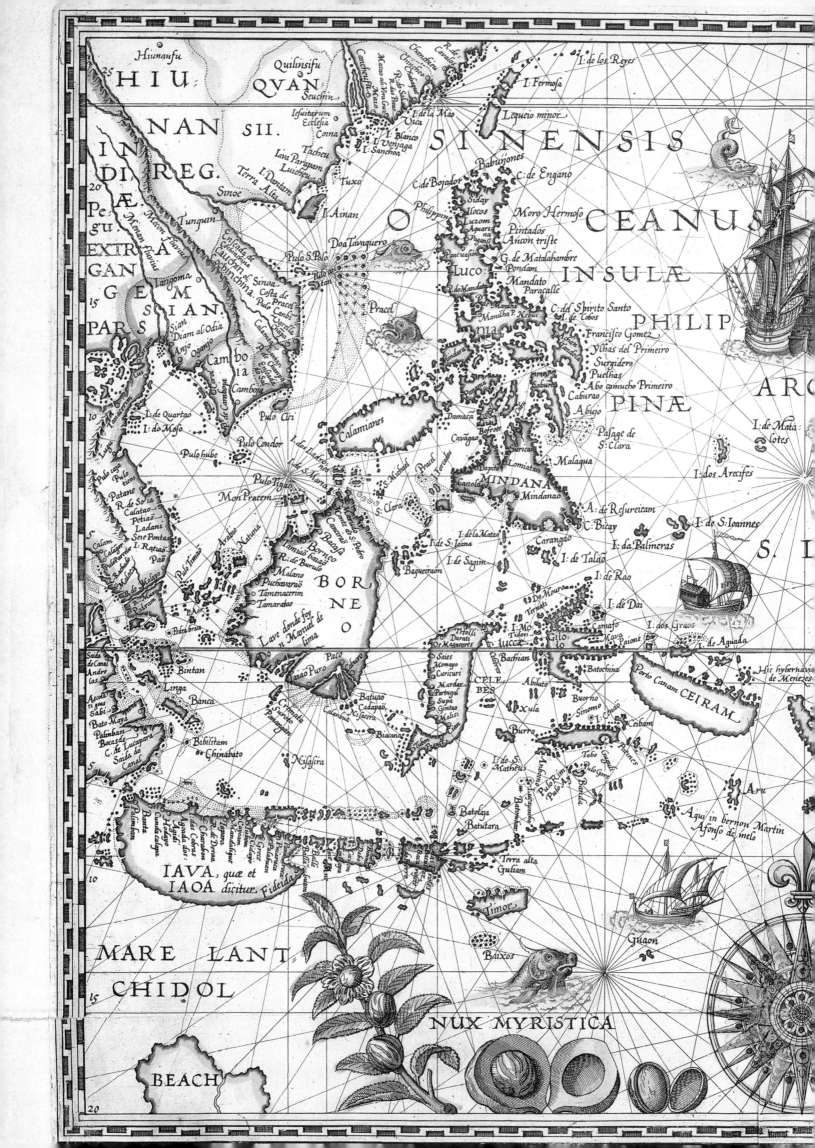

Some confusion resulted from the new information reaching López de Velasco's desk. 'Manila', Spain's new community founded in 1572 on Luzon, is used as the name of the island of Luzon, while 'Luzon' is marked as a city on the island. Another new name found on the map is Mindoro, the island south of Luzon which was discovered in the mid-1560s. But the similarity of the island's name with that of *mandana*, López de Velasco's rendering of 'Mindanao', resulted in his confusing the two. Mindoro is labeled *mindanao* (Mindanao), while Mindanao is *mandana*.

One interesting place name, so poorly engraved as to be barely legible, is *Mira como vas*, designating a small group of isles at about 8° north latitude. This comes from the Legazpi expedition (1565), commemorates the discovery of reefs near Truk in the Marshalls:

> At daybreak, we tacked back to see if we could see the reef, or island, but we could not . . . it is in 8 degrees. The pilot baptized it Mira como vas [= 'watch how you go!] because that is what those passing here should do.

The 'Southeast Asian' Island of Japan

Another curious island mapped by Velasco/Herrera is *Xipon,* off the coast of China and north of Formosa and the other Liu-ch'iu islands. *Xipon* is, of course, Japan, but Herrera y Tordesillas seems to have followed a *Relation of the Islands of the West* by Diego de Artieda, ca. 1570, which did not appear to make the connection, treating Japan as an extension of Southeast Asia. The report states that

> farther north of [the Philippine islands] are others, the nearest to Luzon being called Xipon. We have not seen this island, and what I say about it has been related to us by the Moros who carry on trade with that land.[249]

The 1601 engraved version of the map has *Xipon* partially scratched from the plate, as if an attempt were made to delete it (the extent to which this is obliteration is visible varies among examples of the map, depending on how the plate was inked).

Herrera y Tordesillas' depiction of the Liu-ch'iu islands is also muddled. Of the Liu-ch'iu and Formosa, the same de Artieda document notes that "a little to the east behind [*Xipon*] and China are the islands of Lequios. They are said to be rich; but we have been unable to learn much about them, for I have not seen any one who has been there. For this reason, I conclude that they must be small, and that the people are not much given to commerce." On the nearby mainland, López de Velasco, influenced by Mercator, has marked the Canton River as being the Ganges.

André Thevet's Unpublished *Grand Insulaire* (1580s)

The island world of Southeast Asia was perfectly suited for *isolarios,* or island books. The Southeast Asian islands covered by Benedetto Bodone's opus of 1528, the first printed *isolario* to include such shores, was largely symbolic; its delineations were for the most part hypothetical ones extrapolated from textual accounts. Another printed *isolario* appeared in 1572, by Tomaso Porcacchi, entitled *L'Isole piu famose del Mondo,* but its coverage of Southeast Asia was limited to a reduced copy of the by-then outdated 1548 *India Tercera Nova Tabula* of Gastaldi.

In contrast, a monumental *Grand Insulaire et Pilotage* was undertaken by the French explorer and chronicler André Thevet which — had it ever been published — would have included descriptions and maps of many Southeast Asian islands, including the Philippines. But Thevet, who was already eighty years of age when he announced the *Insulaire* in 1584, never saw the work to completion, whether due to illness, or to political or financial

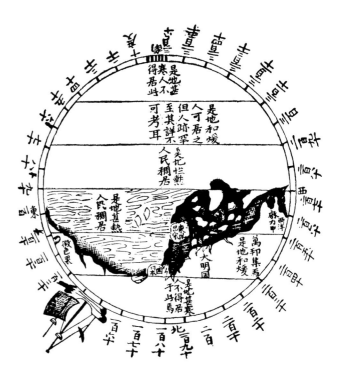

Fig. 91 Eastern Asia and the Pacific, from *Pien cheng-chiao chen-ch'uan shih-lu,* Manila, 1593. Woodcut. [University of Santo Tomas Publishing House, Manila]

troubles. Far more exhaustive than any previous island book, it would have been Thevet's finest creation. Fortunately, fragments of the work are extant in manuscript and engraved form. Typical of Thevet, the maps mix an impressive degree of detail with lapses into fancy and imagination. He claims, for example, to have derived his information about Mindanao from a chart he saw in Lisbon by "Jean de Boheme", which Magellan had consulted before his voyage, a nonsensical fantasy based on Pigafetta's statement that Magellan had been influenced by a chart of Martin Behaim.

Beginnings of Map Printing in Southeast Asia: Manila, 1593

Another important event during this last decade of the sixteenth century was the beginning of the printing of maps by Europeans in Southeast Asia — ironically by Spain, whose disinclination to disseminate maps via the printed medium was second only to that of Portugal. The map was part of a book written in Chinese, entitled *Pien cheng-chiao chen-ch'uan shih-lu* (*Testimony of the True Religion*), printed in Manila in 1593.

The map, a simple woodcut, superimposes a rudimentary delineation of the Pacific upon a zonal map framework (fig. 91). South is at the top. The zonal skeleton differs, however, from the medieval genre in that the equatorial regions are described as being "densely populated by men", instead of uninhabitable. Although the northern temperate region is well-inhabited, the southern (upper) temperate region, though inhabitable, is said to have "very few vestiges of man and they cannot be ascertained with precision." Like the traditional zonal maps, both polar regions are deemed uninhabitable. The diagonal coastline running along the lower left of the map is that of Mexico, with Japan at the lower center and China to the lower right; the large black area to the south and east of China is the 'Eastern Sea of Luzon', dotted with various islands. A cosmographical woodcut map in the same work depicts a geocentric universe, with the earth encircled first by atmosphere, then fire, and eight heavens. Other woodcuts and text argue for the sphericity of the earth, which was a new concept for most of the book's intended (Chinese) audience.

Dutch and German Maps at the Turn of the Seventeenth Century

Following Holland's declaration of independence from Spain in 1584, Iberian ports were closed to Dutch commerce and efforts to find an alternative means of trade with the Indies made the last decade of the sixteenth century a decisive one for Holland's future, expediting her attempts to reach the Indies directly. During this critical period, four people played a particularly crucial role in establishing a Dutch presence in Southeast Asia. They were: Jan Huygen van Linschoten, Petrus Plancius, and the Houtman bothers, Cornelis and Frederik.

Jan Huygen van Linschoten was an early Dutch traveler and one time secretary to the Portuguese archbishop of Goa, a position which allowed him to spend his spare time profitably collecting navigational information relating to the sea route to the Indies. In 1572, when he was about ten years old, the Spanish subdued Haarlem, and the Linschoten family were forced to move to the active seaport of Enkhuizen, where Spanish control was weaker. Despite the wars between Holland and Spain, the two countries still maintained commercial links, and when Linschoten was sixteen he traveled to Spain and Portugal. In 1583 he made the voyage from Lisbon around Africa to southern Asia, where he lived for five years in Goa. Although he never traveled to Southeast Asia, he had a keen interest in the region, and recorded a great deal of information about it while in Goa. His *Reysgheschift,* which was published in 1595, recorded explicit sailing directions that he had garnered from Portuguese *rutters* for entering the Indian Ocean by way of the Cape of Good Hope. But his more famous work is the *Itinerario,* which quickly became a standard text for Indies-bound pilots, and is among our most important sources for Southeast Asia during the sixteenth century. Linschoten's *Itinerario* made details of the formerly mysterious world of the Portuguese Indies easily available to anyone with the dream and the initiative to venture to the East. Furthermore, Linschoten provided the geographic 'key' to unlocking the Portuguese grip on passage through the Malacca Strait: Linschoten advocated approaching the Indies from the south of Sumatra through the Sunda Strait, thereby minimizing the danger of Portuguese notice or reprisal.

Petrus Plancius was a minister of the Reformed Church who was an expert on navigation. Plancius was instrumental in motivating and focusing Dutch energy toward Southeast Asia, and was one of the primary forces in the creation of the V.O.C. (Dutch East India Company) in 1602. He was to Holland what Richard Hakluyt was to England. In the early 1590s, Plancius was a scientific consultant for the pioneering voyage to Southeast Asia being planned by Cornelis and Frederik de Houtman.

Cornelis and Frederik de Houtman were two brothers who were sent by a group of Amsterdam merchants to Lisbon in 1592 in an effort to extract whatever knowledge they could of the Portuguese spice trade. While there they obtained classified sea charts made by the Portuguese cartographer Bartolomeu Lasso. Cornelis commanded the first Dutch voyage to Southeast Asia (1595-97), accompanied by Frederik.

The 'Plancius' Map of ca.1594

In April of 1592, the States General of the Netherlands recorded that they had granted a patent to the Amsterdam publisher Cornelis Claesz to "print . . . such twenty-five special sea charts as he obtained by the direction of Petrus Plancius, but at his own expense, from Bartolomeu Lasso". Such were the beginnings of one of the most fabulous maps of the Indies to be circulated at the close of the sixteenth century (fig. 90). Engraved by Johannes à Doetechum ca. 1594, the map was sold as a loose sheet into the early seventeenth century, and was on rare occasions bound into Linschoten's *Itinerario.*

Lasso's depiction of the Philippines was a vast improvement over earlier maps, and quickly established itself as the standard model for other mapmakers to copy. Of particular note is 'true' Luzon, superbly formed for the time, rather than the hypothetical *Lozon* found on earlier maps. Lasso's depiction of the Philippines was extracted by Petrus Bertius for a small map engraved in 1616, and was loosely followed for a map published by Langenes in 1598 (see page 188, below, for a closer look at Lasso's cartography).

Seram and Irian Jaya are Confused with Each Other

Many mapmakers of the early seventeenth century confused the peninsula which forms western New Guinea (Irian Jaya) with the island of Seram, which lies just southwest of it; the beginnings of that error can be seen on the Plancius map. On some earlier works (for example, Ortelius' *Asiae Nova Descriptio* of 1570, fig. 85), both the island of Seram (Ortelius' *Cera*) and western New Guinea (Ortelius' *Cainan*) are rudimentarily shown. Plancius, however, shows a large island of *Ceiram* (Seram) with a port of *Canam,* thus intertwining Seram and Iranian Jaya. This confusion of Seram and New Guinea was reinforced by Linschoten on his map of 1595 (fig. 92).

Above Plancius' *Ceiram* lie the islands of *Dos graos* (two degrees), and *de Aguada* (watering place), both of which suggest that Lasso tapped Spanish sources. The Marianas are shown in detail and now bear the name *Islas de las Velas* (Islands of the Sails), the original name given the Marianas by Magellan, describing the lateen sails used by the islanders. The pejorative name *Ladrones,* used by most mapmakers, was given afterwards.

Whereas the *Ladrones* lost their pejorative name, another archipelago gained one. The 'garden' islands discovered by Villalobos and dubbed for their "green and beautiful" appearance, already mapped by Ramusio in 1554, now have the disparaging remark 'al la Desaprovechada' affixed to their name, meaning that these are the garden islands of no worth or merit.

Several new islands appear in the Pacific to the north and northeast of New Guinea, derived from the 1565 Legazpi expedition. *De los dos Vesinos* ('Islands of the two neighbors, or residents') was discovered by Arrelano from the Legazpi expedition and was so-named because "there were no other [people] in the islands except two who were fishermen who had come from outside to fish at those islands."[250] Lasso placed these islands at $3^{1}/_{2}°$ north latitude, although Arrelano stated they lie at 9°. Arrelano also discovered *I: de los Nadadores* (Island of the swimmers), named "because [the people of the island] came aboard by swimming when we were over one league from the island . . . they would make the best possible rowers for a galley, according to the tall and well-built bodies they had."[251] Lasso places the islands at a latitude of 3° north, though Arrelano fixed them at $8^{1}/_{2}°$. The Legazpi expedition also discovered the *I: de Paxaros,* or "island of birds", which according to pilot Estéban Rodriguez, "was so small that we think it was uninhabited . . . the whole island was full of birds and for this reason we named it Isla de Pajaros."[252] The island of *Miracomo Vāz* is the "watch where you're going" island we saw earlier on the Velasco/Herrera map. Lastly, the enigmatic island of *St. John (I. de S: Ioannes)* still hovers in its correct position, far to the east of Mindanao at about 6° north latitude.

Various commodities of the Indies are depicted along the bottom of Plancius' map: *nux myristica* is nutmeg, *Caryophilorum Arbor* is clove, and *Santalum fluvum* is sandalwood. The bogus kingdom of *Beach,* unwittingly created by Mercator from a corruption of Marco Polo's *Lochac* (Siam), appears on the lower left (see page 160, above).

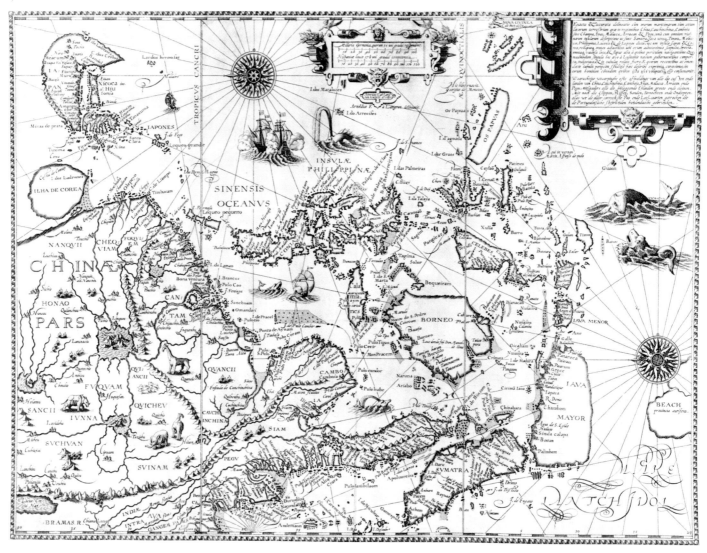

Fig. 92 Southeast Asia, Linschoten, 1595. (38.8 x 52.2 cm)

The Linschoten Map of 1595

While the separately-published Lasso-Plancius map was, in very rare instances, bound with Linschoten's *Itinerario,* another fine map of Southeast Asia, also heavily indebted to Iberian chart makers, was included with all issues of that work. This was Linschoten's own map of the Indies (fig. 92), which, in Linschoten's words, was made "from the most correct charts that the Portuguese pilots nowadays make use of." Those words were fair enough; the map's representation of Japan and Southeast Asia, except for the Philippines, was based on the work of the Portuguese cartographer Fernão Vaz Dourado, while the depiction of China is taken from Barbuda. The Philippines appears to be a variant of the Lasso model, and is most obviously characterized by its peculiar east-west orientation for Palawan. Although the *Itinerario* itself was not published until 1596, the license for the book was granted by the States-General in October of 1594, and this map of the Far East (which appears to have been engraved specifically for the book) is dated 1595 on the plate.

Linschoten labels the island of Seram as *Os Papuas,* reinforcing Plancius' confusion of Seram with the newly-emerging land of New Guinea. His depiction of New Guinea, however, proved influential at a later date; Thomas Forrest, who explored New Guinea in the service of the East India Company in 1774-76, cited the Linschoten

map as proof that the islands of New Britain discovered by William Dampier were one and the same archipelago as the Solomon Islands. He reproduced the New Guinea section from the Linschoten map and compared it with that of Dampier:

> It is to be regretted, that Dampier, who sailed to New Britain in the Roebuck 1699, had not seen Linschoten's map. Such a guide might have induced him to put into harbours which he did not visit, not knowing they existed: for the least additional light to a discoverer may be productive of important consequences.

In southwestern Borneo, both Plancius and Linschoten marked the region of *Laue donde foÿ Don Manuel de Lima (Laue,* which has been visited by Don Manuel of Lima). The only surviving reference to this Don Manuel or his endeavors is a mention, in the 1550s, by the Portuguese historian Fernão Lopes de Castanheda, who refers to Don Manuel being in Malacca in 1537. Notation of Don Manuel's presence in *Laue* is found as early as the 1554 world map of Lopo Homem, and was carried on through the charts of Bartolomeu Lasso and Vaz Dourado to the printed maps of Linschoten and Plancius. The region of *Laue* itself was mentioned by Tomé Pires in the early sixteenth century; although he located it to the east of Tanjonpura, Portuguese chart makers generally placed it further to the west, as do Linschoten and Plancius.

Fig 93 A crudely-engraved, semi-fanciful view of Banten. Giacomo Franco, 1598. (24 x 30 cm)

The rest of Indonesia varies slightly from earlier maps. East of Java, Bali has become *Galle,* and Sumbawa has assumed its old variant guise of *Java Menor.* Though the Plancius and Linschoten maps had succeed reasonably well in portraying the peculiar shape of Halmahera (*Gilolo*), the wild geography of Sulawesi has eluded both cartographers. Sulawesi's wonderful paramecium-like shape, so long as it was the island's own secret, could easily ensnare the most competent navigator. Francis Drake, after leaving Ternate, struck a reef off the Sulawesi coast and only escaped after throwing overboard some of his hard-won cloves in order to remain sea-worthy until the winds reversed. But as a result of such misfortunes, the mapping of Sulawesi began to improve in the seventeenth century.

On the mainland, the mysterious Mekong veers even further to the west than in the Ramusio map of 1554, reducing Siam — if one defines Siam as the tract of land between Burma and the Mekong — to a mere sliver. Ayuthaya, which actually lay on an island in the Chao Phraya (*Menam*) River, has been transferred here to another, shorter river which empties into the Gulf of Thailand a little further to the west of the *Menam.* The city of Ayuthaya is shown twice, once as *Odia,* and, just over the other side of the river, it appears again as *Siam.* Linschoten did not accept the inland sea placed by Lasso in northern Malaya, and corrected his omission of Singapore.

Chiang Mai *(langoma)* is placed as a neighboring city to Ayuthaya, on the bank of the Mekong. Lake Chiang Mai, the mystical lake from which the major rivers of Southeast Asia were believed to originate, has been placed further to the west than on earlier maps (it is the lake straddling the lower border), probably as a result of reports that one or more of the Southeast Asian rivers were connected to the Ganges. Barbuda, upon whose map of China much of the Linschoten's map interior information was derived, believed that the river systems of China originate in several large lakes, and Chinese tradition did in fact hold that the kingdom's rivers originated in a great western lake. Thus Linschoten includes four large lakes in the Chinese interior (more are on the Barbuda map of China, further north), one of which occupies the approximate place where earlier mapmakers (Ramusio and Ortelius) had placed Lake Chiang Mai.

The international history of the Plancius and Linschoten maps gained one more twist in 1598, when they were both re-engraved for the 'other' country vying for a slice of the Indies trade, namely England. Although the English renderings are inferior in execution to the Dutch plates, they nevertheless retain all the detail of the Dutch originals. So, here we have Portuguese and Spanish cartographers relinquishing geographic secrets to the Dutch, which in turn were copied by the English. In short, the four main rival countries in Southeast Asian affairs were all party to some angle of the map, reflecting the intrigue that surrounded mapmaking in those days.

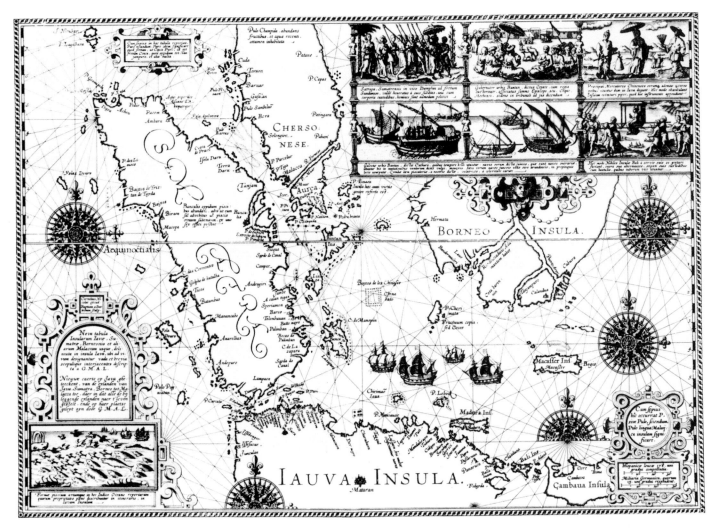

Fig. 94 Southeast Asia, Willem Lodewijcksz, 1598. (38 x 54 cm) [Universiteitsbibliotheek Amsterdam]

The Lodewijcksz Map of 1598

The dreams and labors of Plancius and Linschoten culminated in Cornelis de Houtman's pioneering voyage to Southeast Asia of 1595-97, which initiated Dutch presence in the Indies. The fleet, heeding Linschoten's advice, entered Indonesian waters by way of the Sunda Strait between Sumatra and Java, rather than the Malacca Strait. De Houtman, in fact, carried on board a copy of Linschoten's *Reysgheschift* in manuscript and probably relied heavily on it for sailing directions. Though the financial returns of the voyage were meager, de Houtman nonetheless established commercial relations with the great pepper port of Banten, on the north west coast of Java, near which the Dutch colony of Batavia would soon be founded. The facade of Portuguese invulnerability was quickly eroding; as the great English imperialist Hakluyt observed in 1599, "their strength is nothing so great as heretofore hath been supposed." The de Houtman expedition gave Holland first-hand data about the Sunda Strait, the northern coast of Java, and the island of Bali.

This first Dutch voyage to the Indies reached home in August of 1597. News of it was related in several works, the first being an anonymous account published by Barent Langenes before the year was out. Most important, however, was the *Historie van Indien,* published by Cornelis Claesz in April of 1598. Essentially the log of one of the expedition's participants, Willem Lodewijcksz, this book contained profiles of the coasts of the Sunda Strait, a

plan of the port of Banten, a plan of the town of Banten, a view of the market at Banten, coastal profiles of Bali and Java, plans of Bali's coast (fig. 99), and a general map of Bali (fig. 98). Several engravings depicted scenes from daily life in Indonesia (figs. 56, 5, 6, 97 & 110).

The volume was also supposed to boast a new general map of southern Malaya and the western Indonesian islands; however, in chapter 19 of the volume, we find this note:

> Here follows the chart of Java. But there is no chart. There is the accepted opinion that this mentioned chart was Lodewijcksz's chart. When the merchants saw this chart, the first printed one from this area in such detail, they have forbidden to insert this chart in the log.

The forbidden map, however, was published later the same year as a loose-sheet (fig. 94). It the upper part of the main cartouche, the name Cornelius Nicolai appears, which is the Latinized form of the publisher Claesz, along with that of Baptista van Doetechum, the map's superb engraver. The 'G.M.A.L.' cited in the attribution is the author, Guilielmus M. A. Lodewijcksz.

As the printed volume noted, this was "the first printed [map] from this area in such detail". That it was made good use of by pilots is evidenced by the fact that there are two extant examples of the map in which sailing directions for the Java Sea are printed on the verso. But the Amsterdam merchants' concern for keeping the chart confidential proved futile, since before the end of the

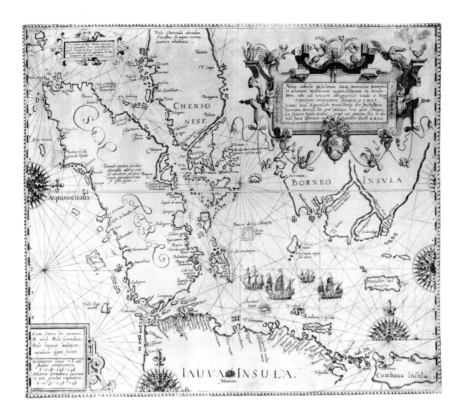

Fig. 95 Southeast Asia, Lodewyckszoon /Theodore de Bry, 1598. Theodore de Bry repackaged and republished chronicles of contemporary voyages to reach a wider audience than the original accounts had. His adaptation of Willem Lodewijcksz' map is particularly important since the original (fig. 94) was banned from the author'own published account, and thus would have otherwise been known only to a small circle of privileged insiders.

year a copy of it was published by the German chronicler, Theodore de Bry, in Part II of the *Petits Voyages,* dealing with the travels of Linschoten (fig. 95). The following year, de Bry published Lodewijcksz's account, together with re-engravings of his views and map of Bali, in Part III of the *Petits Voyages.*

De Bry's rendering of the Lodewijcksz map is typical of his beautiful engraving and aesthetic sense. One flaw crept into the copying process, however: de Bry's latitude markings err by one degree compared with the original. Lodewijcksz' log records that the north coast of Bali lies at 8.5° south latitude, which is very close to the correct figure of 8°. The Claesz /van Doetechum original follows this meticulously, but de Bry's markings are mis-aligned, mistakenly placing the island's north coast at a latitude of about 7° south.

Larger scale than either the Plancius or Linschoten maps of a few years earlier, the Lodewijcksz map focuses exclusively on southern Malaya, Sumatra, Java, southern Borneo, and the islands east of Java through to Sumbawa — the limited region reconnoitered by de Houtman. The map records unprecedented detail along western and northern Java, and a plethora of small islands in the Sunda Strait itself and on the Indian Ocean threshold to it. Entering the region via the waters between Sumatra and Java rather than by way of Malacca and Singapore, the crew reported so many islands on the western side of the Sunda Strait that they had difficulty finding the channel.

Mataram, a city in the interior of Java, is illustrated in vignette. Though the map is conceived in the style of a mariner's chart, which rarely included interior features, Mataram was relevant to the commercial affairs of European mariners. Most of the northern coast of the island had become dominated by the Muslims by about 1535, the Hindus holding on only at the eastern tip. But Muslim control over the coastal region ebbed in the latter years of the sixteenth century as the interior Muslim states of Mataram and Pajang became the new nerve centers of Muslim trade, frustrating the coastal-based European attempts to control Java. For much of the seventeenth century, Mataram, Banten, and the V.O.C. vied for control of Java.

The Mystery of Java's Southern Coast

The question of Java's southern coast is evaded by the map. Map-makers had no available data about the island's southern coast, and traditionally depicted the island as extending much further south than it actually does. Several early sources had set the premise for this. Polo had described it as "the biggest island in the world," which fact he knew from "the testimony of good seamen who know it well," and early European experience did not contradict this. In fact, as we saw earlier, the reports of Varthema reinforced it.

Before the Dutch reached Java, the mystery of the island's southern coast had been on the minds of Portuguese sailors. The Portuguese questioned their northern Javanese hosts about the southern coast, and were told that few good harbors exist on that coast. The Javanese also explained that the mountains which bisect the island along an east-west axis also stifled communication between the northern and southern shores. Although such indigenous geographical information did not necessarily contribute to the erroneously large idea of the island's width, it did make Portuguese exploration of the southern coast a low priority.

Any Arab sources tapped by the Dutch or Portuguese were likely to concur. The fifteenth-century navigational treatise of Sulaiman al-Mahri stated that

> the outer [southern] coast of Java on the west is in a state of ruin and is not inhabited. There is no well-known port there. The ports are all on the east coast ["east" probably equaling 'north' because of the diagonal orientation implied for the islands in the Arab texts].[254]

Two European sources, however, began to provide clues to the mystery of Java's southern coast. Francis Drake touched on southern Java in 1580, providing a reference point about mid-way along the coast. Secondly, the de Houtman expedition itself, sailing west after leaving Bali, set their course for what should have been midway along the eastern coast of Java according to existing charts, and found only open ocean. In the report's words, if the island extended as far south as geographers believed, they would have "sailed through the

Fig. 96 Title page to the account of the first Dutch voyage to the Indies. Cornelis de Houtman, 1598. [Antiquariaat Forum, Catalogue 105]

Fig. 97 Merchants at Banten. Lodewyckszoon, 1598, (Theodore de Bry, 1599) (13.6 x 17.3 cm). Lodewijcksz described Banten as being enclosed by a high, thick red brick wall, with well-guarded gates and watch towers. He observed three streams flowing from the mountains (actually all branches of the Kali Banten River), two of which surround the city to form a natural moat, the third bisecting the town itself. Banten was deemed the best and largest seaport on the island of Java, and a major center of regional trade. The merchant on the left of the illustration is said to have come to Banten from Malacca, having borrowed money from a creditor who was to be repaid in double if the merchant was successful. Both he and the man on the right were eager to trade, particularly for merchandise brought by the Chinese. The figure in the center is said to be dressed in the manner typical of their women.

middle" of it. Thus mapmakers began to narrow the island's latitudinal width and better approximate its southern shores.

The earlier map of Plancius (fig. 90) had depicted the southern region of Sulawesi as segmented from the rest by a narrow neck. Lodewijcksz goes one step further, severing *Macasser* (Makasser) from the main part of the island and placing it as a separate island to the west, between Borneo and Sumbawa. The city on the east coast of the island, *Bogis,* is a reference to the Muslim Buginese of south Sulawesi, renowned throughout the region for their seafaring skills.

Bintan, Lingga, Bangka and other islands to the south of the Malay Peninsula, appear in better detail than on earlier maps, while *Baixos de los Chineses* are marked in the waters between Malay and Borneo, probably referring to low islands frequented by Chinese vessels. The accompanying *China bato* literally means 'rocks of China' or of the Chinese. East of Bali, old errors still remain: Lombok and Sumbawa are shown as a single island.

Lodewijcksz' Bali

Bali was nominally under the rule of Srivijaya until the thirteenth century. Subsequently the island was incorporated, as a vassal state, into the mighty Majapahit empire based in east Java, which cast its political hegemony over the territories that today constitute the modern state of Indonesia. With the collapse of Majapahit at the beginning of the sixteenth century, Bali was left as the last surviving representative of Indonesia's classical Hindu-Buddhist past.

During the monumental upheavals of the sixteenth century, Bali was little-affected by Christian intruders. It was, however, indirectly transformed by the success of Islam on Java. As Islam took hold in Java, many Majapahit Javanese fled to neighboring Bali, making the island a flourishing sanctuary for Hindu arts and traditions, a cultural distinction which it retains to this day.

Lodewijcksz depicts the island of Bali in recognizable form for the first time, both on the general map and on a separate map of the island in his book (fig. 98). The shape of Bali, despite understating the island's east-west breadth, is a superb attempt to portray an island still little known by Europeans. In lieu of more specific information about the interior, simple repetitions of the legend *Palatium Regis* (Royal Palace) denote the island's many kingdoms. The southern peninsula, where most modern-day tourism is focused, has been exaggerated, perhaps following Lodewijcksz's text, which described the southern coastline as a long, high point extending far into the sea.

Off the eastern coast of the southern peninsula (upper left of map) lies a small, unnamed island. This is Nusa Penida (Deer Island), which served as a penal colony for the Balinese king. Lodewijcksz reported that ten or twelve years before their visit to Bali, an attempt had been made to assassinate the king. The rebels were captured and, instead of being executed, were banished to Nusa Penida, where they led a fruitful life in exile.

The Lodewijcksz map of Bali reflect a brief European romance with the island rather than a sustained interest in it. The crew's fuss over Bali was largely a labor of love — they were so impressed with Bali that they wanted to name it 'New Holland'. Two sailors jumped overboard and returned to the island as the ship departed. But although Bali was, by all accounts, abundantly endowed with rice, fish, fruit, and everything else that was necessary for a good life, the Dutch were not sufficiently interested in the island during the seventeenth century to place it under V.O.C. custody. Bali had no adequate sea port on the north coast, and the south coast was not convenient to Dutch routes. Nor was the island a source of trade spices, and it was independent of the Javanese powers at Mataram. These deterrents were a blessing for Balinese civilization but stifled European mapping of the island.

When V.O.C. interest in Bali did blossom, however, it was precisely because the island was not aligned with Mataram. The Dutch built a base on Bali in 1620 but it only survived for a year before being destroyed. In 1633 the Company attempted to recruit a Balinese king to help them in their struggle against Mataram which controlled much of Java's commerce. Even then, Bali managed to remain aloof from European worries, compared to most of its neighboring islands, and its remarkable arts and culture remained little known beyond its shores; it was not until 1841 that Bali finally came under Dutch control.

After the map of the island which appeared in the Lodewijcksz and de Bry accounts, European knowledge of Bali and its geography was frozen in time for well over a century; this 1598 delineation of the island was copied and recopied without improvement well into the eighteenth century by authors such as Renneville and Tardieu. Perhaps the greatest testimony to Bali's relative seclusion from the West is found in François Valentijn's *Oud en Nieuw Oost-Indien* (1724-26); although that work is known for its improved maps of the Indies, its large map of Bali is geographically identical to that of Lodewijcksz.

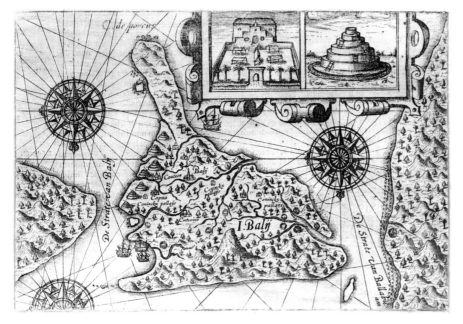

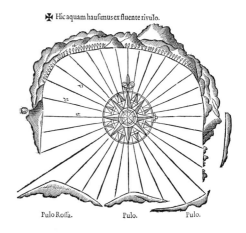

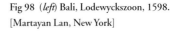

Fig 98 (*left*) Bali, Lodewyckszoon, 1598.
[Martayan Lan, New York]

Fig. 99 (below) Coastal profile of Bali, Lodewyckszoon, 1598. Woodcut. (15 x 19 cm)

Fig. 100 Singapore, Theodore de Bry, 1603 [Lent from a private collection].

Two Early Images of Singapore: de Bry and Hulsius

Though Bali was not high on the Dutch wish list after the de Houtman voyage, the Spiceries most certainly were, and the most proven route to the Moluccas was via the Malacca and Singapore Straits. De Houtman's decision to enter the region through the Sunda Strait was a clever strategy for the moment, but Holland would eventually need to secure the waters of Malacca and Singapore if the V.O.C. were to truly prosper. In 1603, Theodore de Bry published Europe's first close-up glimpse of Singapore, *Contrafactur des Scharmutz* (fig. 100). The map shows Pulau Karimon and other major islands in the Singapore Strait, and the Johor River *(Rio Batasubar)* on the upper left, with the the island of Singapore not yet severed from the mainland.

In stark contrast to de Bry's sophisticated mapping of southernmost Malaya, a naive view of the region, depicting the attempt by a Dutch fleet under Cornelis Matelief to dislodge the Portuguese at Malacca in 1606, was offered by a competitor, the German chronicler Levinus Hulsius, in 1608 (fig. 101). Matelief's attack was performed in concert with the sultan of Johor — for whom, at the moment, avenging a century of Portuguese intrusion was more important than any future concern about the Dutch. The Portuguese prevailed, remaining in control of this coveted 'throat' of Southeast Asian commerce for yet another thirty-five years. Matelief visited Banten, Jakarta, and Makassar, and was able to further Dutch interests in Ambon and Ternate, building a fort in the latter.

Hulsius' map of the Malacca Strait is inverted (northwest-southeast, so that the relative positions of Malacca, Banda Aceh *(Achin)* and Singapore are reversed. This was probably the result of the engraver of the copperplate having transcribed the image directly from the original, instead of reproducing it in mirror image.

Fig. 101 Malacca and Singapore Straits, from Part X of Levinus Hulsius' voyages, 1608; this engraving illustrating Cornelis Matelief's attempt to wrestle Malacca from the Portuguese in 1606. In his published journal, Matelief continued to stress the importance of Malacca to the future of Holland and the V.O.C. After failing to win the entrepôt, he divided his fleet and visited Bantam, Jakarta, Ambon, and Ternate, where he built a fort. He then sailed to China, where he tried intensely to establish trading relations. When he returned to Holland in 1608, he brought with him a five-member embassy from Siam, as well as Cornelis Specx, a Dutchman who had opened a factory in Ayuthaya in 1604.

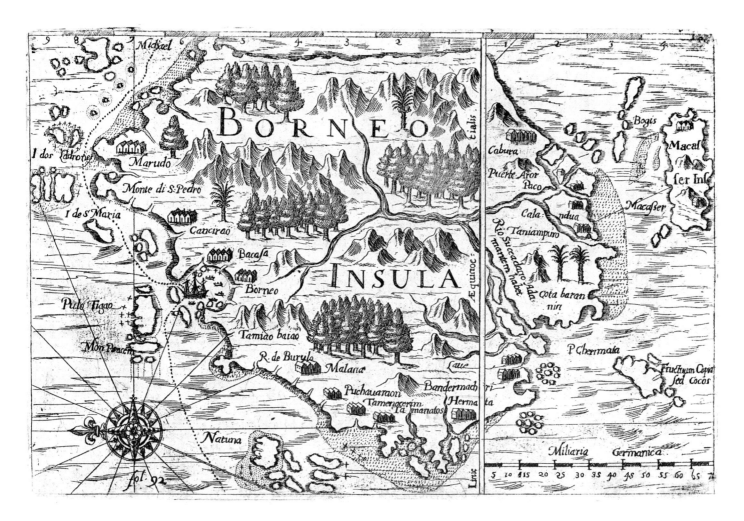

Fig. 102 Borneo, Theodore de Bry, 1602 (14 x 21.5 cm).

The de Bry Map of Borneo

The success of the pioneering Dutch expedition to the Indies under Cornelis de Houtman infused such a fever of confidence in the Netherlands that in 1598, the very year following the fleet's return, five different companies financed the voyages of twenty-two ships to the Far East. One of the fleets, under the command of Olivier van Noort, reached the Philippines, Borneo, and Java by Magellan's route, around South America and across the Pacific, completing the first Dutch circumnavigation. An account of the voyage in Theodore de Bry's *Voyages* of 1602 contained a map of Borneo (fig. 102) recording van Noort's tracks into Brunei Bay in December of 1600.

Geographically, however, the map reflects nothing from the voyage. De Bry's map is a close copy of one accompanying an account of Noort's voyage which was published in 1601, and this model was itself derived from Iberian prototypes already popularized on the Linschoten map of 1595 (fig. 92). It was reused as a separate map in Langenes' miniature atlas of 1598, with some differences in interior rivers and nomenclature, and the same geography was employed again in 1616 by Bertius.

The most striking aspect of this mapping of Borneo is that the island's east coast (top of map) is merely a hypothetical line. This conspicuous void in European knowledge of Borneo was a result of the route taken by early European mariners making the run between Malacca and the Spiceries. After leaving Malacca and passing eastward through the Singapore Strait, they generally went south by the western coast of Borneo, crossed the Java Sea, and followed the northern coast of Java to Sulawesi. This route missed the eastern coast, and thus, as Linschoten reported, "the breadth [of Borneo] as yet is not known nor discovered." Not until the end of the sixteenth century did it become common for mariners to sail around Borneo to the Moluccas, leading to the mapping of the island's eastern shores in the seventeenth century.

Few of the places found on these earliest maps of Borneo were mentioned by subsequent visitors. Some of the settlements recorded by de Bry were cited by the Portuguese historian Fernão Lopes de Castanheda during the 1550s as principal seaports used by his countrymen. One is *Marudo,* marking the bay at the northern tip of the island (upper left), a name which the bay has retained to this day (Teluk Marudu). *Monte di S. Pedro* is Mount Kinabalu.

Another place cited in Portuguese records is *Taniampuro,* or more commonly, *Taiao pura.* This is probably Tandjong Pura, the old capital of Matan, although Matan lies on the west coast while *Taiao pura* was normally placed on the south. The Portuguese knew *Taiao pura* as a source of diamonds; Linschoten, repeating what he had learned in Goa, seems to be describing the fat river system which mapmakers placed by *Taiao pura* when he refers to it as a strait: "there is a straight called Tania pura . . . where many diamonds are found, that are excellent." *Laue,* another Portuguese port-of-call, was mentioned by Pigafetta (though he did not reach it), and could conceivably be the enigmatic *Lochac* of Polo. De Bry follows most of his contemporaries in placing it in the southwest of the island.

Van Noort's ship is shown in the large bay of Brunei, marked *Borneo*. The Dutch captain relied on Chinese pilots in his employ to serve as ambassadors to the sultan, who had to be persuaded that the visitors were not Spanish. The map's depiction of local boats clustered round van Noort's vessel refers to the appearance of almost one hundred proas in the bay on January 1ˢᵗ, 1601. They surrounded the ship and their crews attempted to cut its anchor cable.

The descriptions of the sultanate left by both Pigafetta in 1521 and van Noort eight decades later would both sound familiar to a modern visitor. Pigafetta placed the city a fair distance up the Brunei River, and describes it as built entirely over the water, with wooden houses sitting high on pilings, as much of the city of Brunei remains today. Only the raja and some of his chiefs, Pigafetta observed, live on the land proper. Noort judged the village to lie some three miles upriver, and to comprise about 200 houses, built on a swamp, with its residents traveling from house to house by proa. Noort, like many before him, praised the island's camphor as the finest available.

Although it is generally agreed that the name of the island of Borneo probably derives from *Brunei,* another theory is suggested by the Ibans of Sarawak, who claim that the name derives from *buah nyior,* a Malay term for coconut.²⁵⁵

Van Noort in Albay (Southern Luzon)

En route to Manila, van Noort put into Albay Bay, in the southeast corner of southern Luzon. The view-map of Albay engraved for de Bry's account of the voyage (fig. 103) shows Van Noort's ship anchored in the bay, with the northern tip of Samar in the lower right. The mountains rising in the upper left lead to Mayon Volcano, which soon became an important landmark for sailors, as its near-perfect symmetry and the halo of vapor which often glows at its summit at night, were unmistakable and could be seen from a great distance.

Andreas/Metellus, 'Siam' (1596)

A map published by Lambert Andreas in 1596 was the first to use the name Siam *(Sian)* as a title (fig. 104). Though spelled *Sian* on the map (as it was, for example, on the maps of Juan López de Velasco) 'Siam' is spelled in the conventional fashion in the verso text of the various issues of the map.²⁵⁶ The country's present name of *Thailand* is a hybrid term, the first syllable being Thai, the second, English.

What, then, is the origin of the term 'Siam'? La Loubère observed that "the Name of *Siam* is unknown to the *Siamese*," and noted that "the *Siamese* give to themselves the Name of Tai, or Free, as the word now signifies in their language." He correctly quotes *Meuang Tai* as the full term for Thailand, "for *Meuang* signifies Kingdom." La Loubère believed that *siam* meant 'free' in the Peguan language, and that the name Siam was simply a translation of *tai,* acquired by the Portuguese from the Burmese. Gervaise, a contemporary of La Loubère, speculated that Siam may be "a whimsical derivation by the Europeans" of the capital city, *Sijouthia* (Ayuthaya). Other theories are that Siam may come from the Malay *Siyam,*²⁵⁷ or from the Arabic *hadyia,* meaning 'a gift given to a superior or a teacher of the Koran'.²⁵⁸ Siam may have been a corruption of the Indian word *suna,* a term for 'gold', and thus a reference to the 'golden peninsula'.²⁵⁹ Or it may have been an adaptation of the Persian *Shahr-I Nao,* meaning 'new city', which was originally applied to Ayuthaya when it was founded in the fourteenth century, or Lop Buri, to the north of Ayuthaya. Some early European authors had also used the term *Sornao,* or some variation of it. Back in the early fifteenth century, Conti, probably following the pronunciation used by Muslim sailors, relayed the term *Cernove,* and *Xarnauz* was heard on Vasco da Gama's voyage. Varthema knew Siam as both *Sornao* and *Cini;* Pinto referred to the kingdom as *Sornau* and Siam, stating that the Emperor of *Sornau* is the king of Siam. According to Pinto, *Sornau* "is a province comprising thirteen separate kingdoms, otherwise known as Siam," whose capital is Ayuthaya. As a result, some maps (e.g., Sanches 1641, fig. 80) place a kingdom of *Sornau* (Sanches' *Suinão*) in the interior.

Andreas continues the cartographer's romance with mythical Lake Chiang Mai. The map's configuration is rare, however, in that it places Lake Chiang Mai *(Chianaÿ lacus)* in the correct position of Chiang Mai town, rather than far to the north. Chiang Mai itself *(Iangoma),* as usual, is wildly misplaced, in this case washing up on the Cambodian bank of the Mekong River. Ayuthaya is shown twice, though not in the same two places as on the Linschoten map. Here it appears once as *Sian* in its approximately correct position just north of the apex of the Gulf of Siam, and duplicated as *odia* on the northern part of the peninsula. This aberration was inherited from the 1569 world map of Mercator, and had already been sanctioned by both the *Asiae Nova Descriptio* and *Indiae Orientalis* of Ortelius.

Fig: 103 Albay Bay, in southern Luzon. Theodore de Bry, 1602 (14 x 17.5 cm). In his account of his circumnavigation, Van Noort reported bewilderment at the complexity and vastness of the Philippine archipelago. His vessels are seen here in Albay Bay, where they anchored *en route* to Manila. From Albay, they seized Philippine canoes and a Chinese junk, kidnapping their pilots to guide them through the San Bernandino Straits separating Luzon from Samar. The strait is shown in the lower right of the engraving as the 'strait of Manila'. [Martayan Lan, New York]

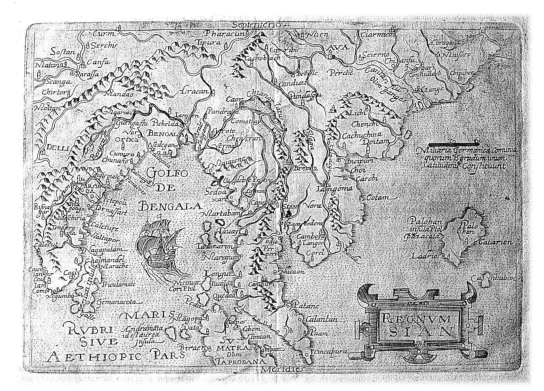

Fig. 104 Mainland Southeast Asia, Andreas/Metellus, 1596. The first to use 'Siam' in the title to designate the Southeast Asia mainland in general, this map first appeared in Giovanni Botero's *Theatrum oder Schawspiegel …* published by Lambert Andreas in 1596. Natalius Metellus, under whose name the atlas also appeared, was probably the author of the map. [Lent from a private collection]

The single prominent feature in the Pacific Ocean is a large *Palohan*, the Philippine island of Palawan. An inscription explains that Palawan was the *Bazacata* of Ptolemy. This opinion was taken from Mercator, who, as with the identification of Borneo as the Ptolemaic *Bone fortune*, was influenced here by his identification of Sumatra as having been *Taprobana*; Palawan lies in the same position relative to Sumatra as *Bazacata* did to Ceylon.

About the time of Andreas' map, Europeans were becoming increasingly common in Siam, where they were accorded liberal privileges. Siam's kings, however, were generally adept at maintaining the upper-hand in their kingdom's affairs, and foreign presence diminished dramatically after a 'Revolution of 1688' considerably down-scaled Ayuthaya's cosmopolitan nature.

The Trans-Peninsular Waterway

The Lodewijcksz map of 1598 depicts the southern end of the Malay Peninsula separated from the mainland by the Muar and *Formeso* (Batu Pahat) 'Rivers', offering a particularly clear portrayal of a trans-peninsular strait found on several maps of this era. We saw the earliest traces of this non-existent waterway on the 1507 map of Ruysch (page 104 and fig. 55), evolving into a far more pronounced indentation on the 1548 map of Gastaldi (fig. 73), and culminating with a true strait on the 1554 map of Ramusio (page 151 and fig. 74).

The origin of this cartographic folly lay in misunderstood trans-peninsular shipping routes.[260] Merchants wishing to transport goods from one coast to the other could either sail around the peninsula, which risked the political entanglements of Malacca, pirates, and the logistics of the voyage itself, or transport the merchandise overland. Although there was no strait piercing the peninsula, there were rivers on one coast whose sources lay close to the sources of rivers flowing to the opposite coast. One could transport the cargo upriver, drag it overland to the source of an opposing river, and float it down to the opposite shore. Malayan traders used such hybrid routes to transport goods from western Malaya to points on the Gulf of Siam, such as the bustling port of Patani, while circumventing the claws of Malacca.

The key to the riddle of the trans-peninsular strait can be found in the work of the chartmaker Godinho de Eredia. In a map of 1613, de Eredia depicts the course of the Muar and Pahang Rivers, which flow to the west coast and east coast of the Malay Peninsula, respectively. He also indicates a track connecting them, which he denotes as *Panarican*. This is a transliteration of a Malay word, *penyarekan*, which is a drag-way or portage. Thus de Eredia was recording how traders, transporting goods between the west and east coasts of the peninsula, would go upriver on the Muar as far as possible, then transport the merchandise overland to the upper reaches of the Pahang, flowing to the opposite coast. Because of the difficulties involved, this method was suitable only for compact, valuable items, such as gold, and the more highly-prized spices.

South of this route, a second trans-peninsular route was formed by the Batu Pahat and Endau Rivers, which were joined by a swampy area in the middles of the peninsula. This route appears to have been the less favored, though, for despite it being shorter in total distance than the Muar /Pahang route, the distance between the two rivers was greater, and it passed to the south of the peninsula's gold fields.

The association of the region with gold was, of course, an ancient one; Ptolemy had recorded a 'river of gold' in Malaya, the *Khrysoanas,* which might have represented the Muar (flowing to the west on the 'leaf' at the bottom of Malaya in figure 32). In combination with other waterways and land connections, the *Khrysoanas* would have provided access to the Malayan goldfields of Ulu Pahang, as well as transport across the peninsula.[261] That the peninsula was traditionally known as a source for gold is also reflected in the old term *Chersonese Aurea*, associating Malaya with gold, which straddles the mainland and insular region on the Lodewijcksz map.

The fake waterway christened by Ruysch and consummated by Ramusio may also have been borrowed from indigenous Siamese or Burmese maps, on which such routes appeared.[262] European mapmakers may have been misled by a literal — and therefore erroneous — understanding of indigenous maps or geographic descriptions which had been meant only to define itineraries.

The Miniature Atlases of Langenes and Bertius

Miniature atlases, being less costly and less cumbersome than those of folio size, were popularized by the indefatigable Ortelius. His diminutive atlas first appeared in 1577 under the name *Spieghel der Werelt,* but is commonly known by its subsequent name, the *Epitome.* As for Southeast Asia, it contained only miniature renderings of the general map from the *Theatrum.* But Southeast Asia was an ideal canvas for a small format atlas, since it could be divided into many islands and regions, each the subject of a map. The small size allowed such individual coverage without the loss of geographic data for the area.

In 1598 Barent Langenes issued a miniature atlas, the *Caert-Thresoor,* which took advantage of precisely this opportunity, according individual maps to the various regions and islands of Southeast Asia. Cornelis Claesz, the Amsterdam publisher who had acquired portolan charts from Bartolomeu Lasso with the help of Plancius, and who had published the Lodewijcksz account of the de Houtman voyage, was involved with the publication.[263] It contained separate maps of the Philippines, Moluccas, Borneo, Java, Sumatra, 'Malacca' (i.e., Malaya), and Arakam (Burma), in addition to the usual general map of *India Orien[talis],* most of them engraved by Peter Kaerius and Jodocus Hondius. Another fine miniature atlas, similar in concept but expanded and with revised maps, was published by Petrus Bertius in 1616.

Langenes and Bertius Maps of the Philippines

The first two published maps that were specifically devoted to the Philippines, and which used the term "Philippines" as their title, were the *Insulæ Philippinae* of Langenes (1598) and the *Philippinæ Insulæ* of Bertius (1616, fig. 105). The Langenes version is taken directly from the Linschoten map of 1595 (fig 92), with its peculiar east-west orientation for Palawan, while the Bertius rendering is derived from the Plancius map of 1592 (fig 90), which in turn was based on Bartolomeu Lasso. Although there is nothing about the two maps that is identical, they nonetheless contain the same general features.

These represent the first tolerably accurate depictions of the archipelago's complicated shores, including Luzon, whose fine port of Manila had quickly become the center of the Spanish empire in the Indies. The only major error in the general outline of Luzon is in the winding peninsular region to the southeast, which on both maps should extend much further than is, in fact, depicted

Lasso portrays Samar accurately for the first time, labeling it on the map as both *Achan,* a name which is sometimes applied to the island's northern half, and *Tandola* (or *Tandaya*), which was actually the name of a ruler of a region of the island. When the Spanish expedition under Legazpi reached Samar in 1565, they asked for the island's name, but their source — who was the nephew of the chief — simply gave the name of his uncle. The maze of islands in between Luzon and Mindanao are still only

Fig. 105 Philippines, Bertius, 1616 (8.7 x 12.5 cm).

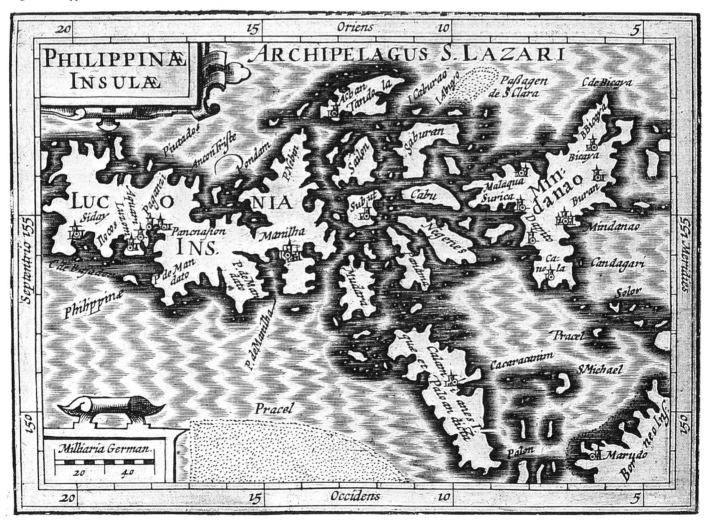

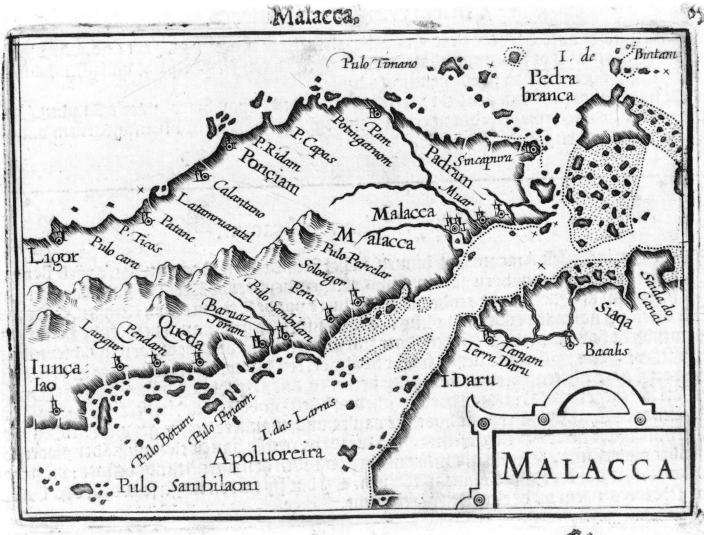

Fig, 106 Malay Peninsula, Langenes, 1598 (8.5 x 12 cm).

crudely represented, though the major islands are nonetheless depicted: *Mindara* (Mindoro), *Panama* (Panay), *Cabu* (Cebu), *Sabunra* (Leyte), *Negoes* (Negros), as well as Masbate (unnamed). On the west, as with many Iberian charts of the time, Palawan has been confused with *Calamianes,* a group of small islands situated between Mindoro and Paragua or Palawan. The mischievous island of *San Juan* does not appear, since it lay (correctly) far to the east on the Lasso model.

Though Bertius omits a few place names found on the Plancius, Linschoten, and Langenes maps, he includes many that are of interest. Off Samar *(Tandola)* lie the *Ylhas del Primeiro Surgidero* (islands of the first anchorage), which refers to the place where Legazpi first anchored in the Philippines. Above it is *Francisco Gomez,* the name being that of a Spaniard who was murdered there while performing a blood ceremony. At the southern end of the island is *Abo camucho Primeiro,* a corruption of *Aboca mucho Primeiro,* which in turn was probably originally notated as *Abocamiento Primeiro,* the first mouth of the channel used by the early Spanish explorers to enter the archipelago. On the west coast of Luzon is *G. de Matalahambre,* the 'gulf of killing the hunger' where a good feast must have been had. More descriptive names follow to the north: *Ancon triste* (sad cove), *Pintados* (painted), *Moro Hermoso* (beautiful Moor), and *C. de Engano* (cape of deceit), this last term being used not infrequently by Spanish mariners.

Langenes' Malaya

Langenes' map of *Malacca* (fig. 106) is derived from the Linschoten map of 1595. It offers a different rendering of the false trans-peninsular strait than that recorded by Ramusio (1556) and Lodewijcksz/de Bry (1598; see pages 187, above). Whereas those maps interpreted the waterway as a strait severing southernmost Malaya from the mainland, Langenes maps a river which originates in the peninsula's mountains (regions of which were known for their gold fields), flowing south until it divides into two branches, one flowing into the Indian Ocean, the other emptying into the Gulf of Siam. The entire system is identified as the Muar River.

Langenes, following Linschoten's lead, gives prominence to *Pedra Branca* Island (Pulau Batu Puteh), off Singapore, which is vastly out of proportion to its size. Known to early Chinese seafarers as *Pai-chiao,* this was an important landmark for mariners in the waters off southern Malaya and Singapore.[264] *Pedra Branca's* significance on the map can perhaps be better understood by Linschoten's description of it:

> From the Cape of Singapura to the hooke named Sinosura eastward, are 18 miles, 6 or 7 from thence lyeth a cliffe in ye sea called Pedra bianque, or white rock, where the ships that come and go to and from China, do often-times pass in great danger and some ships are left upon it, whereby the pilots when they come thither are in great fear, [for there is no alternative to the route].

Fig. 107 *Aracam*, Langenes, 1598. (8.5 x 12 cm)

A corruption of the name is found on the Lodewijcksz and de Bry maps of 1598, which dubbed it *Pedra Breva* (short rock). Langenes' Malaya includes Phuket *(Iunca: Iao)* to the north, Nakon si Thammarat *(Ligor)* on the east, and Perak *(Pera),* another tin-producing regions on the Malay Peninsula.

Langenes' Arakan

The Langenes map of *Aracam* (Arakan; fig. 107) is named after one of the kingdoms which comprised the region of Burma (see also page 149, Gastaldi, above). Arakan was a proficient sea power, shielded from attacks from Pegu by the Arakan Yoma range, and wielded considerable influence over neighboring Chittagong, which it had claimed as a vassal state since 1459. Although early relations between Portugal and Arakan were

marred by mutual distrust, piracy and looting, during the latter part of the sixteenth century Portugal aligned itself with Arakan and Chittagong to subdue Pegu. This alliance led to the tragic devastation of Pegu by the Arakanese-Chittagongese-Portuguese triangle in 1599-1600, just after the publication of the Langenes map. Langenes used Arakan, the mightier of these two Portuguese allies, to denote Burma as a whole.

Bertius' Burma

The Bertius atlas' map of Burma, *Arachan & Pegu,* is geographically unrelated to Langenes' map. but both the Langenes and the Bertius maps are, however, among the few to include the Burmese town of *Macao.* In the mid-1580s, the Englishman Ralph Fitch visited *Makhau* on his way to Pegu, describing it as "a pretty town, where we left our boats or *paroes.*" He stayed the night there and the next morning they took "*delingeges,* which are a kind of coaches made of cords and cloth quilted, and carried upon a stang between 3 or 4 men," by which transport "we came to Pegu the same day."

Bertius' Malacca

Bertius' general map of the Southeast Asian mainland, *Malacca* (fig. 108), is also quite different from that of Langenes. Bertius, following the Lasso model for Malaya disseminated by Plancius, depicts the trans-peninsular river as a true strait, squeezing a mammoth inland sea into the northern part of the peninsula; he does not include the name 'Singapore'. Deviating from Plancius, *Sian* (Siam) is shown as a large island at the mouth of the Chao Phraya River, which was an oft-copied misinterpretation of Ayuthaya's riverine location. As with many other maps of the period, Siam is shown twice, once as the river-island, then again as *Diamal:odia,* which is a corruption of some charts' *Diam, al:Odia,* which in turn was a corruption of *Siam al: Odia* (Siam, alias Ayuthaya).

Fig. 108 The Southeast Asian mainland, Petrus Bertius, 1616 (8.7 x 12.5 cm). Bertius combined elements from different geographic models then in circulation. The huge, island-filled bay to the northwest of Malaya is a misinterpretation of the Mergui Archipelago, a group of many islands scattered off the west coast of Burma. The island at the apex of the Gulf of Siam is an exaggerated interpretation of (correct) reports that Ayuthaya lay on a riverine island. The Mekong River is erroneously oriented straight to the northwest, with Chiang Mai (*Langoma*) on its banks. In southern Malaya, the peninsula is bisected by a strait, which was a misunderstanding of hybrid land and river trans-peninsular routes.

Fig: 109 Sunda Strait, Waghenaer, 1602 (18.5 x 26 cm) [Universiteitsbibliotheek Amsterdam]

Lucas Janszoon Waghenaer and the Transition to Printed Sea Charts and Pilot Books

In the latter part of the sixteenth century, increasing trade within European waters led to an escalating demand for navigational charts. Although as early as about 1539 a Venetian mapmaker, Giovanni Vavassore, sold a sea chart intended for actual onboard use, which was printed by woodblock, the printed medium was not commonly exploited for navigational charts until the end of the century. A Dutch pilot by the name of Lucas Janszoon Waghenaer recognized that a market existed for mass-produced sea charts, and so created the printed *rutter,* or mariner's guide with a full set of sea-charts for a particular region. First published in 1583, it covered the coasts of western Europe. Waghenaer's splendid charts — like many of their manuscript progenitors — tended to exaggerate important features along the coast, a quirk frequently noted about maps of Asian origin as well.

Although the work was a success among academics, pilots found its large format impractical, and so in 1592, Waghenaer introduced a new sea-atlas on a smaller oblong format. Like the earlier work this atlas, the *Thresoor der Zeevaerdt,* was conceived for use within European waters. But, reflecting the widening outlook of Dutch pilots, it also included information on the tracks between China and Japan which Linschoten had obtained from his friend, Dirck Gerritsz.

In May of 1598, a fleet of eight ships under Jacob van Neck and Wybrand van Warwijck left Texel for Southeast Asia, approaching the Indies through the Sunda Strait and establishing contact with the Javanese markets at Banten. Exploring Java, they continued on to Banda, Ambon, and the Moluccas, where they established what were known then as 'factories'. Factories were strongholds which would collect merchandise between the visits of Company vessels, thus avoiding the scarcity and high prices that a sudden demand would create. Descended from the *fondachi,* houses maintained by Genoese and Venetian merchants in North African and Middle Eastern ports, the first factory in Southeast Asia had been built by the Portuguese in Ternate. Van Neck's factories became the foundation of the Dutch spice trade.

Van Neck returned in July of 1599 with a bounty of pepper from Banten. Unlike the de Houtman voyage, this expedition was a commercial windfall, and so provided a strong incentive to lure other Dutch fortune-seekers to the Indies. It provided an inspiration for the English as well, proving as it did that defying the Iberian monopoly in the Indies trade could be richly rewarded.

Acknowledging the newly invigorated interest in the region and the accompanying need for navigational aids, Waghenaer tapped the van Neck voyage to add a chart of the Sunda Strait region to the *Thresoor* in 1602 (fig. 109). That Waghenaer would include such a chart in what was essentially a pilot-guide to European seas reflects the fact that Holland was buzzing with fever for the Southeast Asian spice trade. Waghenaer's chart was a brilliant production, presenting for the first time a detailed, large-scale chart of a strategic parcel of Southeast Asian coasts. The chart was impressively accurate for the day, despite its imperfections, and was engraved with Waghenaer's usual finesse and elegance.

Early Dutch voyages such as those of de Houtman and van Neck, which braved the Sunda Strait to avoid the Portuguese stranglehold on Malacca, sailed eastwards after clearing the strait to the north. They skirted the emporiums along the northern coast of Java, and headed directly for the Moluccas and other important spice markets. As a result, Waghenaer had insufficient data to chart the coast of Sumatra north of the strait, which veers erroneously to the northeast (the upper right-hand corner of the map), rather than due north.

Waghenaer, however, thought like a chart-maker serving the needs of sailors, rather than the publisher of static maps. He treated his charts as inherently imperfect, evolving creations. By drawing Sumatra's eastern shores with less definition than other coasts, he indicated, in his own map-jargon, his relative uncertainty here. Secondly, once his sources confirmed that the information he provided was indeed faulty, he deleted it. The inaccuracy of the chart's southeast Sumatran coastline must have been confirmed within a few years after the map's initial publication in 1602, as by 1609 those shores (upper right on the map) had been erased from the plate. Such meticulous attention and updating stands in great contrast to the stagnation which dominated Dutch published maps for much of the seventeenth century.

The prominent cape at the lower center of the chart is the northwest tip of Java. This cape forms the western shoulder of the large bay of Banten, the great spice mart of Java and a focal point of the earliest Dutch voyages. Soundings are marked in the bay, as well as along the tricky neck at the northern end of the strait and the waters approaching it. The Sumatran port of *Sumor,* at about the center of the map on the west side of the cape, is still the name of a town (Sumur) on the east side of the cape, and the large island

at the mouth of Banten's bay, Waghenaer's *Pulo Panjan,* is still known as Pulau Panjang. In the strait itself, *Pulo Carcata* represents the first mapping of Krakatau by its modern name. Krakatau, the active volcanic island whose eruption in 1883 was the most violent in modern days, became well-known to European sailors; in 1681 a V.O.C. vessel under Elias Hesse sailed close to Krakatau and reported desolate charred forests with smoke visible for many miles, the devastation of the eruption there the previous year. The island seemed calmer four years later, in 1685, when the first French embassy to Siam skirted the island. As recalled by Guy Tachard, "we made many Tacks to double the Isle of *Cacatoua* (so called because of the white Parrots that are upon that Isle, which incessantly repeat that name)."

The Sunda passage was not without problems; de Houtman's crew complained that the western side of the straits had so many small islands that it was difficult for them to find their way through. Other pilots noted how sensitive the strait was to winds and currents; in 1685 the first French mission to Siam traversed the strait and found that

> all our attempts [to reach Java] were unsuccessful because the wind was too weak, and Currents too strong in the middle of the Channel. That which causes the Currents, is because the water that for several months has been forced into the Streights by the South and South-west-winds, which reign commonly from the Month of March to September, set out again impetuously during the other six Months of the year.

In terms of published mariner's charts of the Indies, after this chart of 1602 the golden age of Dutch printed maps suddenly, for the moment, ended. Not entirely coincidentally, in the very same year of 1602 the *Vereenigde Oost-Indische Compagnie,* or 'United East India Company', was born.

Fig. 110 View on the little island of Panaitan (Prince's Island), at the southern entrance of the Sunda Strait, with the strait itself visible in the background. Lodewyckszoon described the Panaitan islanders as excellent seamen, swimmers, and fishermen. Many vessels crossing the strait from the south found the island to be both a navigational hazard to be avoided, and at the same time a welcome landfall for provisioning, if it could be anchored at safely. When the French embassy to Siam reached the strait in 1685, they wasted about four days dancing with the island before favorable winds allowed them to slip past it, despite being desperate for food and rest. From de Bry, 1599, based on the Lodewyckszoon view of the preceding year (13.7 x 17.5 cm).

The Role of England

During the first half of the sixteenth century, the English public had no published materials in their own language describing overseas discoveries. The publication of Richard Eden's *A treatys of the newe India* in 1553, a translation of part of the fifth book of Sebastian Münster's *Cosmographiae* of 1550, began the process of correcting that gap. Two years later, Eden published *the Decades of the Newe Worlde or West India,* a translation of writings by Martyr, Pigafetta, and Maximilian, and in 1577 he offered the English public a translation of Ludovico di Varthema's *Itinerario.* Eden's texts were often mediocre, but nonetheless pioneered vernacular reports about Southeast Asia for the English public.

Far more material in English appeared during the last decade of the century, and in 1598 the text of Linschoten's *Itinerario* was translated into English, dramatically expanding England's knowledge of the Indies. Both the Plancius map of Southeast Asia and the Linschoten map of the Far East were re-engraved in London for the new volume, which was given to the captains of the early voyages of the English East India Company. Thus Linschoten's *Itinerario*, which had been intended to help pioneer Dutch trade with the Orient, also contributed greatly to an English presence in Southeast Asian waters.

In fact, while England and the Low Countries were historically seen as rivals in the Indies trade, there was a great deal of intercourse between the two peoples on an academic level during the concluding decades of the sixteenth century. England became a temporary home for such continental figures as Jodocus Hondius and Theodore de Bry. England's greatest proponent of overseas exploits, Richard Hakluyt, who had avidly studied the compilations of the Italian chronicler Ramusio, met with Abraham Ortelius in 1577, the year that Drake began his dramatic circumnavigation of the globe. Hakluyt also corresponded with the Belgium mapmaker Gerard Mercator, an interesting link in that Hakluyt's colleague Edward Wright later refined the projection pioneered by Mercator in 1569.

Mercator and Wright

The greatest English contribution to the cartography at this time was carried out by Edward Wright. Wright was largely responsible for the revival and ultimate acceptance of the Mercator projection, which was a breakthrough for navigators because it allowed them to plot a 'straight' line on the earth's surface as a straight line on a chart. First brought to wide attention by Mercator on his large world map of 1569, the mathematical problems of constructing the projection, and perhaps the visual distortion it created, dampened its acceptance.

The idea for the projection may date back to the Portuguese earlier in the century. During the 1530s Portuguese sailors, whose voyages to the Indies pushed existing charting techniques beyond their practical limits, pressed their chief hydrographer, Pedro Nunes, for better solutions. Nunes was aware of the problems of the convergence of meridians toward the poles, of magnetic variation, and of the need for a method of charting which would translate a 'straight' line on the earth as a straight line on the map.[265] A work he published on the subject in 1537 probably influenced Mercator.

Three decades after the projection had been formally inaugurated by Mercator, Edward Wright, realizing that it would be an invaluable asset to pilots, revived the concept and devised mathematical tables for constructing charts that allowed mariners to plot their course as a straight line which corresponded to a true compass bearing. Jodocus Hondius was privy to Edward Wright's research, the manuscript of which was completed perhaps by 1592. When the latter was finally published in 1599, Wright, in his preface, accused Hondius of having plagiarized his technique.

Jodocus Hondius and Francis Drake

Two beautiful new maps of Southeast Asia were created by the Dutch mapmaker Jodocus Hondius for an enlarged issue of the Mercator *Atlas* of 1606. One covered the Indian Ocean region (*India Orientalis,* fig. 111), and the other focused on the islands (*Insulæ Indiæ Orientalis, Præcipuæ, In quibus Moluccæ celeberrimæ sunt,* fig. 112). For their artistic character and engraving style, these splendid maps better represent the concluding chapter of the sixteenth century rather than the dawn of the seventeenth. Geographically, Hondius largely followed the Bartolomeu Lasso pattern, adopting many of its idiosyncrasies, such as the *Enseada de Cochinchina* (Bay of Cochinchina = Gulf of Tongkin) with an exaggerated rendering of the Red River Delta. Hondius adds, however, a bit from his own notebook.

The *Insulæ Indiæ Orientalis* is one of few maps to show any trace of Francis Drake's presence in Southeast Asia. Hondius had spent several years in London, having fled there in about 1583 to avoid religious persecution. At that time, Francis Drake's flamboyant circumnavigation was still fresh news, and Hondius (as we know from other maps he made) became well-acquainted with the voyage. Drake is believed to have taken with him a sea chart prepared by a Lisbon maker, possibly Vaz Dourado. To this manuscript chart and other printed materials, he added Iberian sea charts and pilot books, plundered *en route.* Like Magellan before him, his only predecessor in circumnavigation, Drake's first landfall in Southeast Asia was some point in Micronesia (the precise landfall is calmly disputed, although Palau is the most likely candidate). Unlike Magellan, he had set off across the Pacific from some point in North America (that landfall is hotly disputed). From Micronesia he continued west to Mindanao, then sailed southeast in search of the Spiceries. He picked up two native fishermen in canoes in the seas somewhere northeast of Sulawesi, who guided him to the Moluccas. Leaving the Moluccas filled with spices and the precious spoils of earlier plunder in South America, Drake attempted to navigate the tricky waters leading to the clearer seas to the south, but ran aground on a steep reef off Sulawesi. Three tons of cloves, among other valuables, were dumped overboard to lessen their weight, but nothing seemed to help them from what appeared to be inescapable disaster until the strong winds reversed, freeing them from the reef. Hondius seems to recognize the fact that the eastern Sulawesi coastline was far more complex than current charts showed. Although he does not know what configuration to give the island, he 'ripples' the eastern shores of Sulawesi in a more pronounced fashion than Lasso, acknowledging that there were significant, though unknown, undulations in the coast, such as that which befouled Drake.

Drake and his crew eventually found their way out of the maze of islands in the Molucca and Banda Seas, assisted by local sailors. After crossing into clear ocean by way of Timor, Drake made landfall somewhere on the southern coast of Java, probably in the vicinity of *Tjilatjap* (Cilacap). Drake is the first European known to have made landfall on the southern coast of Java.

Hondius notes this important event on the *Insulæ Indiæ Orientalis.* He draws the southern coast of Java only as a dotted (hypothetical) line, save for a port at about the coast's midpoint, which is shown with a solid line. He inscribes *Huc Franciscus Dra. appulit* on the port, thus recording that Drake had landed there. A separately-published world map of Hondius, which was probably published in Amsterdam in about 1595, contained insets depicting Drake's south Java landing, his reception by the king of the Moluccas, and his near-catastrophic accident on the reefs near Sulawesi.

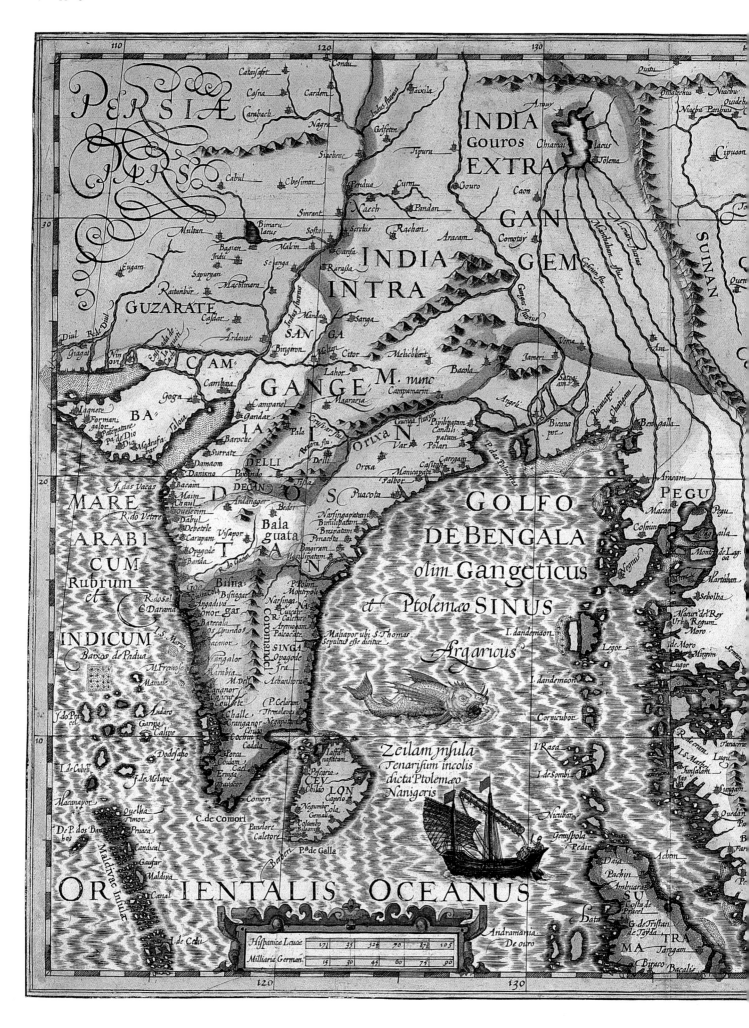

PERSIÆ
PARS

GUZARATE

CAM
BA

MARE
ARABI
CUM
Rubrum
et
INDICUM

DECAN

Baxos de Padua

Maldivac Insulæ

ORIENTALIS OCEANUS

INDIA
Gouros
EXTRA
GAN
GEM

INDIA
INTRA

SAN   GA

GANGE M. nunc
IA
N   Campamarin

ORIXA N

DECAN

Balaguata

Bisna
gar

Coromandel

GOLFO
DEBENGALA
olim Gangeticus
et Ptolemæo SINUS

Argarious

Mahapor ubi S. Thomas
Sepultus esse dicitur

I. dandemaon

Legor

I. dandemaon

Cornucubor

Zeilam insula
Tenarisum incolis
dicta Ptolemæo
Nanigeris

I. Rasa

I. de Sombi

Nicubar

Genuspola
Pedir

Andramania
De ouro

PEGU

Negrais

SU

TRA

MA

| Hispanica Leucæ | 17½ | 35 | 52½ | 70 | 87½ | 105 |
| Milliaria German. | 15 | 30 | 45 | 60 | 75 | 90 |

Fig:111 Mainland Southeast Asia, Jodocus Hondius, 1606 (35.5 x 48.6 cm). Hondius shared the classic view of Southeast Asia' river systems, mapping five river,s from the Mekong westwards, as originating in a Himalayan lake associated, if not in Hondius' mind, with the kingdom of Lan Na in what is now northern Thailand. Note also the depiction of the Mergui Archipelago, off the Burmese portion of the Malay Peninsula, as an islan-studded sea, and the exaggerated representation of the island on which the Siamese court of Ayuthaya sat (the red island at the top fo the Gulf of Siam).

As a youth, Hondius showed a talent for drawing and calligraphy, and later developed a fine reputation as an engraver. He also studied mathematics, Greek,and Latin, and the Lutheran faith. The Duke of Parma offered Jodocus the opportunity to continue his studies in Rome, but he declined the offer, probably to avoid religious conflict. Ironically, the Duke of Parma subsequently captured Ghent, where Hondius was living, and many Protestants were forced to flee, Jodocus among them. He resettled in England, where he carried on as an engraver and instrument-maker, as well as a makcr of maps and portraits. He returned to the Continent in 1593 and set up shop in Amsterdam. Here he was successful as a maker of large wall maps and of globes, for which he had obtained, in 1597, the privilege from the Staten Generaal. In 1604 he acquired the copperplates from the Mercator Atlas, which he began publishing two years later, supplemented by his own, more up-to-date, maps. Two of his new contributions covered Southeast Asia: this map focusing on the mainland, and one devoted to the islands (fig. 112).

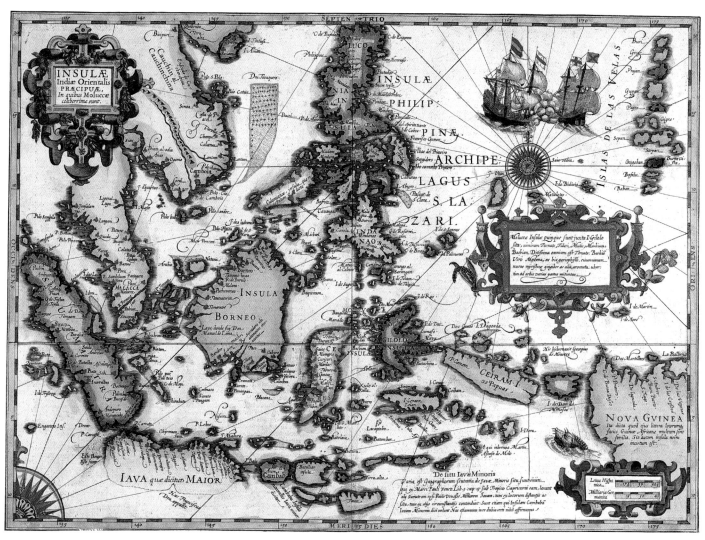

Fig: 112 Insular Southeast Asia, Jodocus Hondius, 1606. (34.5 x 47.5 cm)

Hondius marks the island east of Bali as *Cambaba* (Sumbawa), though it should actually be Lombok; like Linschoten, he identifies it as having been the *java minor* of old. The islands further east are abbreviated even more than on the Lasso model. One new term is *Buena vista* (good view) to denote the island of Tinian, north of Guam, possibly taken from one of the Spanish charts pillaged by Drake. Due west of Guam, just to the right of the compass rose, is the island of *Saia vedra,* a reference to the Spanish captain Saavedra, who sailed to the Philippines from Mexico in 1527; use of his name to denote an island is also found on the Herrera y Tordesillas /López de Velasco map (fig. 84) as the island of *Sahavedra.*

The Search for a Northern Route to the Indies
Both of the two known sea routes to the Indies — to the east, around Africa, or to the west, through the Magellan Strait — were long and difficult. The belief that a better and more direct route existed through or above America remained strong throughout the sixteenth century. By clutching to this theory, the English, and to a lesser extent the Dutch, probably delayed their entrance into Southeast Asia.

The northern route to the Indies was first pursued in the late fifteenth century by entrepreneurs from northern Italy sailing under the English flag. In 1496 John Cabot, a native of Genoa but of Venetian citizenship, was granted permission by Henry VII of England for a northern voyage to 'Cathay'. Sailing west the

following year, he reached northeast Canada, but like his countryman Columbus, believed he had reached Asia. In a second attempt, Cabot hoped to reach Japan.

In the early sixteenth century, with the realization that a continental landmass blocked westward access to the Far East, some explorers hoped to pierce the continent via a waterway. In 1524 England's rival across the Channel, France, sent the Florentine navigator Giovanni da Verrazano "to penetrate [the New World] to those blessed shores of Cathay." Three years later, in 1527, Robert Thorne wrote a letter to Henry VIII encouraging the exploration of a northern route to the Moluccas, which would circumvent America to the north.

Several sixteenth century geographers believed in such an expedient passageway to the East, among them the important theoretician Gemma Frisius. Frisius helped codify the methodology for determining longitude by time, more than two centuries before Harrison's clock made the method possible, and was a defender of the theories of Copernicus. In about 1537 Frisius was involved in the production of a globe which depicted a "strait of the three brothers by which the Portuguese attempted to sail to the Orient & Indies & Moluccas," probably referring to the Portuguese/Azorean Corte-Real Brothers.[266] A cordiform map of the world made by Frisius in 1544 for the *Cosmographia* of Peter Apianus (fig.113) illustrated the sort of passage to the Indies that theoreticians and entrepreneurs envisioned.

Gemma Frisius taught a student named John Dee, who soon became a prominent English geographer and a strong proponent of the northern route to the Indies. Dee regarded the Orient with a mystical awe, believing it to be both a repository of the occult arts and the resting place of Biblical treasure. He considered the East to be the true source of all mystical knowledge, as well as wealth, wisdom and true faith. While fastidiously keeping himself abreast of the latest geographic news and theories, he tried to reconstruct King Solomon's tracks to Ophir by using the Judeo-Christian Bible as a literal travel log. He encouraged his countrymen to try to find a northern route to the Indies but they only met with failure and in 1583 Dee ceased his study of geography. Any inclination he may have had to resume his pursuit of the Indies was squelched when Dee's library and workshop were destroyed by an angry mob who distrusted his obsession with the occult.

Ortelius and Linschoten also considered the possibility that the Dutch might benefit from a northern route to the Far East and Spiceries, and explorers such as Willem Barentsz and Henry Hudson devoted their lives to the scheme, with both of them finally perishing in the attempt. Barentsz made three voyages between 1594 and 1597 in search of a route to the northeast of Europe, round the top of Asia. Hudson, on the other hand, pressed the route to the northwest, above America, making four attempts between 1607 and 1610. English auspices financed Hudson's first, second, and fourth, voyages, while the third was undertaken in the service of the Dutch East India Company. The French explorer Samuel de Champlain followed in 1612, hoping to penetrate North America via a mammoth waterway, and even the early English voyages to Virginia kept alive the hope that the region might reveal a water passage to Asia.

Fig. 113 World map, Gemma Frisius, 1544. Gemma Frisius (i.e., Gemma 'of Friesland'), physician, mathematician, cosmographer, and cartographer, was a proponent of the theory that navigable seas north of America would allow easier access to the Indies than the long southern route through the Strait of Magellan. He illustrated this belief on this world map of 1544, constructed on a truncated cordiform (heart-shape) projection. The map is probably a reduction of one he is known to have made in 1540, but which is not extant. Access to the markets of Southeast Asia even proved important to his career as a physician; in 1545 he was sent to Brussels to treat the Emperor, which he accomplished with "wood of the Indies." Gemma was influential in the development of navigational and surveying apparatus, and a work he published in 1530 detailed theories about the determination of longitude which would be proven correct after the development of the chronometer in the eighteenth century. With "a very finely made clock which does not vary with change of air", he believed, one could determine longitude "even if dragged off unaware across a thousand miles".

Part IV
Companies and Colonization

Chapter 14

The Advent of The East India Companies

The English East India Company

Although England flirted with Southeast Asian commerce as early as the Netherlands did, English imperialism in Southeast Asia was less swift. For the couple of decades following the circumnavigation of the indefatigable, ruthless Francis Drake (1577-80), English voyages were generally disappointing. Thomas Cavendish was initially successful, sailing through the Magellan Strait and on to the Philippines in 1588, acquiring a large Chinese map plus notes on navigating in Eastern waters, then continuing to Java before finally completing his circumnavigation of the globe just as Francis Drake was defeating the Spanish Armada. Four years later, however, he died while attempting to sail the Magellan Strait a second time. In 1594 Richard Hawkins sought the Orient via the Magellan Strait but ended up a prisoner of the Spanish in Peru, and two years later Benjamin Wood successfully opted for the African route but was later shipwrecked off Burma.

The English East India Company was chartered on the last day of 1600, and it was not long before the Company had established factories in Banten (Java), Ayuthaya, and Patani, and was conducting limited trade with Cambodia and Cochin-China. Early experience in Indonesia had taught that textiles were the preferred medium of exchange for pepper and spices, so textiles were first purchased with silver in India — the drain of silver from England was a major gripe of Company critics.

During the Company's early years, its pilots were dependent upon foreign maps and sailing directions. But just as the Dutch had first gained experience by sailing with the Portuguese, so one of England's pioneers, John Davis, learned from the Dutch by sailing on Cornelis de Houtman's second expedition to the Indies in 1598. By 1614, the Company had begun giving lectures on navigation, and recognizing that a successful East India Company could not rely indefinitely on foreign maps and documentation for its geographic data, Edward Wright was assigned the task of reviewing and compiling data from the Company's own pilots. This attempt "to examine their journals and mariners and perfect their plats" may have benefited charts made by and for Company pilots, but had little discernable impact on indigenous English cartography.[267]

England's early entrepreneurs in the Indies were beset by hardship. In 1623, a notorious act of violence by Dutch East India Company agents against English, remembered as the 'Massacre of Ambon', dampened English enthusiasm for adventures in the Indies. The combination of the Ambon tragedy and poor market conditions in the same year precipitated the Company's withdrawal from the Spice Islands, Company officials deciding that it would be more profitable to shift their focus to India instead. However, outposts which had previously been established on the Southeast Asian mainland, at Patani and Ayuthaya in 1612 were maintained, with difficulty, until 1684.

In 1606, the English public was offered an English-text edition of the Ortelius *Theatrum,* with its general maps of the Asian continent and Southeast Asia prepared in 1570. An English edition of Mercator's *Atlas* appeared in 1636 with the Hondius map of mainland Southeast Asia, prepared in 1606, and an updated map by Jansson of the islands, introduced in 1633, as well as a map of the Moluccas. England finally produced its own world atlas with *The Prospect of the Most Famous Parts of the World,* by John Speed, with maps dated 1626, although the atlas was not actually issued until the following year. Its coverage of the Far East was limited to a map of the full Asian continent, which was based on Dutch models, and even engraved by a Dutchman, Abraham Goos. A map of Southeast Asia, also copied from Dutch models, was added to the atlas for the last edition in 1676, while a miniature edition of the Speed atlas appeared in 1646, which put the acquisition of an atlas within the reach of more people. But the publishers of atlases in seventeenth-century London, never targeted an international audience. Whereas atlases produced in the Low Countries were commonly issued with their text translated into several European languages, those produced in England appeared only in English.

The monumental *Pilgrimes* of Samuel Purchas (1625) carried extensive accounts of various voyages and included important maps of America and India. As regards Southeast Asia, however, it contained only restrikes of the maps engraved by Hondius for the Mercator *Atlas Minor* of 1607 (reduced versions of figs. 111 & 112). Purchas himself brought the inadequacy of the old Hondius map of the Indies to the attention of his readers. "This [map] of

Hondius," he advised, is "mean and obscure enough, but somewhat more than nothing." Purchas had apparently secured a much better chart of Southeast Asia with the intent of including it in the book, but refrained from doing so because of his "doubt" about "displeasing", insinuating some political consideration:

> I had another far better [chart] sent out of the Indies, but partly the cost, and partly doubt to displease have detained the publishing . . . if I had been able to have given thee also those Draughts, thou shouldest have had them.

The V.O.C. and Early Amsterdam Publishers

In the final years of the sixteenth century, several rival commercial organizations in Holland competed with each other for the lucrative Indies trade. This drove up the prices for Southeast Asian commodities at their source, while lack of cooperation between the competing outfits denied them the benefits of shared geographic knowledge.

Recognizing this inefficiency, conflicting Dutch aspirations for the Indies trade were consolidated, in 1602, into the single capitalistic behemoth — the Dutch East India Company (the *Vereenigde Oost-Indische Compagnie,* or V.O.C.), which became one of the most successful and self-perpetuating commercial organisms ever invented. The Company, which was virtually a surrogate government in itself, was afforded prerogatives normally associated with states, not commercial enterprises. In the name of Holland it could make treaties, occupy lands, declare war, raise armies and designate their officers. In contrast to England, Holland had already established a formidable presence in the field of map publishing before her dive into the Spice trade, and her entry into Southeast Asian commerce was far quicker and more decisive than that of her rival across the North Sea. Petrus Plancius, who a decade earlier had acquired the Bartolomeu Lasso charts from the de Houtman brothers and had helped mastermind their pioneering Dutch voyage to the Indies, was one of the Company's primary architects and became its first mapmaker.[268]

The incredible inertia of the V.O.C. encouraged the mapping of the region as regards producing its own manuscript charts, but stifled progress in published maps. Thus, the history of the Dutch mapping of Southeast Asia in the seventeenth century follows two parallel trajectories, being the history of charts made by and for V.O.C. pilots and cartographers on one hand, and of decorative maps designed to please lovers of pretty books on the other. Presentation, more than content, became the lure of the published maps.

Blaeu, Hondius, and Jansson

At the time the V.O.C. was founded, Willem Blaeu was an emerging Amsterdam mapmaker. In 1608, Blaeu solidified his career by publishing spectacular *carte à figures* wall maps of the world and continents (fig. 136), probably inspired by Jodocus Hondius, and in the same year copied Waghenaer's *Thresoor* (see page 190, above) to create his own pilot-guide, the *Het Licht der Zee-Vaert.* In its execution, the coastlines of Blaeu's *Zee-Vaert* were more 'literal' than the stylized contours of Waghenaer's earlier charts, thus initiating a greater break between the conventions of Western and Asian mapmaking. Blaeu's *Zee-Vaert* was to be divided into four books, the last of which was to comprise Africa, America, and the East Indies. But this section, which would have been the only one to include Southeast Asia, was never published. Nor did Blaeu's *Zeespiegel,* a new enlarged pilot-book introduced in 1623, cover any part of the Indies.

Willem Blaeu's first map of Asia was his wall map of the continent of 1608. Nine years later a folio-size reduction of the map was engraved which was at first sold as a loose-sheet, and later became the general map of Asia in the Blaeu atlas. One might have imagined that this would have been the flagship map of an atlas produced by the mapmaker to the Dutch East India Company, yet it was republished without improvement from its first appearance in Blaeu's 1634 *Theatrum* through to editions of the *Atlas Major* of the 1660s. The cost of a new plate in itself does not explain the lethargy, since many of the map's more obvious failings, such as Java's hypothetical southern coast, the rudimentary eastern shores of Borneo and Sulawesi, and the identification of Irian Jaya as Seram, were known to be faulty and could have been remedied by inexpensive reworking of the same plate, just as Waghenaer had rubbed the erroneous coast of Sumatra from his chart of the Sunda Strait to serve his far narrower audience.

In the meantime, the Mercator *Atlas,* to which Jodocus Hondius had added a pair of maps of Southeast Asia in 1606 (figs. 111 & 112), continued to evolve. The elder Jodocus died early in 1612, and later that year Elisabeth, his daughter, married Jan Jansson, who became inducted into the family's publishing business along with Jodocus' widow and their two sons, Jodocus Jr. and Henricus. By the 1620s, the family had published a number of maps as loose-sheets, among them a detailed chart of the Moluccas which located all five principal members of the group with reasonable accuracy. Ternate, Tidore, Motir, and Makian are still a negligible bit too far south, while Bakian is shown by an inset map without coordinates identified. In 1629, Jodocus Jr. died and these plates roughly 40 in number   were sold to Blaeu, a transaction that the Hondius family quickly lamented. Blaeu changed the plates' attribution from Hondius' name to his own, and began publishing many of the maps, including that of the Moluccas, in his *Atlantis Appendix* of 1630. The Hondius-Jansson family, needing to replace the plates to remain competitive, commissioned two engravers to prepare 36 new plates within 18 months. These were to be "accurate and fine, yes, finer and better and not less in quality" than those they lost to the Blaeus.[269] One of these new plates was the Moluccas chart (fig. 114).

In 1630, Hondius issued a new general map of Southeast Asia, entitled *India Orientalis Nova Descriptio,* and beginning with the Hondius' *Atlantes Majoris Appendix* of 1633 — which appeared, coincidentally, in the same year that Blaeu was appointed official cartographer to the V.O.C. — the firm published its copy of its own (now Blaeu's) chart of the Moluccas. At the same time, the Blaeu atlas included the ex-Hondius Moluccas chart, which remained the only large-scale map of any part of Southeast Asia published by that firm.

In 1635, Blaeu added to his atlas a new general map of Southeast Asia, *India quæ Orientalis dicituret insulae adiacentes* (fig. 115). This fine map represented a marked improvement over the depiction of Southeast Asia on the map of the Asian continent, which continued to be published for another three decades. Significantly for the day, the new map was constructed on the Mercator projection, although its scale was too small for it ever to have been intended for navigational use. One interesting difference between the older continental map and the new portrayal of Southeast Asia is in the depiction of Irian Jaya (western New Guinea). The continent map, following late sixteenth century models, shows Irian Jaya as a large island east of Halmahera, labeled *Ceiram,* as well as true Seram, which lies below Halmahera. The new map of Southeast Asia shows a more correct outline of New Guinea and at the same time deletes the erroneous 'Seram' designation.

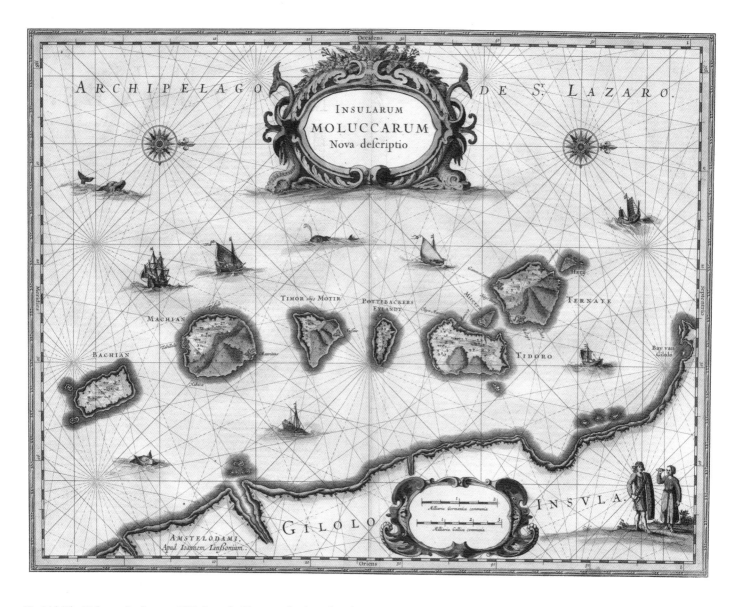

Fig. 114 The Moluccas, Jan Jansson, 1633. From the Mercator atlas. (38 x 49 cm)

A proof state of the *India quæ Orientalis* is known, lacking the cartouches, which was probably available as a loose-sheet map, but was also included in the 1634 German edition of the atlas. In 1636, Henricus Hondius published a virtual twin of Blaeu's Southeast Asia map, differing only in its omission of one place name in the coastline of northern Australia in the bottom center of the map, and in its less elaborate cartouches.

Luxuriously produced on fine paper and often placed in exquisite bindings or even fitted into their own fine furniture, Blaeu atlases were a popular acquisition for the wealthy — they were quite expensive — and were often chosen as gifts which were certain to impress even the most indulgent recipients. Although the engraving quality (and arguably their aesthetic sensibilities) rarely equaled that of many maps produced in the Low Countries in the late sixteenth century, Blaeu maps fulfiled their function as objects to be marvelled at in themselves. They were designed to entertain and enhance the glamour of their owners, and in this they were exceedingly successful.

The V.O.C. and Cartographic Secrecy

Like Portugal and Spain before her, Holland, alias the V.O.C., saw no reason to share its geographic booty with rival contenders for the riches of the Indies, and passed laws to safeguard its cartographic

secrets. By systematically guarding the geographical information collected by its pilots, the Company helped to protect its trade monopolies in the region.

The first dramatic example of how the V.O.C. dealt with perceived threats to geographic intelligence came when the Company was still in its infancy. The trick of entering the South China Sea through the Sunda Strait rather than the Malacca Strait, touted by Linschoten and successfully used by the early Dutch voyages, had served its purpose. Dutch pilots now looked to secure both ends of the Pacific by mastering the Malacca/Singapore Straits on the west side, and the Magellan Strait on the east. In 1614 the V.O.C. sent a fleet of six ships, under the command of Admiral Joris Spilbergen, to enter the Pacific via the Magellan Strait, which was the only known entrance into the great ocean at its eastern end, and over which the Company claimed exclusive commercial rights. Spilbergen had a successful but unremarkable time, enhancing his country's military prestige by defeating the Spanish in skirmishes along the Pacific Coast of South America, before reaching the Philippines (figs. 116 & 117) and then reinforcing the V.O.C.'s grip over the Moluccan trade.

In 1615, while Spilbergen was still engaged on his mission, Jacob Le Maire and Captain Willem Schouten set sail under the auspices of a group of merchants headed by Jacob's father Isaac. The expedition

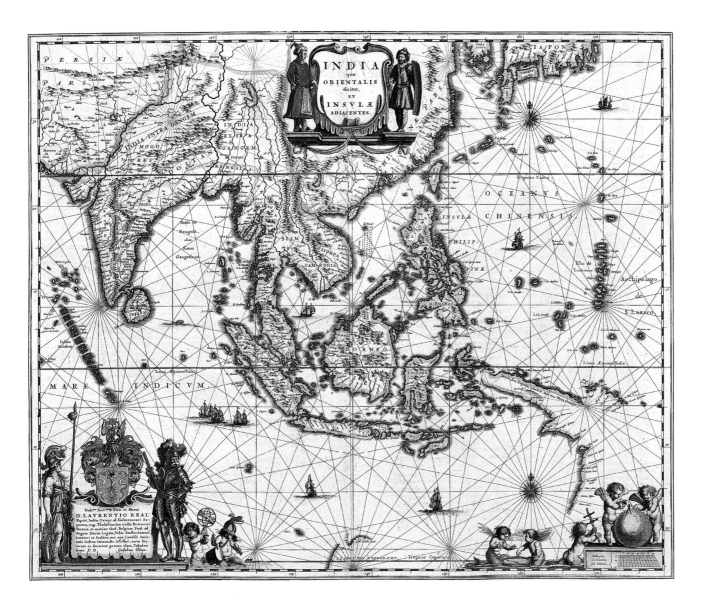

Fig. 115 Southeast Asia, Blaeu, 1635. [Paulus Swaen Old Maps Internet Auction]

Fig. 116 View of Manila Bay, from
Joris van Spilbergen, *Speculum
orientalis occidentalisque Indiae
navigationium*, 1619 (14.5 x 21 cm).
Spilbergen reached Manila Bay early
in 1615, where he pirated Spanish
vessels for nearly a month until a
concerted Spanish offensive ousted
him from Philippine waters. This
view of the bay appeared in the
published account of the circumnavigation which also included a
record of the voyage of Le Maire and
Schouten. Spilbergen's text added
little to current knowledge about the
Philippines, but in it he advocated the
Dutch seizure of the archipelago as a
means of securing the Moluccas.

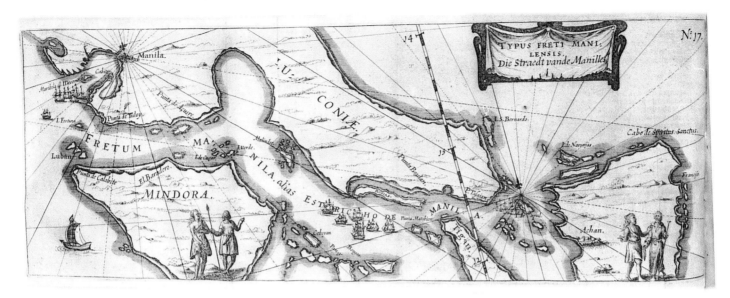

Fig. 117 J. van Spilbergen, *Oost ende West-Indische Spieghel . . .* 1619. Map of the strait separating Luzon from Mindoro. (15.2 x 43.5 cm) [Antiquariaat Forum, Catalogue 105]

sought to discover an alternative route to the Spiceries by taking a gamble on the theory, that had originated with Francis Drake and was currently being flaunted in intellectual circles, namely that Tierra del Fuego was a group of small islands, not part of *Terra Australis*. Le Maire and Schouten sailed south of Tierra del Fuego and rounding Cape Horn, dismissed the century-old assumption that the Strait of Magellan was bounded on the south by a continental landmass. In crossing the Pacific, they made numerous discoveries, principally in the Tuamotu and Tongan groups, and along the coast of New Guinea. Upon reaching the V.O.C.'s outpost in Batavia, however, the local Dutch officials did not take kindly to Le Maire's discovery that Tierra del Fuego was not a part of *Terra Australis*, for it effectively negated the Company's monopoly as it was then defined. The crew was incarcerated for infringing on the V.O.C.'s monopoly, and Le Maire, having led one of the greatest of Dutch exploratory missions, was shipped back to Holland in chains and died *en route*. The V.O.C. at first attempted to squelch all knowledge of Le Maire's new-found strait; an injunction prohibiting Blaeu from revealing the strait was enforced in July of 1617. By August of the following year, however, the futility of trying to keep the strait secret was acknowledged, and the ban was lifted. The existence of such a strait had, in fact, been already mapped hypothetically by cartographers influenced by Drake, such as Jodocus Hondius and Edward Wright — and their works, in turn, had been part of the evidence which influenced Le Maire.[270]

The V.O.C.'s policy of secrecy was more effective when it came to Holland's own published maps, especially since the Blaeu firm dominated map publishing from about 1630 to 1670, during which period Willem Blaeu, and after him Joan Blaeu, were the V.O.C.'s chief cartographers. There were 'leaks', however. The Company's vessels carried numerous manuscript charts with them, which were stored rolled in a metal container. But these charts were occasionally 'lost' when an officer would slip one to an outsider for cash. To tighten security, V.O.C. authorities printed ledger sheets on which any charts carried on the vessel had to be itemized and signed for.

There were other circumstances under which charts sometimes vanished from Company holdings. It is known that some wealthy, influential private Dutch individuals were secretly entrusted with copies of V.O.C. charts for their personal libraries. Ostensibly, these were to be for the enjoyment of their eyes only, but there was no way to safeguard against abuses of this confidentiality.

The question of the longitude and latitude of Java offers an illustration of the discrepancy between published maps of the Indies and V.O.C. charts of the latter seventeenth century. According to the records of the first French embassy to Siam in 1685, those who were familiar with eastern waters were well aware of the inaccuracy of the island's placement on printed maps. When the commanders of the embassy's fleet, which was proceeding to Ayuthaya via the Cape of Good Hope and the Sunda Strait, attempted to follow the charts which had been supplied to them, they were advised not to by members of their crew. According to Guy Tachard, one sailor "assured us that the Isle of Java was ill placed in the common maps, and that it was about an hundred leagues nearer the Cape, and much more to the leeward than was believed."

Abbé de Choisy revealed that it was the Dutch members of the crew, hired because of their familiarity with the Indies, who warned that the standard maps were faulty: "If the Dutch have told us the truth . . . the cape of the island of Java is 100 leagues further west than is marked on the maps."

The Dutch sailors' claims were soon proven correct. What was common knowledge among Dutch seafarers had not yet been incorporated into the published charts. They "discovered a great Coast of Land, and . . . found it to be the Isle of Java," although they thought themselves to be "far from it."

> This made us observe that that Island lies much more to
> the West, and by consequent is nearer by threescore
> Leagues to the Cape of Good-Hope than it is marked in
> the Geographical Maps.

Another example, involving latitude, is all the more persuasive since it is relatively easy to determine latitude with a reasonable degree of accuracy. This occurred when the French embassy attempted to make a landfall on the small island of *Mony* (Christmas Island) before reaching the Sunda Strait to allow their many sick crew members to recuperate. Since the island was mapped at a latitude of 8° south on published charts, they believed themselves to be lost when they passed an island at about 10°. When they later told their story to a Dutch official in Batavia, they were shown a great chart "which placed that isle just in ten degrees eleven minutes south latitude . . . exactly as we found it." The island, indeed, is centered at about 10° 30′.

Portugal Relaxes its Secrecy

As the V.O.C. increased its stranglehold on Southeast Asian commerce, the original European master of the Indies, Portugal, had all but lost the empire she had enjoyed a century earlier. One effect of this was that Portugal's reluctance to share its geographic secrets softened. Although Portugal never dived into the map publishing field in any meaningful way, in the early seventeenth century, with Portuguese commerce in Southeast Asia in decline, Portugal allowed some of its pilot guides to be printed.

There were several reasons for this new outlook. Portugal's loss of most of her Indies empire rendered the secrecy of her charts less important. Information from Portuguese *roteiros* had already been made accessible to any European pilot at the end of the sixteenth century, courtesy of Linschoten, who had published some of this material in his *Reysgheschift* in 1595. Finally, Lisbon had been required to share its Indies data with Spain as a result of agreements between the two nations' temporarily united crowns, making further breeches in confidentiality much more difficult to control. Lastly, the Portuguese themselves had attempted the privatization of their Indies trade with the establishment of the Portuguese India Company in 1628, but the experiment was short-lived and was aborted five years later.

The stunning Portuguese portolan chart of 1641 in figure 80 comes from this period of Portuguese maritime decline. It is signed in the lower left by Antonio Sanches, a cartographer who worked in Lisbon from about 1623 to about 1642, which is the date of the atlas in which the chart appeared, *Idrographisiae Nova Descriptio.*

The Trend Toward Printed Sea-Charts

Robert Dudley

By the middle of the seventeenth century, the Blaeu *status quo* was being undermined. The first serious challenge came from an English expatriate in Italy, Robert Dudley. Dudley was born in 1574 from an affair — or secret marriage — between the Earl of Leicester and the widow of Lord Sheffield. Although in his youth Dudley enjoyed all the privileges accorded to children of royalty, in 1603-05 his continued attempts to prove his 'legitimacy' brought about the wrath of the Countess of Leicester, who accused him of conspiracy and defamation. He fled England for France, leaving behind his (third!) wife but taking with him his cousin Elizabeth, disguised as a page. In France the two declared themselves Catholic and were married. They continued to Italy, where they asked the Grand Duke of Tuscany for protection and passed their lives in exile in the region of Florence. Dudley's atlas, *L'Arcano del Mare* (*Secret of the Sea*), was published toward the end of his life, in 1646.

Dudley's furthest experience at sea was a voyage to Guyana in 1594. As a youth he had wished to travel to the Pacific, but this aspiration was nixed by the Queen. His interest in the Far East is indicated by his backing of Benjamin Wood's 1596 expedition to Southeast Asia which ended in shipwreck on the Burmese coast. Ultimately exiled in Italy, a country with no serious designs on Asian exploits, Dudley drew his geographical data from various sources, though not necessarily the most current ones. Whether or not one takes literally the claim of the engraver, Lucini, that he was

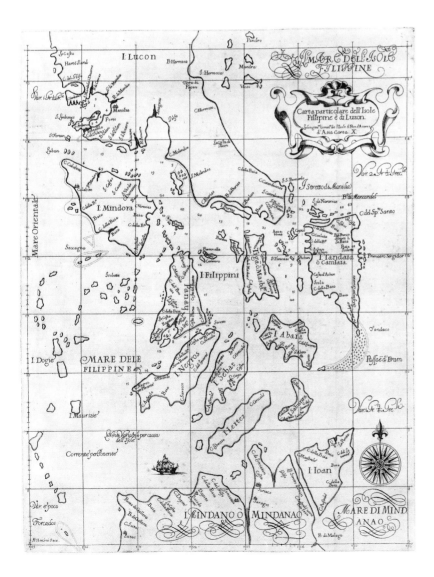

Fig: 118 The Philippines, Robert Dudley, 1646 (35.5 x 48 cm). Though smitten with wanderlust as a youth, Dudley never reached the East. Writing if his "great desire to discover new countries," he wished to sail to "India and other parts to which navigation should take him." Queen Elizabeth, however, "would not allow such a mere youth to break his maiden lance in an enterprise requiring so much knowledge of the world, in which many veteran Captains had fared so ill." Elizabeth's fuss, indeed, proved wise. "Not being able to take the desired voyage to China," Dudley continued, he "sent ships and men there under the command of Captain Wood," who was the shipwrecked off the coast of Burma.

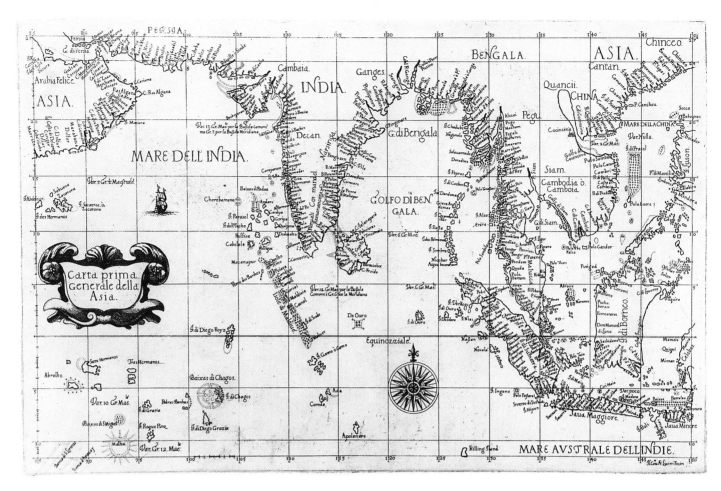

Fig. 119 Southeast Asia, Dudley, ca. 1665. (24.5 x 38.2 cm) [Richard B. Arkway, New York]

"for twelve years sequestered from all the world in a little Tuscan village" consuming "5000 lbs of copper" to produce the plates for the atlas, it was certainly a monumental undertaking which was a long time in preparation. His charts are remarkable for their large scale, their indication of magnetic variation, and their elegant, unembellished execution. Most importantly, Dudley made consistent use of the Mercator projection throughout the atlas.

Roughly conceived by Portuguese and Spanish theorists early the previous century in response to the needs of their pilots, the projection was first applied by Mercator in 1569, and formalized by Edward Wright thirty years later. It became popular among chartmakers in the French port of Dieppe, on the English Channel, by about 1630, and was used by Blaeu on his printed chart of the Indies in 1635. Although Dudley's own understanding of the projection was imperfect, he was instrumental in promoting its use. In championing the projection Dudley, in a sense, represents the culmination of one of England's great *theoretical* contributions to mapmaking on the eve of her awkward entry into the world of printed nautical charts.

Two Dudley charts are illustrated here. The *Carta particolare dell' Isole Fillipine è de Luzon* (fig. 118), is one of four relating to the Philippines, and focuses on the region from southern Luzon through to northern Mindanao. This chart offers another look at the island of *St. John* (the triangular-shaped *I. Ioan,* lower right), which had begun life in Micronesia over a century earlier and subsequently migrated to northeastern Mindanao on the López de Velasco charts of ca.1575-80 (page 170, above). The island here has truly 'become' the northeastern corner of Mindanao, separated from the mainland by a strait, *Li Stretto d'Ioan.* Its northernmost

tip lies within a degree of the correct latitude of the northernmost point of Mindanao, suggesting that *St. John's Island* is in this instance based on true Mindanao. However, whereas much of the nomenclature for the rest of Mindanao's coastlines is indigenous or Spanish-given, *St. John's* nomenclature is entirely generic: 'The point', 'Bay', 'Cape of the Bay', 'Southern point', and the like, indicative of its genesis at the cartographer's desk rather than via the explorer's log.

An inscription in the seas to the southwest of Negros warns pilots that the islands cause variable winds. The fact that all of Dudley's soundings are in the northern region reflects the typical routes taken by Spanish vessels. Magnetic variation is calibrated in each corner of the chart, except in the southwest, where Dudley believed the error to be negligible, and thus has marked *Var. e poco* (the variation is little) — the great English scientist Edmund Halley reached the same conclusion at the turn of the next century, finding that the line of zero declination runs east of Palawan and through Sulawesi. In his youth, Dudley had been particularly interested in magnetic variation because he accepted the then popular (but erroneous) theory that it could be used to determine longitude, and he tried to employ it in such a way on his voyage to America in 1594.

Abaia, in the waters between Mindanao and Manila, is Leyte, here named for *Abuyog,* a section of the southeastern coast of the island. Leyte is then duplicated and triplicated to the south with *Sabunra* island, a name of unknown origin used by Bartolomeu Lasso (and those who followed Plancius' charts) to denote Leyte, and *Leitez,* naming an island which is in the place of Bohol. *Sebat* is Cebu, *Tandaia* is Samar, while the remaining major islands are identified by their modern names.

A second edition of the *Arcano* was published in 1661, with most of the charts unchanged except for the addition of a volume number below the title (the Philippine chart illustrated here is the first state; "L.°6.°" appears on the second issue). A series of smaller charts was also engraved by Lucini which have been provisionally dated at about 1665. The smaller, oblong atlas for which these reduced charts appear to have been intended was never completed, however. Of the several charts known to be extant, one is of the Indian Ocean and Southeast Asia (fig. 119). Lucini, apparently fond of cartography, also engraved a board game which used maps of the various parts of the world (detail, fig. 89).

Jan Jansson

Dudley's *Arcano del Mare* was the first printed sea-atlas of the world, and it was a radical work in many respects. In 1650, four years after the *Arcano* was first issued, Jan Jansson published his own sea-atlas of the world, the first such work from Holland. It formed a section entitled *Waterwereld* (waterworld) from Volume V of his *Novus Atlas*. Jansson's work is a conservative effort compared to Dudley's, containing far fewer maps and being composed on a far smaller scale, while boasting none of Dudley's ingenuity. But for the layman weary of the stale maps available from Amsterdam publishers, and interested in an attractive, manageable sea-atlas, Jansson's *Waterwereld* had much to commend it. The 1650 issue included a chart of the Indian Ocean, the Gulf of Ganges, and the Pacific Ocean; in 1657 it was expanded to include charts of Java, Sumatra, and Borneo.

The map of Sumatra details the Malacca Strait, Singapore, and the western shores of the Malay Peninsula, at a moment when the V.O.C.'s sights were particularly well focused on the region. The political intrigue, which often spanned the strait, was heightened when the Dutch seized Malacca from the Portuguese in 1644. The V.O.C. had ready markets in India and Europe for the vast reserves of tin found in Malaya, but the tin deposits were under the control of the north Sumatran sultanate of Aceh. In an attempt to secure the tin for themselves, the Dutch, in around 1650, entered into an alliance with Aceh's perennial Malayan enemy, Johor, disrupting ports in north Sumatra and instigating the revolt of Aceh's vassal states.

Jansson's chart of Borneo (fig. 120) shows some improvement in the island's east coast (bottom), which remained 'blank' on many maps of Southeast Asia still being published at that time. The *Waterwereld's* map of Java also shows a similar improvement in detail, but not in the island's location, which is representative of the disparity between published Dutch charts and the knowledge of Dutch pilots, which we examined earlier (page 204, above). Jansson placed the westernmost tip of Java at 134° east of the Cape Verde Islands, which is about 5° further east than its true position, depending on precisely what point in the Cape Verdes was imagined as the prime.

Fig. 120 Borneo, Jan Jansson, 1657. (42.2 x 52.7 cm)

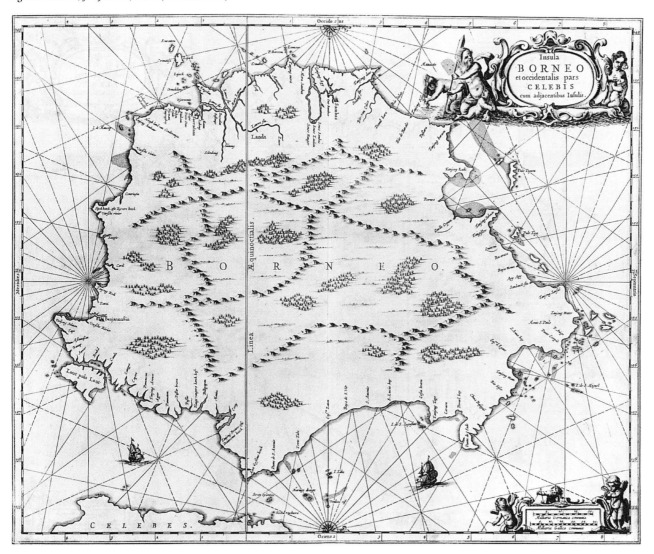

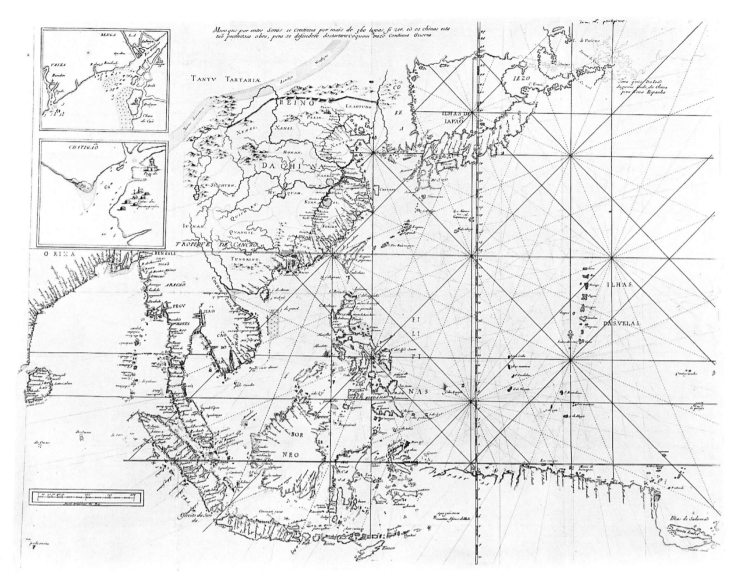

Fig. 121 Southeast Asia, Melchisédech Thévenot, 1664. (51 x 68.6 cm)

Melchisédech Thévenot

The French chronicler Melchisédech Thévenot published an engraved rendering of a Portuguese portolan chart in 1664, which was the very year that the French East India Company was formed (fig. 121). The map appeared in the second part of Thévenot's great compendium of voyages, *Relations de divers Voyages,* a work which was to France what Hakluyt's writing had been to England and Ramusio's had been to Italy. Thévenot, who is believed to have invented the spirit level, based his handsome map on a portolan chart of 1649 by the royal cosmographer of Portugal, João Teixeira. The eighteenth century cartographer J. N. Delisle believed that the original chart came into Thévenot's possession after a French ship had pirated it from a Portuguese carrack. Tradition also suggests that the map had actually been published in Lisbon — which would have been a noteworthy event in itself — in 1649, although there is no corroborative evidence for this and no such map is known.

Parts of the portolan original are attributed to a Portuguese merchant in Macao by the name of Don João da Gama, who after sailing by a northern route from China to Mexico, reached Acapulco and was promptly arrested by Spanish authorities. Most of the chart, however, probably reflects the Portuguese *padrão* of the time.

A New Crop of Charts from Amsterdam.

Joan Blaeu, who had succeeded his father as mapmaker to the V.O.C., did nothing to improve his firm's published maps of Southeast Asia, despite the lead taken in 1657 by his competitor, Jansson, with his *Waterwereld.* At about the same time, several new Amsterdam publishers set up shop who were specifically interested in sea-atlases. Each published a general, small-scale map of the Far East, encompassing the entire region from Tasmania to Japan. Most are similar in geography, and represent some improvement in the existing maps of Blaeu and Jansson.

Probably the first of these was by Arnold Colom. Arnold and his father, Jacob, both produced sea-atlases with a general chart of Southeast Asia, but it was the son's *Atlas Marinus,* which appeared first (ca. 1654); the elder Colom's *Atlas of Werelts-Water-Deel* was not published until 1663. Arnold happened to be a tenant of another Amsterdam mapmaker, Nicolas Visscher, who had also made a general map of the Indies. Ironically, in 1663 Arnold was forced to surrender his copperplates to Visscher to defray overdue rent, and as Arnold's plates were never used afterwards, Visscher probably melted them down to re-use the copper for his own maps.

Arguably the finest of Amsterdam's sea-atlases was one produced by Hendrik Doncker. First published in 1659, by the following year the atlas contained a fine general chart of the Indies, which was mated with another chart extending to the east African coast, so that together they formed a chart of the full Indian Ocean and southwest Pacific. The map was modified and re-engraved in several stages, and in about 1705 a new general chart of the Indies was added, along with a larger, folding chart of the region of Java, Sumatra, and Borneo.

Johannes van Loon included a general map of Southeast Asia in his sea-atlas *Klaer-Lichtende* of 1661. The same plate was used in a 1666 sea-atlas entitled *Nieuwe Water-Werelt* by Jacob and Casparus Jacobsz (who were also known as 'Lootsman', a term they adopted to avoid confusion with other Amsterdam publishers of the same name). Jacob Lootsman added a chart of the Moluccas to their atlas in 1681.

One of the better-known Amsterdam publishers of sea-atlases was Peter Goos. Peter was the son of Abraham Goos, an engraver who had worked for various people, including the London mapmaker John Speed. His *Zee-atlas*, first published in 1666, contained a beautifully engraved chart of Southeast Asia. Goos' chart was given a new cartouche and restruck by the London mapmaker John Seller (fig. 123), and also copied by Frederick de Wit in 1675; de Wit's engraving is slightly cruder than Goos', but boasted a more decorative cartouche. De Wit's plates were later acquired by Louis Renard, who inserted his imprint and restruck them for his *Atlas de la Navigation* of 1715 and 1739. The same plates then came into the hands of R. & J. Ottens, who published them again, with their own imprint, in 1745.

Johannes van Keulen, one of the last map publishers to set up shop in Amsterdam in the seventeenth century, issued his own sea-atlas with a general chart of the East Indies in 1680. Enterprising and prolific, in 1693 he purchased a stock of plates from his neighbor, the chart-maker Henrik Doncker who, after 45 years in business, wished to retire. The van Keulen firm became a veritable dynasty, remaining in existence into the nineteenth century, and achieving its greatest importance in the early- to mid-eighteenth century, during the time Johannes II was mapmaker to the V.O.C.

The van Keulens entered the stage at a fortuitous moment. It was during the early years of the firm that the Dutch East and West India Companies finally loosened their grip on cartographic material relating to the West Indies, Brazil, and West Africa. As a result, the charts in the fourth and fifth parts of van Keulen's *Zeefakkel* (1683-84) contained information from newly declassified V.O.C. charts of these regions. Maps of Southeast Asia, however, remained off limits. Not until the early eighteenth century, when the V.O.C. was in decline and its foreign competition gaining ground, did the V.O.C. allow the publications of maps which tapped its cartographic resources for Southeast Asia.

The Rise of Printed Charts in England:
Seller, Fisher, and Thornton
As the V.O.C. enjoyed its heyday, England made its awkward entry into the field of printed sea charts. In 1671 King Charles II granted a privilege to John Seller for a thirty-year monopoly for the printing of *The English Pilot* and *Sea Atlas*. Seller was an instrument maker and teacher of navigation — and, curiously, a Baptist who had earlier aroused the wrath of the English royalty for his association with Nonconformists. Seller's atlas, first published around 1675, consisted largely of re-worked or re-engraved Dutch plates. His map of Southeast Asia (fig. 123) was simply Peter Goos' 1666 map of the region, with an English title substituted for the Dutch and Goos' cartouche replaced with a vignette of a palanquin, the 'land ship' from China, which had been quite a curiosity in Europe.[271] Yet Seller slyly announced on the title page to his atlas that his charts were based on "the latest and best discoveries, that have been made by diverse, able and experienced navigators, of our English nation." Seller was duly criticized for this expediency, critics not missing the opportunity to make the pun that he was just a 'seller'.

In 1678, Seller produced a miniature atlas entitled, appropriately, the *Atlas Minimus,* which included separate maps of continental and insular Southeast Asia. These maps, although of little cartographic interest, at least reflect some individuality. The map of peninsular Southeast Asia refers to the region as India *within* the Ganges, rather than the traditional perspective as being India *beyond* the Ganges. This, of course, would be the view of someone approaching the continent from the Pacific Ocean, rather as if Seller were indulging the ghosts of Drake and Cavendish to boast of English circumnavigatory glory.

Risking bankruptcy on his main atlas project, Seller collaborated with two other chart-makers, William Fisher and John Thornton, to produce *The English Pilot*. The inclusion of Thornton, who later became hydrographer to both the East India and Hudson Bay Companies, gave the work greater credibility. For the remainder of the seventeenth century the atlas had to battle with financial crises and unstable partnerships, but the tradition of the printed English sea-atlas had been established.

The new sea atlas was conceived in four parts, of which *The Third Book* was to deal with "Oriental navigation." Seller never saw the book to completion, but produced a fragmentary edition in 1675. Another edition was attempted in 1701, as *The Sixth Book,* just after the expiration of Seller's 30-year copyright privilege.

It was not until 1703 that John Thornton, son of the original collaborator, saw *The Third Book* to fruition. The atlas' title page states that its 62 charts were "collected for the general benefit of our own country-men," and spanned the coastlines of the British Isles, Africa, India, Southeast Asia, and China. Its information was "gathered from the experience and practice of the English and Dutch navigators; the like never yet published," and it included tables of magnetic variation, of longitude, and latitude. This compilation of sea charts, extensive sailing directions, and numerous coastal profiles became the standard reference of English ships making the long haul to Asia by way of the Cape of Good Hope, and underwent 12 editions through 1761. The 1703 issue contained a general chart of the Indies, one of the Ganges Delta (including the Burmese kingdom of Arakan), a chart of the Sunda Strait and Western Java, of Eastern Java and the Bali Strait, of the Malacca/Singapore Straits, of southern Borneo, and a chart of eastern Australia "and parts Adjacent," which extended northward through Java and southern Borneo. While most of the charts made no claim of being original, the one of the

Thus sheweth the high-Land of *Pattany*, bearing South West by West four leagues off.

Fig. 122 Profiles of the coast of Patani, an important kingdom on the Malayan shores of the Gulf of Siam, probably the Langkasuka referred to in ancient texts. It is now part of Thailand. From *The Third Book of The English Pilot*, 1703 and subsequent. (13.5 x 3 cm)

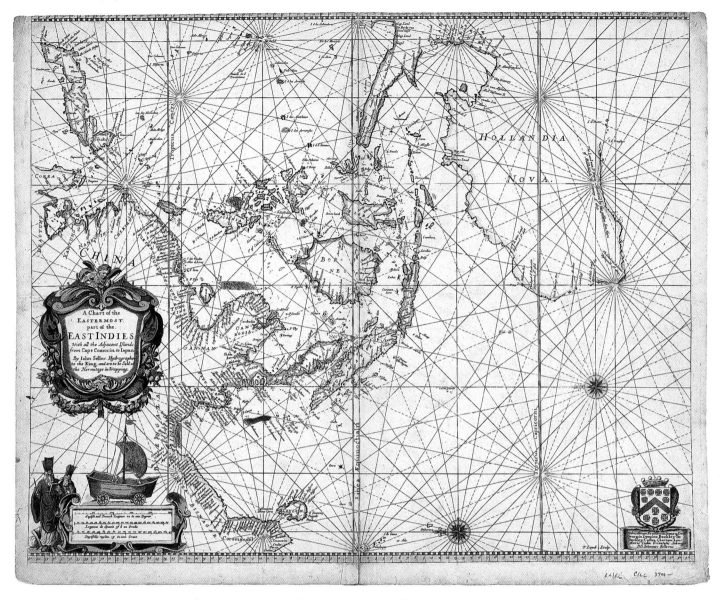

Fig. 123 Southeast Asia, John Seller, 1675. (43.5 x 54 cm) [courtesy of Martayan-Lan, New York]

Ganges Delta (which extended east as far as the Martaban River in Burma) claimed to be "Taken From a Draught Made upon the Place, by the Agents for the English East India Company Never Before made Publique." That chart is also atypical in that it bears a dedication to officials of the East India Company, suggesting that some relationship existed between Thornton and the Company. Subsequent issues of the book added some new charts. One of Borneo was added in 1711, of part of Java in 1716, and of Mergui (Burma) in 1743.

The Third Book continued the practice of including numerous profiles of coasts in the sailing instructions, such as those of the coast of Patani (Malaya) illustrated in figure 122. These were generally reproduced by woodcut, not copperplate engraving, even though engraving had long ago virtually replaced woodcuts as the singular method of map printing. The reason woodblocks were still used for the coastal profiles is that they were set within text; the woodblock could be printed together with the typeset, since both are relief printing; an engraving, however, would have had to have been printed as a separate register.

The kind of information offered by the *Pilot* made it truly a *merchant* pilot's guide, with even the occasional bit of advice on how to conduct trade being included. In Arakan, for example, the merchant is advised that although "the trade is profitable, [you] must trade on board, [since] they will cut you off if they can"; in the Nicobar Islands, on the other hand, "the people [are] civil and kind with good looking after, but trust them no further than you can see them."

What was the source material for charts of Southeast Asia in *The English Pilot*? In December of the same year *Book III* was published, 1703, two French ships intercepted an English East Indiaman, the *Canterbury*, and took a bounty which included seventeen portolan charts.[272] These charts were brought to the authorities in Paris the following year, and are believed to be the same set as a collection of seventeen vellum charts by Thornton which now reside in the Bibliothèque Nationale. All but one of the charts are signed by Thornton and bear a date of 1699 or 1701. The printed charts of Southeast Asia in *The Third Book* appear to have been derived from these Thornton portolan charts (or other renderings of the same charts). Yet, the actual data Thornton used for his manuscripts seem to have come not from English sources, but rather from Dutch manuscript sources, most notably Blaeu. However, Blaeu's charts were unpublished and were closely guarded by the V.O.C., so that the question of the sources for *The English Pilot,* now becomes: How did England manage to gain access to V.O.C. charts whose publication was forbidden even in Holland?

In fact, Thornton's immediate source was probably the English East India Company itself. Although the younger Thornton, unlike his counterpart Blaeu, is not known to have held an official post with his country's East India Company, *The Third Book's* map of the Ganges is signed by him as "Hydrographer to the Honble East India Company." This suggests that in some capacity *The Third Book* was produced with the Company's patronage, and was probably intended as a standard guide for its pilots. So how, then, did the Company acquire Dutch charts? We saw earlier that copies of V.O.C. charts were illegally sold, stolen, and surreptitiously slipped to favored private Dutch citizens. It has also been suggested that the Dutch and English East India Companies may have shared information for a brief time at the end of the seventeenth century, concerning regions in which their spheres of influence did not at that moment conflict.[273]

Regardless of the circumstances of its creation, *The Third Book* was well-used by East India pilots for navigation in Asian waters, and this evidences a major difference between the English and the Dutch approach. Whereas Holland, during its heyday, never printed a pilot guide for its Indies-bound pilots, the English Company's pilots relied heavily on the printed pilot book. And whereas the English East India Company allowed its pilots a fair amount of freedom of personal judgement during the voyage to the Indies, the V.O.C. dictated precisely the course to be taken. The Dutch system supplied its pilots with confidential manuscript charts which could be constantly improved as better data was collected, but which they were required to surrender once on shore. The English approach invited more open participation in the Indies enterprise, but lacked an efficient mechanism to pool the accumulated experience of its sailors. By the eighteenth century, with the specter of secrecy waning and perhaps the belief that the rate at which charts would become obsolete was slowing, the concept of a standard, printed guide to Asian seas made sense, just as such guides to European waters had been commonplace for over a century.

Evidence of feedback from East India Company pilots is found in *The Third Book,* suggesting that charts from Company voyages were consulted. For example, the rediscovery in 1681 by the vessel *London* of the dangerous Tryal Rocks, named for an East Indiaman that was wrecked on them in 1622, is recorded on one. Pilots, such as those guiding the first French embassy to Siam in 1685, carefully steered clear of the Tryal Rocks.

Manuscript Sea-Atlases
Manuscript reproduction persevered longer for sea-atlases of the East Indies than for those of the West Indies. While Seller, Thornton, and other English cartographers were pioneering their country's trade in printed charts, others continued in the established tradition of producing manuscript maps. Notable among these, and of particular interest for Southeast Asia, was William Hack. Hack's output was prodigious — so prodigious that one might envision him competing against the threat of mass-produced engraved maps. In about 1700 he produced multiple copies of an atlas of Asian shores, three copies of which are extant.[274] The atlases contain 90-100 maps each, drawn on paper, with remarkable uniformity. These atlases are perhaps the first to methodically record such a large extent of coastline in such detail and on such a large scale. The influence of Dutch charts is less evident in Hack's work than in Thornton's.

William Dampier
One of the most intriguing characters to roam — and map — Southeast Asian waters at this time was the English adventurer William Dampier. Dampier took to the seas at age sixteen, sailing to various destinations, among them Asia, for the next decade. In 1679, at the age of 27, he left England for Jamaica, where he remained for a year. When on the return voyage the crew mutinied, Dampier opted to join the rebels. For nearly six years, Dampier and his comrades pirated Spanish settlements in South America, twice crossing the Isthmus of Panama. In 1686, he joined a band of pirates who had commandeered a vessel and crossed the Pacific to the Philippines, Sumatra, and Tongkin. In 1688, tiring of the scavenger's life, he convinced the pirates to leave him on one of the Nicobar Islands. Finally — after twelve and a half years, and ten or more ships plus native vessels — Dampier reached England again in 1691.

The enigma of Dampier is that in great contrast to the mercenary nature of his fellow pirates, Dampier can be seen as one of the first of a new breed of more 'enlightened', carefully observant, and impartial collectors of information who subsequently joined the great scientific expeditions of the eighteenth century . He kept an accurate record of a variety of geographic data — the nature of shorelines, shoals, winds, and currents; he proposed that conventional charts underestimated the breadth of the Pacific Ocean by 25° latitude and that the Indian Ocean was being proportionally overstated; and he tried to compile evidence to show that ocean depths were greater along steep mountainous coastlines and shallower off low-lying shores.

Dampier also described the nature and customs of the peoples and civilizations he came across, being careful not to speculate about that which he did not know, and always stating when his information came from sources which he could not himself confirm. There is even a hint of the coming glorification of the Noble Savage when, having left himself stranded on Nicobar shores, he notes that he had never encountered the cannibals which are purportedly to be found there and speaks of the basic goodness of people. "I am of the opinion," he states, "that there are no people in the world so barbarous, as to kill a simple Man that falls accidentally into their hands . . .". We will hear more of this scholarly rogue in later pages.

Isaac Newton and Edmund Halley
Other extraordinary Englishmen of this period with some effect on the mapping of Southeast Asia include Isaac Newton and Edmund Halley. Newton correctly proposed that the sphere of the earth is flattened at the poles, rather than flattened at the equator (as the French scientific establishment of the day believed) or perfectly spherical, while Edmund Halley introduced new themes to world mapping by attempting to record the trade winds and to chart magnetic variation. The latter's map of the trade winds, which was first published in the *Philosophical Transactions* of 1686, is rich in Southeast Asian wind data.

In 1698-1700 Halley made two remarkable voyages in the Atlantic Ocean recording compass variation. His first so-called 'isogonic' map (1701) recorded magnetic variation in the Atlantic Ocean from a latitude of about 50° north to roughly 50° south, but none whatsoever for the Indian or Pacific Oceans. The following year, Halley extended his chart to Southeast Asia by using data, according to text accompanying his chart, "collected from the comparison of several journals of voyages lately made in the Indian Seas," specifically from William Dampier and East India Company navigators. Halley determined that "the *Variation* [in the Indian Ocean] is altogether westerly," increasing toward Madagascar but then decreasing "till you arrive on the east-coast of *China,* or the *Philippine-Isles.*" A line indicating zero variation runs just east of Palawan and cuts through Sulawesi; Halley had insufficient data to record variation further east. He also understood that magnetic variation was not a static measurement, but rather that "there is a perpetual, tho' slow Change in the *Variation,* almost every where," and that this would make future calibrations necessary.[275]

Fig. 124 da Vignola /Rossi, Indies, 1688 (44.5 x 58 cm) [Martayan Lan, New York].

Italian and French Maps of the Later Seventeenth Century

Despite the paramount role played by Italy in the publication of maps throughout the sixteenth century, few maps of any consequence were produced by her city-states during the following century until Dudley's *Arcano del Mare* appeared in 1646. In 1652, possibly motivated by Nicolas Sanson's new collection of maps, the Congregation for the Propagation of the Faith (*Propaganda Fide*) hired G. B. Nicolosi to compose an atlas which was eventually published in 1660 as the *Dell' Hercole e Studio Geografico*. It contained maps of the continents, each on four sheets. Southeast Asia appeared with sparse detail despite the relatively large scale of the map. Irian Jaya *(Terra d. Papui)* is rendered an island separate from New Guinea, following an error which began in the 1590s which confused Irian Jaya and Seram, and continued in the seventeenth century owing to the uncertainty as to whether or not Irian Jaya was connected to the mainland of New Guinea (and whether New Guinea, in turn, was connected to Australia). The error was retained on an otherwise improved map of the Indies issued in the 1690s by the Roman publisher Giacomo de Rossi based on a 1683 model of G. Cantelli da Vignola (fig. 124), which was also used by Vincenzo Coronelli.

France had been publishing important maps since the fifteenth century, but had not produced any formal atlases of the world (excepting those compiled in Strassburg in the early sixteenth century) until Nicolas Sanson plunged into the commercial map market in the 1650s. Sanson published a quarto-size atlas of Asia in 1652, *L'Asie en plusieures cartes,* followed by a folio map of Asia in 1654, which was subsequently included in his atlas in 1658. Geographically, his maps of Southeast Asia were competent but unremarkable. Like the maps of Nicolosi and da Vignola, they were similar to the standard Dutch models of the day, varying among themselves most conspicuously in their depictions of the smaller Indonesian islands east of Java, and differing from the better Dutch models in interpreting Irian Jaya as an island. France's true *forte* at this time was in the exploration of the New World, not Asia. Whereas revolutionary maps of America, composed from first-hand French experience, had been published in the accounts of French explorers since the early years of the seventeenth century, the gem of French charts of Southeast Asia, that of Melchisédech Thévenot (1664, fig. 121), was actually an engraved rendering of a Portuguese portolan chart. Significant French contributions to the mapping of Southeast Asia only began with Louis XIV's embassies to the Siamese court at Ayuthaya in the 1680s.

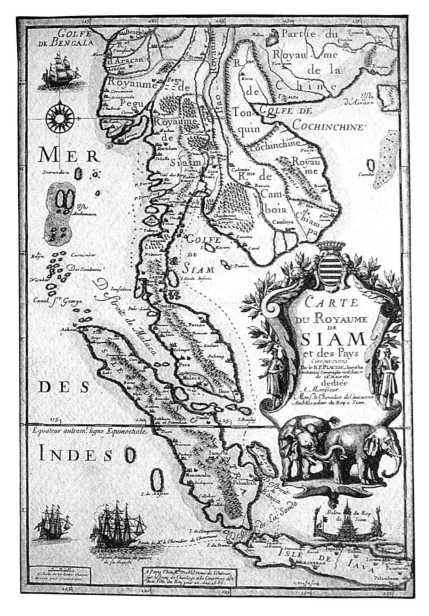

Fig. 125 Map of Siam, le Père Placide, 1686. Placide incorporated data from the French embassy into older geographic models. The single greatest inspiration for Placide's map was the observation, by King Narai and French envoys, of an eclipse of the moon in Lop Buri on 11th December, 1685 (see fig.128). Narai was no mere passive onlooker to this event; he enquired why the part of the moon that was eclipsed was still visible, what the time was in Paris, the possible uses of the astronomical observations made in concert, and "had a mind to look in a Telescope twelve feet long, [which] we presently carried to him. He suffered us to rise and stand in his Presence, and would look in the Telescope after us; for we must needs set it to its Point when we presented it to him:. Although that evening must have seemed the beginning of a close cultural and commercial relationship between the two kingdoms, the situation deteriorated precipitously during 1687. Distrust of European intentions, fostered by intrigue pivoting around a controversial figure named Constance Phaulkon, and a ruthless power struggle upon Narai's death in 1688, ushered in a period of relative isolationism which endured for the remaining years of the Ayuthaya period. Lop Buri continued to serve as a point for celestial observations after the 1685 eclipse; on the eve of the 1688 *coup de état*, a runner sent to Lop Buri found Phaulkon, his wife, and a group of Jesuits observing a lunar eclipse. Placide's map was re-published from the original plate as late as ca. 1775, by Dezauche. (47 x 33 cm) [Lent by a private collection] .

Although French merchants had sailed from Dieppe for Southeast Asia in the 1520s, and had also made a second attempt to join in the Indies trade at the turn of the seventeenth century, France's own East India Company was not chartered until quite late — 1664. Initially,. the Company's strategy was to use the African route and establish a colony in Madagascar, but this outpost proved cumbersome and was soon abandoned to the missionaries. A bold and expensive attempt to seize the Dutch trade ended in failure. so France then focused her attention on Indochina and Thailand, where the existing foreign strongholds were less impenetrable. More so than for the Dutch or English, religion afforded the paramount pretext for French intervention in Southeast Asia. French embassies were sent to Siam shortly before the kingdom effectively closed its doors to foreigners in 1688 and their records provide much of what is known about the region from that period.

A Lunar Eclipse is used to Determine the Longitude of Siam
Paris in the latter seventeenth century had become a center for the sciences, including cosmography and the determination of longitude. With scientific, commercial, political, and religious aims all in mind, France embarked upon a series of embassies to Southeast Asia, the first setting sail for Siam in 1685, with Chevalier de Chaumont as envoy. Soon after the embassy's return to France

in 1686, five accounts were published by its participants, and maps inspired by the expedition were composed by Le Père Placide, geographer to Louis XIV, in 1686 (fig. 125), and by Jean Baptise Nolin in 1687, based on the geography of Coronelli (fig. 126). Two of the accounts are especially useful here, that of Father Guy Tachard, a Jesuit missionary who returned to Siam in 1687 and again in 1698, and Abbé de Choisy, a fascinating character who had the rank of co-ambassador and who brought with him a reputation for transvestism, seducing young women, and then dressing them up as men. Add to that the humorless, impossibly rigid Chaumont, and the embassy must have been an interesting bunch indeed. To put this voyage, its maps and its scientific endeavors into a time perspective, we can note that soon after they departed, on 21st March, sailing south through 13° latitude in the north Atlantic, Johann Sebastian Bach was born in Eisenach, Germany.

According to the account written by Father Guy Tachard, the mission aspired, among other things, to "correct the geographical maps, facilitate navigation, and raise astronomy to its perfection." Following a theory currently being tested by the French scientific community, the expedition planned to use eclipses of the moon to determine the longitude of the Cape of Good Hope and Siam. Father Tachard tells us that the French members of the expedition brought "sea-charts of the course we were to follow", that they

Fig. 126 Siam, Vincenzo Coronelli / Jean Baptiste Nolin, 1687. (45 x 61 cm)
[Roderick M. Barron]

had previously copied from charts "in the Kings Library, which had served in other voyages". Choisy notes that these were transcribed by Jean de Thévenot, who had been in the Indies from 1663-67. Other charts were supplied by Bénigne Vachet, a missionary who was part of the expedition and who had previously been in Siam, from 1668-73. Early in the voyage, before the failings of these charts became apparent, Choisy reported that the pilots found the charts supplied by Thévenot to be good, and those from Vachet to be excellent.

The expedition was endowed with a veritable laboratory. For the necessary celestial observations they brought "several large telescopes of 12, 15, 18, 25, 50, and eighty Foot." For the all-important task of keeping time during the lunar eclipses they also lugged along "two repetition-pendulum-clocks," and for various other experiments and measurements they carried "lodestones, microscopes, several thermometers and barometers, and the tubes and machines that serve in making experiments of vacuity," as well as two "machines of Romer", a type of orrery, "one of which represents the motion of the planets, and the other the eclipses of the sun and moon." And to help them with their calculations of longitude they brought "tables of the satellites of Jupiter, which have been made with so great labour." Not a spartan enterprise, they also brought trumpets and a violin.

Once in the seas off southern Africa, they found that the star charts they had been supplied with were faulty. Although the tossing of the ship prevented them from using their instruments

Fig. 127 The town of Lop Buri (Siam), Nicolas Bellin, 1747 (1757). (19 x 31.5 cm)

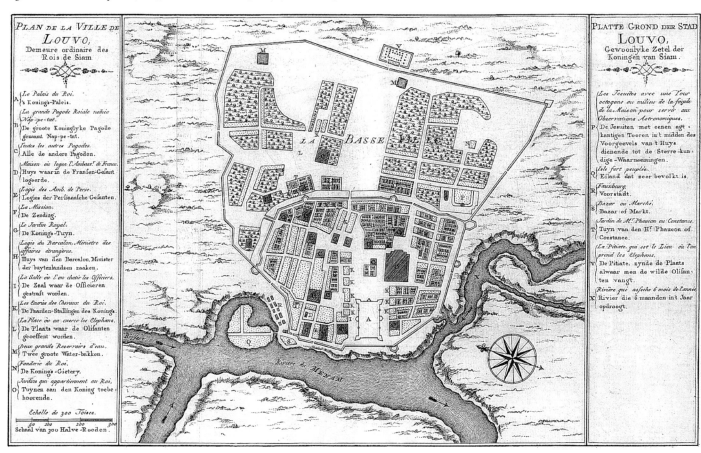

FIG: XXVI

Palais de louvo d'ou le Roy de Siam Observe l'Eclypse de Lune

Fig. 128 King Narai and the French envoys observate an eclipse of the moon in Lop Buri on December 11, 1685, from Tachard's *Voyage to Siam*, 1688. Narai was enchanted by European science and scientific instuments. [D. Ande, Bangkok]

to construct a new star chart properly, Tachard boasted that they "made a new one only by the eye' which though not "so exact as such works require", was nonetheless "not so defective" as the star chart they had brought from France. But the truly extraordinary event of the whole enterprise was the observation of the lunar eclipse by two teams working in concert, one in Paris under Jean Dominique Cassini, the eminent scientist of the Paris Observatory, the other in Siam. By recording the difference in local time between the two locations when the eclipse occurred, they hoped to establish a more accurate figure for the longitude of Siam.

The eclipse was observed from Lop Buri (*Louvo*), which was a town to the north of Ayuthaya and was used by the King of Siam as a country retreat (fig. 127). With their native hosts looking on, the French scientists assembled their equipment at "a royal house called *Thlee-Poussonne,* a short league from *Louvo* eastwards, not far from the forest where the king was hunting elephants". There, on December 11th, King Narai joined the French visitors in observing the heavens through a twelve-foot telescope (fig. 128). On his map of Lop Buri, Bellin marked with the letter 'P' (center-right) the actual site where the observations of the lunar eclipse were made. Narai's guests probably showed him the Milky Way, Saturn's belt, and Jupiter's moons (Choisy recorded that they demonstrated these to the Dutch while at the Cape of Good Hope). Tachard, who witnessed the Lop Buri observation, did not detail the procedures used, but Choisy gave his own account of the methods employed when he participated in an earlier observation at the Cape of Good Hope, *en route* to Siam on June

4th. According to Choisy's summary, the number of degrees the moon moves during a period corresponding to the difference in local time between two points on the earth's surface — in this instance Paris and *Louvo* — when the "occultation" (as he called it) of the moon occurs, equals the difference in degrees of longitude between the two places.

"The King expressed a particular satisfaction," Tachard wrote, "seeing all the spots of the moon in the telescope, and especially perceiving that the type of map that was made at the observatory of *Paris,* agreed so well with it." The technology was a novelty for Narai, but the prediction of celestial events was not; this same eclipse had been predicted to within about a quarter-hour by a "Bramen astrologer" in Lop Buri, although he misjudged its duration.

How accurate were their results? At the Cape of Good Hope, local time was determined to be 74 minutes ahead of Paris, and the moon moved $18\frac{1}{2}°$ during that 74 minute period, and thus the longitude of the Cape was determined to be $18\frac{1}{2}°$ east of Paris, slightly closer to the true figure of 16° than most published maps of the time. In Siam, the observations placed *Louvo* at 121° 2′ longitude east of the eastern tip of the Cape Verde Islands, as compared to the actual figure of 118′, an error of about three degrees.[276] This was reasonably close for those pre-chronometer days, though it is not clear how dependent the figure was on the observation of the eclipse. The limitations of their time pieces, and the fuzzy nature of the lunar shadow, kept them from achieving the accuracy necessary for reliable ocean travel, and thus the race for a means of determining longitude remained a

Fig. 129 Simon de la Loubère, Siam, 1688 (1693) (26 x 13.4 cm). Composed by Jean Dominique Cassini from the surveys of a French engineer, M. de la Mare, who traveled up the Chao Phraya River, and from unidentified Siamese geographic "memorials".

European obsession until John Harrison's chronometer was tested in 1761-62. Choisy, in his often self-depreciating account, joked that he might collect the 100,000 crown reward then being offered by the Dutch for a solution to the longitude problem, were it not for a "tiny but insurmountable difficulty" with his own theories about it.

In contrast to V.O.C. data, the French figures were quickly circulated in Tachard's *Relation of the Voyage to Siam.* In Ayuthaya in 1690, Engelbert Kaempfer, in the employ of the V.O.C., had learned of the Jesuit's calculations, noting that the longitude of Ayuthaya is "according to the common maps 138, but according to the late observations of the Jesuits 120 degrees."

Tachard acknowledged that they were still largely ignorant of the regions that lay to the north: "The provinces that lie up in the country towards the north are but little known, and our geographical maps mark not their situation and limits well." Three years later, in 1688, the embassy headed by La Loubère gleaned what they could about the lands to the north, with the result that the map of the kingdom included in the account of their voyage (fig. 129) finally placed Chiang Mai in roughly its proper position, that is, far to the north and slightly west of Ayuthaya. Though plotted at a latitude of just below 25°, which is about 650 kilometers north of its true position at just below 19°, this still represented an improvement over the traditional placement of Chiang Mai in the Himalayas or on the banks of the Mekong. The map of Asia from the miniature edition of John Speed's atlas (1646) represented the extreme opinion, shoving Chiang Mai *(Iougama)* to the northeast of the Caspian Sea and northwest of the Great Wall.

European Missions to Vietnam

Italian, Portuguese, and French Jesuits began organizing missions to Vietnam in the early seventeen century. Alexandre de Rhodes, a native of Avignon who worked for two decades in East Asia, was sent to Tongking in 1627 by his superiors in Macao. Having already learned the Vietnamese language during a previous stay in the country, he established a mission in Hanoi and claimed to have enjoyed quick and substantial success in conversions (though William Dampier reported that many of the converts reported by Christian missionaries were poor souls who came to the Europeans for rice "in the scarce times"). Rhodes, however, was expelled three years later, largely because the king's wives and their families, who feared what the future would bring if the Christians' edicts against polygamy were accepted, pressured the king to banish him. In 1650, a year after returning to Rome, Rhodes published an account of Vietnam which contained a map of the northern part of the country (fig. 130). Its title invokes the ancient term 'Annam', meaning 'pacified south', a reference to its subjugation under Chinese rule from about 111 B.C. to 939 A.D. The map reflects the image of northern Vietnam as having numerous short rivers, what Choisy described in 1685 as "so short and numerous that no one has given them names."

Another interesting map of Vietnam was composed about this time by Daniel Tavernier, who made eleven or twelve voyages to Tongking during the 1640s from outposts in Indonesia. Daniel died in 1648, but his better-known brother, Jean Baptiste, published Daniel's map in 1679 (fig. 131) as part of an addendum to an account of his own voyages to Asia, which had brought him considerable wealth from diamond trading. The map and its various inscriptions, like its author's account of Tongking, contains both accurate and fanciful information.

Dampier left us his impressions of heading upriver toward Hanoi (*Kecio* on Rhode's map, *Checo* on Tavernier's):

> As we advanced thus up the River, sometimes rowing and sometimes sailing, we had a delightful prospect over a large level fruitful Country. It was generally either Pasture or Rice-fields; and void of trees, except only about the Villages, which stood thick, and appeared mighty pleasant at a distance. There are many of the Villages stand close to the banks of the Rivers, encompassed with Trees on the back side only, but open to the River.

Dampier described the city as a crowded, but well-ordered metropolis divided into 72 sections, each with a wide main avenue that had many smaller side streets. It was the main market for the Red River Delta. While Rhodes' map is devoted to Annam

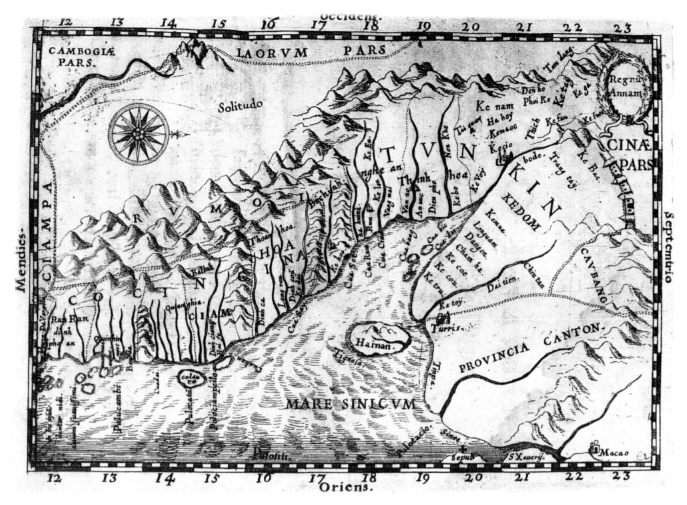

Fig. 130 *Regnŭ Annam*, an early separate map of Vietnam, from A. de Rhodes' *Relazione de' felici successi della Sante Fede nel Regno di Tunchino*, 1650. [Antiquariaat Forum, Catalogue 105]

(northern Vietnam), Tavernier's map covers the full Indochinese coast. Off his very distorted Mekong Delta (lower right), Tavernier has identified the Condore Islands as the *Tortues* after the turtles which were so abundant there. Dampier spoke at length about the green turtles found on Condore and its neighboring islands, and theorized that they might migrate as far as Ascension Island.

After the English shut down their factory in Hanoi in 1697 (the Dutch closed theirs in 1700), the English looked to Pulo Condore as an alternative. Dampier had suggested Condore as a base; stopping at the island in 1687, he noted that "Pulo Condore is the principal of a heap of Islands, and the only inhabited one of them", adding that "these Islands lie so near together, that at a distance they appear to be but one island."

Dampier went on to describe the physical features of the people of Condore, whom he seemed to find appealing, and commented on the freedom of the island's women. But he also had plans for his country there. "These Islands," Dampier explained,

> lie very commodiously in the way to and from Japan, China, Manila, Tunquin, Cochinchina, and in general all this most Easterly Coast of the Indian Continent . . . Any Ship in distress may be refreshed and recruited here very conveniently . . . it might also be a convenient place to usher in a Commerce with the Neighboring Country of Cochinchina, and Forts might be built to secure a Factory; particularly at the Harbour, which is capable of being well fortified.

Dampier included a chart of Condore in his account since, he explains, its geography is little-known. In 1702, as he envisioned, the East India Company did in fact establish a factory on Pulo Condore, but it survived only a few years. In hindsight, Alexander Hamilton, who reached Southeast Asia just after Dampier and remained until 1723, noted that Condore had been "a bad choice for a colony, that island producing nothing but wood, water, and fish for catching."

Coronelli's Ayuthaya

During the 1690s, Vincenzo Coronelli composed a map of the Indies which geographically is identical to that published by Rossi in 1688 (fig. 124). He also produced a fine map of Ayuthaya (fig. 132), the capital of Siam and the most international and cosmopolitan of mainland Southeast Asian cities at that time. Ayuthaya had been established as the kingdom's capital in the middle of the fourteenth century, supplanting the old capital of Sukhothai. Soon after Ayuthaya's founding, a canal was dug joining two river branches which effectively rendered the city an island, tapping the natural river system to create a moat for defense and transportation.[277] Today Ayuthaya stands on a peninsula, the rivers having dried and silted up since the destruction of the city by Burma in 1767.

Joost Schouten, who lived in Ayuthaya for several years in the service of the V.O.C. during the 1620s and 1630s, reported that the city "is situated upon the River Menam, in a little round island, encompassed with a thick stone wall, about six English miles

Fig. 131 Vietnam, by Daniel Tavernier. From J. Chr. Wagner, Delineatio Provinciarum Pannoniae . . . 1684-86. (16.5 x 21 cm)

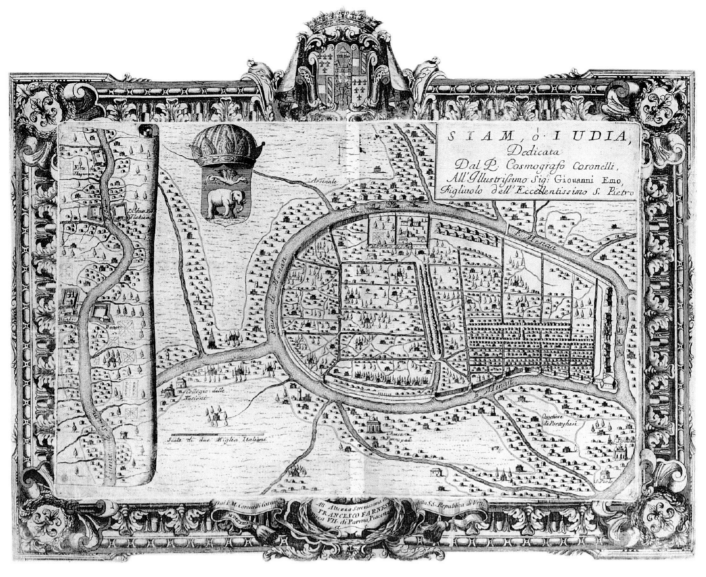

Fig. 132   Ayuthaya, Vincenzo Coronelli, 1696 (borders added by Farnesi, 1698). The inset on the left shows the Chao Phraya River south of Ayutthaya; at the midway point is Bangkok, marked on both sides of the river.

round." He added that "the suburbs are on the other side of the river, closely builded, and full of temples and cloisters, lying in a flat and fruitful country."

Gervaise, who visited the city a half century later, reported that the king was having a new wall built, as the old one was "falling into ruin", and indeed the city's fortifications were undergoing modifications under the guidance of a French Jesuit. The city was designed around a series of intersecting canals and streets which connected various districts within the walls. At the top (north) end of the island on Coronelli's plan there is a double-square enclosure which, Gervaise explained, is the royal palace "enclosed by a double wall of brick, which is always very well maintained." Within these walls are kept the royal elephants, including the mystical, venerated 'white' ones. Both the king's apartment and that of his wives and the princess-queen looked out onto large, well-tended gardens with walks that

> are intersected by little streams, which make everything
> fresh, and the gentle murmur of their waters lulls to sleep
> anyone who rests on the evergreen grass on their banks.

Outside the island, in the lower right, Coronelli has indicated the quarters for the Portuguese, who in 1512 had become the first

Europeans to trade directly with Siam, and had set up their own station in the city in 1602. Nearby are two Christian churches, the Santos Domingo (Dominican Order) and St. Paul (Jesuit).

Jacques de Bourges, resident in Ayuthaya for a year from 1662-63, noted one benefit of Ayuthaya's being comprised of islands:

> as the city of Siam is divided into several islands by canals
> forming the river, care is taken to place each nation on an
> island or in a separate quarter, so that there are few
> quarrels which are often caused by the mixture of
> nationalities who have natural antipathies.[278]

Only the French were permitted to take up residence within the city walls. Though not marked by Coronelli, a comparison of his map with that of La Loubère would place the French quarters in about the six o'clock position on the oval form of the city. At eight o'clock one finds the *Collegio delle Nationi,* which was a French school for boys. At twelve o'clock, on the river outside the palace grounds, Coronelli has indicated a few of the magnificent royal barges which so captivated European visitors to the city. These barges are seen in better detail in an engraving from Tachard's record of the 1685 French Embassy to Siam (fig.133). A description from Jacques de Bourges' account follows:

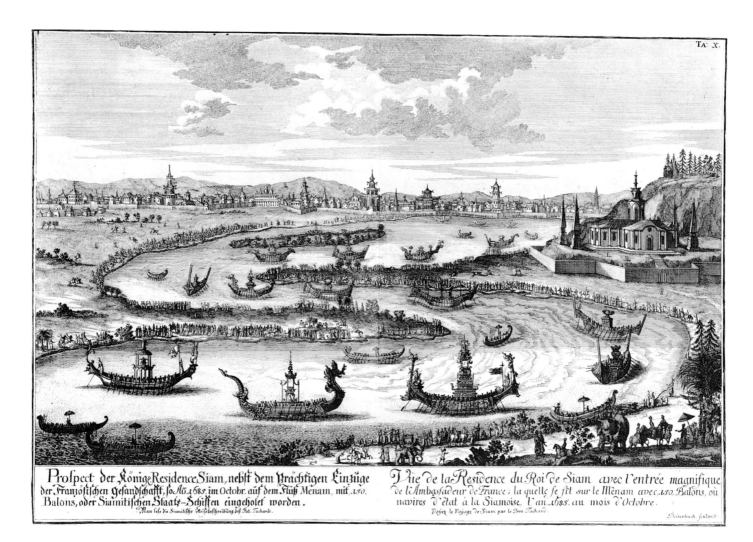

TA. X.

Prospect der Könige Residence Siam, nebst dem prächtigen Einzüge der Französischen Gesandschafft, so Ao.1685. im Octobr. auf dem Fluß Menam, mit .150. Balons, oder Siamitischen Staats-Schiffen eingeholet worden .
"Man lese die Siamitische Reisebeschreibung des Pat. Tachard.

Vüe de la Residence du Roi de Siam avec l'entrée magnifique de l'Ambassadeur de France ; la quelle se fit sur le Menam avec .150. Balons, où navires d'état à la Siamoise, l'an .1685. au mois d'Octobre .
Voyez le Voyage de Siam par le Pere Tachard.

Heissenbach sculpsit

Fig. 133 View of vessels at Ayuthaya, from the voyage of Guy Tachard. "There are no words sufficiently magnificent to express what I have just seen," wrote Abbé de Choisy in 1685. "The procession numbered more than 200 barges . . . The King's barges were rich and handsome above imagining. There were in most of them 150 oarsmen with paddles gilded overall." The king, covered with jewels, dined at "a country house he has a good two leagues from here," and then "all the barges lined up, each according to the rank of its occupants . . . It is something to watch, the speed with which they move up the broad, swiftly flowing river." Engraving from a book of architectural plates by Johann Bernard Fischer von Erlach, 1725. (29.5 x 42 cm)

When [the King of Siam] goes by land he is carried on an elephant in a howdah [tower] resplendent in precious stones and the number of his suite reaches 10,000 persons. But the procession still more grand is that which is made on water, because of the great throng of galleys which accompany him, numbering more than 300 or 400. They are all gilded inside and out, and each carries thirty or forty oarsmen, some of whom also have their arms and shoulders gilded These oarsmen cut through the waters with extraordinary speed, and the banks of the river of Siam echo with the sound of the waves caused by the passage of the oars.

The King's galley sparkles with very fine gold which goes even down below the waterline; above it is a magnificent throne. The King appears there dressed in precious robes, covered with a crown of solid gold decorated with fine diamonds. Two golden wings hang from this crown onto the King's shoulders. All the lords and officials of the crown follow the King in their covered barges, each decorated according to their power, their means and their offices. The banks are lined up with great crowds of people who fill the air with shouts of joy. The ruler, in order to appear no less pious than powerful, does not omit on such ceremonial occasions to visit some famous temple and make magnificent presents to the priests who officiate the devotions.

Holland in the Latter Part of the Seventeenth Century

By the mid-seventeenth century most of the Indonesian Islands, except for Borneo, had come under V.O.C. control. Even Ternate had become Dutch turf when the Spanish relinquished that island in 1663. Having thus consolidated its supremacy in the commerce of the Indies, in 1664 the Company effectively declared itself a sovereign state:

the places [the Company has taken] in the East Indies ought not to be regarded as national conquests, but as the property of the individual merchants who have the right to sell them to whomever they wish, be it to the King of Spain or to another enemy of the United Provinces."[279]

HEDENDAEGSCHE
HISTORIE;
of
TEGENWOORDIGE
STAET VAN ALLE
VOLKEREN
Twede Deel.

TE AMSTERDAM
By ISAAK TIRION MDCCXXX.

Fig. 134 Titlepage by Isaac Tirion, 1730 (10 x 16 cm). The dream of Dutch omnipotence in Southeast Asia is bluntly portrayed in the title page to a 1730 atlas of Isaac Tirion. Subservient, groveling "natives" happily offer symbolic riches of their lands to a crowned semi-deity figure representing the all-mighty V.O.C, whose insignia is displayed on the shield. The gifts relinquished by the dark "savages" include assorted spices, gold, and ivory, but most importantly a rolled wall map of Southeast Asia itself. By offering up the map to the Company, the Southeast Asians symbolically relinquish their lands to the V.O.C., just as the Javanese prince had surrendered a map to his Mongol conqueror in 1293, symbolizing his abdication of sovereignty. [Lent from a private collection] .

This was a yet another reworking of the idea of Southeast Asia on the map. At the turn of the sixteenth century, Europe first saw the region as perhaps rightfully belonging to Christendom, then to conquering nations, then to chartered 'companies' which functioned as semi-autonomous states. Finally, in this last instance, the V.O.C. has come to see the Indies as the property of the Company itself or even of its individual merchants who were not accountable to their home country; each merchant of influence in the Company was in effect deemed a potential sovereign power.

The pivot of Holland's Southeast Asian empire was Batavia (fig. 135). Just as Ayuthaya was considered the most grandiose 'native' metropolis in Southeast Asia, Batavia was the most spectacular 'European' city, set against a back drop of dense forests, with white walls which shone in the sunlight, and streets lined with coconut palms, mango, and citrus trees. Abbé de Choisy, who visited Java in 1685, wrote that Batavia "is like all Dutch towns: white houses, all the streets between two canals, handsome green trees, well-paved paths for the gentry, the middle of the roads well sanded." The nerve center of Batavia (and of the entire V.O.C.'s Indies operation) was its castle. The castle lay at the east end of the harbor, and had walls extending to the sea.

Visitors consistently reported that the harbor was full of ships and smaller vessels, while the streets were lined with restaurants and shops, including (we learn from Abbeé de Choisy in 1685) a bookseller.

At the dawn of the eighteenth century the Dutch, from their headquarters in Batavia, controlled Java, Formosa, Makassar, and the Moluccas, with indirect control over many of the other Indonesian islands. The Philippines and Flores were closed to the Dutch, while an English fort remained in Sumatra despite Dutch opposition to it. The V.O.C. approach to controlling the vast and varied islands of Indonesia was by delegating responsibility to native overlords. In general, the Dutch only wielded the Bible when 'necessary', leaving the Indonesians to their own ways when this did not affect commercial concerns. On those islands whose infrastructure the V.O.C. was able to harness, the Company's disinterest in indigenous affairs helped to preserve Indonesian cultural traditions intact through these centuries. A telling example is the fact that although Dutch traders frequently traveled between Batavia and the Javanese courts at Yogyakarta and Solo (Surakarta), and in doing so passed close to Borobudur, there is no record whatsoever of these colossal ruins until the British arrived in the early nineteenth century.

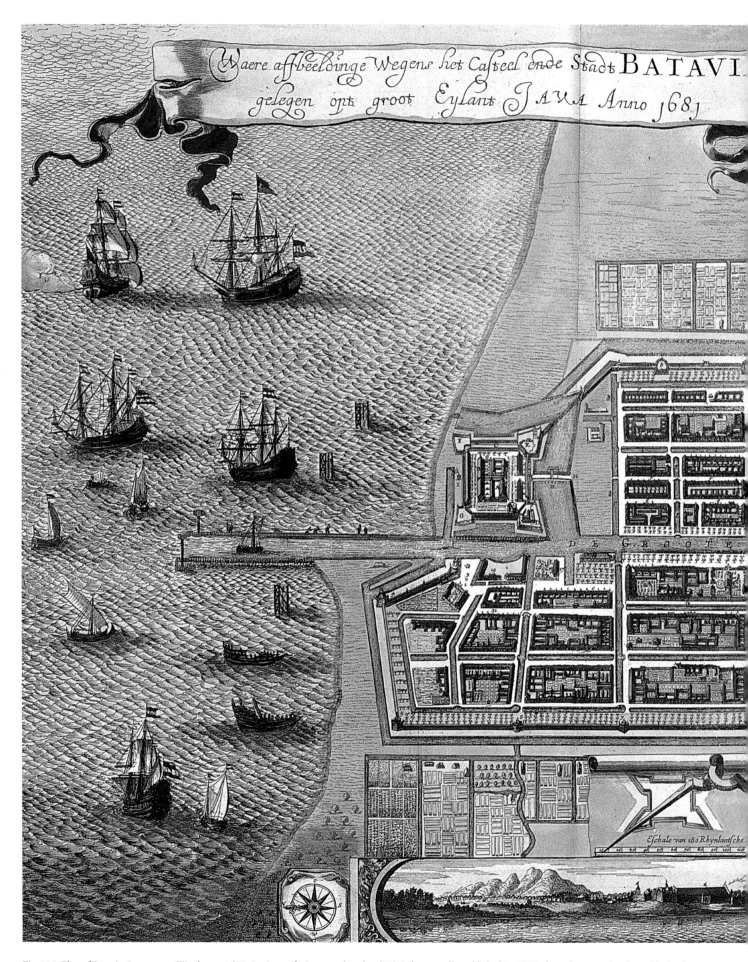

Fig. 135 Plan of Batavia, Jansson van Waesberge, 1682. A reissue of a Jansson plate dated 1652 (but actually published in 1657), from the same plate but with the date reworked to read 1681, and in turn based on a 4-sheet view of 1650 by Clement de Jonghe. (41 x 52cm.) [Edwin Rahardjo, Jakarta]

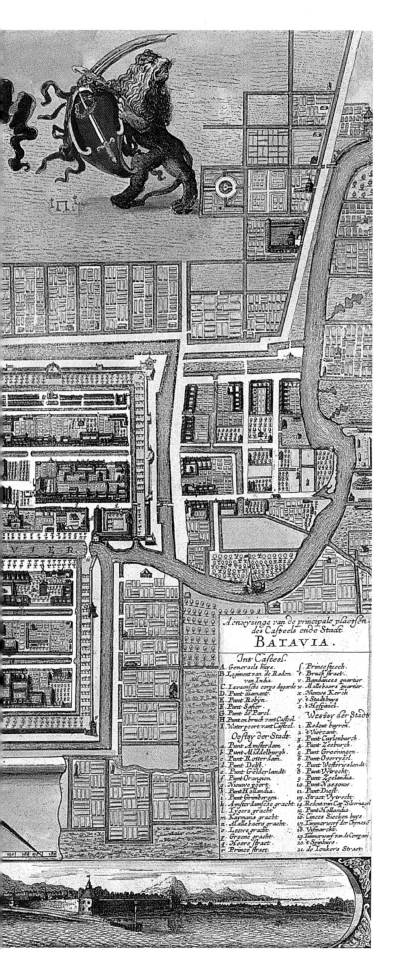

Following page: Fig. 136 A 1679 copy of Willem Blaeu's wall map of Asia (1608) by Alexis-Hubert Jaillot (this is a re-issue of 1685). [Martayan Lan, New York]

The V.O.C.'s policy was conceived as an efficient and expedient mechanism for controlling numerous islands dispersed over a vast sea, and did not reflect any particular Dutch regard for Indonesian civilization. Calvinism did come to play a role when the V.O.C. decided that the islanders would be less likely to rebel if they were lambs of the same flock. Efforts made to bring the Muslim, Catholic, and 'pagan' population of the Moluccas under the Calvanist helm to avoid rebellion, had quite the opposite effect, in that it led the sultan of Makassar, which was an international spice mart and a bastion of Islam in the region, to prepare for war against the Dutch. Unable to persuade the sultan to see things their way, the Dutch attacked Makassar in 1666, eventually forcing the sultan into an agreement in 1669 which granted the V.O.C. a monopoly on Makassar trade and a fortress on the island.

Dutch Wall Maps

Among the grand works which had established Willem Blaeu as a master at his craft in the early part of the century were his wall maps. The genre was not new, however. Magnificent *mappaemundi* had adorned the walls of the wealthy, the State, and the Church at least since medieval times, and maps reflecting new discoveries and voyages of exploration were designed by top cartographers and painted by artists on the walls of the privileged. Prominent mapmakers such as Waldseemüller, Gastaldi, Mercator, and Plancius had all made large maps which would have been unwieldy anywhere except on a wall. Blaeu popularized (though he did not invent) a new aesthetic of wall maps, the *carte à figures* (fig 136), whose borders were decorated with vignettes of natives in colorful costumes and other exotic scenes, a genre which was copied throughout the seventeenth century by fellow Dutchmen and by Italian imitators.

During the latter part of the seventeenth century, the aesthetic element of wall maps evolved, keeping pace with contemporary developments in the field of painting. Mapmakers, no doubt influenced by the Dutch artists of the day, came to see the map not only as a canvas for the elegant recording of cartographic data, but also as a landscape for artistic portrayal of other kinds of images. The latter were often allegories based on classical figures, but in other instances they were more relevant to the map itself, attempting to represent life in the part of the world depicted on the map, even if only in an idealized fashion. Southeast Asia offered a particularly fertile source of images for these highly-embellished maps. The region had become intimately entwined with Dutch life, and its people provided exotic subjects for map artists to portray.

Two exceptional examples of such works are those of Hendrik Doncker of ca. 1664 (fig. 2), and Peter Goos of ca. 1680 (fig. 137). They share a similar geography which is typical of the period. Insular Southeast Asia is nearly identical on the two, with the exception of the island of Luzon, while the mainland reveals some differences in the Gulf of Tongkin, northwest Malaya, and in the placement of the major mainland kingdoms.

Maps like these were much like the portolan charts individually crafted by manuscript chartmakers, but they were embellished so as to appeal to wealthy, land-bound patrons and engraved on copper so that they could be produced in quantity. From the copperplate, impressions were made both on paper and on vellum. While paper was less expensive and better suited to inclusion in an atlas, vellum offered far greater durability as a loose-sheet map, and was also particularly well suited for sea voyages. The fact that geographic knowledge had advanced to the stage that it was now relatively stable made such expensive maps more commercially viable, since they could be produced with greater confidence that they would not quickly become unmarketable.

NOVVELLE DESCRIPTION DE TOVTE L'ASIE.

CARTE DE L'ASIE
Nouuellement dressée sur les Memoires des meilleurs Geographes de nostre temps, et distinguée suiuant les Royaumes souuerainetez, et Principales parties qui se nomment aprent

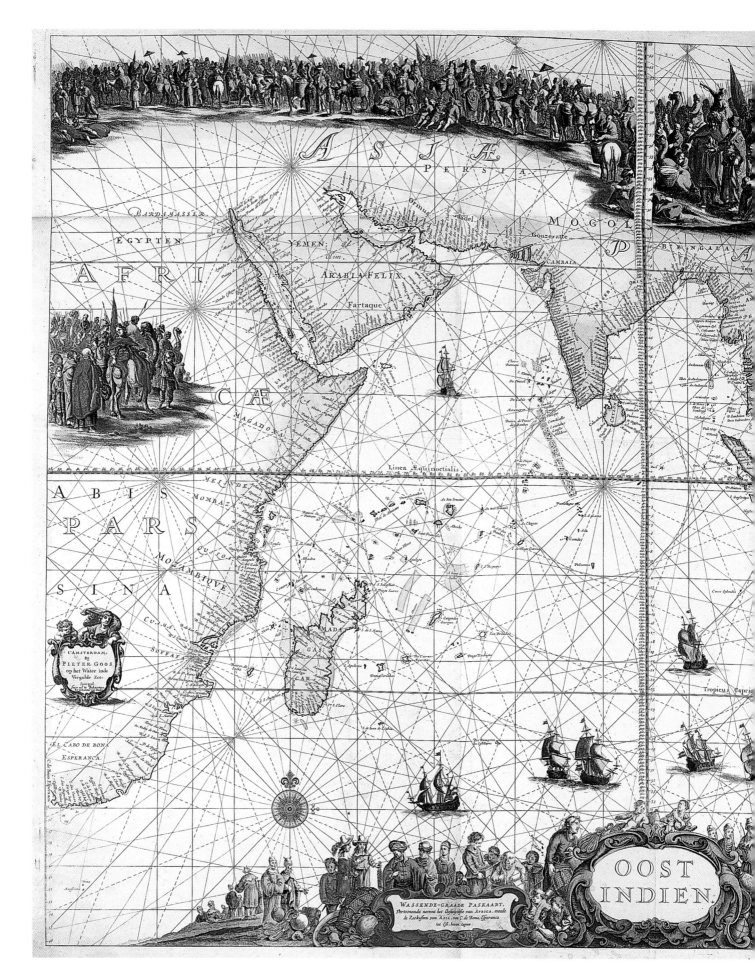

Fig. 137 Southeast Asia, Abraham Goos /Van Keulen, 1680 [Robert Putman, Amsterdam].

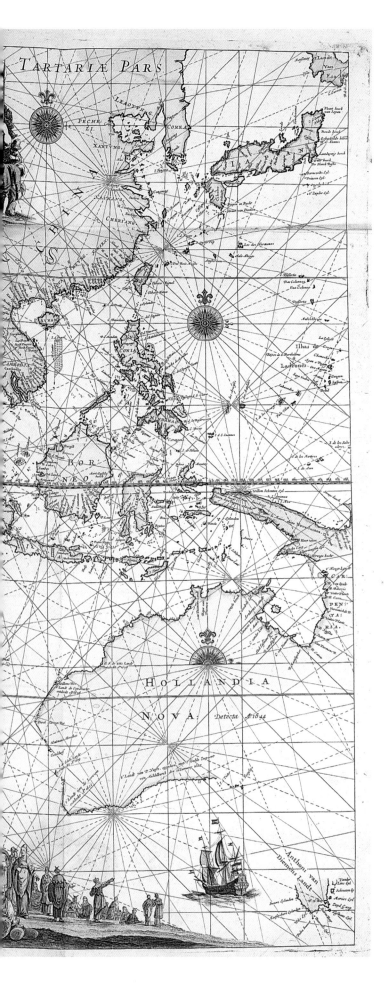

The End of the Blaeu Legacy

In early 1672, in the midst of the boom in printed sea-atlas in Holland, the Blaeu publishing firm suffered a tragic fire that destroyed its premises, equipment, and many of its copper-plates. Joan Blaeu died the following year. His heirs continued the business on a limited basis while the inventory and plates that had survived the fire were sold in a series of auctions. A final auction in 1695 liquidated the remaining holdings and ended what was perhaps the most famous map publishing house of all time. The Blaeus had a profound effect on the course of commercial map-making, producing spectacular wall maps, atlases, and globes, and in their inner circles they compiled data from V.O.C. pilots. Although Willem Blaeu's *India quæ Orientalis* of 1635 was a fine general chart of Southeast Asia for its day, neither Willem nor Joan Blaeu had a serious impact on public knowledge of the Indies. Given that Lucas Waghenaer, whose pilot book inspired Blaeu to produce his own, had managed to publish a chart of the vital Sunda Strait region back when the Dutch grasp of the Indies was in its infancy, Blaeu's failure to parallel even that achievement is a thunderous silence indeed.

Spain Looks To Micronesia

The latter part of the seventeenth century was a rather different period of transition for the Spanish in Southeast Asia than for the Dutch, English, or French. Although the authorities in Manila and Madrid had no serious thoughts of expansion to the mainland, the various archipelagoes of Micronesia were beginning to attract their attention. Micronesia was ideally situated as a way-station for the Mexico-Philippine voyage, and there was no serious contest for them from other Western powers. These islands, however, promised little in material gain, their only obvious commodity being heathen souls in need of redemption.

Spanish encounters with the islands of the Marianas, Carolines, and Marshalls began with the first trans-Pacific Spanish voyages. Later, the English circumnavigations of Drake and Cavendish skimmed the islands, and by the early seventeenth century Dutch vessels approaching the Moluccas by way of the Pacific used Guam as a stopover to replenish supplies. They continued to provision galleons operating between Manila and Acapulco, but played only a minor part in Europe's evolving image of Southeast Asia. The English geographer Robert Morden reflected a typical attitude in his *Geography Rectified* (1688), writing that "I find nothing memorable [about the Ladrones], except their names in some maps."

In 1662, a Jesuit named Diego Luís de Sanvítores, stopping at Guam on his way to Manila, decided that the salvation of the islanders' souls was his destiny. The *Ladrones* themselves, apart from the value of their location for vessels crossing the north Pacific, must have seemed like a disappointing prize compared to the 'rich' lands which had brought the Spanish to the Indies in the first place, and the authorities in Manila reacted coolly to Sanvítores' proposal to establish missionary stations there. But within the same year, Spain lost its grip on the Moluccas and the southern Philippines, making expansion to Micronesia a more attractive option. When Sanvítores beseeched the help of Madrid and Rome, his pleas were relayed to King Philip IV and Mariana of Austria. In 1667, permission and funds for a mission in the *Ladrones* was approved by the Spanish court. Philip died three months later, leaving Mariana as the queen-

Following page: Fig. 138 Joan Blaeu II, 1697, South China Sea Manuscript on vellum. Made for the V.O.C. (72 x 107 cm) [Universiteitsbibliotheek Amsterdam]

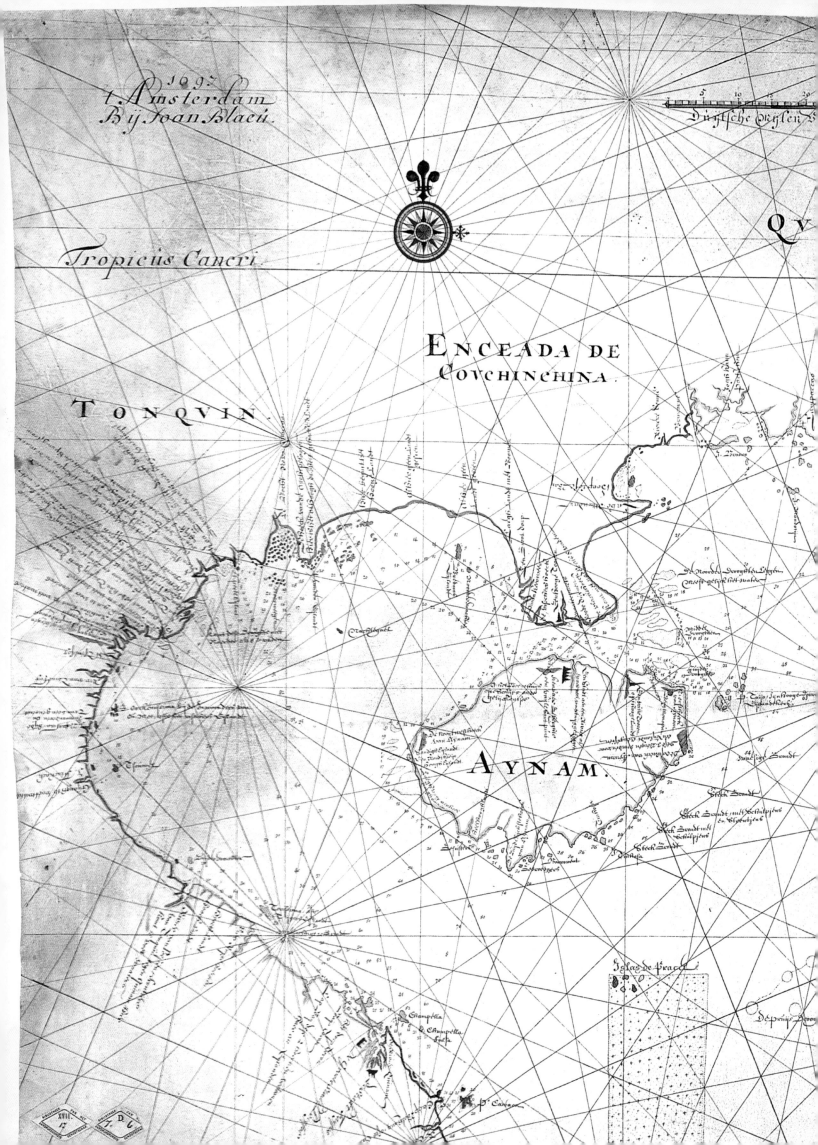

t'Amsterdam
Bij Ioan Blaeu.

Tropicus Cancri

ENCEADA DE
COVCHINCHINA.

TONQVIN.

AYNAM.

QV

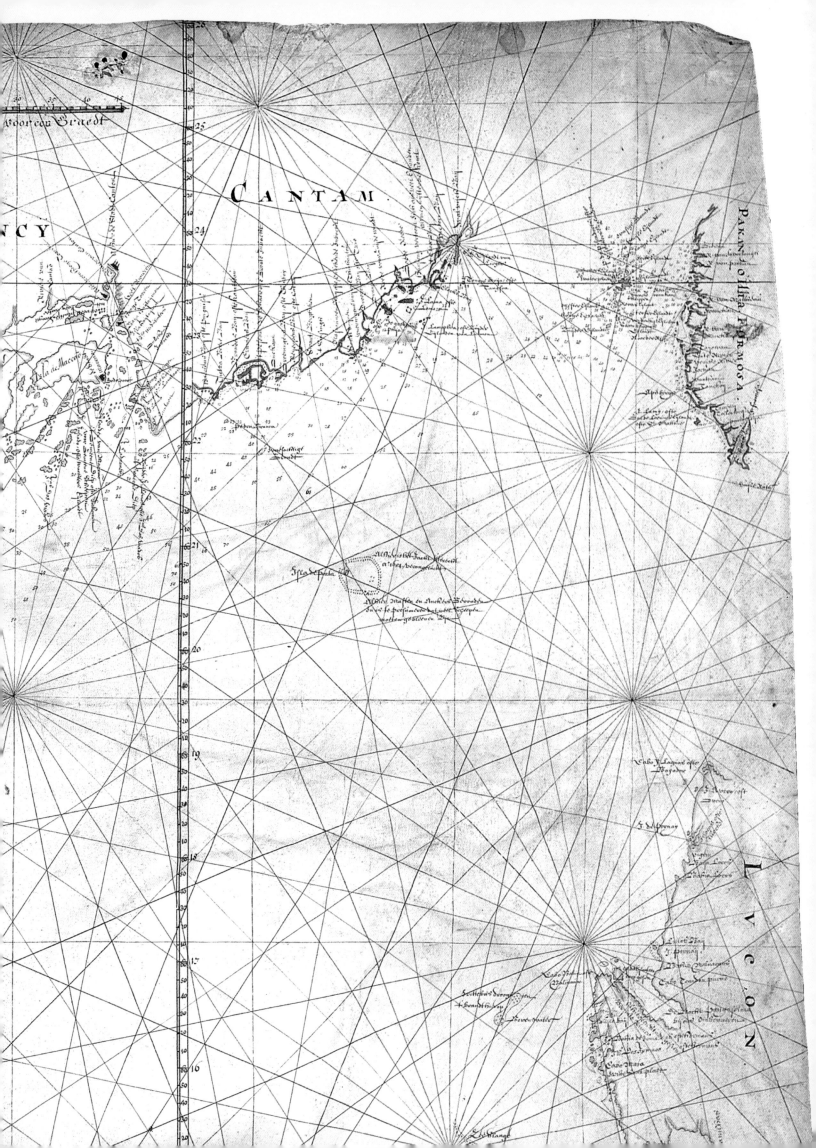

regent, and Sanvítores proposed renaming the archipelago after her, an appellation which has survived to this day — the Marianas.

Thus began Spain's turn from the fabled points to the west and south of her Philippine empire to the 'poor' islands to her east — islands which nonetheless can conjure up the poetry of Paradise as easily as the mainland and Indonesian kingdoms that Spain had failed to attain. The next quarter-century, however, was hardly paradisiacal in the islands. Much of the Chamorros population succumbed to epidemics resulting from imported diseases, as well as from the violence which ensued when the people rebelled against the Spaniards' obligatory 'conversions', 'baptisms', and social mandates contrary to their culture.

This epoch, however, put the Marianas on the map more prominently than they had been before. Whereas the traditional itineraries in the Indies were acknowledged for their purported riches or their strategic location, the Marianas become the subject of prolific writings because of the martyrdom of its Jesuits. These reports began well before the survivors of the resistant Chamorros finally surrendered to the Spanish in 1695. For the Jesuits who lamented the failure of the missions on the Southeast Asian main-land, the Marianas had become a symbol upon the map — they showed that with the proper military and financial support, even the most recalcitrant and obstinate folk could be 'converted' to Christianity. Maps of the archipelago were made in 1671 by Alonso Lopez, which were used in an account by Charles Le Gobien published in 1700.

"These islands," the *Philosophical Transactions* reported in 1721, "are situated between the Molucca's, the old Philippines, and the Marianas. They are eighty-seven in number, and make up one of the finest archipelagos in the East." The engraved map of Micronesia included in the volume (fig. 13) was composed from indigenous information gathered from the islanders themselves, who arranged stones to form a map of the archipelago, and then gave the number of days' sailing time around each, and number of days' sailing time between the islands (see page 41, above). Ironically, Palau *(Panlog)* and Sonsorol *(Sarol),* our two candidates for the original, real island of *San Juan,* by coincidence now lie immediately to the northeast of Mindanao, in the very position long occupied by the bogus *San Juan.*

Seven years later, another Jesuit letter contained a map of Micronesia which focused on the "New Philippines", and introduced the modern name *Carolines,* as given by Cantova in 1722 (fig. 139) It depicts Yap (upper left circle) and Palau (lower left circle) with reasonable accuracy, but, by unwittingly combining two reports about Ulithi, duplicates that archipelago, once as *Egoi* (elliptical shape) and again as *Lamürrec-Ulee* (large circle). In latitude, and in longitude relative to Yap and the southernmost point of Samar (Philippines), the western Ulithi is more accurate. Among the islands in the western part of the map, lie Truk (Chuuk) and Ponape (Pohnpei).

Germany

While German mapmakers of the sixteenth century, such as Andreas, can be grouped together with the cartographers of the Low Countries as regards their general content and style of their work, a more individual genre of mapmaking emerged in Germany in the early eighteenth century. A thematic map of Southeast Asia by Heinrich Scherer, ca. 1700 (fig. 140), purports to show fauna of the region, including a Bird of Paradise to the east of the Philippines. The creature is shown without wings or feet, for according to Linschoten, they "keep themselves continually in the air, without lighting on the earth, for they have neither feet nor wings, but only head and body, and the most part tail." Ortelius depicted the bird on his 1567 map

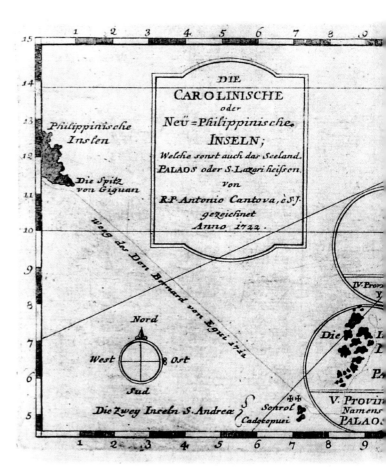

Fig. 139 The Caroline Islands from *Allerhand so lehr-als gesit-reich Brief, Schrifften und Reis-Beschreibungen ...,* vol.3, 1728. Longitudes are calibrated from teastern Samar in the Philippines. (14.5 x 39cm)

of Asia, and a southern constellation noted by Vespucci (in South America) and Corsali (in Southeast Asia) was named after it.

Below the Bird of Paradise is a flying fish, which according to Linschoten "fly by great flocks together, two or three fathoms above the water, and fly in that manner at the least a quarter of a mile, until their wings or fins be dry, and then they can fly no longer." The reason for their flight, Linschoten believed, was to escape from predatory fish. Some of Scherer's fauna is utterly alien to the region. In Vietnam he places a two-humped camel — an error committed even by Gerard Mercator on his Ptolemaic map of Southeast Asia, 1578 — and in the waters below Java he puts a narwhal.

Scherer prominently shows zero magnetic declination on large rose compass, perhaps having noted the comments about negligible declination on the Philippine chart of Robert Dudley, or, more likely, from Edmund Halley's determination that the isogon of zero deviation ran east of Palawan and through Sulawesi.

Johann-Baptist Homann, a prodigious publisher of atlases, typified German mapmaking eighteenth-century. His style is characterized by a strongly engraved map area with a bold wash of color and elaborate, etched cartouches, left uncolored. Interestingly, one of his early depictions of Southeast Asia is found on a map of Holland. In the early 1700s he published a map of Holland and its extra-territorial possessions. Maps of Southeast Asia and the New Netherlands, plus views of Batavia and New York, are all featured as insets. Curiously, these insets were removed for later issues of the map — the New Netherlands map and the New York view were probably deleted because England had succeeded to the city, and the Batavia view and Dutch Indies map rubbed out with them.

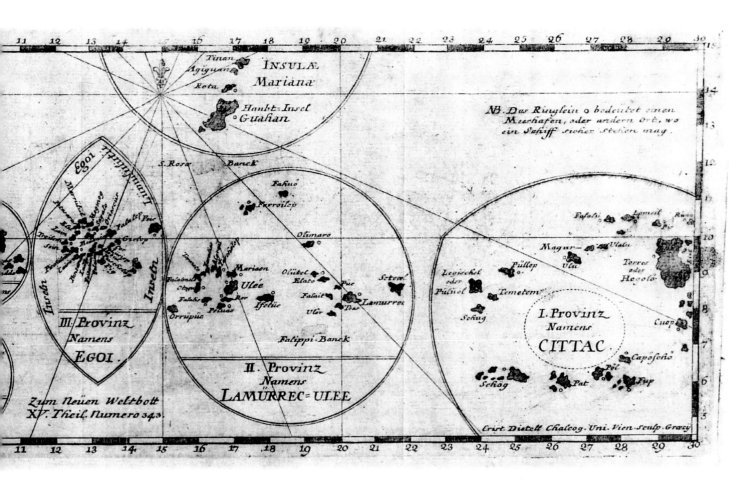

Fig. 140 Southeast Asia, Heinrich Scherer, 1700 (22.7 x 35.5 cm).

Chapter 15
The Eighteenth and Early Nineteenth Centuries

Despite Western tendency to view the history of Southeast Asia since 1500 as the history of European involvement in the region, the influence of the West was for the most part superficial up until the middle of the eighteenth century. At that time, the political map of Southeast Asia consisted of some forty major kingdoms or states, each possessing varying degrees of autonomy or vassalage in relation to one or more of the others, plus a great many other smaller political units. The various wars, rebellions and conquests which continued between and within the nations of Southeast Asia were seldom reflected in European maps, unless they affected alliances or schemes with which Europe was actually involved. These alliances apart, Southeast Asian life and culture remained little affected by Europe, save for some of the smaller Indonesian islands where the interests of the V.O.C. had compelled the Dutch to supplant the indigenous infrastructure, as well as Portuguese Timor, and the northern Philippines, which had truly become a Spanish colony.

In the early eighteenth century, the V.O.C. began to relax its policy of secrecy as regards its storehouse of geographic data relating to Southeast Asia. The earliest indication of this new openness came in 1719, with a superb chart of Java published by Henri Châtelain in his *Atlas historique*. This large-scale map is substantially more accurate and detailed than any previously published map of the island; even the southern coast of Java, which had long been a weakness in European knowledge, is now well-defined. Nine years later, in 1728, the V.O.C.'s hydrographer, Johannes Van Keulen II, published a chart of Java on two sheets (fig. 1); though more elaborate in execution, it is geographically identical to the Châtelain.

At the beginning of the eighteenth century, one could still purchase archaic maps of Java, such as the one by Peter Vander Aa illustrated in figure 141, which had been frozen in time for more than a century. Even after the publication of the Châtelain and Van Keulen charts, a history of the Dutch Indies by Renneville continued to identify Java as *Java Major* and Sumbawa as *Java Minor*. Furthermore, the map offered no clue as to the delineation of Java's southern coastline, while Lombok was omitted altogether.

François Valentijn and Johannes Van Keulen

A comprehensive mapping of Southeast Asia is found in the work of François Valentijn, a minister who devoted most of his life to the employ of the V.O.C. In 1685 Valentijn, then age 19, was sent by the V.O.C. to Ambon as a Minister to the East Indies, where he remained for a decade. After another ten years (1705) Valentijn, now married and with five children, returned to the Indies with his family. The following year Valentijn was Army Chaplain on an expedition in eastern Java, but suffered health problems and requested permission to return to Holland. This was denied, however, and he remained in the East until 1714. Finally back home, Valentijn composed his monumental work, *Oud en Nieuw Oost-Indien,* which was published between 1724 and 1726.

The *Oud en Nieuw Oost-Indien* was created both from the voluminous journals Valentijn had amassed during his two stays in Southeast Asia, as well as from his own research, correspondence, and from previously unpublished material secured from V.O.C. officials. The work contained an unprecedented selection of large-scale maps and views of the Indies, many of which were superior to previously available maps. Some of his maps, however, parroted older models which had long grown stale — an example is the chart of Bali, which repeats, on a larger format, the configuration published in 1598 by Lodewijcksz and de Bry. We will look at two of Valentijn's most interesting maps, one of the mainland, and one of the islands.

Valentijn's Menam River (Thailand)
Perhaps the most intriguing map in Valentijn's work is one devoted to the Chao Phraya River region of Thailand, and neighboring upcountry regions (fig. 142). The grandest of early maps of the river and adjacent territories, its stylized, pictographic quality, and its seeming lack of regard for uniform scale, all suggest that it may have been inspired by an indigenous map.

The map's title, *De Groote Siamse Rievier Me-nam,* continues the old confusion about the term *menam*, which is not the name of the river as such, but simply the Thai word for 'river' (literally, 'mother of waters'). Gervaise had already brought this error to light in his 1688

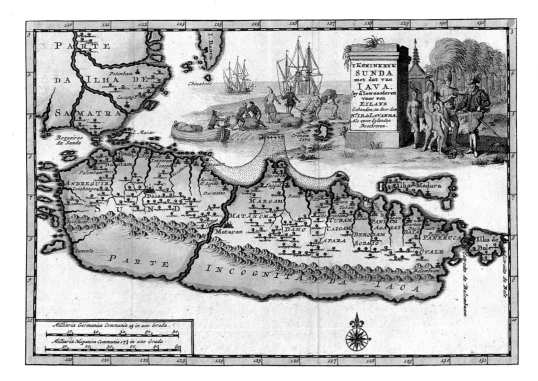

Fig.141 Java, from Peter Vander Aa's atlas entitled *Atlas nouveau et curieux des plus celebres itineraires,* 1714. Note that the southern coastline of Java is still designated as unknown, and that an imaginary strait bisects the island into two parts.

book about Siam. "The authors of earlier accounts of the kingdom call the great river of Siam the Méenam, but this is because they are ignorant of the language of the country, for in Siamese *Méenam* simply means a river . . . since the people believe that the rivers actually create their own water, they call them by this name, which means 'mother of waters'." 'Menam', however, continued to be used even into the nineteenth century by British envoys in Siam.

What was a trip along the river like in those days? The French missionary Jacques de Bourges left this description of his journey up the *Menam* from the Gulf of Siam to Ayuthaya (marked as the large walled city in the center of the map), in 1662:

> We traveled on the river with three small boats covered with palm leaves, each of which had three men to man it. These boats are normally made of one piece, of large tall trees hollowed out by fire, and which are a good twenty feet long. They are very appropriate for these rivers, which are swift, and barred by many rapids; that is why the boats, being exposed to rude blows, would be quickly smashed if they were made of several parts. As trees of this thickness are lacking in France, it would be difficult to copy the skill of these boats, which have this convenience that they cost little to construct. We gave for each about twelve écus. One has to prepare one's food and sleep in the boats, it being too dangerous to leave them and to stop on land, because they comprise uninterrupted forests full of lions, boars, tigers, and other carnivorous beasts. We went upstream with much difficulty, because of the rapidity of the flow and the waterfalls blocking its passage. Then the boatmen get into the water, to shift the boats by brute strength, some pulling them with ropes, others pushing them with long poles, yet others almost carrying them on their shoulders, so difficult is it to fight the impetuous waters, which pass between the rocks and rush like our millraces.

Valentijn's map depicts the river almost as a series of mazes, as indeed it appeared to visitors like Gervaise. A local, mental map of the Chao Phraya River was important, Gervaise wrote, for "the meanderings of the river . . . [divide it] up imperceptibly into a labyrinth of countless branches, in which, if one does not have the secret, one is in danger of getting lost."

Bangkok

Bangkok is marked in the accompanying key as no. 12, where the western (upper) and central branches of the river emptying into the Gulf of Siam meet. Given its position on the main waterway of the country, it is easy to understand why Gervaise declared Bangkok to be "the chief key to the kingdom of Siam," and why it became the capital after Ayuthaya was destroyed. The Dutch, in fact, had previously tried to exploit this 'key' back in 1664, when the V.O.C. blockaded the lower Chao Phraya in an attempt to exact a monopoly from Ayuthaya on the trade in hides. Gervaise offers a premonition of the court's ultimate move to Bangkok, noting that it is "assuredly the most important place in the kingdom of Siam, for it is the only place anywhere on the sea coast that could offer some resistance to enemy attack."

La Loubère reported that Bangkok lies about seven miles up the *Menam* river from the sea, and that it is called *Fon* by the Siamese. He confesses that he does not know where the name Bangkok came from, but correctly observes that the names of several Siamese towns begin with *ban*, as it means 'village'.

Kaempfer passed through the city in 1690 e*n route* to the V.O.C. factory just south of Ayuthaya. He noted that "the banks above Bangkok are pretty well inhabited, and stock'd with houses and villages." Nearby, the V.O.C. had a "Dutch Habitation and Store-House called Amsterdam" which was two leagues from the sea and "built according to the fashion of the Country on piles of *Bambous.*" A fort erected earlier by the French at Bangkok had been demolished.

In 1821, four decades after Bangkok had become the kingdom's capital, the British naturalist George Finlayson described Bangkok as a modest metropolis. Although the city stretched for a fair distance along the Choa Phraya River, "the houses rarely extend more than one or two hundred yards from the river, and by far the greater number of them are floating on bamboo rafts secured close to the

Fig 142   Chao Phraya River Valley, upcountry through Laos, Valentijn, 1724-26. (29.5 x 73.5 cm)

bank." Finlayson adds that "to every house, floating or not, there is attached a boat, generally very small, for the use of the family [while] the few streets that Bangkok boasts are passable on foot only in dry weather."

What of the region that lay further upstream, between Bangkok and Ayuthaya? "Betwixt *Bancock* and *Siam*," reported the first French embassy in 1685,

> you meet with a great many Aldees or Villages, that almost every where border the River. These Villages are no more than a great many Huts or Hovels raised upon high Pillars, because of the Inundation. They are made of Bambous, which is a Tree whose Timber is much used in that Countrey. The Trunk and great Branches serve for making Pillars and Joysts, and the small Branches to makes the Walls and Roof. Near the Villages are the Bazars or floating Market-places, where the Siamese, who go up

or down the River, find their Victuals ready drest; that's to say, Fruit, boyl'd Rice, Rack (which is a kind of Strong Water made of Rice and Lime) and some Ragousts after the Siamese Mode, which a French-man could not taste.

On the northern (right-hand) extreme of the Valentijn *Menam* map one finds the town of *Lipü*, which is *Louvo*, or Lop Buri, some 155 kilometers north of Bangkok. Gervaise aptly described this town as being "to Siam what Versailles is to France." Being situated on a plateau, Lop Buri was raised above the level of the flood waters that invariably plagued the region during the rainy season, and in this respect served as a fine country retreat for the king at that time of year. "Its situation is so agreeable," wrote Gervaise, "and its air so pure that no one ever leaves it without regret." The map also depicted, albeit out of scale, various upcountry towns and regions about which European visitors knew little more than what could be gleaned from native sources. Some of the many towns indicated by

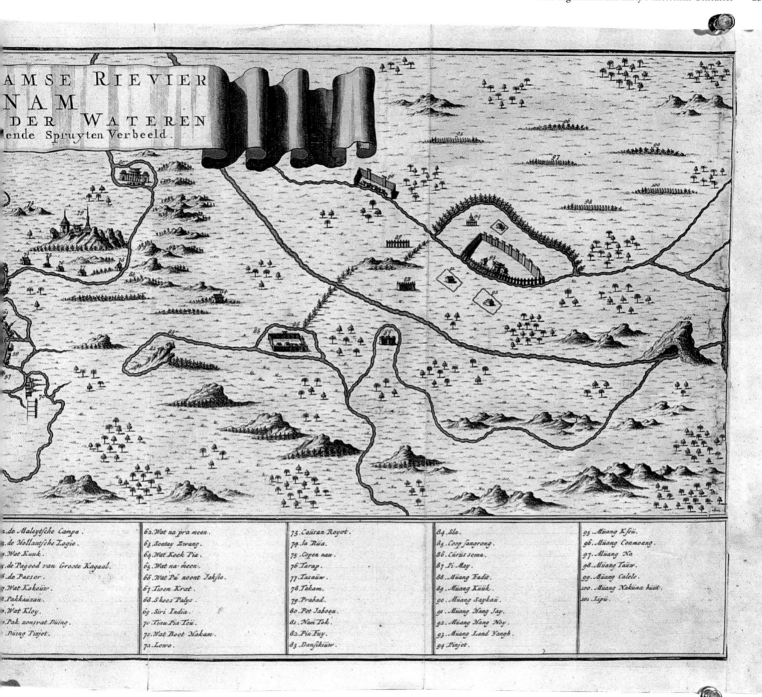

..do Maleytsche Campo .	62.Wat na pra meen .	73.Caŭsan Royot .	84.Sla .	95.Müang Ksoŭ .
. de Mollantsche Logie .	63.Sontay Ziwang .	74.la Rŭa .	85.Coop sangrong .	96.Müang Conmoang .
.Wat Kuuk .	64.Wat Koek Pia .	75.Copen nau .	86.Cŭrŭs sema .	97.Müang Na
. de Pagood van Groote Kagaal .	65.Wat na meen .	76.Tarap .	87.Pi May .	98.Müang Taŭw .
.de Passer .	66.Wat Pŭ noont Jakste .	77.Tasaŭw .	88.Müang Fadit .	99.Müang Calele .
.Wat Kokoŭw .	67.Toen Krat .	78.Takam .	89.Muang Kuŭk .	100.Müang Nakŭaz hŭit .
.Pakkauzan .	68.Skoes Palys .	79.Prabad .	90.Muang Saykaŭ .	101.Ligŭ .
.Wat Kloy .	69.Siri India .	80.Pet Jaboen .	91.Müang Hang Jay .	
.Pak zonsvat Düing .	70.Tiau Pia Teŭ .	81.Nam Tok .	92.Müang Hang Noy .	
.Düing Pinjet .	71.Wat Boet Hakam .	82.Pia Fuy .	93.Müang Land Yangh .	
	72.Lewo .	83.Danskiŭw .	94.Pinjet .	

the key are as follows: (79) *Prabad* = Saraburi; (80) *Pet Jaboen* = Phetchabun; (86) *Cürüs sema* = Nakhon Ratchasima (in Isaan, northeast Thailand); (93) *Müang Land Yangh* = Vientiane (Laos).

Valentijn's Map of the Bandas

Tight control over Banda was important to the V.O.C. for the same reason that controlling Ambon, Seram, and the Moluccas was — namely to protect the Company's effective monopoly on the trade in nutmeg and cloves. The V.O.C. attempted to secure exclusive markets both by monopolizing *and limiting* the sources. Having calculated that their market could be supplied from the clove production on Ambon alone, it set about destroying all the clove trees on Seram, though happily without complete success.[280] This policy was more effective, however, on the small island of Pulau Run, where the V.O.C. succeeded in extirpating all the nutmeg trees that grew there, thereby ensuring that the English, who had

been allowed to remain on a tiny coral islet just to the north of the island could not profit from their fruit. Valentijn (fig. 143) shows both Pulau Run (*Poelo Rhun,* the westernmost island on the map), as well as the small coral islet occupied by the English. English traders competing for cloves in Ambon had been eliminated in the notorious massacre of 1623; by the time of Valentijn's tenure in the Indies, the English had given up on Banda as well.

We saw earlier how the V.O.C. generally found it advantageous to leave indigenous societies intact so long as the needs of the Company were served. It was more efficient to tap into the culture's existing infrastructure rather than drain Company resources by establishing and maintaining a local administration of its own. For this reason, the interior of many Indonesian islands remained, for the most part, relatively outside the sphere of Dutch influence since the indigenous social structures could continue to be relied upon for communication and the transport of goods. But Banda was one of

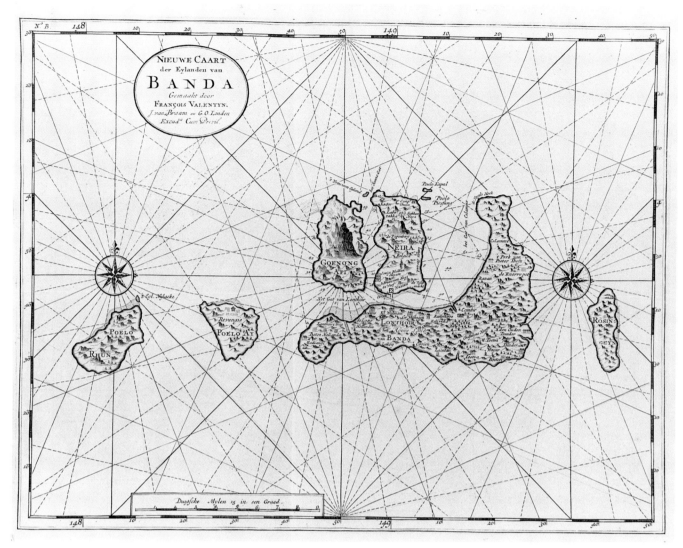

Fig.143 Banda, Valentyn, 1724-26. (31 x 40.7 cm)

the exceptions. The Bandanese, unwilling to cooperate with V.O.C. demands for exclusive rights to their nutmeg, were effectively erased from the map in 1621 — those who had not either fled or been killed were forced into labor. The V.O.C. then divided the islands into plantations which were leased to former Company employees who had stayed on after their term of service was completed. The Company gave them slaves to work the land, and in turn the Company bought their crops at fixed prices.

The Bandas first appeared on maps as early as the *mappamundi* of Fra Mauro (ca. 1459), who took his information from Conti. The Portuguese reached Banda in 1512 and set about building a fort in 1529 on Neira, though it was never completed. They opted instead to obtain nutmeg from intermediaries in Ternate and Tidore. Barros commented on the flat, horseshoe shaped coast of the largest island of Banda, which he said was covered with shiny nutmeg trees that had a sweet smell. Earlier European visitors noted that vessels from Java, Malacca, and China frequented Banda's harbor.

Sailors also remarked on how the three principal members of the archipelago, Valentijn's *Lonthoir Banda* (Great Banda), *Neira* (Bandaneira), and *Goenong Api* (Gunung Api), were clustered together and formed an inland sea with an especially fine harbor. The last of these, the archipelago's active volcano, was one of a number of Indonesian islands which shared the same Malay name, which means 'fire mountain', and is not the same volcanic island as noted by Gastaldi in 1548 (fig. 73) or Laurie & Whittle in 1798 (fig.

113). One particularly violent eruption of the Bandanese Gunung Api, which probably occurred in 1629, catapulted so many rocks and boulders into the waters between Gunung Api and Neira that the channel was no longer navigable by sailing ships.[281] The Bandanese themselves had their own cosmology to explain such tumultuous disruption of the islands, believing their archipelago to lie on the horns of a great ox which caused earthquakes by shaking its head.

Johannes van Keulen

In 1726, the same year that the publication of Valentijn's work was completed, Johannes Van Keulen II took up the position of Hydrographer to the V.O.C. During his tenure, which lasted until his death in 1755, charts based on the Company's 'secret atlas' found their way into print. These maps were sometimes added to the usual van Keulen *Zee-Fakkel*, or bound, without text, as a separate volume. Johannes' two-sheet map of Java (1728, fig. 100), tapped V.O.C. data which had already been used by Henri Châtelain nine years earlier. V.O.C. knowledge of the southern coastline of Java was extended in 1739, the year in which Paulus Paulusz, then head cartographer in Batavia, led an expedition along its southern shores.

Johannes' paramount achievement was the publication, in 1753, of Part VI of the *Zee-Fakkel*, a sea-atlas and pilot-guide for navigation in the East Indies (see the chart of Singapore and vicinity, fig. 144). By this time, however, both the V.O.C. and Dutch cartography were

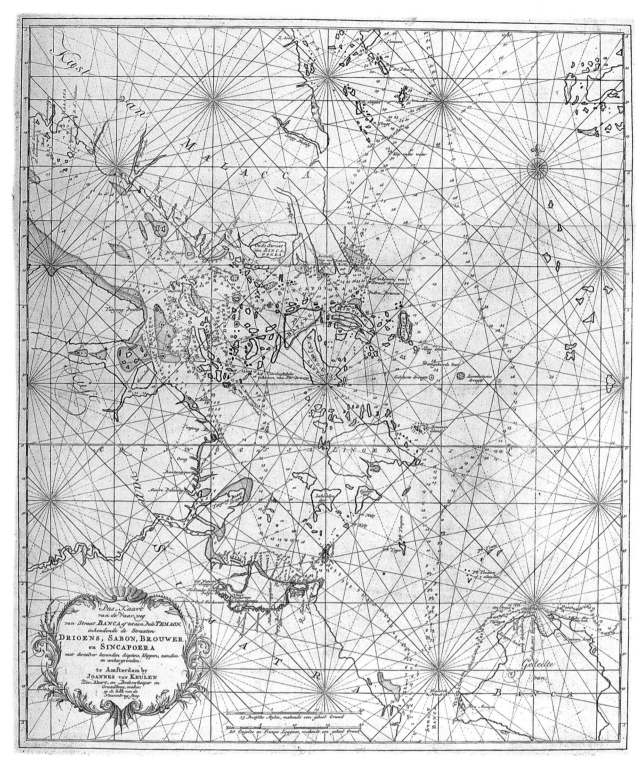

Fig. 144 Singapore, Van Keulen, *ZeeFakkel*, 1753. [Geoff Edwards, Jakarta]

in decline. For the very reasons that the V.O.C. no longer feared sharing its geographic secrets, the V.O.C. and Holland no longer dominated map publishing or the spice trade.

Part VI of van Keulen's the *Zee-Fakkel* and Valentijn's *Oud en Nieuw Oost-Indien* have both been cited as being among the possible sources used by the seafaring Bugis of Sulawesi, when compiling their own charts of the regional seas. Regardless of whether or not the Bugis obtained copies of the actual printed Van Keulen and/ or Valentijn volumes or not, they clearly consulted late seventeenth or early eighteenth century Dutch charts which were in the same cartographic family.

England and France

As the prime of the V.O.C. passed, and the works of François Valentijn and Johannes Van Keulen II were disseminated, English pilots in the Indies were becoming increasingly aware of the inadequacies of their 'bible', *The Third Book* of *The English Pilot*. The fatal flaw, here, was one of management, for when Samuel Thornton died in 1715, the work was acquired by the firm of Mount & Page, which lacked any true connection with the East India Company. Diligence could have overcome this obstacle, had Mount & Page possessed the courage to invest the considerable time and capital required to rework or re-engrave their copper-plates as newer data became available. But the initiative was not there, and the demand for better material far outpaced the rate at which improvements and additions were made.

The ultimate price for this lethargy was enormous — faulty charts were the suspected cause of many shipwrecks the East India Company suffered in Asian waters. That the pilots who used *The Third Book* were keenly aware of its shortcomings is reflected by the fact that many surviving examples have numerous manuscript annotations, corrections, and additions.

Jean-Baptiste d'Après de Mannevillette

A key inspiration for reforming the state of printed sea charts came from a French navigator by the name of Jean-Baptiste d'Après de Mannevillette (1707-80). The son of a captain in the service of the French West India Company, Mannevillette, like William Dampier, made his first sea voyages in the West Indies. In about 1735, Mannevillette set about collecting charts of the coasts of Africa and the Southeast Asian islands, a task which culminated in a request to the Compagnie des Indes in 1742 to publish a sea atlas. The Compagnie agreed, and the L'Académie des Sciences supported the work, which appeared in 1745 under the title *Le Neptune Oriental.* The new sea atlas quickly found its way into the pilot cabins of ships of several nations, and its 22 charts were immediately recognized as being superior to all previous printed maps of Southeast Asian coasts.

One interesting aspect of Mannevillette's methods was that he sometimes attempted to correct inaccuracies on existing printed charts, once these errors had been discovered, by pasting a small slip of paper, printed with the corrected data, over the faulty portion of the chart. In this way the existing stock of charts could be easily brought up-to-date, while the prohibitive cost of engraving a new copperplate did not prevent new and better information from being incorporated, once it became available. Mannevillette's charts are occasionally found with these small, inconspicuous corrections or improvements, a reflection of his intention that these charts could indeed be relied upon for actual use at sea.

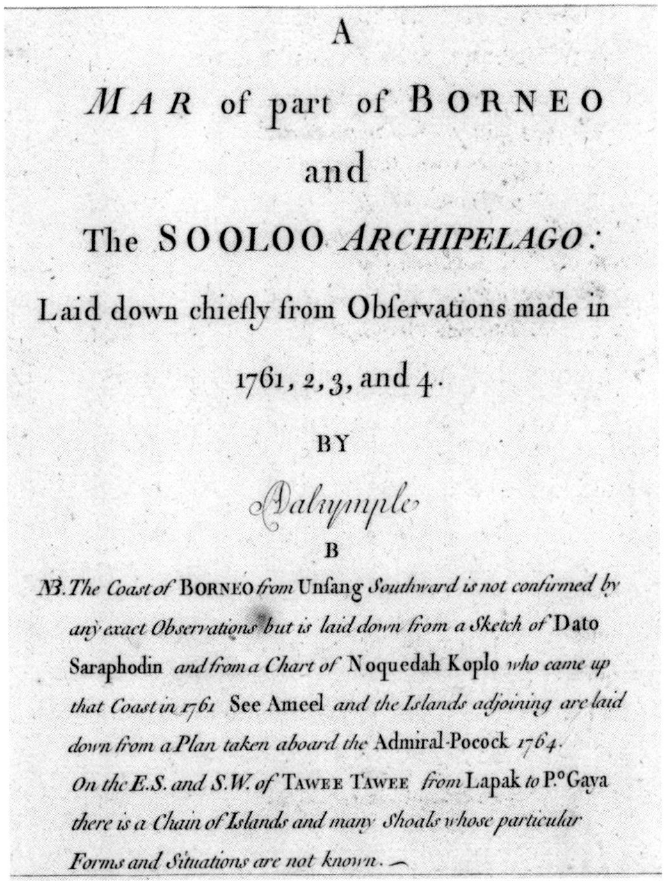

One individual who was keenly aware of the *Neptune's* benefits was William Herbert (1718-95), an Englishman who traveled to India in about 1748 as a purser's clerk. Back home in the early 1750s, Herbert set up a map and print shop on London Bridge, and in 1758, with the encouragement of the East India Company, he introduced a new pilot guide, the *New Directory for the East Indies.* Herbert gathered superior sources to those used in *The Third Book,* making full use of Mannevillette's *Neptune Oriental,* and also consulting the navigator William Nicholson and the cartographer Samuel Dunn. He acknowledged his reliance on Mannevillette on his title page, which must have further emphasized the inferiority of the charts from the *Third Book* of Mount & Page's *English Pilot.* Herbert's opus saw six editions through 1780, and along with the *Neptune Oriental* sounded the death knell for The *Third Book* of *The English Pilot.*

Mannevillette, meanwhile, never stopped collecting and compiling data, and in 1775 an enlarged edition of his *Neptune Oriental* was published which now contained 63 charts. Toward the end of his life, Mannevillette became a good friend of the British navigator and cartographer Alexander Dalrymple (1737-1808), who shared his indefatigable obsession with collecting marine surveys and charts.

Alexander Dalrymple and Sulu Sources

Alexander Dalrymple was Mannevillette's British soul-mate, a chronic compiler of coastal charts with long experience in Southeast Asian waters. In 1779, he convinced the directors of the East India Company what the V.O.C. had known for nearly two centuries — that it was absolutely essential that data from its pilots and captains be systematically collected and analyzed. This renewed sense of purpose led to the establishment of the Hydrographic Office by the British Admiralty in 1795, which published its first map five years later under Dalrymple's leadership. The Hydrographic Office's charts, originally available only to the Royal Navy, were offered to the general public beginning in 1821.

While sailing in the Indies, Dalrymple solicited data from Malay seamen to supplement his own surveys. Some of Dalrymple's published charts from 1770 contain inscriptions identifying which coastlines were composed from indigenous sources, and even citing those sources by name as cartographers in their own right. For example, a note on a chart of Borneo states that the coast from *Unsang* southwards had not been "confirmed by any exact Observation but is laid down from a Sketch of Dato Saraphodin and from a Chart of Noquedah Koplo who came up that Coast in 1761" (fig. 145). Elsewhere on the chart, Dalrymple notes that certain features "marked with points, are well determin'd from my own Observations," while others are based on "Sketches I receiv'd from the Sooloos, but chiefly from the information of Bahatol an intelligent old pilot."

Balambangan as Key to the Indies

Dalrymple believed that Balambangan (northern Borneo) would be England's best base for trade and empire in insular Southeast Asia, and composed careful charts of the region from both his own, and indigenous, sources (fig. 146). In 1769, writing in praise of

Fig. 145 Northeast Borneo and the Sulu Sea, Alexander Dalrymple, 1770.

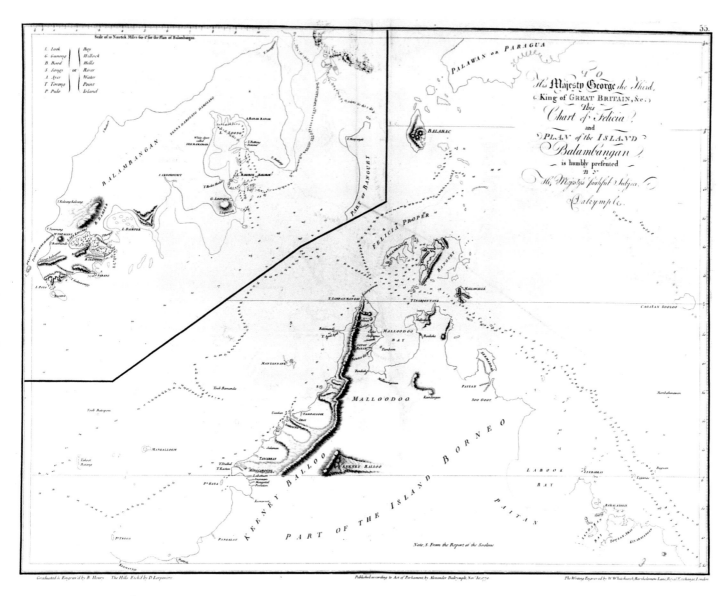

Fig. 146 Balambangan, Alexander Dalrymple, 1770. (47 x 61.5 cm)

Balambangan's advantages, he noted that few islands

> enjoy one harbour so excellent as either of those this island possesses. They are both land-locked, and capable of admitting the largest ships . . . Balambangan, situated at the north point of Borneo, and south of Palawan, is the most convenient place for the inhabitants of the eastern and western coasts of these islands to bring the productions of their country.

Due to these considerations, he proposed that Balambangan be England's center of power in the islands, just as Batavia was to Holland and Manila was to Spain:

> . . . on a general and particular view, [I] found Balambangan exquisitely adapted for a capital to the oriental Polynesia, as strickly central, both in distance and conveniency of access.

It was to this end that Thomas Forrest was sent to the Indies in 1774-76, to explore the many islands that lay to the east as far as New Guinea. Ultimately, Dalrymple's dream of a British Batavia in Borneo came to naught and another half century was to elapse before the founding of Singapore.

Laurie & Whittle

One of the finest commercial publishers of sea charts in the late eighteenth and early nineteenth century was the London firm of Laurie & Whittle. Following the practice that had been gradually established during the eighteenth century, Laurie & Whittle emendated their charts with updates, corrections, comments, explanations, worthy opinions — any kind of information which might help a pilot reach his destination safely. We will look at four of their charts, spanning 1794 to 1807, each depicting a region focused on by British pilots.

Indonesia assumed a particular importance for the British at the close of the eighteenth century and the beginning of the nineteenth, as a result of what at first sight might seem to be an irrelevant series of events — namely the French Revolution and the subsequent wars with France which only ended with the final defeat of Napoleon at Waterloo in 1815. When the French occupied the Netherlands in 1795, Prince William of Orange escaped to England and invited his hosts to take over any of the Dutch colonies scattered around the world, if only so as to prevent them falling into the hands of the French. The English were already neighbors of Holland's vast Southeast Asian empire, and the East India Company consented, occupying Malacca in 1795, and taking Java in 1811.

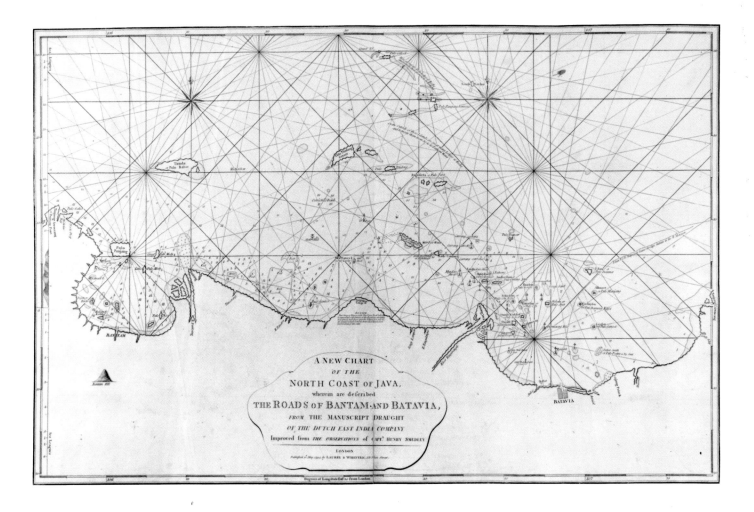

Fig. 147 Northeast Java, Laurie & Whittle, 1794. (56 x 87 cm)

The 1794 chart of the north coast of Java from Banten to Batavia (fig. 147) is, as the title acknowledges, derived from V.O.C. manuscripts and emendated with more recent English reports. Among the notes to pilots is a warning that the shoal surrounding 'Men Eaters Island', which was a danger point when coasting between Banten and Batavia, is larger than is mapped, and that the beacon on the island "is down of late years."

A chart of the eastern Indonesian islands (figs. 148, 149) was published four years later, on the 12ᵗʰ of October 1798. It incorporates improvements gleaned from the observations of a Captain Robert Williams, made in 1797, and also records the voyages of both James Cook and William Dampier. Dampier had sailed through these seas on a government-sponsored expedition to Australia, having shaken off the reputation of his pirate days and winning the respect of the scientific community. In 1699, after leaving Australia, Dampier explored the coastline of Timor and the waters round about. "At the South-West end of *Timor*, is a pretty high Island, called *Anabao*," Dampier explained in his published account. This is Semau Island and the chart labels it as such, but also adds that it is *Dampier's Anamabao.* "It is about ten or twelve Leagues long," Dampier continued,

> and about four broad; near which the *Dutch* are settled. It lies so near *Timor*, that 'tis laid down in our Draughts as part of that Island; yet we found a narrow deep Channel fit for any Ships to pass between them." North of Timor, "by a good Observation we found that the South-East point of *Omba* lies in Latitude 8 d. 25 min. In my Draughts it's laid down in 8 deg. 10 min.

Dampier was fascinated by the highly active volcanic island of Gunung Api, sitting by itself in the Banda Sea to the north of Timor, which had first been described by Europeans as early as de Abreu and Serrão in 1512. Dampier records that on the 27ᵗʰ of December 1699,

> we saw the burning Island, it lies Latitude 6 deg. 36 min. South; it is high, and but small. It runs from the Sea a little sloping towards the Top; which is divided in the middle into two Peaks, between which issued out much Smoke: I have not seen more from any Volcano. I saw no Trees; but the North side appeared green, and rest look'd very barren.

Laurie & Whittle mark it as "the burning island" on their chart, and include an inset view taken from an English vessel which passed by in August 1797, the year before the chart was compiled.

Thomas Stamford Raffles, Singapore, and Penang

The modern city of Singapore was the brainchild of one Thomas Stamford Raffles. Born at sea off Jamaica, and a scholar of the Malay language and Malay culture generally, Raffles certainly knew the chart of the Singapore Straits by Laurie & Whittle (fig. 150). When it was published in 1799, Raffles was only eighteen years old, but had already been in the employ of the East India Company for four years.

The chart tapped the experiences of various pilots, noting, for example, that a Captain Elmore had navigated the strait in 1784 and investigated the question of whether it was better to sail north or south of Pedra Branca. The chart explains that whereas some pilots believed there was no navigable passage to the south, Elmore had reported that the southern passage was in fact superior.

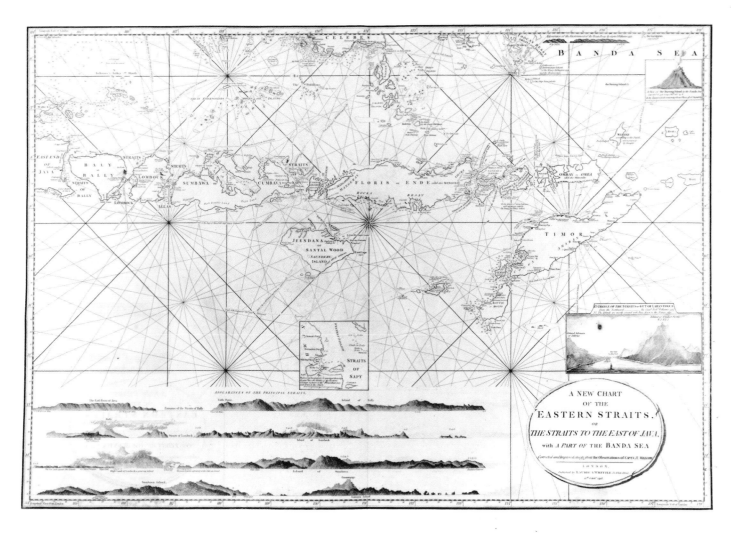

Fig. 148 Eastern Indonesia, Laurie & Whittle, 1798. (59.8 x 87.7 cm)

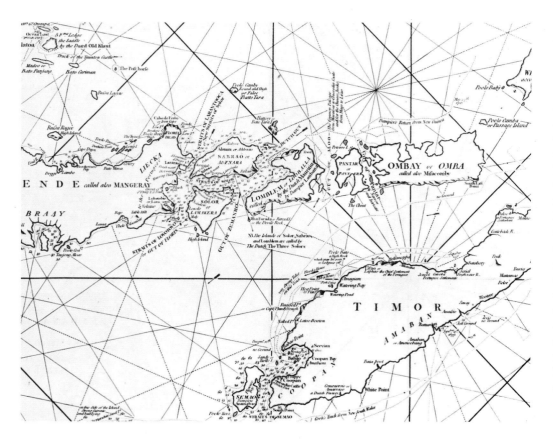

Fig. 149 Detail from the Laurie & Whittle chart of Eastern Indonesia (1798), showing Timor and the small neighboring islands of Semau and Roti. The chart records soundings and other data, warnings about navigational hazards and other useful advice for vessels sailing along the coast of Timor. The tracks of William Dampier and James Cook, both of whom reached Timor sailing north from Australia, are plotted. When Dampier skirted the island in 1699, it marked his return to Southeast Asia as a formal explorer, rather than as a pirate. Cook put into Timor in 1770, on the first of his three voyages around the world, shortly after nearly losing the *Endeavour* on a coral reef off Australia. From Timor, Cook sailed to Batavia where the *Endeavour* was refitted for the long voyage back to England. This was the captain's only experience in Southeast Asian seas.

Raffles' experiences in Southeast Asia did not begin until 1805, when he was appointed assistant secretary to the governor of Penang, then known by the British as 'Prince of Wales' Island'. This was the same year that a British lieutenant, named Thomas Evans, was busy surveying the waters between Penang and the mainland, recording depths, as well as rocks and shoals. Evans' information was subsequently used in a grand map of the island published by Laurie & Whittle in 1807 (fig. 151). Using a technique that was first proposed by Galileo in the early seventeenth century, the precise longitude of the Admiral's House, which lay on the northern coast of the island's easterly peninsula, was ascertained (as an inscription explains) "by the mean of three eclipses of Jupiter's 1st Satellite," despite the wide availability of the chronometer by this time, which had rendered such methods obsolete. The figure that was achieved was in fact highly accurate: 100° 23′ 39″ east of Greenwich, which is only about 4′ too far to the east.

The map marks the British settlement of George Town in red on the eastern tip of the island. Fort Cornwallis, which defended the approach to the harbour, is indicated at the water's edge by a red 'x' and is also illustrated in a detailed view on the right-hand border of the map. Below the fort one can see the customs house, while to the left of the fort are situated the Admiral's House and Governor's House where, if we could zoom in on the map, we would find Raffles.

Raffles and Java

In 1811, the British occupied Java at the invitation of the Dutch to forestall the possibility of Napoleon's forces using the island as a base from which to launch an attack on India. The liberal-minded Raffles, who judged Dutch rule to be oppressive and believed that British stewardship of the Indies would be more humane, was appointed Lieutenant-Governor of Java. It was during this interlude of British rule that Europe first became aware of the mighty ruins of Borobudur (fig. 8) near the city of Yogyakarta in central Java, which had apparently remained invisible to the Dutch despite two centuries of their presence on the island. Raffles, keenly interested in Indonesian antiquities, ordered that a survey be made of the Borobudur site, apparently the first mapping of an Indonesian classical ruin.

In general, however, the lack of official surveys of the Javanese interior could not be quickly alleviated, causing some problems for Raffles' attempts to reform what he perceived as the injustice of Dutch colonial policy. Although he instituted a new land allotment system for Javanese villagers, which returned a degree of autonomy to the Javanese, the dearth of local maps or cadastral records meant that the headman of each village could manipulate the system of land tenure according to his own whims and advantage. Meanwhile, Raffles found his relatively liberal policies, and the cost of implementing them, severly criticized, and was soon recalled to England. Java was subsequently returned to the Dutch at the conclusion of the Napoleonic wars.

Fig.150 Singapore Straits, Laurie & Whittle, 1799. (42.5 x 58.5cm)

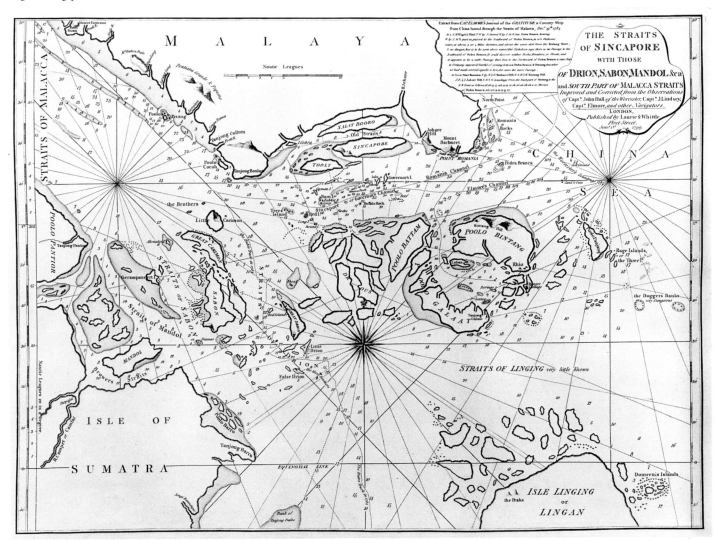

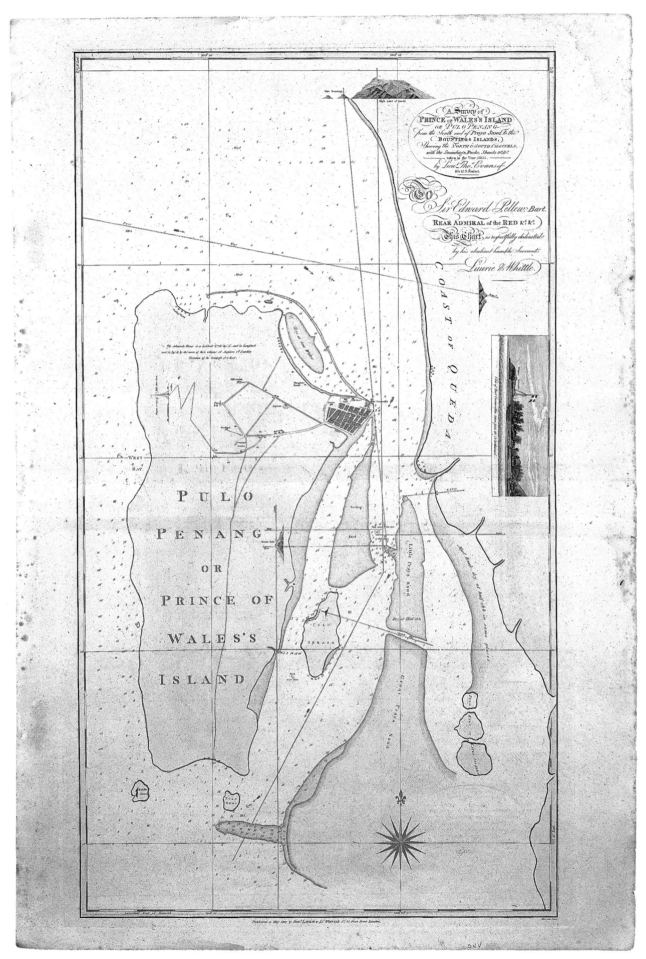

Fig. 151 Pulo Penang or Prince of Wale's island, Laurie & Whittle, 1807.

Founding of Singapore

In 1818 Raffles was appointed Lieutenat-Governor of Bencoolen, a British settlement on the southwest coast of Sumatra. Angered by Dutch attempts to monopolize trade in the islands, he conceived his masterpiece, the founding of Singapore. Raffles pressed the East India Company to establish a base at Singapore which, at the time, was home to a Malay fishing community of about 500 people.

Curiously, over a century before Raffles founded Singapore, the Scottish captain, Alexander Hamilton was offered the island as a gift. In 1703, Hamilton stopped at Johor, where the king "made me a present of the island of Singapore." Hamilton declined the gift, explaining that it was "of no use to a private person," but clearly understood Singapore's significance, noting that it was "a proper place for a company to settle a colony on, lying in the center of trade, and being accommodated with good rivers and safe harbors, so conveniently situated that all winds served shipping both to go out and come into those rivers."

Raffles focused on Singapore's unique position, predicting that British control of the island would "eventually destroy the spell of Dutch monopoly," for "what Malta is in the West, that may Singapore be in the East." In a treaty signed by Raffles in 1819 with the supposed sultan of Johor, Singapore was ceded to the East India Company. The Dutch, however, protested that they had won the Riau Archipelago, which included Singapore, in an 1818 treaty with the prior sultan. Nevertheless, the English settlement soon became an indelible feature of the island.

Singapore's location, as well as the stability insured by the presence of the British military, proved attractive, and British banking, shipping, and insurance firms from India and Sri Lanka opened branches there. The island's excellent harbor, free trade, and security brought thousands of new settlers, particularly Chinese, making Britain's position there even stronger. Only five years after the founding of British Singapore, an agreement was signed with the Dutch which essentially divided Southeast Asia, outside of the Philippines, into two spheres of influence. Britain was proclaimed the master of the mainland, while Holland was guaranteed their traditional realm, the Indonesian islands.

British law gave the East India Company a monopoly on the China-England trade. Independent entrepreneurs, however, soon began exploiting a loophole in this arrangement, declaring that their goods, regardless of ultimate origin, were being shipped out of Singapore rather than China. Finally, in 1833 the East India Company lost its monopoly altogether.

The Singapore-China Route and the Paracels

Portuguese mariners became familiar with the Paracel Islands in the early sixteenth century, when they first sailed northeast from Malacca along the coast of China. But whereas islands such as Tioman and Condore were safe havens on the route between China and the Indian Ocean, the Paracels were a menace. Dampier noted their danger, and shortly afterwards Alexander Hamilton remarked that the Paracels are "a dangerous chain of rocks, about 130 leagues long,

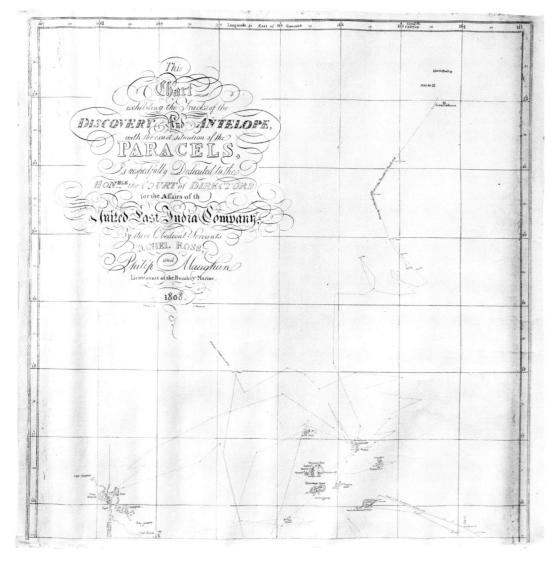

Fig: 152 The Paracel Islands, a manuscript copy of James Horsburgh's printed chart, ca. 1810 (61.5 x 61.5 cm). This manuscript copy was probably made in the Indies onboard ship or at port where additional copies of the published map were not available. The diagonal coastline in the lower left is Da Nang, Vietnam. Macao is marked toward the upper right of the map, Canton is plotted in the upper border of the map, and Hainan lies in the area taken up by the extensive title. The various islands comprising the Paracels are in the foreground.

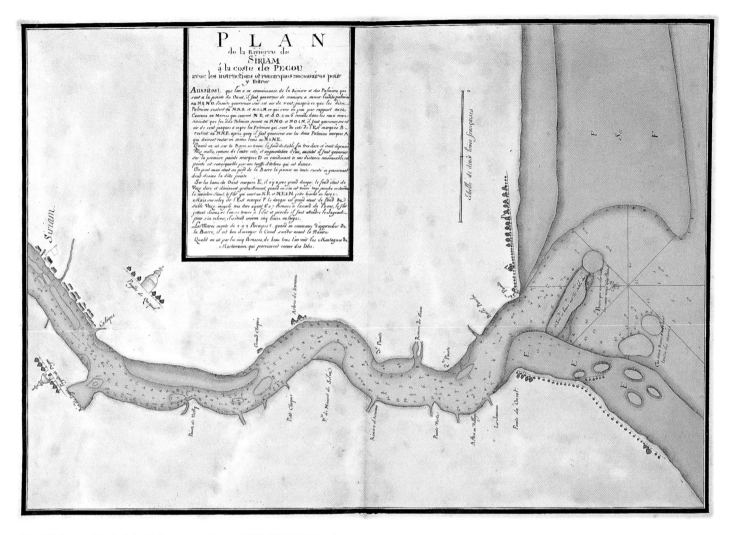

Fig. 153 Burma, Dépôt de la Marine, manuscript, ca. 1770. (49.7 x 71.9 cm)

and about fifteen broad, and have only some islands at each end. There are several inter-currents among those rocks, but no known marks to keep clear of dangers by." Hamilton noted that fear of shipwreck on the Paracels kept pilots close to the Indochina coast.

In 1808, a chart of the Paracels was printed and published by James Horsburgh, Hydrographer to the East India Company, and was carried onboard English vessels making the run between China and Singapore. But the example illustrated here (fig. 152), is a manuscript map, not a printed map. Although printed charts had by this time largely replaced manuscript maps as the principal reference of pilots and captains on routine voyages, the manuscript medium continued to serve as an expedient method of reproduction while at sea or at a distant port. When additional copies of a map were needed *en route*, it would be copied by hand. This is an example of one such production, meticulously copied from the printed chart (or another MS copy of the engraved chart), probably aboard a British ship in the Indies.[282] Even the name of the engraver, a seemingly superfluous feature for the copy's purpose, has been retained.

France and England Compete on the Mainland
Jean-Baptiste D'Après de Mannevillette raised the standards of printed sea charts in his native France as well as in England. Other French mapmakers of the eighteenth century had a marked impact on the mapping of many mainland regions, notably Jean Baptiste d'Anville. In the meantime, French commercial interests were enduring rough times.

In 1719 the French East India Company had been reorganized as the Compagnie des Indies, placing commerce with the Americas, Africa, and Asia under one helm. Its first head was the infamous John Law, concocter of the 'Mississippi Scheme' which collected funds from investors while flooding France with paper money from Law's own bank, which had in the meantime become the country's official bank. That particular 'bubble' burst in 1720. The Compagnie lost its slave trade with Africa a decade later, and then in 1731, it lost its trading place in Louisiana. Finally, in 1736, the Compagnie lost its coffee trade with America. In the wake of these commercial disasters the Compagnie's Asian operation became all the more important.

France's interests in southern Asia consisted of two very different endeavors. There was the Compagnie's venture in India, which continued to be quite profitable at a time when its exploits in America and Africa soured. The other preoccupation, however, was the dream of finding a practical river route through which to penetrate Southeast Asia to China.

Britain was a constant threat to France's India operations. In 1751 British troops succeeded in confining the French to southern India, though the Compagnie operated profitably from there for another decade. The death blow for the French, however, came with the British capture of Pondicherry in 1761, which ultimately led to the suspension of the Compagnie's operations eight years later. The loss of India made France look to the eastern mainland with renewed fervor.

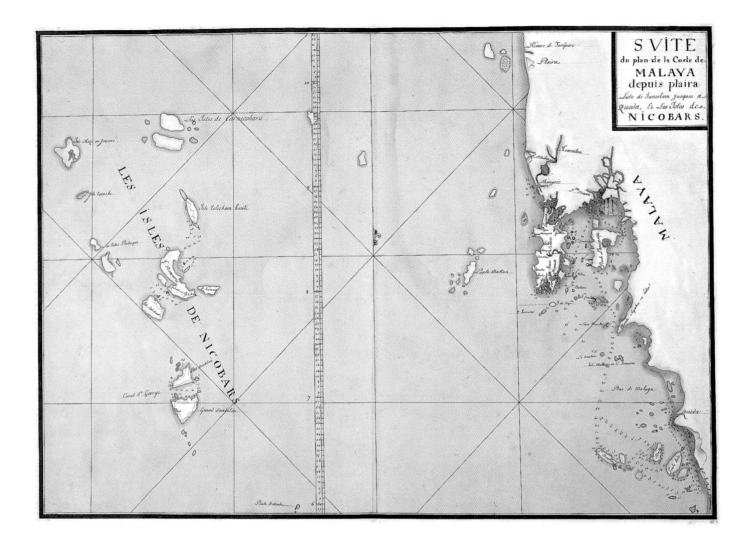

Fig: 154 Map of Phuket and the Nicobars, Dépôt de la Marine,
manuscript, ca. 1770. (50 x 71 cm)

British expansionists in India viewed Burma as a cultural, economic and political extension of the subcontinent, and on the strength of this, argued that the control of India justified the control of Burma. Thus the fact that the French were poking about Burma at this time — using maps like the one illustrated in figure 153 — was of major concern to the British. In 1802, after leading two expeditions into Burma to investigate the feasibility of establishing trade between India and China via Burma, a British captain named Michael Symes reported that despite the potential profits of such a scheme, it was of more immediate concern to spoil French interests in the region.

Dépôt des Cartes et Plans de la Marine

To a large extent, French pilots sailing Southeast Asian waters could rely on the various printed sea atlases being produced in France and England, but for more detailed or current data they would acquire manuscript charts from such sources as the Dépôt des Cartes et Plans de la Marine. Two charts by the Dépôt de la Marine (as it was commonly known) dating from ca. 1770 are illustrated here. They both show potentially profitable regions which France pursued at the end of the eighteenth century. The first, *Plan de la Rivierre de Siriam àla coste de Pegou avec les instructions et remarques necessaires pour y entrer* (fig. 153) is, as the title explains, a map of the Syriam (Rangoon) River on Pegu coast of Burma.

This Dépôt de la Marine chart contains navigational instructions and advice for entering the river, guiding the pilot from its wide entrance as far upstream as Rangoon and Syriam, where it narrows and divides. Tricky tides made accurate maps of the river's entrance especially important; Alexander Hamilton, visiting Syriam in the early eighteenth century, noted that "if by accident a ship bound to Syriam be driven a league or two to the eastward of that river's mouth, a strong tide carries her on hard sands till she sits fast on them, for anchors are of no use to stop them, because of the rapidity of the current."

At low tide, with the ship stranded on the sand, "the shipwrecked men walk on the sands toward the shore for their safety, for the sea comes back with so much noise, that the roaring of the billows may be heard ten miles off."

The chart illustrates two 'pagodas' (temples) as landmarks. The *Pagode de Digon* (Rangoon) had often been commented upon by early European visitors. In the mid-1580s Ralph Fitch traveled "about two days' journey" from Pegu to *Dagon,* where there is a temple which "is of a wonderful bigness, and all gilded from the foot to the top." The English merchant was so taken with the town of *Dagon* that he considered it "the fairest place, as I suppose, that is in the world." Just upriver from *Dagon* is Syriam, which Fitch also found to be "a good town," with "a fair port into the sea, whither come many ships from Mecca, Malacca, Sumatra, and from diverse other places."

A century and a half later, Syriam was described by Hamilton as being the only Burmese port then open to trade. It was

> built near the river's side on a rising ground, and walled round with a stone wall without mortar. The governor, who is generally of blood royal, has his lodgings in it, but the suburbs are four times bigger than the city.

Hamilton also described the temples: "There are two large temples near Syriam, so like one another in structure that they seem to be built by one model", and indeed they appear nearly identical on the map.

The temple marked on the chart as *Pagode de Quiquel*

> stands about six miles to the southward [of Syriam], called Kiakiack, or the God of Gods temple. In it is an image of twenty yards long, lying in a sleeping posture . . . The temple stands on an high champain ground [open country], and may easily be seen, in a clear day, eight leagues off.

The temple was thus a useful point of reference for pilots navigating the river, just as the Hmawdin temple on the high ground of Cape Negrais was an important landmark for pilots off the coast. The chart's other temple, *Pagode de Digon,* "stands in a low plain, north of Syriam, about the same distance, called Dagun." Whereas the doors and windows of the *Pagode de Quiquel* were always open, and anyone could enter, the *Pagode de Digon* was closed to all but monks.

The second manuscript from the Dépôt de la Marine, *Suite du plan de la Coste de Malaya depuis plaira l'isle de Janselam jusques a Queda, Et Les Isles de Nicobars* (fig. 154), charts the western coastline of the Malay Peninsula in the vicinity of the island of Phuket, and the waters east through to the Nicobar Islands. *Janselam,* or more commonly *Junkceylon,* is the port and island of Phuket. The name *Junkceylon* possibly came from the Malay term for 'tempestuous cape', *ujung salang.* It was an important port for vessels bound for points west of Thailand, despite its draft being relatively shallow. Gervaise described the mainland harbour as "one of the best among those where art has had to be used to improve upon nature." Hamilton cautioned, however, that the island's trade was stifled by the pirates who infested the many islands along the coast between Phuket and Mergui.

Spain and the Philippines

Through most of the eighteenth century, the northern Philippines remained the only part of Southeast Asia which had been extensively colonized. William Dampier passed through in 1688 and described the Spanish city of Manila as lying

> at the foot of a ridge of high Hills, facing upon a spacious harbour . . . The Houses are large, strongly built, and covered with Pan-tile. The Streets are large, and pretty regular; with a Parade in the midst, after the Spanish fashion . . . The Harbour is so large, that some hundreds of Ships may ride here: and is never without many, both of their own, and strangers . . . the Chinese are the chiefest Merchants and they drive the greatest trade; for they have commonly 20 or 30, or 40 Junks in the Harbour at a time, and a great many Merchants constantly residing in the City.

The mapping of the complex archipelago had gradually improved during the seventeenth century, and reached its maturity in the early eighteenth century with a large chart of the islands, compiled by Francisco Díaz Romero and Antonio de Chandia, published in 1727. The map far surpassed any previous map for its exhaustive detail; it also included informative inscriptions of both political and navigational interest, and plotted the sailing route of the Manila Galleon between Manila and Acapulco.

Pedro Murillo Velarde

The most important and famous map of the Philippines, however, was that of Father Pedro Murillo Velarde, a native of southern Spain who had been sent to the Philippines in 1723 (fig. 155). His map was a truly joint Spanish-Philippine effort, supplemented by vignettes drawn by a Tagalog artist, Francisco Suárez, and engraved and printed by another Tagalog, Nicolás de la Cruz Bagay. That the map was a cooperative Spanish-indigenous effort is particularly interesting given that Father Murillo had bitterly criticized both the religious and secular elements of the entrenched Spanish colonists for their maltreatment and exploitation of native Filipinos.

Alejandro Malaspina

It is often noted that during the latter part of the eighteenth century a more 'enlightened' attitude prevailed in the West, one that nurtured a new breed of exploratory voyages dedicated to scientific research rather than mere exploitation or trade. This is not to suggest that earlier travelers from Europe and elsewhere had not expressed a similar curiosity about the places they visited. Nor were the more 'sophisticated' travelers of the eighteenth and nineteenth centuries any less inclined to help themselves to their share of plunder. Nevertheless, these later voyages did increasingly provide a more thorough, methodical account of the people and places they visited. William Dampier, ironically a pirate, was a pioneer of such careful observations, as were, in different ways, the French embassies to Siam. The most famous of the eighteenth century scientific expeditions were British and French — the three circumnavigations of James Cook and the ill-fated voyage of La Pérouse — but there was also a less well-known Spanish voyage, that ranks along side them.

In 1789, Alejandro Malaspina, a native of Italy who had gained the respect of the Spanish authorities for his skill in fighting the British at Gibraltar, led two vessels from Cadiz on a planned circumnavigation of the world — the names of his ships are noted on the chart illustrated here (fig. 156): *Descubierta* (discovery) and *Atrevida* (the daring one). Before leaving, he consulted with scientists about the chronometer and other matters related to the expedition, including his sailors' health — like Cook, Malaspina was a pioneer in preventing scurvy and other ailments of long sea voyages. And while he carried all the obligatory armaments, he also brought along a harpsichord.

Following an important and lengthy exploration of the American Northwest, Malaspina crossed the Pacific in 1792, reaching Guam and the Philippines, before continuing on to Macao, Australia, and New Zealand. While in the Philippines, he visited the ports of Palapag, Sorsogon, and Manila, and made cartographic surveys of the islands of Luzon, Mindoro, Mindanao, and Negros.

Rather than circumnavigating the globe as originally planned, Malaspina touched on the recently-founded colony of New South Wales and then returned east across the Pacific, arriving back in Cadiz in September of 1794. Although the expedition lacked the romantic appeal of a circumnavigation, its bounty of scientific data was enormous, and its surveys in the Philippines provided for a fine, new map of the archipelago. A detailed chart of the strait between Luzon and Samar appeared as an inset, and a separate map charted Sorsogon Bay and part of northeast Samar (fig. 156).

Fig. 155 The Philippines, Pedro Murillo Velarde, 1734 (32 x 49 cm). In the Albay region of southern Luzon, Velarde has represented the volcano Mount Mayon, which was an important landmark for sailors. The illustrations by Suárez are, left column, top-to-bottom: Chinese residents of the Philippines; African residents (four on left) and Indian residents (two on right); *Mestizos,* or mixed Spanish-Filipino (three on left), then a Moluccan and a Japanese. The next illustration, fourth down on the left, is a political caricature of Spanish officials whose arrogance Murillo — and presumably his illustrator friend Suárez — condemned. Commoners and "mountain savages" appear in the background. The next-to-last vignette features a Persian, a Mongol, and an Indian of Malabar; while the bottom vignette depicts a typical street scene, the woman in the foreground selling guava from her basket, two children carrying bamboo vessels and a crab, and people making music and dancing.

The right column of illustrations, top-to-bottom, show a farmer with his plow and water buffalo, another man with a carabao and bamboo sled, various predators such as crocodiles and snakes, and (under the dwelling in the background) a woman pounding rice with a *luzon,* from which, Murillo believes, the island of Luzon gets its name (the Tagalog word for a mortar used to grind food is *lusóng*). Next is an inset map of Guam, in the Marianas; then a map of Manila; then a rural scene with a man cutting bamboo and various animals; a map of the Fort of Zamboanga follows; and finally Cavite, in Manila Bay.

But the map of the Philippines was not published until 1808, fourteen years after Malaspina's return. Nor was any credit given to Malaspina or his colleagues; the chart simply acknowledges that it was based on observations *levantada . . . por los Comandantes y oficiales* of the vessels *Descubierta* and *Atrevida* in 1792-93. What happened? Why did the map omit the authors' names?

In November of 1795 King Carlos IV, whom history records as having been a weak and impotent leader, issued a royal decree praising Malaspina. But a mere eight months later, the commander was arrested, confined, and later jailed in the prison castle in the bay of La Coruña, eventually to be banished from Spain. Malaspina's transgression was probably two-fold. The explorer had returned from his voyage espousing various 'liberal' ideas regarding the treatment of the people of overseas colonies. According to the Italian ambassador to Spain, who was a friend of Malaspina, his experiences during the voyage "made a deep impression on [his] passionate soul as to what is best for his fellow creatures, thinking it the absolute necessity for a radical change in the system of govern-ment and for legislation in the colonies"; one of the queen's ladies-in-waiting who supported his political ideas, named La Matallana, was also expelled from the royal court. Secondly, Malaspina may have had a romantic involvement with the daughter of the Duke of Parma, who happened to be Carlos IV's queen — hardly the way to stay on good terms with one's king.

In 1797, while Malaspina was in prison in La Coruña, the *Direccion Hidrografica* was created, with Felipe Bauza y Canas, an artist and cartographer who accompanied Malaspina, as deputy. Bauza was therefore in a position to promote the publication of materials from the expedition. But the much-expected account of the Malaspina voyage and accompanying maps never materialized, and when, in 1808, the *Direccion Hidrografica* finally published Malaspina's general map of the Philippines, *Carta General del Archipelago de Filipinas. Levantada en 1792 y 93 por los Comandantes y oficiales de las Corbetas de S.M Descubierta y Atrevida, durante la Campaña,* the name 'Malaspina' did not appear — it was as if the explorer had been purged from the memory of history.

The official shunning of Malaspina was successful; even otherwise exhaustive modern-day encyclopedias acknowledge the name 'Malaspina' only as the name of a glacier in the American Northwest, without a word of the person after whom it was named.

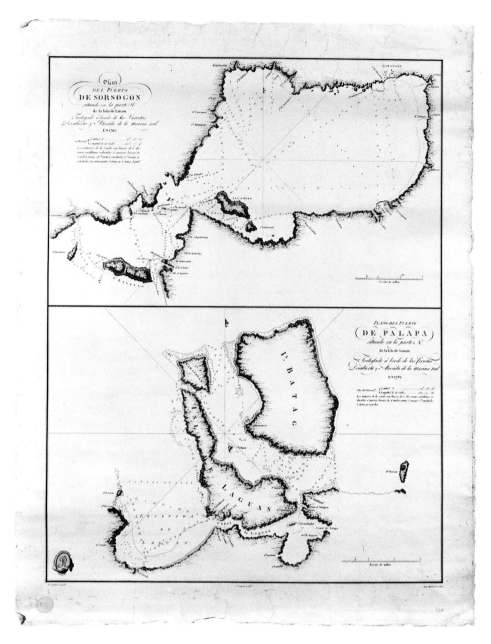

The Spanish crown effectively robbed Spain of a great explorer, visionary, cartographer, and doctor. Only in recent years, with the second centennial of the voyage, have scholars and the media in Spain begun restoring the memory of Malaspina.

For determining latitude and longitude, the expedition made astronomical observations at Manila, on the northeast of Samar, and on Zamboanga. These astronomical records from the voyage were published by José Espinosa y Tello, a member of the expedition, in 1809. The full account of the voyage was not published until 1885.

European Maps in Southeast Asia

Rudimentary European cosmographical principles were imported to Southeast Asia at least as early as 1593, with the printing of a European text in Chinese by the Spanish in Manila (see page 176, above). Transmission of far more sophisticated European science is found by the mid-seventeenth century in the cosmopolitan kingdom of Makasser (Sulawesi), whose well-educated, wealthy ruler, Karaēng Pattingalloang, lusted after the geographical offerings of the Blaeu firm of mapmakers.

In 1644, as part of sandalwood negotiations, Pattingalloang had the V.O.C. authorities in Batavia secure for him a Blaeu wall map of the world, plus Blaeu's four-volume atlas, a telescope, a burning-glass, and a globe. The globe, which was of huge proportions (approximately 130 centimeter diameter), had to be specially made and did not reach the prince until February of 1651.

The court of Makasser, in fact, appeared to be avid students of Western science in general, and geography in particular. In 1647 the French Jesuit, Alexandre de Rhodes, was received by the chief minister of the Makasser court and reported that his host was

> exceedingly wise and sensible, and apart from his bad religion, a very honest man. He knew all our mysteries very well [and] had read with curiosity all the chronicles of our European kings. He always had books of ours in hand, especially those treating with mathematics, in which he was quite well versed. Indeed he had such a passion for all branches of this science that he worked at it day and night.[283]

There is further confirmation of Makasser's fascination with Western cosmology a decade later, in 1657, when Francisco Domingo Fernández Navarette found himself stranded on the island

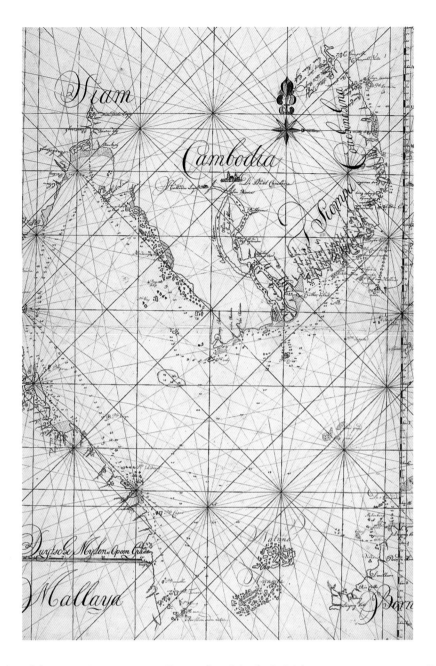

Fig. 157 An anonymous V.O.C. chart covering the South China Sea from north Borneo through to Vietnam and Malaya, made in Batavia in the early eighteenth century. Although Cambodia had been a target of the V.O.C. since 1620, serious Dutch infiltration into the country came only when Japanese commerce with Cambodia was halted after 1635. The V.O.C. established bases in Phnom Phen to carry on the trade abandoned by the Japanese, operating irregularly until about 1670, when the country's political instability proved too great a burden. That Europe remained largely ignorant of Cambodia until relatively modern times is reflected by the fact that reports of the true grandeur of Ankgor were not held as credible until the nineteenth century. (100.5 x 66 cm) [Paulus Swaen Old Maps Internet Auction]

while attempting to return to Europe from Manila. In Makassar, Francisco was entertained by the son of Karaēng Pattingalloang, who showed him his father's books, maps, globes, and clock.[284]

In the mid to later eighteenth century, when Dalrymple was consulting Sulu charts for his own maps, we also have a record of Bugis traders acquiring British maps. The Bugis mariners were in fact keenly interested in European maps, as we know first-hand from the British navigator Thomas Forrest. Forrest sailed extensively in Southeast Asian waters between 1774 and 1776 compiling charts of the Moluccas and Borneo. The Bugis, Forrest wrote, "are fond of sea charts, [and] I have given many to certain [pilots] for which they were very grateful." Forrest, who was proficient in Malay, stayed with the sultan of Mindanao for several months in 1774, during which time he drew a large map of the world on wood for his hosts. The published account of his adventures (1779) included an engraving of a nuptial scene in the sultan's royal chambers, where Forrest's map was said to be hung. An image of the map was not included in the original engraving of the scene, but was subsequently added to later editions of the book, where it is shown hanging prominently on the wall, high above long shelves of variously-shaped vases. If the illustration is

accurate, Forrest's map already incorporated the discoveries of James Cook's first circumnavigation of the world, which was completed in 1771.

'European' Maps Made in Southeast Asia

The maps Forrest made for the Mindanao court and Bugis acquaintances were 'European' maps made in the East Indies. Map production was, in fact, an ongoing enterprise of Europeans in Southeast Asia. Nearly all such maps were in manuscript form, made both for the local use of settlers and for export to the mother country, where they would serve to improve official government charts of the region. Such Asian-Western charts are rare, since not only are manuscript maps inherently scarce, being produced individually, but in addition they suffered the wear of ship-board and local use, as well as a hot, moist climate. For the Dutch, the principal center for chart-making in Southeast Asia was Batavia; figure 157 illustrates one such anonymous Batavian work.

Chapter 16

The Nineteenth Century and The Mapping of The Interior

At the dawn of the nineteenth century, despite three centuries of European presence along the shores of mainland Southeast Asia, most of the interior still remained unknown and shrouded in mystery. In the words of Lieutenant James Low, an Englishman involved with the mapping of Southeast Asia at that time, "existing Maps of the World exhibit [Siam, Cambodia, and Laos] nearly as a blank."[285] Even the mighty, protective kingdoms of China and Japan had long ago yielded to the West far more than had the hinterland of the Southeast Asian mainland. Yet it was the reputed hiding place of considerable wealth, and its river systems promised a practical water route to China.

The frustration felt by some nineteenth-century Europeans by the refusal of the Southeast Asian interior to yield her secrets made the land into a conqueror's *femme fatale*. As a colonial officer of British Malaya in the late nineteenth-century wrote, "Sphinx-like she propounds riddles which few can answer, luring us onward with illusive hopes of inspiring revelations, yet hiding ever in her splendid, tattered bosom the secrets of the oldest and least amply recorded of human histories."[286] There was an urge to dominate this mysterious and little-known region: "The East baffles while she fascinates us: fascinates us because she baffles," the innuendo continued.

At the same time, purported philanthropical concerns ostensibly warranted the exploration and mapping of the Southeast Asian hinterland. "Men in high station and particularly those in India," a British official in Malaysia wrote in the early nineteenth century, "should never lose sight of" the fact that better maps of the Southeast Asian interior "will afford the best means for enabling us to diffuse among our ruder and less fortunate fellow creatures the blessings of useful knowledge and the comforts of civilized life."[287] Similar sentiments were voiced by the French in Indochina. In the mid-nineteenth century Francis Garnier, most famous explorer of Indochina's rivers, boasted of "France, the arbiter of Europe, making use of her influence only for the happiness and for the moral improvement of the peoples."[288] Thus, the mapping of the Southeast Asian interior was necessary for the European powers' sense of 'manifest destiny', to spread its cultural and material ways among Southeast Asia's peoples, and to save the vulnerable from unworthy conquerors, since the English and French nations each deemed their protectorships to be in the best interests of the indigenous peoples of the region.

Indochina

Garnier cited the British legacies of the Opium Wars in China and the decay of Indian civilization under British rule as proof that Cambodia, Laos, and Vietnam should welcome French protectorship to spare their kingdoms similar misery. "Go to India," he decried, "visit this country ruined and impoverished by the plunderings of the English Company." And of the Opium Wars, "who was not disgusted when [England, for opium profits, declared war on China to force the] Emperor to . . . sanction the poisoning of three hundred millions of men?"[289]

It was dreams of wealth, however, that continued to be the principal reason why Europeans were interested in obtaining an accurate cartographic record of Indochina and its rivers. Auguste Pavie, a major explorer of Indochina and the Mekong in the last decade of the nineteenth century, was still able to speculate that ancient Cambodia was none other than King Soloman's Ophir. But in the absence of any hard evidence that Indochina still harbored such mythical wealth, French fortune-seekers became intoxicated instead with the fantasy that its rivers might provide waterborne conduits of trade through the "wild countries" to "the fabulously wealthy regions of Western China," which, they believed, "the Celestials kept jealously guarded from prying European eyes."[290]

These ideas can be traced back a long way in the cartographic record. The 1641 portolan chart of Antonio Sanches (fig. 80), for example, shows a huge waterway in the vicinity of the Pearl River, in Guangdong, southern China, running wide and relatively straight all the way through to *quançi* (Kwangsi), where there is a large lake that would evidently provide easy docking facilities and travel in the region. From that same lake two other limbs of the river continue west to *quacheu* (Kweichou), while other branches could take trading vessels still further north into the interior of China. In the nineteenth century, while the British viewed the rivers of Burma as potentially promising waterways leading to the Chinese interior, French explorers looked upon the Mekong and the Red River in the same light. Of these two rivers, it was the immense Mekong, known as the 'River of Cambodia', which seemed to be the best-positioned waterway with an abundant supply of water to provide access to the far north.

Jean-Marie Dayot and John White

In the expectation that the Mekong would become a grand highway to the interior of southern China, an attempt was made to provide a river link between the Mekong and Saigon. A canal was therefore cut connecting tributaries of the Mekong with the Dong Nai River, one of the principal river approaches to the city. When an expedition from the United States, under John White, reached Vietnam in 1819, the commander reported that

> from the western part of the city, a river or canal has recently been cut, (indeed it was scarcely finished when we arrived there,) twenty-three English miles long connecting with the branch of the Cambodia [Mekong] river, by which a free water communication is opened with Cambodia.

Saigon itself was "within a few miles of the head of the ship navigation of that branch of the Don-nai river, on which it is situated." The 1823 account of the White expedition included a detailed map of the Dong Nai and its new canal (fig. 158), since an accurate mapping of the tricky system of waterways was a key to Saigon-based interests hoping to tap the potential of the Mekong. The map was based on surveys carried out in 1791 by Jean-Marie Dayot, "hydrographer to the King of Cochin China," updated to include White's own observations and the new canal. As White

sailed across a bay called *Nga-bay* or *sete-bocas* (seven mouths) in the waterways near Saigon, he wrote that

> the prospects from this noble basin, though possessing few features of the sublime, were beautiful and romantic. Lofty and venerable trees crowned the points formed by the effluence of the several streams, which, branching out in various directions, like so many radii from a center, presented to view long vistas, fringed on each side with foliage of different shades of verdure, while their polished surfaces reflected, with chastened beauty, the varied tints of the impending forests.

The Wall in Vietnam — Vander Maelen's *Atlas Universel*, 1827

A close view of Indochina at the beginning of France's intense exploration of the region can be found in the maps of Philippe vander Maelen, whose *Atlas Universel* of 1827 was the first atlas — aside from Coronelli's bound volume of globe gores — to employ a uniform scale throughout. The relatively new medium of lithography was used for the atlas' four hundred maps. Brilliant but impractical, its production would not have been viable had the author not inherited considerable capital from his parents' flourishing soap business.

Fig. 158 *Map of the River of Don-Nai. from Cape St. James to the City of Saigon.* The *Donnai* [Dong Nai] River in Vietnam, with Saigon in the upper left-hand corner. John White, 1823. (43.5 x 57.5 cm) [Edward Lefkowicz, Providence]

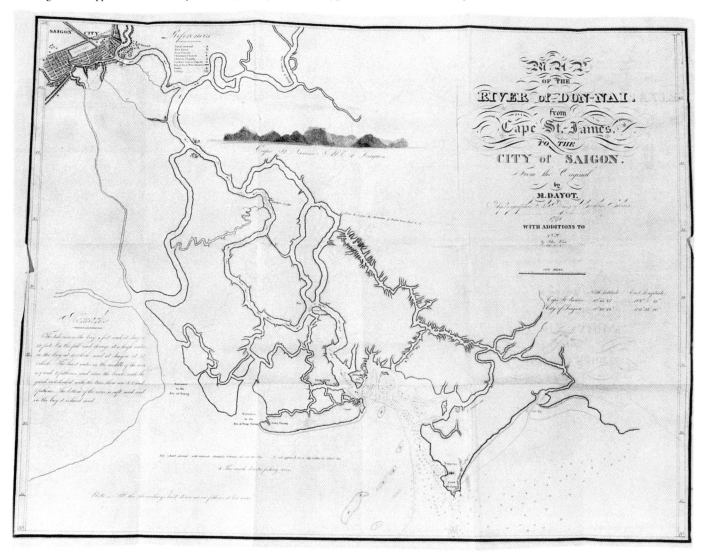

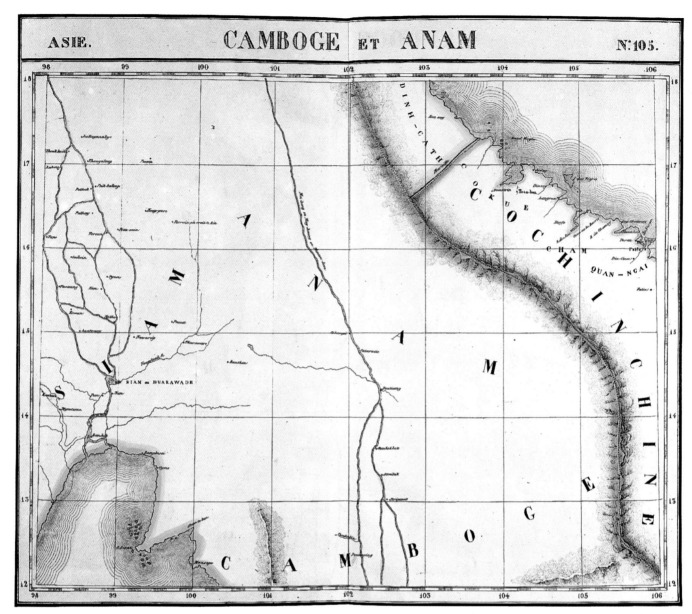

Fig. 159 Indochina, Philippe vander Maelen, 1827. (46.8 x 54.7 cm)

Fig. 160 "The wall that separates the two kingdoms" of Vietnam is marked on this map based on prototypes issued by Jean Baptiste d'Anville in the 1730s. Solomon Bolton, 1755.

The atlas' map of *Camboge et Anam* (Cambodia and Annam, fig. 159) depicts the region "nearly as a blank" (as James Low had complained), despite the opportunity for inland detail afforded by its large scale, had reliable data been available. One unmistakable feature, however, is the wall *(Muraille)* running from the coast, near Dong Hoi (not marked), to the Annam Mountains. This wall was originally constructed by China in about 1540 to mediate a conflict between rival Vietnamese factions, the Trinh and Nguyen.[291] William Dampier, who visited Vietnam in 1688, related that the wall was the only permanent fortification left standing after the incessant fighting in the country, but despite such reports, the wall was not recorded on printed maps until the French geographer, Jean Baptiste d'Anville, popularized it in the early eighteenth century (fig. 160). Placed at a latitude of about 17° north, it lies eerily close to the line drawn at the Geneva Convention of 1954 to temporarily divide Vietnam until general elections, which were scheduled for 1956, could be held (the elections never took place).

Cambodia, Caught Between Vietnam and Thailand

Vander Maelen provides a focus on Cambodia during the height of competing claims for vassalage from the kingdom by its mighty neighbors, Thailand and Vietnam. Seven hundred years earlier, in the heyday of the Khmer empire, the realm controlled by Angkor had extended well into what is now Thailand and Vietnam. When

Angkor began to decline in the fourteenth century, Sukhothai and later Ayuthaya extended their control over Cambodia, ultimately claiming it as a tributary kingdom. By the seventeenth century Vietnam was making demands upon Cambodia as well, and 1834 saw the beginning of fourteen years of open warfare between Siam and Vietnam to decide who should possess Cambodia. In 1847 Cambodia, now 'independent' again, yet also, once again caught between her two powerful neighbors, was in ruins.

The amorphous nature of the indigenous 'map' of Southeast Asia is marvelously painted by the king of Cambodia's plea to his neighboring kingdoms at the war's end. He understood that because of his land's geographic position, Cambodia would never be left at peace as long as it was vassal to only one of its giant neighbors. Thus the king begged to be symbolically 'conquered' by both nations: "Please let me be subjected to the merit and power of both great kingdoms [Siam and Vietnam], so that my people can live in peace and happiness."[292]

The Mystery of the Mekong

In 1858-60, the naturalist Henri Mouhot traveled far north into Siam, Cambodia, and Laos, where he reached Luang Prabang and gained a rudimentary picture of the course of the Mekong. The 1863 map of Southeast Asia by Edward Weller (detail, fig. 161) is typical of this period, correctly depicting the Mekong veering to the east at the capital city of *M. Udong* (Phnom Penh). But neither the Mekong's dramatic turn to the north just east of Vientiane, nor its sharp doubling-back to the west above Luang Prabang, the river's northernmost limit of navigability, were as yet recorded (Luang Prabang is Weller's *Muang-luang Prabang*; *muang* = city).

These additions to the map came shortly after Mouhot's time, courtesy of the energetic Doudart de Langée and Francis Garnier, whose expedition of 1866-68 finally brought Europe detailed knowledge of the upper Mekong and the region's mountainous frontier with China. The material from Garnier's expedition was

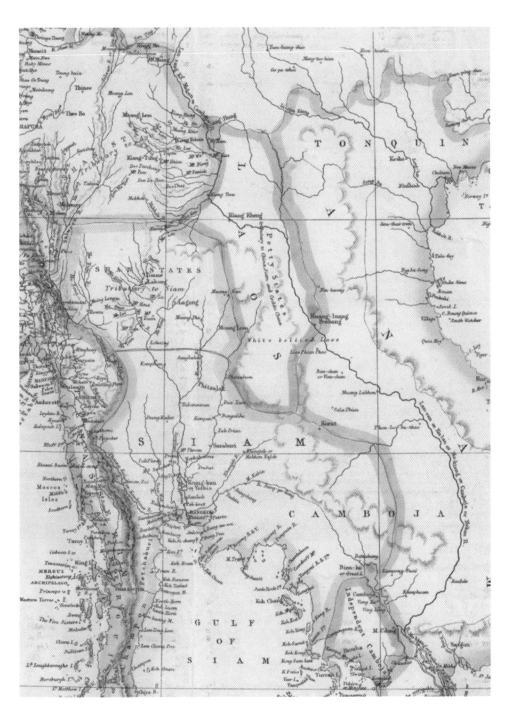

Fig, 161 Southeast Asian mainland (detail), Edward Weller, 1863 (45 x 30.5 cm). Despite the clear boundaries inferred by the colored lines, Weller tries to indicate the subtleties of Southeast Asian political alignments of the day by marking certain areas as "petty states" and indicating their kingdom's tributary status. The area outlined in purple, which was peopled by various little chiefdoms, is interesting is this regard. The fact that the Mekong ran through their collective territories, and that they lay in the cracks between Siam and occupied Indochina, meant that within the next few decades these groups were literally squeezed out of existence as Siam and France pushed in from opposite sides. Burma boasts the greatest detail in the map, a result of Britain's studious mapping of their exploits there. Although the Lan Na kingdom (center left) was still independent, its close alignment with Siam led Weller to include it in the same wide orange outline, separating the two former rivals with a thin red line. Its capital, Chiang Mai (here *Zimmé*), whose location was long left to guesswork, is finally mapped accurately.

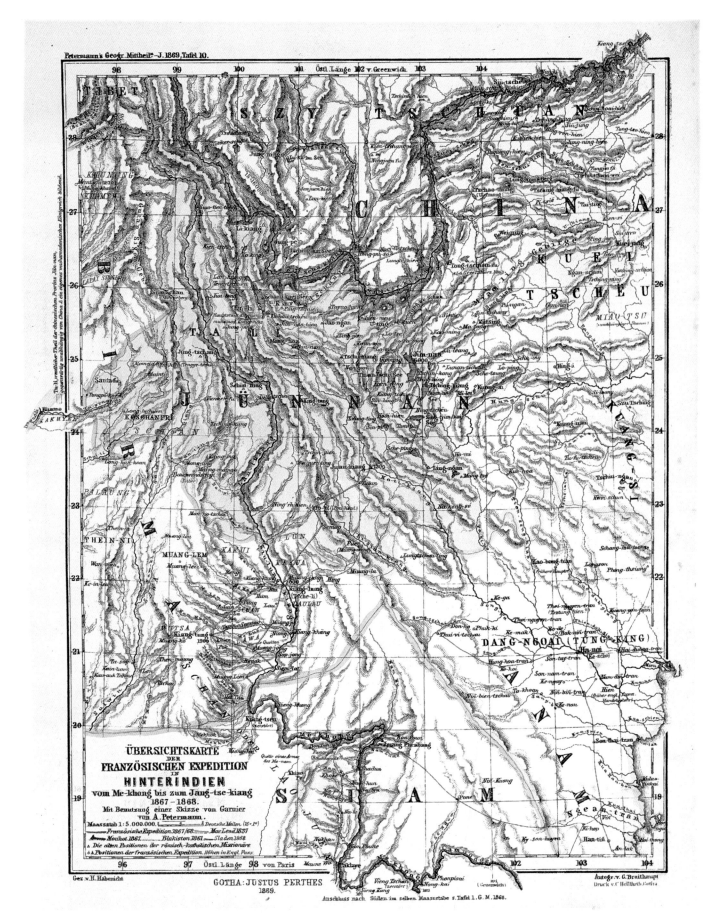

Fig 162 Northern Indochina, recording the course of the Mekong as mapped by Garnier. Justus Perthes, 1869. (26 x 20.5 cm)

quickly tapped by the the German mapmaker Justus Perthes for his *Übersichtskarte der Französischen Expedition in Hinterindien vom Me-khong bis zum Jang-tse-kiang 1867-1868* (fig. 162), which appeared in 1869, the very year after the expedition was completed.

Garnier described Luang Prabang as being the capital of Laos,

> picturesque and pleasant to the view, [with numerous houses] arranged in parallel lines around a small central hillock, which, like a dome of verdure, rises above the mass of gray thatched roofs . . . on the summit a tat or dagoba elevates its sharp arrowy pinnacle above a belt of trees, so as to form a landmark for all the surrounding country.

The red roofs of pagodas were "vividly defined against the sombre green vegetation," and there was a riverine city of huts erected upon rafts, "connected with the capital city itself by zigzag paths, shining like white ribbons in the distance." The river itself was a bustling highway for boats of all sizes, where "the hum of voices mingles confusedly with the murmur of the stream, and the whisper of the palm-trees [waving] their feathery crests upon its smiling and fertile banks."[293] Upriver from Luang Prabang, Garnier found that the Mekong "narrowed considerably, and resumed its wild and romantic aspect."

Many stretches of the river were impassable, forcing the pioneers to carry their boats and supplies overland. After considerable hardship (de Langée died *en route*), Garnier and company succeeded in reaching Yunnan via the Mekong, but the trials of the Mekong had been so daunting that they did not attempt to retrace their path. Instead, the expedition, led by Garnier, returned to Saigon by the Yangtze River and China Sea. Exceedingly difficult and utterly impractical, the great river did not fulfil their dream of a grand Indochinese seaway to the Celestial Kingdom.

The expedition proved that what Engelbert Kaempfer had gleaned from the Thai, concerning Laos, nearly two centuries earlier, was correct. While staying in Ayuthaya in 1690, Kaempfer was told that travel on the upcountry rivers

> is very inconvenient and troublesome traveling thither by Land, on account of the high mountains, and on the River, by reason of the rocks and Catarracts, so that the Vessels, or Prows as they call them, are built in such a manner, that they can be taken to pieces, and carried over the eminences in order to pursue the Journey by water.

Yet, although Francis Garnier left behind his hopes for the Mekong, his dream of a water channel to China remained. He now believed that of the five great rivers of Southeast Asia — Irrawaddy, Salween, Chao Phraya, Mekong, and Red — it was the last of these, still virtually unknown to the West, by which "the plethora of the riches of Western China would one day flow into a French port." Moreover, it would be the Red River basin and the region of Tongkin, with its principal city of Hanoi, which would be "a necessary sequel to the establishment of the French flag in the six provinces of Cochin-China." Garnier was killed in Hanoi in 1873 during the French capture of the city.

The usefulness of the Red River *south* of Hanoi, for access to Chinese commodities via the Gulf of Tongkin and South China Sea, had in fact been a consideration to the Vietnamese a thousand years earlier, when a capital was founded near the modern city of Hanoi. A tablet dating from 870 preserves the tradition that "the rousing power of thunderbolts" had caused the sea "to form a channel, where boats can pass in safety, with the deep sea stretching out peacefully, a highway of supply for our city" — the genesis of the Red River.[294] Hanoi was symbolically considered the 'Dragon's Belly' *(Long-do),* the spiritual center of the kingdom.

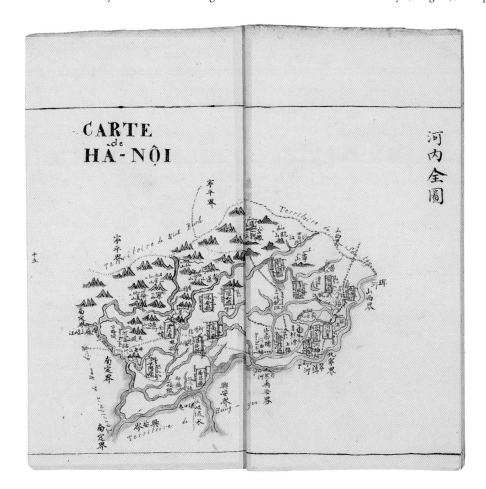

Fig. 163 (*left*) Hanoi and environs, from an anonymous Vietnamese atlas. Manuscript, ca.1870. [Rodolphe Chamonal, Paris]

Fig. 164 (*below*) Cover to the anonymous Vietnamese atlas illustrated on the left.

Fig. 165 Manuscript map of a 20 x 35 km parcel of land in the Indochinese interior, made for the Ministry of Indigenous Affairs, 1874-75. Found among the papers of the French explorer F. J. Harmand, and signed by Dr. Nansot. (27.5 cm. x 39 cm) [courtesy of Cornell University Library]

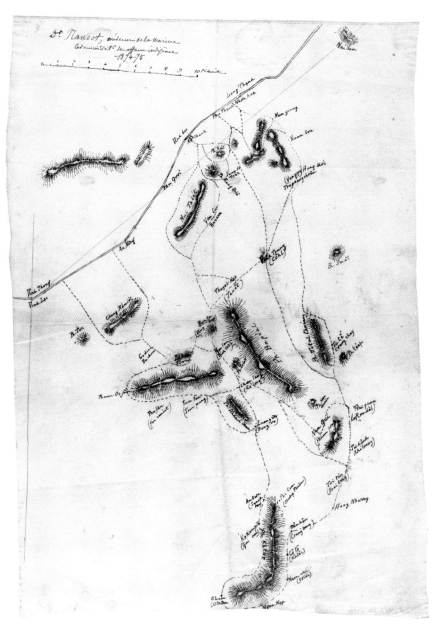

French-Vietnamese Itinerary Atlas

An example of the new, imposing French presence in Vietnam is reflected by a manuscript atlas of the country entitled *Hoàng Việt Dú Dô* (*Atlas of Vietnam*) from which the map of Hanoi illustrated in figure 163 is taken. The map dates from ca. 1870, about the time Hanoi was established as the capital of French Indochina. Though Asian in the style of its execution, it is nevertheless one of the earliest Vietnamese atlases with nomenclature in both French and Chinese calligraphy. The map perfectly exemplifies the Vietnamese concept of their country as *Non nuóc,* meaning 'mountain and water', with the river systems around Hanoi being drawn out of scale , while the inland mountains rise in a stylized fashion in the northwest, as if in a Chinese painting.

Meanwhile, French expansionists continued to explore the Indochinese interior and the upper reaches of the Mekong River. In early 1877, François Jules Harmand set out to explore the Boloven Plateau and the river valleys in Central Laos, hoping to establish a route from Bassac on the Mekong to Hué on the Vietnamese coast. A manuscript preserved among his reports is illustrated in figure 165.

To realize his vision of a French Laos, Harmand believed that the indigenous people must first be replaced. In 1880 he wrote that "it is necessary first that the Laotians be eliminated, not by

violent means, but by the natural effects of competition and the supremacy of the most fit." To accomplish this, he envisioned shipping "Annamites" (the Vietnamese), once subjugated, "to colonize to our profit a large part of [the Mekong Valley in Laos] where they will rapidly supplant the debris of the decrepit races which inhabit it."[295]

After Harmand, Auguste Pavie conducted a series of arduous explorations through central Indochina in 1886 to 1895. Traveling by elephant, raft, horse, and foot, Pavie managed to complete an unprecedentedly thorough survey of Laos. The resultant map from his field work was issued in 1899 and remained the model for the mapping of the region for the following two decades. The lure of the Mekong as a shuttleway to China had not been entirely forgotten, a new generation of proponents believing that it could serve as such if a viable pass around the river at Khone were found. Even Laos itself was still idealized as a region of vast wealth — in anticipation of Pavie's expedition, the Commercial Geographic Society of Paris referred to the country's "incalculable riches".

The petty-states of Laos could not resist French expansion. But Siam's monarchy, keenly aware that France was scooping up as much of the 'unclaimed' territory of Indochina as possible, sought to expand its sovereignty to the eastern shores of the river, and began a game of geographic chess with the French.

Burma and Thailand

Of the various causes behind the Burmese-British conflicts which led to the British taking of Burma, one is of interest to us here, namely a clashing of alien geographic concepts which exacerbated the quagmire. Burma claimed some allegiance from the kingdoms of Assam, Manipur, and Arakan, but dismissed culpability for any raids that those kingdoms might make into lands already taken by the East India Company in India. Britain, with its Western concept of political space, could not understand the Southeast Asian concept of frontier zones and indistinct sovereignty. King Bagyidaw of Burma did not think in terms of precise borders, but rather regarded Assam, Manipur, and Arakan as essentially autonomous regions which were simply expected not to act against Burmese interests. He did not perceive national borders in the clear, absolute terms of the European interlopers, but instead envisioned borders as approximate and permeable. But hindsight aside, King Bagyidaw made the worst possible decision in response to the escalating vanities between himself and the British: he invaded Bengal. Thus ensued the Anglo-Burmese war of 1824-26.

The British dispatched forces up the river to Prome and demanded that Arakan and Tenasserim be ceded to the Company. When the King refused, the British advanced further north, ultimately forcing Bagyidaw to give up Tavoi, Mergui, and Tenasserim. By having secured these strategic lands, the East India Company occupied the entire vital coastal strip from India to Siam, effectively strangling Ava. Though no further advances were made for another twenty-six years, the British colonial empire of Burma had been born.

Sidney Hall, Burma and Environs, 1829

The 1829 map by Sidney Hall entitled *Birmah, with part of Anam and Siam* (fig. 166), records Burma during its transition to a British colony, being published three years after the British conquest. Hall locates a striking number of places along the Irrawaddy River, north through Mandalay (*Amarapura*), in considerable detail, even further to the north, evidencing the degree to which the British had by this time penetrated Burma. British maps of the river as far as Mandalay had improved markedly as a result of a surveyor by the name of Lieutenant Woods, who accompanied a British embassy to Ava in 1795 led by Captain Michael Symes. The river journey to Mandalay had become well-known to British expansionists, since Burma's reluctance to grant them lavish trading privileges brought them repeatedly to the then-capital.

The land to the east of Burma remains, for the most part, a mysterious and unexplored region on the Hall map. In the mountains of the Burma-Siamese border there is a section marked off as "Extensive Hilly Tract occupied by the Karaen." Hall is referring to the Karen, one of the so-called 'hill tribe' people who live in the western and northern mountains of Thailand and its frontier with Burma. Of these groups, the Karen may have lived in the mountains of western Thailand (at this time still Lan Na) for hundreds of years, while the Hmong, Akha, Lisu, Yao, and others began settling in northern Thailand from Burma and Laos around 1880.[296] Although the Karen are rarely mentioned on maps, they were known by Europeans since the sixteenth century. Camões, repeating heresay, described the Karens as "savage tribes inhabiting the remoter hills who eat the flesh of their enemies and cruelly tattoo their own with red-hot irons," perhaps associating them with the feared *Gueos*.

The Rebirth of Chiang Mai

Continuing east into the kingdom of Lan Na, marked both as Laos and Annam, we find Chiang Mai present and designated by its modern name, an early testimonial to the rebirth of the region. At the close of the eighteenth century Chiang Mai, in a gradual state of decline since its capture by the Burmese in the sixteenth century, was virtually deserted, and had been nearly forgotten by the cartographer's quill (although it does appear on, for example, the map of Siam in La Loubère, fig. 129). The major population centers of what is now northern Thailand were in ruins, their people scattered about the countryside. In 1775 — the very year Siam re-opened itself to the West — the royal court of Chiang Mai moved to a camp south of Lamphun.

The fateful event that placed Chiang Mai back on the map was, ironically, the Burmese devastation of Ayuthaya in 1767. That attack made clear to both Siam and Lan Na that it would be in their mutual interest to join forces against Burma. Following the sacking of Ayuthaya, the capital of Siam was re-established further to the south on the east bank of the Chao Phraya River at Thonburi. There it remained until 1782, when the capital was moved again, downriver to the west bank at Bangkok. In 1797 the king of Lan Na re-entered Chiang Mai, protected by an alliance with Siam. During the next several years the combined forces of Siam and Lan Na successfully fought off repeated Burmese attacks; Burma, in the meantime, had become increasingly preoccupied with the British.

The Siam - Lan Na alliance directly led to Chiang Mai, Chiang Rai, and the other northern states ultimately becoming part of Thailand. Chiang Rai, lying close to the frontiers of both Burma and Laos, was not formally re-founded until 1844 and thus does not yet appear on Hall's map. The present borders of northern Thailand, Burma, and Laos are the result of treaties with France and England drafted between 1894 and 1904 (see 1904 map of Chiang Mai, fig. 168).

The British Tap Native Sources

Whereas European maps traditionally contained a distance scale calibrated in one or more types of European miles, Hall has marked his scale with English miles and "Roé-ning of Siam", the "roé-ning" presumably being derived from *nueng roi,* the number one hundred in Thai — it may have equaled a league.[297] This attempt to indicate Thai linear measurement is but one piece of evidence of British interest in gleaning what they could from their indigenous colleagues. In 1795, Francis Hamilton, resident in Burma for eight months as an aide to his country's first ambassador to the country, collected, studied, and supervised Burmese mapmaking. While Hamilton was interested in Burmese maps for their scholarly interest, he was also gratified that they "tended to throw much light on the geography of what . . . is called the *Farther Peninsula of India*." No doubt this was especially interesting, coming at a time when Britain was vying for possession of Burmese soil. Among the maps acquired, for example, was one of Lan Na, which in 1795 was still sought by Burma and thus a likely object of British designs.

Another British mapmaker interested in native maps was James Low. In Penang (Prince of Wales Island) in 1824 Lieutenant Low presented the island's British authorities with a map of Siam, Cambodia, and Laos which, according to an accompanying letter, he had first completed in 1822 "from a large mass of original, and I firmly believe, authentic Native Plans, and Itineraries, and oral information."[298] Low's map contained considerable topographic data but failed to meet its author's promise that it was accurate "in all its principal outlines and features." Soon after the map was issued, Low became involved with the First Anglo-Burmese War.

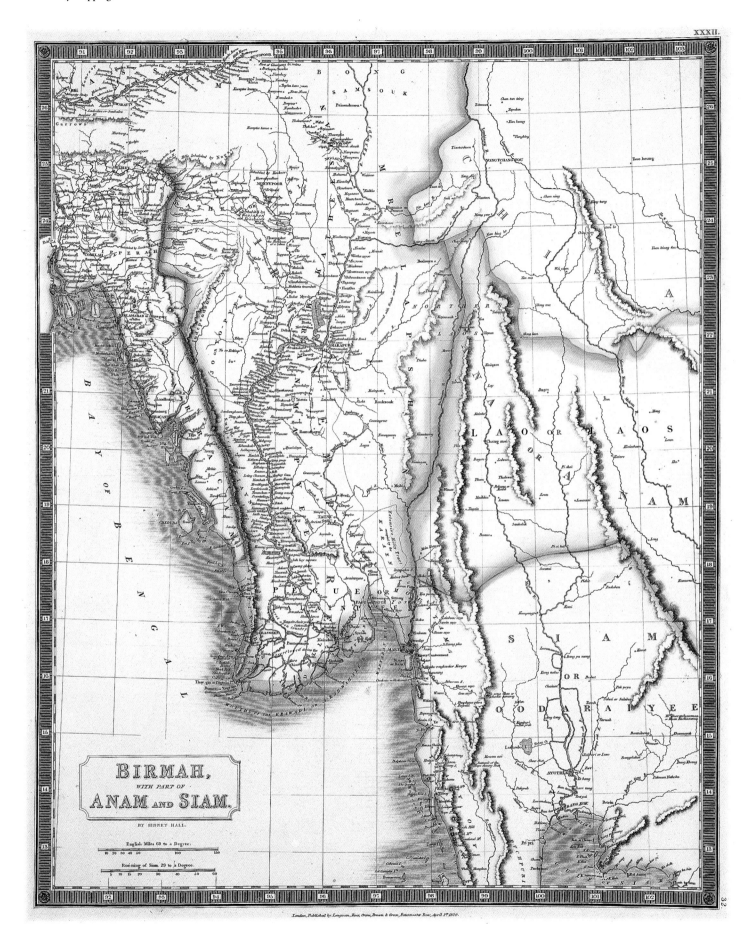

Fig. 166 Burma and Thailand, Sidney Hall, 1829. (42 x 51 cm)

Fig. 167 A lithographed map card printed by the Donaldson Brothers of New York between 1891 & 1893. The map of Siam was numbered 91 in a series of 50 cards numbered 51-100, which were included in 1 lb. bags of Arbuckle coffee. (7.3 x 12.3 cm)

The Mapping of the Mekong and the Fate of Thailand

In 1830 Low produced a new map of the region while residing in Calcutta.[299] This work offered some modest improvements in its rendering of coastlines as surveyed by John Crawfurd, an enterprising British envoy sent to Bangkok in 1821, as well as some shores charted by Low himself. The map's main shortcoming, however, was that it perpetuated the erroneous European view of the Kingdom of Siam as being much narrower (from east to west) than it actually was by placing the Mekong River too far to the west. Since the Mekong was seen as the boundary between Siam and her neighbors to the east, ignorance of the river's course transposed what was actually eastern Siam into Laos and Cambodia. Low's reinforcement of this error probably influenced the course of events in the colonial struggles of nineteenth century Southeast Asia.

There are two reasons for this. Firstly, had the true extent of Siam been understood by the British authorities in India or French colony-seekers in Indochina, the conquest of Siam would have been a higher priority, both as a material prize in itself and to avoid conflicts with Siam in the future. The second reason is the way in which Siam figured into British-French accords to divvy up mainland Southeast Asia. While the British were subduing Burma, the French were beginning to consolidate their claims in Indochina. Britain and France agreed to use Siam as a buffer between them — a seemingly reasonable accord if the negotiators believed that Siam was a slender piece of land. Thus Siam was envisioned as a frontier zone, or buffer sate, by Britain and France. Much later, this same geographic principle can be found in the 'DMZs' (demilitarized zones) of Korea and Vietnam.

The Weller map shows the kingdom of Lan Na as the southern 'Shan States', newly incorporated into Siam as tributary lands in an alliance to counter Burma. The more northern Shan States are demarcated as tributary to Ava (Burma). The border separating the two kingdoms on the north is drawn straight at a latitude of 20° north, the vicinity in which the modern irregular border now lies. Chiang Mai is mapped as *Zimmé,* which was a common

nineteenth century English rendering of the Burmese pronunciation of the name. Vientiane, the modern capital of Laos, had been overrun and largely destroyed by Siam in 1827, and appears only as the village of *Ban-chan or Vien-chan* several miles from the Mekong.

British negotiations with the Siamese court proved at least as frustrating for the British as those with Burma. Britain wanted to establish a clear, Western-concept boundary between her spoils in Burma and the land claimed by Siam, but Bangkok did not understand why it was necessary to define a boundary at all, and when pressed to do so it could only offer a traveler's description of a vague 'frontier' rather than a border. Rivers and mountains, of course, were part of the guide posts, but so were "teak forests, mountains upon mountains, muddy ponds where there were three pagodas, Maprang trees, three piles of stones," etc.[300] The British negotiators were further confounded when Siam repeatedly showed no interest in actually going into the field to identify and plot their claimed boundaries, leaving it to the British to do so if they really felt it was necessary.

Shortly before Weller's map there was a conflict in a Karen village in the mountains between Chiang Mai and Ava — a fairly blank area in the map. The British were surprised to find the people of Chiang Mai rather indifferent about the trouble, saying that the Karen town "belongs to no one." In British eyes, of course, a town could not belong to 'no one', but even to this day, the various hilltribes along the Thai-Burma frontier live largely aloof from national identity, and many who live within Burmese territory have violently resisted Rangoon's jurisdiction over their lives. In their concept of geographic space, they live in 'the cracks' between larger neighbors — their home does not lie on the 'map' of any nation.

The northwestern border of the blank area on the Weller map is the Salween River. There were many teak forests along the banks of this river and in the region of Lan Na (much of the blank area), which the British were keen to acquire. Here the British found themselves in an even more unfamiliar conceptual 'territory' than

Fig. 168 Map of Chiang Mai, ca.
1904. According to a note at the
bottom, written by a military officer
in 1972, the map was "made in 1904
when General Prince Nakhon Caisi
Suradet traveled to the northern
provinces." [reproduction by H.
Penth (1986) of a copy in the
National Museum, Chiang Mai,
Thailand. The original is preserved in
the Map Division, Royal Thai Army,
Bangkok.]

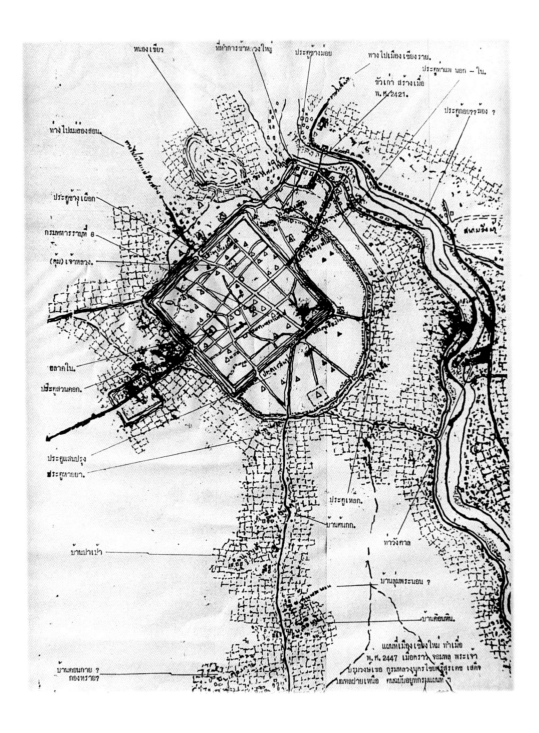

before, since the northern regions were more fragmented than those they had already experienced in Burma and Siam proper. Various villages were accountable to one or more kingdoms and the people who lived along the frontier with what had now become British Burma could not understand why they could no longer cross a certain invisible line (Burma's British-defined boundary) to continue trading, hunting or fishing as they had previously done without regard to the soil's 'ownership'. Just as the Siamese authorities had suggested to the British that they go ahead and plot the proposed boundaries without them, so Chiang Mai showed no interest in mapping the Salween River, but encouraged the British to do so themselves. As a result, between 1847 and 1849, British surveyors mapped the Salween in detail, hence the improvement in the portrayal of that river between the Hall map of 1829 and the Weller map of 1863.

Weller, however, was aware of the fact that many little kingdoms lay in the cracks between the major Southeast Asia powers. Along

the west bank of the Mekong in the north-northeast of Thailand we find "Petty States tributary to China, Siam and Cochin China". The northern 'Shan States' (now the Shan region of Burma) are marked as being tributary to Ava, while the southern Shan States (Chiang Mai) are tributary to Siam. Cambodia is ambiguously marked overlapping Thailand and Vietnam. Below it lies the remains of 'Independent Camboja', which in the earlier part of the nineteenth century served as a buffer state between Vietnam and Thailand.

Transition to the Modern Era

Britain and France mapped Burma and Indochina on their own terms, since they were usurpers in those regions — they simply introduced Western mapmaking as a part of the prerogative of conquest. Siam, in contrast, remained independent. Nevertheless, the desire to modernize the kingdom's infrastructure, as well as the contest to establish the most favorable borders possible in the changing political map of the region, led the kingdom to import Western cartography.

The acceptance of Western mapping techniques in Thailand accelerated under the reign of King Mongkut (1851-68). Although Mongkut was already familiar with Western maps, it was French imperialism in Indochina that provided a catalyst for Siam's own "Western" mapping of its kingdom. After France had effectively conquered Vietnam, it began to move inland, penertrating the numerous little chiefdoms which lay in the blurred expanses that separated Siam from Vietnam and Cambodia. In 1866, Mongkut learned that French explorers were surveying areas along the Mekong and, understanding the political implications of such mapping, hired a Dutch surveyor to lead a team to the Mekong to map the region for the Thais.

This initiated a contest which effected the demise of the ancient "mandala psychology" of Southeast Asian geo-politics, in which power and sovereignty radiated out from the center of a kingdom. Both France and Siam were rushing to establish precedence in what had always been the lands-in-the-cracks — those regions occupied by chiefdoms and hill tribes which may have formally honored one or more of the large kingdoms with tribute and tokens of allegience, but which remained essentially autonomous. Suddenly, these remote enclaves were pressured to align themselves with either Siam or France, to include their community on one side or the other of an imaginary line defining political ownership. This geo-political polarization, by definition, meant the ultimate demise of these same minor kingdoms.

Siam also hired foreign technicians, though not necessarily professional surveyors or cartographers, to carry out the mapping entailed by modernization projects. The desire to build roads, railways, and telegraph lines all required that modern surveys be undertaken. Even Auguste Pavie, later to become famous as a determined French explorer of the Indochinese interior, worked in Siam during 1878-79, directing the construction of a telegraph line from Bangkok to Battanbang (now in Cambodia), which linked up with the French line connecting Saigon.

In 1880, new impetus came from Siam's western colonial neighbor. Britain had been conducting extensive triangulation surveys in India and Burma, but needed to enter Siam to complete the process. Many in the Siamese court viewed the British request for triangulation in Bangkok with distrust, especially since two of the specified marking points were important religious sites. But King Chulalongkorn agreed to the proposal when it was suggested that triangulation be conducted by British engineers to construct a map of Siam itself.

This, in turn, led to other British-led Siamese mapping projects in the early 1880s. These spanned the kingdom and were diverse in purpose: a telegraph line was surveyed and constructed linking the Siamese town of Tak with Burma; a map of the Chinese district in Bangkok was made which facilitated more efficient collection of a tax on Chinese residents; a map of the frontier between Chiang Mai and neighboring Thailand was drafted to settle a dispute over a woodcutting tax; and a map was composed of the boundary between Pattani, a Malay state under Thai control, and a neighboring Malay state under the British. However, the British surveyors claimed that many Thais, high-ranking and commoner alike, were suspicious of their activities and impeded their surveys. But in 1882 a school for cartography was establish by Prince Damrong, and three years later Siam established the Royal Survey Department, effectively integrating Western mapmaking as an official part of Thai culture.

At the same time that Britain and France vied for control over mainland regions and Western mapping techniques were adopted by independent Siam, shifts in land usage were also causing dramatic changes in the "map" of both continental and insular Southeast Asia. One crop which had a particularly visible effect on Southeast Asia was rubber. Rubber, was a substance that had little practical application until scientific advances of the nineteenth century developed processes which enabled it to be made into a stable substance relatively unaffected by temperature, It was originally available only from unreliable and expensive sources in South America, but during the last two decades of the nineteenth century rubber was transplanted to Southeast Asia where it was found to grow profitably. This coincided with a dramatic surge in worldwide demand for the material, and soon rubber plantations covered vast areas of Vietnam, Cambodia, Malaya, Sumatra, and Java, usurping existing crops and clearing large tracts of previously uncultivated land and jungle. In turn, the considerable investment these plantations required further solidified colonial control over these regions.

By the early decades of the twentieth century, the colonial grab-bag and Siam's posturing had set the foundation for the political divisions we have today, and the continued energetic exploration of the interior had exposed the region's major geographic secrets. But neither Southeast Asia, nor cartography itself, is static, and the mapping Southeast Asia's land, people, resources, ocean floor, and a myriad other aspects, has continued through to our day.

Endnotes

1. The Pali Text Society's *Pali-English Dictionary*. Thanks to Hans Penth for this information.
2. *History of Cartography* 2/2, p. 479.
3. Williams, p. 25.; Kenneth Hall p. 256. For a thorough analysis of current research see Charles Higham, *The Bronze Age of Southeast Asia*. Cambridge University Press, 1996.
4. Kenneth Hall p. 6.
5. Majumdar, p. 20.
6. Lore about the *naga* was not confined to Southeast Asia proper. Antonio Pigafetta, who supplemented his first-hand record of the Magellan expedition with hearsay to provide a fuller image of the East, mentions a 'serpent' called the *nagha* into which the king of China would enter with his six principal wives so that the people, who looked on them through a window in the *nahga's* breast, might view them. See Paige, *Pigafetta*, p. 143.
7. Keith Weller Taylor, p. 1.
8. Wheatley, *Golden Khersonese*, pp. 253-254.
9. Excepting northern Vietnam, eastern Java, Bali, and Lombok. See Reid p. 8, Steinberg *et al*, p. 14.
10. See Mantle Hood and James Brandon's article about Southeast Asia in Stanley Sadie (ed.), *The New Grove Dictionary of Music and Musicians*, volume 17, pp. 762-779. London, Macmillian Publishers Limited, 1980.
11. Dawn Rooney of Bangkok has kindly brought to my attention the fact that, although women are almost always associated with throwing or coiling pots and decorating their form, there are references in inscriptions found in Southeast Asia in which men quarried the clay and handled the firing of pots.
12. See Reid, ch. 4. That the origin myth of Funan civilization, two thousand years ago, traces Funan roots through a woman — the tradition of the *Naga* princess — suggests that the role of females in Southeast Asia was even more extensive before the Indianization of the region.
13. Except for the Chinese-influenced Vietnamese Court. For Micronesia, see Lévesque, Volume 1, e.g., p. 466. Early Spanish records of life in Micronesia, the eastern periphery of Southeast Asia, also records the dominant role of women. However, attempts to correlate literacy and gender in Southeast Asia are tenuous; some commentators reported a higher literacy rate among women than men, notably in the Philippines, though this is disputed.
14. Tradition records that this woman, a Mon princess named Jam Thewi from Lop Buri (central Thailand), ascended to her reign in 662 or 656, but modern historians believe this figure to be a century too early. See Hans Penth, *A Brief History of Lan Na*, p. 3. The *rishi* did not deliberately select a woman, but rather did not care about the new ruler's gender (thanks to Dr. Penth for clarifying this point to me).
15. Thailand's Queen Suriyothai and Lady Mo; Vietnam's Trung sisters.
16. Kenneth Hall, p. 9.
17. Clifford p. 195.
18. Tibbetts p. 53.
19. SarDesai pp. 66-67.
20. Jacques de Bourges, in Michael Smithies, *Journal of the Siam Society*, Volume 81, Part 2, pp. 113-129.
21. For much of this section on Indian thought, I am indebted to Deepak Bhattasali.
22. These Jain concepts find a modern counterpart in the physicist's concept of an 'inflationary universe' as a solution to the dilemma posed by thermal equilibration and finite time horizons (courtesy of Richard Casten).
23. Tachard, p. 284.
24. The European concept of a mountain partition is found, for example, in the the anonymous mid-seventh century *Ravenna Cosmographia*. The *Traiphum* quotation comes from Reynolds and Reynolds, *Three Worlds*.
25. See Hans Penth, 'Buddhist Literature of Lan Na ...', pp. 43-81.
26. Winstedt, in Wheatley, *Golden Khersonese*, p. 51, note 5.
27. Covarrubias, *Island of Bali*, New York: Knopf (1937), p. 7.
28. See Reid p. 126.
29. Funan is not mentioned in any Southeast Asian annals, save for one Sanskrit inscription in southern Vietnam, which is believed to date from the third century, and may contain a reference to the kingdom. Sar Desai states that the so-called Vo-canh inscription, the earliest known Sanskrit inscription from Southeast Asia, refers to Funan, but this interpretation is based on geographical proximity. It does not mention the kingdom by name. Thanks to Dawn Rooney for this information.
30. Kenneth Hall, p. 48, Chandler, *History of Cambodia*, p. 14; also Majumdar, p. 12. A Roman medal unearthed in Oc-èo bears a portrait of Marcus Aurelius, Roman emperor of the Antonine dynasty, which would date it at the second century.
31. Smithies, *The Siamese Embassy to the Sun King*, of unproven authorship but possibly a translation (from French and in turn from Thai) of the memoirs of Kosa Pan. The source for the material was Jean Donneau de Vizé, 'Voyage des Ambassadeurs de Siam en France,' in Mercure Galant, Paris, 1687. See also Michael Smithies, 'The Travels in France of the Siamese Ambassadors 1686-7,' in *Journal of the Siam Society*, Vol. 77, Pt. 2, 1989, pp. 59-70. Thanks to Mr. Smithies for his correspondence with the author on the subject, September, 1998.
32. Sunthorn Bhoo, p. 9.
33. *History of Cartography* 2/2, and figure 1.14.
34. Quoted from Kenneth Hall, p. 48. In the same work Hall (p. 59) discusses the view, held by some, that Funan cannot truly be classified as a 'state', though for the purposes of this volume it certainly qualifies as such.
35. For Angkor, see Osborne, p. 25, and Eleanor Mannikka, *Angkor Wat: Time, Space, and Kingship*. Honolulu: University of Hawaii, 1996. For Pagan, see *History of Cartography*, 2/2, p. 738.
36. Quoted from Sternstein, 'The London Company's Envoys Plot Siam,' in *Journal of the Siam Society*, Volume 81, Part 2, 1993, p. 20.
37. Thanks to Dr. Hans Penth for this comment on palm leaf writing.
38. Hans Penth, correspondence, 1997.
39. For the batik map, see *History of Cartography* 2/2, p. 697 (which dates it at 'presumably nineteenth century', and p. 771, which dates it as "likely . . . not much later than 1800." For batik and tattooing having been related, see Reid, p. 77.
40. Frédéric Lutké, *Voyage Autor du Monde, exécuté par order de Southeast Asia Majesté l'Empereur Nicolas 1st* . . . Paris, 1835. There is no corroboration of his report or his interpretation of what he saw. Thanks to David Woodward for kindly tracking down this Lutké reference for me, which was alluded to in P. D. A. Harvey, *Topographical Maps*, p. 29. Translation courtesy of Maurizio Favretto.
41. See Stanley Sadie (ed.), *The New Grove Dictionary of Music and Musicians*, 1980, Volume 13, pp. 128-154, especially p. 151.
42. Kenneth Hall p. 138.

43. Chandler, p. 14. Thanks to Dawn Rooney for clarifying this information.

44. In Burma, *nat* spirits were assimilated into the Mahagiri spirit of Mount Popa. See Kenneth Hall p. 5.

45. Wheatley, *Golden Khersonese,* p. 65.

46. Covarrubias, *Island of Bali,* p. 6.

47. Paul Mus, *Indian & Indigenous Cults in Champa,* pp. 31, 43-45.

48. *History of Cartography* 2/2 pp. 480-81.

49. David Chandler, *The Land and People of Cambodia,*, p. 66-67.

50. The analysis of this map is by Hans Penth, personal correspondence.

51. See Hans Penth, 'Inscription 1.2.2.1 Wat Phra Sing — Restoring a Buddhapada,' and 'Inscription 1.2.2.1 Wat Phra Sing — The Buddhapada of Wat Phra Sing.'

52. See A. B. Griswold, pp. 173-221.

53. Mus, pp. 20-21.

54. *History of Cartography* 2/2 pp. 480-82.

55. See John K. Whitmore, 'Cartography in Vietnam,' in *History of Cartography* 2/2, p. 482.

56. Hans Penth, personal correspondence, October 8, 1997. See T. Grimm, 'Thailand in the light of official Chinese historiography. A Chapter in the *History of the Ming Dynasty*'. *Journal of the Siam Society,* 49.1, 1961, p. 3.

57. The tentative dating according to Hans Penth, based on the text (personal correspondence). The partial translation by Ahngsana Suárez.

58. This map is analyzed by Larry Sternstein, in 'The London Company's Envoys Plot Siam', *The Journal of the Siam Society,* Vol. 81, Part 2, 1993.

59. Varthema's record is unclear. He claims he sailed from the Moluccas (probably Ternate) 200 miles due *south* to *Bornei'* an island he described as being "somewhat larger than [Molucca] and much lower." Thus Varthema's *Bornei* may have been Buru.

60. According to the Portuguese historian Armando Cortesão. See Winter, *Imago Mundi* VI, 1950.

61. See Phillimore, pp. 175-79; also *History of Cartography* 2/2 p. 692.

62. Sheet five from the nine-sheet map of Tibet, published by du Halde, 1738-41.

63. *Edinburgh Philosophical Journal,* 1821-1824, and *Edinburgh Journal of Science,* 1824.

64. Matthew Paris exaggerated rivers on his map of England; Bordone did so on his plan of Venice.

65. Thanks to Hans Penth for all the information in this paragraph (personal correspondence, October 8, 1997).

66. *The Chiang Mai Chronicle,* pp. 121-122. Thanks to Hans Penth for this reference.

67. Michael Wright of Bangkok kindly assisted with the analysis of the *Traiphum.*

68. E.G.R. Taylor, p. 128.

69. See *History of Cartography* 2/2.

70. See Helen Wallis, 'The Influence of Father Ricci on Far Eastern Cartography,' p. 41.

71. See Ishihara, Akira, and Howard S. Levy, *The Tao of Sex an Annotated Translation of the Twenty-eight Section of The Essence of Medical Prescriptions (Ishimpo).* Yokohama: General Printing Company, 1968 (1969), p. 29.

72. Knapp, p. 177

73. Hsu, p. 96.

74. Kenneth Hall p. 42.

75. *Chen-la* was not a united state, but a loose term to refer to many realms and princedoms (see Chandler, *History of Cambodia,* p26); possible identities of *Fo-lo-an* are discussed by Wheatley, *Golden Khersonese,* pp. 68-70.

76. Wheatley, *Golden Khersonese,* pp. 62, 63, 110, 255.

77. Kenneth Hall p. 244.

78. Quirino, p. 3.

79. Quirino p. 5.

80. Wheatley, *Golden Khersonese,* p. 92.

81. Hsu, p. 110.

82. Wheatley, *Golden Khersonese,* p.93, states that the average for the east coast is 40 miles to the inch, west coast 30 miles to the inch, Singapore 12 miles to the inch.

83. *History of Cartography* 2/2, p. 53.

84. Wheatley, *Golden Khersonese,* pp. 101-102

85. Most of the extracts from Arab texts quoted here are from Tibbetts.

86. See Wheatley, *Golden Khersonese,* p. 210, and Tibbetts, p. 1; however see also Lore Crusades p. 50.

87. Tibbetts pp. 100-118.

88. The Mergui estuary is suggested by Wheatley (*Golden Khersonese*), the islands off Malaya, somewhere between Phuket and Kelang, by Tibbetts (p. 128).

89. Some scholars believe that there were two *Waq-waqs,* one being Madagascar, the other, which is often described as a province of China, being Sumatra. Japan has also been associated with this land, though probably erroneously. See Hudu al-'Alam, *'The Regions of the World', A Persian Geography 372 A.H.C982 A.D.* Japan has also been proposed as the identity of *Waq-waq,* though this is almost certainly wrong. Thanks to Thomas D. Goodrich for this reference and other information on *Waq-waq.*

90. Hudu al-'Alam, *'The Regions of the World', A Persian Geography 372 A.H.C982 A.D.* The period in the place name *M.qys* denotes a letter that is unknown (Semitic writing often ignores vowels) or undecipherable. Thanks to Thomas Goodrich for this information.

91. Quoted from Thomas D. Goodrich, *The Ottoman Turks and the New World,* Otto Harrassowitz, 1990, p. 104. Professor Goodrich comments that "one is startled by the realization that [an artist's rendering of the *Waq-waq* tree] with its naked women-fruit is the first printed picture to appear in the prudish World of Islam" (p. 58).

92. Tibbetts (p. 180) believes *Jab*a, *Salahit,* and *Harang* to have been at the southern end of the Malacca Strait (near Singapore).

93. *Encyclopedia of Islam,* p. 930.

94. Ibn Battuta, p. 278. I agree with Wheatley and Tibbetts, who identify *Mul-Jáwa* as Java (see *The Golden Khersonese,* p. 226); Yule believed it to be "somewhere on the coast of the Gulf of Siam" (*Marco Polo,* II, p. 349).

95. Some scholars believe that Aristaeus may in fact have had knowledge of peoples from eastern Asia, but this is difficult to substantiate. The 'Hyperboreans' may have been the Chinese, or otherwise an entirely mythical people. Ctesias of Cnidus wrote a history of Persia and India. Silkworms were apparently known in Greece at this early date, but they may have come from Asia Minor rather than China.

96. For example, in Marcelo de Ribadeneira, *Historia de las islas del archipiélago Filipino y reinos de la gran China, Tartaria, Cochin-China, Malaca, Siam Camboge y Japon,* Barcelona, 1601; and Fr. Domingo Navarrete, *Tratados Historicos* (1676). See John Villiers notes in Gervaise, p. 241.

97. Thanks for Dr. Hans Penth for clarifying the definition of the 'white' elephant, and for pointing out that the concept (a misnomer) of a white elephant in the way now understood doubtfully existed at this time.

98. See Majumdar, p. 12, and Wheatley, *Golden Khersonese,* p. 10.

99. According to Schoff, This may refer to the Chinese state of Ts'in.

100. Gervaise, p. 161.

101. Kimble pp. 120, 122.

102. Rubruquis, explored the northern and western shores of the sea, but learned of the southern and eastern shores from a

friend, Andrew of Longjumeau. See Kimble, p. 138.

103. Kimble p. 168.

104. The Ebstorf world map, ca. 1235 (destroyed in the Second World War).

105. Alternate identities for the Ganges are the Indus, Pison, or Danube; alternate for the Nile is the Gihon; alternate for the Tigris is the Hiddekel. See Wright, *Geographical Lore*, p. 72.

106. From Ortelius' 1567 map of Asia, Quoted from Schilder, *Monumenta Cartographica Neerlandica II*, pp. 76-77.

107. See Pauline Moffitt Watts, 'Prophecy and Discovery: On the Spiritual Origins of Christopher Columbus' 'Enterprise of the Indies',' *American Historical Review*, Vol. 90, February, 1985.

108. See Harry Kelsey, 'Ruy López de Villalobos and the Route to the Philippines,' in *Terrae Incognitae*, Volume XVII, 1985, p. 39.

109. Though the name given the metal by Choisy, *tambac*, appears to be *tem-baht* which, on the contrary, means pure gold. Either his hosts had used *mai tem-baht* ('*mai*' = ' not'), or he misidentified the metal.

110. Using 158 meters as the length of a *stade*. The length of a *stade* is not known with certainty; some authorities cite a figure of 148-158 meters, others of 185 meters.

111. Wright, pp. 136-137.

112. Wright, pp. 79-80.

113. British Library, additional manuscript 10049, f. 64.

114. Jean Fernel, ca. 1530. See Lach Volume II, book 3, p. 415.

115. Lucini, in J. T. Leader, pp. 121-22.

116. Kimble, p. 132.

117. Wright, p. 234.

118. Wright, p. 272.

119. The exact date is disputed; Polo's text says 1294, but he is aware that Gházán had become king of Persia in October of 1295; Polo, however, stated that he had become king in 1294.

120. Francis Wood, *Did Marco Polo Go to China*, Westview Press.

121. There are, in fact, extant manuscripts comprising text and maps which purport to be handed down from Marco Polo's three daughters. According to documents supposedly by his daughter Fantina, "in the map of all Asia, made by my father Marco, these islands were according to the Saracens, VIII hours, CXV meridian degrees," but the islands were too savage a place for Marco. Another map was "Marco Polo's drawing of India, Tartary and the many islands he explored," and as a result of this map, "the Great Khan honoured him." The daughter Bellela wrote that Marco was entrusted to deliver a message to "Fusin, Queen of females in the Extreme East," and that along the voyage he met a Saracen who "gave Marco Polo his navigation chart."
The manuscripts also record that a notary was entrusted with some of these documents in 1556. The curious notary translated an inscription which had been dedicated to Marco on a gold tablet, which stated that Marco took "possession of the many islands scattered in the most distant Orient and previously never heard of," and that Marco "used the famous nautical skills of the Arabs."
These documents are at best later copies of 'legitimate' originals, with some significant changes, or old forgeries. A modern forger would doubtfully have detracted from their believability by recording territories which "appeared to be chambers of love," and inserting titillating descriptions of their father's visits to islands of women. The accompanying maps, to their credit, do not include these, suggesting that the text may have been wild tales told to children for fun. But the maps do contain depictions of what appear to be Korea and Alaska, about which Polo knew nothing.
(See Leo Bagrow; and John Black, 'Marco Polo Documents Incorporated in the Felicitation Volumes of Southeast Asian Studies,' in *Journal of the Siam Society*, Volume 70, Parts 1 & 2, January-July, 1982.)

122. Data from Polo's journey through Southeast Asian waters are also found on the so-called Medici atlas (or Laurentian portolan chart), once believed to date from 1351, but now thought to date as late as the early fifteenth century. The Dulcert chart of 1339, which survives in 2 sheets, may, according to some scholars, have had one or two eastern sheets which are now lost.

123. By Pegolotti. See Latham, p. 15.

124. The surviving map is a copy probably made at the same time. See *History of Cartography* 1, p. 315.

125. Clifford p. 14-15.

126. Kimble, p. 53.

127. Kimble, pp. 54-55, and Tibbetts, p. 11.

128. Wright, p. 271.

129. Kimble pp. 214-216.

130. See Lach Vol. I, p. 106-7.

131. Maximilian quoted from Lévesque, *History of Micronesia* I, p. 267.

132. Eden, p. 222.

133. Other theories for the 'three Indias' were (1) Afghanistan/West Pakistan, India, and east of the Ganges; (2) northern Indian subcontinent, southern Indian subcontinent, Ethiopia/East Africa; (3) 'land of pygmies', 'facing the land of the Medes', 'ends of the earth, ocean on side, realm of darkness on the other'. See Peter Forbath, *The River Congo*, p. 27.

134. Interestingly, if coincidentally, an inlet called 'mouth of the dragon' is marked on the shoulder of South America on some charts from the early sixteenth century; South America, in turn, was envisioned as an extension of Southeast Asia.

135. António Galvano, *The Discoveries of the World*, edited by Vice Admiral Bethune, Hakluyt Society, London, MDCCCLXII, pp. 66-67.

136. There are, however, dubious theories citing evidence of early European knowledge of South America, for example, the Walsperger map of 1448 bears an inscription on a "Southeast Asian" subcontinent describing "giants who fight against dragons," a reference which has been correlated to the "giant" Patagonian people later described by Pigafetta as inhabiting southern South America.

137. Some commentators (e.g., Paige, p. 20, and Ravenstein, p. 38) believe that the map Pigafetta referred to was not by Martin Behaim, but rather was a post-Columbian map, possibly by Schöner. Although it is true that Schöner records a strait between South America and an austral land, probably the result of an early encounter with the Rio de la Plata, there is no reason to doubt Pigafetta's attribution of the map to Behaim. The Martellus/Behaim concept of a Southeast Asian subcontinent fit perfectly into the expedition's scheme and the interpretation of America and the Pacific which grew from it.

138. *History of Cartography* 2/1 pp. 261-2

139. Tibbetts, figure 4. Thanks to Professor Tibbetts, who confirmed this information and his opinions in a letter to the author, November, 1995.

140. Letter to Columbus from Jaume Ferrer, 5 de Augusto, 1495. Quoted from Nunn, p. 29.

141. Nunn, p. 38.

142. The Juan de la Cosa map is dated 1500 but parts of it may have been added ca. 1508. It is in the *Museo Naval*, Madrid.

143. Quoted from Nunn, p. 39.

144. The theory that the New World was connected to Asia began with Columbus' initial landfall in South America and was primitively mapped as such by Juan de la Cosa on his chart of ca. 1500-08.

145. Harrisse, p. 524.

146. A modern exception is Quirino, who still identifies Taprobana as Sumatra (see Quinno pp. 2, 72).

147. Chapter XXXII, p.198. Although known as Sir John Mandeville, he was doubtfully a knight, did not travel beyond Western Europe, and may not have been English Some scholars insist that his name was a pseudonym for either Jean d'Outremeuse or Jean de Bourgogne, both of Liege. Others argue that the name was genuine, that a novelist named John Mandeville was born in St. Albans in the late thirteenth century and passed much of his life on the Continent, and that his book was only intended as a travel romance rather than as a factual account. The book, perhaps completed by 1356, was in any event commonly construed as a legitimate travel log.

148. Quoted from Tibbetts p. 210. Richardson (p. 2) states that the transference of *Taprobana* to Ceylon was "doubtless due to the vastly reduced size of the Ptolemaic Taprobana on the Cantino [map of 1502], and the inscription on Sumatra proclaiming it to be the largest island in the world." This is an interesting point, but in fact the transference of the name happened before Cantino and before the correction to the size of Ceylon.

149. Barros, *Decade IX*, Chapter 5.

150. Martín Fernández de Enciso, *Suma de Geografia,* Seville, 1519. See, Richardson, p. 13.

151. Tibbetts p, 206.

152. A. Cortesao, in Pires, p. 262

153. Abu'l-Fida described *Zaiton* as being Shanju, as Chüan-chau was known to early Portuguese traders. See Kimble, p. 138.

154. Tibbetts p. 98.

155. The error of the Polo manuscripts reading 'Java' instead of 'Champa' probably originated in the so-called Latin "Z" manuscript. In that text, the letter 'm' was omitted from *Çamba,* being spelled *Çamba,* with Java being spelled *Çava,* the two then being confused with each other.

156. Condore's mate, *Sodur* (usually *Sondur* or *Sandur*)? was probably just another isle of the same group; Condore is in fact today also known as Côn Sòn. But some scholars have been frustrated by the fact

that many early Arabic writers, including al-Idrisi and Ibn Sa'id, cite an island called *Sundur Fulat,* which was said to lie *after* Champa on the sea route from the Middle East to China. This is presumably far to the north of Condore, which Polo would have passed *en route* to Champa from China, not during the voyage south from Champa. Tibbetts, and Paul Wheatley, *Lochac Revisited,* pp. 91-92.

157. Thanks to Dr. Hans Penth. Lop Buri in Pali is *Lavo.* See also Paul Wheatley, *Lochac Revisited,* p. 94. The name *Lochac* has sometimes been equated with *Langkasuka,* the ancient port on the northeast coast of the Malay Peninsula, but etymological considerations, as well as Polo's emphasis on the inaccessibility of the place, favor the identification of *Lochac* as the inland region of Lop Buri rather than the port of *Langkasuka.*

158. Paul Wheatley, *Lochac Revisited.*

159. Thanks to Dr. Hans Penth for the suggestion that Lama = Lan Na.

160. Information on Leardo (1448) and Behaim (1492) according to Ravenstein, p. 30.

161. Thanks to Hans Penth for this idea. Richardson (p. 2) states that the Ruysch's *Bocat* is a form of Polo's *Lochac,* which if true would mean that *Lochac* is repeated, since it already appears as *Loac* on the Ruysch.

162. Quirino identifies Pigafetta's Ceylon as Panaon; Lévesque states that it is a part of Leyte called Selani.

163. In the various editions of his rendering of the *Cosmographia* of Peter Apianus, 1524 and subsequent.

164. The vestige of the erasure was discovered by Dr. McGuirk on one of the three known copies of the first surviving state of the map, which was in the possession of the author. See Donald L. McGuirk, Jr., 'Ruysch World Map: Census and Commentary', in *Imago Mundi,* Vol. 41, 1989.

165. See Harry Kelsey, "Ruy López de Villalobos and the Route to the Philippines," in *Terrae Incognitae,* Volume XVII, 1985, p. 39.

166. Richardson, pp. 17-19.

167. The placement of the Seven Cities in Southeast Asia is also interesting in that the Cities were associated with the mythical island of Antilia which, according to one theory based on linguistic analysis of toponyms and in feature analysis, was the island of Formosa, knowledge of which had reached the Mediterranean via Arab pilots. See Robert

H. Fuson, *Legendary Islands of the Ocean Sea,* Pineapple Press, Sarasota FL, 1995, and John Logan Allen (ed.,) A New World Disclosed, University of Nebraska Press, 1997.

168. Akhbar al-Sin wa'l-Hind and Ibn Khurdadhbih, in Tibbetts, pp. 26 & 29.

169. The unique surviving example of the world map from the 1513 *Geographia* with the name 'America' is in the collection of the John Carter Brown Library, Providence. See Henry N. Stevens, *The First Delineation of the New World and The First Use of the Name America on a Printed Map,* London, 1928.

170. This is the title set on the 1522 edition; it was not part of the map woodblock itself, and so varied with each edition.

171. Some versions of Polo do not specify the direction southeast.

172. Tibbetts pp. 97 &155.

173. Thanks to Dr. Hans Penth for this information.

174. Wheatley, *Golden Khersonese* pp. 55-56, 105-106

175. Wheatley, *Golden Khersonese* pp. 47, 56-57; Wheatley (p. 53, etc) suggests that *Ko-lo* was in the northwest of Malaya.

176. Thorne's map, however, is only known from a woodcut copy made by Hakluyt in 1582.

177. Mentioned in a Chinese text by Chao Ju-kua, Commissioner of Foreign Trade at Ch'üan-chou. See Wheatley, *Golden Khersonese* p. 65.

178. The Ramusio version has '500 miles'. Yule-Cordier Vol. II, pp. 359, 362.

179. *Bocat* has also been interpreted as a corruption of Polo's *Lochac.*

180. *Cobebe* could not be the city of Kobe at this date.

181. Hidegard Binder Johnson, *Carta Marina world geography in Strassburg,* 1525, p. 27.

182. Kimble p. 240.

183. Bordone received permission to publish a map of the world and a map of Italy from the Venetian senate in 1508, though neither is known to be extant. See Karrow, p. 89.

184. Martín Fernández de Enciso, *Suma de Geografia,* Seville, 1519. See W. A. R. Richardson, p. 13.

185. Yule-Cordier, ii, pp. 404-405.

186. The map by Bellin in Prevost's *Historoire Générale des Voyages.*

187. The first issue (1532) is distinguished by the place name 'ASIA' in large upper-case letters, while in subsequent issues the name is in smaller upper-case letters.

188. A similar method had been used for the maps in the 1511 edition of Ptolemy's

Geografia, by Bernard Sylvanus, with the separate register of nomenclature printed in red (see fig.43).

189. Lach I, p. 540

190. Linschoten left vivid descriptions of such commodities as benzoin and aloe. Benzoin, "by reason of the sweet smell, comforts the heart, the head, and the brain, it cleanses the head from all superfluous humors, sharpens the wit; being smelled unto, it is good to be used when diseases begin to go away … it grows much in the land of Siam, in the land of Sumatra, in the Islands of Javas, and the countries of Malacca … when the tree is young, then it yields the best benzoin, which is blackish of color, and of a very sweet smell." Linschoten goes on to say that the best benzoin comes from Siam and Malacca, and is exported to China, Portugal, and western Asia. A lower grade is grown in Java and Sumatra. As for aloe, Linschoten explains that "The Lignum Aloe … is most in Malacca, in the Island of Sumatra, Cambodia, Siam, and the Countries bordering on the same: the trees are like Olive trees, and somewhat greater: when it is cut off, it smells not so well, because it is green, for the dryer it is, the better it smells."

191. Lach I, p. 525, and Ramusio, Vol. I, p. 164.

192. The only sixteenth century Spanish printed renderings of of Asia are on a map of the world on two sheets, compiled by López de Gómara in 1552-52, and a map of the Asian continent by Hernando de Solis which may have been engraved in 1598 (this date appears on the corresponding map of America). But even this map is not known to have been published until 1602; furthermore, it was simply a copy of the map of Ortelius.

193. Lévesque, *History of Micronesia* 1, p. 131.

194. The book's privilege states that it was "sold in Paris at the shop of Simon de Colines, official booksellers to the University of Paris, established in Rue Saint Jean de Beauvais, at the Sign of the Golden Sun.

195. The figure is 179.20° in the printed account. The Ambrosiana MS cites 171.20°. The actual figure is about 80° east (280° west).

196. The publisher of the 1552 edition of the *Cosmographia* struck coordinates around the map by means of separate blocks apparently cut specially for the purpose, though these were arbitrary and had no geographical value.

197. During the period from 1508 to 1548, copperplates were used for only a few examples of separately-published maps.

198. There are some minor differences between the Gastaldi 1548 map and its larger 1561 re-engraving by Ruscelli, for example, Palawan is *polagua* in 1548, *Polaguan* in 1561. There are similar differences between the Ramusio woodcut of 1554 and copperplate of 1563; Palawan is *Paloba* on the woodcut, and *Paloban* on the copperplate; Sarangani is *Saragan* on the woodblock, *Sarangan* on the copperplate; Sarangani or Balut is *Candigar* on the woodblock, *Candigan* on the copperplate. The only substantive difference between the two Ramusio renderings is the absence of Lake Chiang Mai's namesake city on the copperplate. Finally, the Olgiato re-engraving of the Gastaldi/Forlani map is essentially identical, except that Olgiato neglects to engrave some nomenclature, such as the Moluccan islands of Tidore and Bacan (comparison with the original Forlani supplement reveals that the islands are there, but unidentified). Olgiato's plate also lacks evidence of indecision on the matter of Samar's occupying the same place as Guam, since he merely copied Gastaldi's conclusion.

199. For the identities of the various Philippine islands visited by Magellan and others, I have generally deferred to the analysis in Volume 1 of Rodrigue Lévesque's *History of Micronesia* series, and secondly to my own research and Quirino's *Philippine Cartography.*

200. Some historians (e.g., Lach) believe Maximilian's *Inuagana* to have been Guam and Rota. Lévesque argues, more convincingly, that *Inuagana* was in the Philippines, and that the Maximilian account simply omitted the Marianas.

201. Lach, I, p. 640.

202. From the narrative of Vicente de Nápoles, quoted from Lévesque, *History of Micronesia* I, p. 513. The events are ascribed to 29th December, 1517, but as Lévesque points out the author has confused his dates, since the Epiphany falls on 6th January. He also suggests that the *Isole del Rey* may be Ulithi, in the Caroline Islands to the east of Yap (Lévesque, *History of Micronesia* I, p. 487); Helen Wallis, in *Exploration of the South Sea*, suggests it is Wotje in the eastern Marshall Islands.

203. According to Lévesque, they set foot upon the island of Likiep in the Marshalls. See *History of Micronesia* I, pp. 570 &

617. Helen Wallis had reached the same conclusion.

204. According to Lévesque, these 'garden isles' of the Marshalls are Wotho. Helen Wallis identified them as Kwajalein.

205. Both Lévesque and Wallis identified the *Matelotes* as the island of Fais, in the Carolines. The account quoted (Lévesque, *History of Micronesia* I, p. 585) dates the encounter as 23rd January, the feast day of St. Ildefonso, after whom the Augustinians wanted to name the island; Lévesque (*History of Micronesia* I, p. 581) gives the actual encounter date as 21st January.

206. Stevenson, *Atlas of Portolan Charts facsimile of manuscript in British Museum.*

207. Rodrigues Lévesque, personal correspondence, 2nd January, 1998.

208. Lévesque, *History of Micronesia* I, p. 259. Quirino identifies Pigafetta's Ceylon as Panaon; Lévesque states that it is a part of Leyte called Selani.

209. Limawasa was depicted as a large island possibly because it was mistakenly combined with Leyte, or perhaps arbitrarily so because of its importance in the course of Magellan's voyage.

210. "Siagu" in the printed French account. Rodrigues Lévesque (correspondence) transcribed it as *Siani* or *Siain.*

211. Records cited from Lévesque, *History of Micronesia* I, p. 158.

212. Rodrigues Lévesque, correspondence.

213. Lévesque translation, *The Philippines*, p. 46.

214. Lévesque identifies the island as Panglao (*The Philippines,* p74), while Quirino assumes that it was the same island known today as Negros; the coasts of either Negros or Panglao could have been skirted when sailing to the southwest of Bohol. Negros was also known as *Buglas.*

215. Ramusio, on a map of the Western Hemisphere published in Volume 3 of the *Navigationi* (1556), reflects the notion of an easy passage from Mexico to the Spiceries by depicting a chain of islands linking them. From the western shores of Mexico, Ramusio maps the Pacific islands of *S. Tomas, Rocca partida,* and *S. Bartolameo,* forming stepping stones to the Marianas, Philippines, and Moluccas. The map is entitled *Universale della Parte del Mondo Nuovamente Ritrovata,* 1556 (1st state), 1565 (2nd state), and 1606 (3rd state). Mendes Pinto, in searching for the identity of the "Silver Islands", spoke of "what I saw in the Moluccas in the reports presented by the Castillian General Ruy López de Villalobos to Dom Jorge de Castro, then captain of our Ternate fortress" (Pinto, p. 300).

216. Lévesque, *History of Micronesia* I, p. 131.

217. The earliest European map to show Japan is the 1459 world map of Fra Mauro; the earliest European printed map to show Japan is the Contarini-Roselli world map of 1506.

218. Esteban Rodriguez, in Lévesque, *History of Micronesia* II, p. 94.

219. Thirty-five were purchased, one was "corrected", but it is not clear whether the corrected needle was one of the thirty-five. Records cited from Lévesque, *History of Micronesia* I, p. 158.

220. A travel report Andres de San Martin had been writing was later taken by a Portuguese in Ternate, Edward Rezende, who wrote a treatise based on it, though San Martin's original has not survived.

221. The French edition uses the word "brasses", which Paige (p. 108) has translated as "arm lengths", but is probably "fathoms" (*brazas* rather than *brazos*). Thanks to Rodrigue Lévesque for this point.

222. Lach, I, p. 507.

223. Wheatley, *Golden Khersonese*, pp. 303-304.

224. According to Ravenstein, a *Menam* River appears on the so-called 'Paris' reproduction of the Behaim globe of 1492 (fig. 48). Extensive and improper 'restoration' of the globe during the nineteenth century makes it difficult to know whether certain place names were originally on the globe, but if the Menan did in fact appear in 1492, Behaim's sources are not known.

225. The exception is Pinto, who mentions a river called the *Tinacoreu* which flows from Chieng Mai Lake; this river has been identified as the Red River.

226. "Cochinchina" in the text, but this seems to be an error, as he ends up in India.

227. Thanks to Dr. Hans Penth for the information in this paragraph (personal conversation, July, 1996).

228. Yule identifies *Sindafu* as Ch'êng-Tu Fu, the capital of Sze-ch'wan. See *Polo*, II, p. 37.

229. Lach II, pp. 463-4

230. Unrelated to the incarnation of Southeast Asian nomenclature in *Terra Australis* following Polo, Varthema, and Mandeville, a misunderstanding of the Portuguese voyage reported in the German newsletter *Newe Zeytung auss Presillg Landt* of 1514 (see page 234, above) may have also placed an isle of Southeast Asia in the southern continent under entirely different circumstances, earlier in the century, under the name

'Brazil'. It is possible that the the vessel's reported track around the "Cape of Good Hope" was correct, and that the "land of brazil" was a reference to a Southeast Asian island which was rich in brazil-wood, probably Sumatra. Sumatra is indeed roughly six hundred miles from Malacca, as the newsletter states of Brazil. Brazilwood was well known in Europe by the later medieval period, and is mentioned, for example, by Chaucer. Both Marco Polo and Friar Odoric spoke of brazilwood in Southeast Asia, and Polo even carried some seeds back to Venice from the Sumatran kingdom of *Lambri* (though the Venetian climate was not suitable for them). It has also been suggested that, if the theory is correct, the 'continent' which the expedition skirted to the south may have been Australia. This idea was proposed by James Enterline in his article, 'The Southern Continent and the False Strait of Magellan,' *Imago Mundi* XXVI, 1972. Also thanks to Mr. Enterline for his correspondence with me on the topic in May of 1996.

231. See Andrew Sharp, *Discovery of Australia*, 1963; and W. A. R. Richardson, 'Jave-la-Grande: a case study of place-name corruption', International Cartographic Association, 12th International Conference, Perth, 1984, *Technical Papers*, Vol. 2, 221-248.

232. Quoted from Harrisse, p. 486.

233. Nunn, p. 38.

234. Hall. The Virginia place-name *Wingandekoa* was used on Ortelius' *America Sive Nova Orbis, Nova Descriptio* of 1587.

235. Prior to that, the Portuguese dominated in the Canaries, and in fact had temporarily won official sanction for their claim with the bull *Romanus Pontifex* of Pope Nicholas V in 1454; furthermore, Portugal's King Afonso forfeited any claim to being sovereign of Leon.

236. Treaty quoted from Lévesque, *History of Micronesia* I, p. 74.

237. The copperplate for the *Asiae Nova Descriptio* was replaced in about 1575. The two plates are virtually identical; the only noticeable change regarding Southeast Asia is *La Farfana*, where in the 1575 plate *farfana* is all lower case. Copies of the *Indiae Orientalis* without verso text are known, suggesting that the maps were also available separately and in made-to-order atlases.

238. Translation from Schilder, *Monumenta Cartographica Neerlandica II,* p. 64

239. Lévesque, *History of Micronesia* I, p. 593. Lach I (p722) records the earlier recorded European encounter at 1582.

240. Yule-Cordier II, p. 406.

241. Lévesque, *History of Micronesia* I, p. 585. Helen Wallis identified Arrecifes as the island of Yap in the Western Carolines.

242. Lévesque, *History of Micronesia* II, p. 84. According to Lévesque, La Barbuda is the island of Mejit in the Marshalls.

243. Friars Buzeta and Bravo describe it in their geographic dictionary of 1850. See Quirino, p. 71.

244. From the account of a Genoese pilot. See Lévesque, *History of Micronesia* I, p. 324

245. From António Galvão account. The account actually states "36 degrees", but is presumed to be an error for "6 degrees". See Lévesque, *History of Micronesia* I, p. 329.

246. The Genoese pilot mentioned "about 5 degrees more or less," while Galvão recorded "36" degrees, which is surely a typographical error, presumably for 6 degrees.

247. Without *St. John* island: *Partie de la Mer du Sud Comprise entre les Philippines et la Californie, d'après la Carte Espagnole trouvée sur le Galion pris par l'Amiral Anson en 1743 …*; with *St. John* island: *Partie de la Mer du Sud Comprise entre les Philippines et la Californie, d'après une autre Carte Espagnole communiquée à La Perouse dans Southeast Asia relache à Monterey …*

248. Captain Arellano, in Lévesque, *History of Micronesia* II, p. 191.

249. Lévesque, *History of Micronesia* II, p. 379.

250. From the account of the voyage by Captain Don Alonso de Arellano, in Lévesque, *History of Micronesia* II, p. 186. Lévesque believes the island to be Kwajalein.

251. From the account of the voyage by Captain Don Alonso de Arellano, in Lévesque, *History of Micronesia* II, p. 186. Lévesque believes the island to be Lib.

252. Lévesque, *History of Micronesia* II, p. 85-87. Lévesque believes the island to be Jemo.

253. Many thanks to Professor Günter Schilder of the University of Utrecht for this translation and other information about the Lodewijcksz map (personal correspondence, July 22, 1998).

254. Tibbetts pp. 216-217. Al-Mahri seems to be speaking of Java, rather than Sumatra.

255. R. T. Fell, *Early Maps of Southeast Asia*, Oxford University Press, 1988, p. 49.

256. Sian (or Xian) is in fact a region in the northern interior of China, but it was

little known to Europeans until a supposed Nestorian monument was discovered there in about 1623-1625. This monument was of great importance to European missionaries in China, because if it was authentic and if in fact it recorded early Christian activitiy in China, it would counteract the charge by Chinese authorities that the new religion had no precedent in Chinese culture. The spelling of Sian for *Siam* was doubtfully a confusion with the Chinese province.

257. Lach I, p. 524.

258. Rebecca Catz, in Pinto, p. 550, ch19, n. 5.

259. De Campos, p15. This is related to the Thai *Su'wan,* a poetic word for gold.

260. This was solved by Wheatley, see 'A curious feature on early maps of Malaya', pp. 66-72.

261. See, Wheatley, *Golden Khersonese* p. 151

262. See *History of Cartography* 2/2, pp. 829-30.

263. It has been theorized that the atlas was conceived by Cornelis Claesz, and that the work had first appeared in 1597. See Koeman, Vol. II, p. 254.

264. Wheatley, *Golden Khersonese* p. 96.

265. Hodgkiss, p. 108.

266. Quoted from Karrow; Karrow tentatively associates the voyage with that of Sebastian Cabot.

267. Quoted from Verner & Skelton, p. v.

268. There were other, relatively minor companies which dabbled in the Indies trade as well, such as the Danish East India Company, which was formed in 1616. By the later part of the century, the Danish company established outposts at Banten, Makassar, and in the west Borneo port of Sukadana, and Danish vessels would be found trading in various Indies ports, often as interlopers. But the Danish East India Company remained a relatively minor figure in Southeast Asia seas, and had no significant impact on the mapping of the region. The Company endured in various forms until 1800.

269. Koeman, Vol. II, p. 345.

270. Drake claims to have speculated about the insularity of the Tierra del Fuego when blown south of the Magellan Strait, though some contemporary sources suggest that Drake may have learned of its insularity from Spanish charts stolen *en route.*

271. Abraham Ortelius (1584) and Cornelis de Jode (1593) both depicted the palanquin on their maps of China, and de Bry illustrated it in his *Voyages* (1599). By the seventeenth century it had attracted enough attention in Europe that a Dutch prince had one built for him, which Blaeu illustrated in his town book.

272. See Verner & Skelton.

273. England's William III, who reigned from 1689-1702, was Stadtholder of the United Provinces. See Mlle M. de la Roncière's theory in Verner & Skelton,

274. One of the three atlases is dated 1700 and is dedicated to Lord Somers, while the other two are undated. See Thomas R. Smith, 'Manuscript and Printed Sea Charts;, in Norman J. W. Thrower (ed.), *The Compleat Plattmaker,* p. 86.

275. A Jesuit in Milan, Chrsitoforo Borri, is believed to have made a manuscript chart indicating magnetic variation with straight lines (ca. 1630, but not extant), and Athanasius Kircher worked on one in the 1640s properly indicating the isolines as curved. See Chapman; and Thrower, *Maps & Civilization,* pp. 97-100, 275.

276. This matched a reading French scientists had taken from a lunar eclipse in Siam in early 1682. Earlier geographers generally used the Canaries as the prime meridian (an inheritance from Ptolemy). Most seventeenth-century mapmakers used the Cape Verde Islands as the prime, the influence of the negotiations over the *línea de demarcación.* The earlier Dutch makers, such as Mercator and Jansson, usually used a point toward the west of the archipelago, while Cassini and his colleagues used the eastern tip of the Cape Verde Islands, as did many Dutch geographers of the latter part of the century.

277. Dawn Rooney, Ayuthaya, the ancient capital of Siam on European maps.

278. Jacques de Bourges, in Michael Smithies, pp. 113-129 (quote from p. 121).

279. Lach III, p. 110.

280. Lach III, p. 68.

281. Lach III, pp. 1433-35.

282. Thanks to Dr. Andrew Cook, Map Archivist, India Office Records, The British Library (exchanges over MapHist, November, 1996), for identifying the printed version of this manuscript and for explaining its likely origin.

283. Lach III, p. 1445.

284. Lach III, p.1446.

285. Lieutenant James Low in an 1824 letter to the Secretary to the Government of Penang (then Prince of Wales Island). See Sternstein, *'LOW' Maps of Siam,* p. 132.

286. Clifford, pp. 1-14.

287. Extracted from a letter by W.E. Phillips in the Straits Settlements Factory Records, 15 April, 1824, pages 442-445; quoted from Larry Sternstein, *'LOW' Maps of Siam,* in *The Journal of the Siam Society,* Vol. 73, p. 135.

288. Clifford, pp. 131-133.

289. Francis Garnier, in Clifford pp. 133-134. Ironically, the letter from which this was extracted was supposedly written by Garnier when he was about twenty years of age, which would place it three years into the Second Opium War, in which France joined England in forcing China to open its doors to the drug.

290. *The French in Indochina,* p. 1.

291. SarDesai, p. 73.

292. As quoted in Winichakul, p. 85.

293. *The French in Indochina,* p. 40.

294. Taylor, p. 252.

295. Quoted from Martin Stuart-Fox, *A History of Laos,* Cambridge University Press, 1997, p. 47.

296. Penth, *Lan Na,* p. 39.

297. See Lotika Varadarajan, 'Glimpses of Seventeenth Century Currency and Mensuration in Siam,' p. 207, note 13, in *Journal of the Siam Society,* Volume 83, Parts 1&2, 1995.

298. Quoted from Sternstein, *'LOW' Maps of Siam,* pp. 132 & 155.

299. This second map is entitled *A Map of Siam, North Laos, Martaban, Tennasserim and Part of the Malayan Peninsula . . . .* See Sternstein, *'LOW' Maps of Siam,* p. 153, etc.

300. Quoted from Winichakul, pp. 67-70.

References

[anonymous], *The Periplus of the Erythræan Sea / Travel and trade in the Indian Ocean by a Merchant of the First Century.* Translated from the Greek and annotated by Wilfred H. Schoff, A.M. Longman, Green, and Co., 1912.

[anonymous], *The Chiang Mai Chronicle.* Translated by David K. Wyatt and Aroonrut Wichienkeeo, Silkworm Books, Bangkok, 1995.

[anonymous], *The French in Indo-China.* Edinburgh: T. Nelson & Sons, 1884 [Reprinted, with a preface by Dean Meyers. Bangkok, White Lotus Co. Ltd., 1994].

al-Alam, Hudud, *The Regions of the World.* A Persian Geography, 372 A.H.C982 A.D. Translated and Explained by V. Minorsky. London: Luzac & Company, 1937. Anghiera, Pietro Martire d', see Martyr

Bagrow, Leo, 'The Maps from the Home Archives of the Descendants of a friend of Marco Polo,' in *Imago Mundi V*, 1948.

Battuta, Ibn, *The Travels of Ibn Battúta.* Translated and selected by H.A.R. Gibb; Edited by Sir E. Denison Ross and Eileen Power. New York: Robert M. McBride & Co., 1929.

Bevan, W. L., & Phillott, H. W., *Mediæval Geography.* London, 1873 [Amsterdam: Meridian Publishing & Co., 1969].

Bontekoe, William, *Memorable Description of The East Indian Voyage 1618-1625.* Hoorn: Issac Williamsz, 1646 [London: George Routledge & Sons, Ltd., 1929].

Camões, Luis vaz de, *The Lusiads.* Translated by William C. Atkinson. London: Penguin Books, 1952 (but this edition n.d.).

Campos, Joaquim de, 'Early Portuguese Accounts of Thailand,' in *Journal of the Siam Society*, 32 (1). Bangkok, 1940: 1-27.

Chandler, David, *History of Cambodia.* Second edition, updated. Westview Press, 1996.

— *The Land and People of Cambodia,* New York: Harper Collins,1991.

Chapman, S., Edmond Halley as Physical Geographer, and the Story of his Charts, *Royal Astronomical Society Occasional Notes,* Number 9. London: June, 1941.

Choisy, Abbé de, *Journal of a Voyage to Siam 1685-1686.* Translated by Michael Smithies. Kuala Lumpur: Oxford University Press, 1993.

Chou Ta-Kuan, *The Customs of Cambodia,* Translated into English from the French version by Paul Pelliot of Chou's Chinese original by J. Gilman d'Arcy Paul. Bangkok: The Siam Society, under Royal Patronage. Third edition, 1993.

Clifford, Hugh, *Further India being the Story of Exploration from the Earliest Times in Burma, Malaya, Siam, and Indo-China.* London: Lawrence and Bullen, Ltd., 1904 [Bangkok: White Lotus Co., Ltd., 1990].

Conti, Nicolò de', *The Most Noble and Famous Travels of Marco Polo together with the Travels of Nicolò de' Conti.* Edited from the Elizabethan Translation of John Frampton . . .by N. M. Penzer. London: Argonaut Press, 1929.

Cook, Andrew S., 'Alexander Dalrymple's A Collection of Plans of Ports in the East Indies (1774-1775): A Preliminary Examination,' in *Imago Mundi*, Vol. 33.

Cosgrove, Denis, 'Mapping New Worlds: Culture and Cartography in Sixteenth-Century Venice,' in *Imago Mundi*, Vol. 44, 1992. pp 65-89.

Crone, G. R., *The Discovery of the East.* New York, St. Martin's Press, 1972.

Dalrymple, Alexander, *A Plan for Extending the Commerce of his Kingdom, and of the East-India-Company.* London, 1769.

Dampier, William, *A New Voyage Round the World . . .* London MDCXCVII / Voyages and Descriptions MDCXCIX.

de Bourges, Jacques, 'Jacques de Bourges (ca. 1630-1714) and Siam,' by Michael Smithies, in *JOURNAL OF THE SIAM SOCIETY, Volume 81, Part 2.*

Dilke, O.A.W. & Margaret S. Dillke, 'Sir Robert Dudley's Contributions to Cartography', in *The Map Collector,* issue 19, June, 1982.

Eden, Richard, *The Decades of the Newe Worlde or West India.* London, 1555 [Readex Microprint Corp., 1966].

Finlayson, George, *The Mission to Siam, and Hue the Capital of Cochin China, in the years 1821-2.* London, 1826 (The Siam Society, Bangkok, 1988).

Forrest, Thomas, *A Voyage to New Guinea, and the Moluccas, from Balambangan.* London: G. Scott, 1780.

Gervaise, Nicolaus, *The Natural and Political History of the Kingdom of Siam,* Paris, 1688. Trans. John Villiers. White Lotus Co., Ltd., 1989.

Goodrich, Thomas D., *The Ottoman Turks and the New World,* Otto Harrassowitz, 1990.

Griswold, A. B., 'The Holy Land Transported,' in *Paranavitana Felicitation Colombo,* 1985, pp 173-221.

Hakluyt, Richard, *Hakluyt Voyages.* Selected and edited by Richard David. Boston: Houghton Mifflin Company, 1981.

Hair, P. E. H., 'A Note on Thevet's Unpublished Maps of Overseas Islands.' in *Terra Incognitae* 14, 1982, pp. 105-116.

Hall, Joseph, *Mundus Alter et Idem (The Discovery of New World),* London, 1605, translated from the original Latin by John Healey, ca. 1609. Cambridge: Harvard University Press, 1937.

Hall, Kenneth R., *Maritime Trade and State Development in Early Southeast Asia.* Honolulu: University of Hawaii Press, 1985.

Hamilton, Alexander, *A Scottish Sea Captain in Southeast Asia 1689-1723.* Michael Smithies (ed.). Chiang Mai: Silkworm Books, 1997.

Harley, J. B., & Woodward, David (ed.), *The History of Cartography, Volume One,* Cartography in Prehistoric, Ancient, and Medieval Europe and the Mediterranean. Chicago & London: The University of Chicago Press, 1987.

— *The History of Cartography, Volume Two, Book One, Cartography in the Traditional Islamic and South Asian Societies.* Chicago & London: The University of Chicago Press, 1992.

— *The History of Cartography, Volume Two, Book Two,* Cartography in the Traditional East and Southeast Asian Societies. Chicago & London: The University of Chicago Press, 1994.

Harrisse, Henry, *The Discovery of North America.* Amsterdam: N. Israel, 1969 (reprint of London-Paris, 1892).

Heidel, William Arthur, *The Frame of the Ancient Greek Maps.* New York: American Geographical Society Research Series No. 20, 1937.

Herodotus, *The Histories.* Trans. Aubrey de Sélincourt; revised with introductory notes by John Marincola. London: Penguin Books, 1954 (1972).

Hodgkiss, Alan, *Understanding Maps.* Kent: Wm Dawson & Son Ltd., 1981.

Hsu, Mei-Ling, 'Chinese Marine Cartography: Sea Charts of Pre-Modern China,' in *Imago Mundi*, Vol. 40, 1988.

Hudu al-'Alam, 'The Regions of the World', A Persian Geography 372 A.H.C982 A.D., translated and explained by V. Minorsky, University Press, Oxford, 1937.

Jamsai, Sumet, 'The Reconstruction of the City Plan of Ayudhya,' in The Journal of the Siam Society, 1970. Bangkok.

Johnson, Hidegard Binder, *Carta Marina world geography in Strassburg,* 1525

Kaempfer, Engelbert, *A Description of the Kingdom of Siam,* in *The History of Japan.* London: 1727 [Bangkok: White Orchid Press, 1987].

Karrow, Jr., Robert W., *Mapmakers of the Sixteenth Century and Their Maps.* Chicago: Speculum Orbis Press, 1993.

Keane, John, *The Evolution of Geography.* London: 1899 [Amsterdam, Meridian Publishing Co., 1976].

Kennedy, Victor, 'An Indigenous Early Nineteenth Century Map of Central and Northeast Thailand,' in *In Memoriam Phya Anuman Rajadhon: Contributions in Memory of the Late President of the Siam Society,* pp. 315-348. (ed. Tej Bunnag and Michael Smithies). Bangkok: Siam Society, 1970.

Kimble, George H. Y., *Geography in the Middle Ages.* London: Methusen & Co. Ltd. 1938.

Knapp, Bettina L., *Women in Myth,* State University of New York Press, 1997.

Koeman, C., *Atlantes Neerlandici*. Amsterdam: Theatrum Orbis Terrarum, 1967.

— *Joan Blaeu and his Grand Atlas*. London: George Philip & Son Limited, in cooperation with Theatrum Orbis Terrarum Ltd., Amsterdam, 1970.

— *The Sea on Paper*. Amsterdam: Theatrum Orbis Terrarum, 1972.

Lach, Donald F., *Asia in the Making of Europe, Volume I: The Century of Discovery. Book One*. Chicago and London: The University of Chicago Press, 1965.

— *Asia in the Making of Europe, Volume I: The Century of Discovery. Book Two*. Chicago and London: The University of Chicago Press, 1965.

— *Asia in the Making of Europe, Volume II: A Century of Wonder. Book One: The Visual Arts*. Chicago and London: The University of Chicago Press, 1977.

— *Asia in the Making of Europe, Volume II: A Century of Wonder. Book Two: The Literary Arts*. Chicago and London: The University of Chicago Press, 1977.

— *Asia in the Making of Europe, Volume II: A Century of Wonder. Book Three: The Scholarly Disciplines*. Chicago and London: The University of Chicago Press, 1977.

Lach, Donald F., & Edwin J. van Kley, *Asia in the Making of Europe, Volume III: A Century of Advance. Book One: Trade, Missions, Literature*. Chicago and London: The University of Chicago Press, 1993.

— *Asia in the Making of Europe, Volume III: A Century of Advance. Book Two: South Asia*. Chicago and London: The University of Chicago Press, 1993.

— *Asia in the Making of Europe, Volume III: A Century of Advance. Book Three: Southeast Asia*. Chicago and London: The University of Chicago Press, 1993.

— *Asia in the Making of Europe, Volume III: A Century of Advance. Book Four: East Asia*. Chicago and London: The University of Chicago Press, 1993.

La Loubère, Simon de, *A New Historical Relation of the Kingdom of Siam*. London: 1693.

Leader, J. T., *Life of Sir Robert Dudley*, pp.121-22. Florence: G. Barbèra, 1895 (Amsterdam, Meridian Publishing Co., 1977).

Lessa, William A., *Drake's Island of Thieves*. Honolulu: The University of Press of Hawaii, 1975.

Lévesque, Rodrigue, *The Philippines, Pigafetta's Story of their Discovery by Magellan*. Gatineau, Quebec: Lévesque Publications, 1980.

— *History of Micronesia, Volume 1, European Discovery*. Gatineau, Quebec: Lévesque Publications, 1992.

— *History of Micronesia, Volume 2, Prelude to Conquest*. Gatineau, Quebec: Lévesque Publications, 1992.

Linschoten, John Huyghen van, *The Voyage of John Huyghen van Linschoten to the East Indies*. The English translation of 1598, from the original Dutch edition of 1596. Two volumes; volume one edited by Arthur Coke Burnell, volume two by P. A. Tiele. London: The Hakluyt Society [reprinted New York: Burt Franklin, 1970].

Macrobius, Ambrosius Theodosius, *Macrobius' Commentary on the Dream of Scipio*. Translated with an Introduction and Notes by William Harris Stahl. New York: Columbia University Press, 1952.

Majumdar, R. C., *Ancient Indian Colonization in South-East Asia*. First published 1955; University of Baroda Press: 1972.

Maggiolo, Vesconte (attributed), Atlas of Portolan Charts, Facsimile of Manuscript in the British Museum, Edited by Edward Luther Stevenson. New York: The Hispanic Society of America, 1911.

Mandeville, Sir John, *The Travels of Sir John Mandeville. (with: the journals of Johannes de Plano Carpini, Friar William de Rubruquis, and Friar Odoric)*. London: MacMillan and Co., Ltd., 1900 (1915).

Marques, Alfredo Pinheiro, *A Cartografia Portuguesa do Japão*. Lisbon: Impresa Nacional-Casa da Moeda, 1996.

Martyr, Peter, *The Decades*.

Motte, Benjamin, *The Philosophical Transactions From the Year MDCC. (where Mr. Lowthorp ends) to the Year MDCCXX. Abridg'd and Dispos'd under General Heads*. London, 1721.

Münster, Sebastian / Richard Eden, *A treatyse of the newe India*. London: 1553 [Readex Microprint Corporation, 1966].

Mus, Paul, *India Seen From the East; Indian & Indigenous Cults* in Champa.Monah Papers on Southeast Asia Number Three. Translated from the French by I. W. Mabbert. Edited by I. W. Mabbert and D. P. Chandler. Centre of Southeast Asian Studies, Monash University, 1975.

Neale, Frederick Arthur, *Narrative of a Residence at the Capital of The Kingdom of Siam*. London: Office of the National Illustrated Library, 1852.

Nunn, George E., *The Columbus and Magellan Concepts of South American Geography*. Glenside: privately printed, 1932.

Odoric, Friar, *The Journal of Friar Odoric*. (see Mandeville).

Olschki, Leonardo, *Marco Polo's Precursors*. John Hopkins Press, 1943 [New York: Octagon Books, 1972].

Osbourne, Milton, *Southeast Asia an Introductory History*. Chiang Mai: Silkworm Books, Sixth Edition, 1995.

(Pali Text Society), *The Pali Text Society's Pali-English Dictionary*. London: Luzac & Co., Ltd, 1966.

Parks, George Bruner, *Richard Hakluyt and the English Voyages*. New York: American Geographical Society Special Publication No. 10, 1928.

Penth, Hans, *A Brief History of Lan Na*. Chiang Mai: Silkworm Books, 1994.

— 'Buddhist Literature of Lan Na on the History of Lan Na's Buddhism", in *Journal of the Pali Text Society*, Vol. XXIII, 1997, pp. 43-81.

— Phanphen Khrüathai, Silao Ketphrom, 'Inscription 1.2.2.1 Wat Phra Sing — Restoring a Buddhapada,' and 'Inscription 1.2.2.1 Wat Phra Sing', in Hans Penth *et al., Corpus of Lan Na Inscriptions, Vol. 2, King Kawila Inscriptions A.D. 1791-1814*. Chiang Mai 1999.

— The Buddhapada of Wat Phra Sing; in *Corpus of Lan Na Inscriptions*, Vol. 2, King Kawila Inscriptions A.D. 1791-1814. Chiang Mai, Social Research Institute, Chiang Mai University, 1998, pp. 35-52, and pp. 251-274.

Phillimore, R.H. 'An Early Map of the Malay Peninsula,' in *Imago Mundi* 13, p175-79 (1956).

Pigafetta, Antonio, *Relation du Premier Voyage Autour du Monde par Magellan*. Edition du Texte Français d'Apres les manuscrits de Paris et de Cheltenham, par J. Denucé, 1923.

— *The Voyage of Magellan*. Translated by Paula Spurlin Paige. New Jersey: Prentice-Hall, 1969.

Pinto, Mendes, *The Travels of Mendes Pinto*, edited and translated by Rebecca D. Catz. The University of Chicago Press, 1989

Pinto, Fernão Mendes, *The Travels of Mendes Pinto*. (Lisbon, 1614). Edited and translated by Rebecca D. Catz. Chicago and London: The University of Chicago Press, 1989.

Pires, Tomé, *Suma Oriental*, translated and edited by Armando Cortesão. Hakluyt Society, 1944 (Kraus reprint, 1967).

Plano Carpini, Johannes de, *The Voyage of Johannes de Plano Carpini*. (see Mandeville).

Pliny the Elder, *Natural History*. Translated with an introduction and notes by John F. Healy. London: Penguin Books, 1991.

Polo, Marco, *The Travels of Marco Polo*, annotated and translated by Henry Yule, revised by Henri Cordier. London: John Murray, 1920 [reprinted, three volumes bound as two, by Dover Publications, New York, 1993].

— *The Travels of Marco Polo*, translated and edited by William Marsden. New York: Doubleday & Co., 1948.

— *The Travels*, translated and with an Introduction by Ronald Latham. Middlesex: Penguin Books, Ltd., 1958 (1978).

Ptolemy, Claudius, *The Geography*. Translated and edited by Edward Luther Stevenson. New York: New York Public Library, 1932 (Dover Publications, 1991).

Quirino, Carlos, *Philippine Cartography*. Amsterdam: N. Israel, 1963 (Reprint edition, 1969).

Ramusio, Giovanni Battisa, *Delle navigationi et viaggi.* Venice, 1550 & 1554 (vol 1); 1556 (volume 3); 1559 (vol. 2).

Ravenstein, E. G., *Translation & Commentary on Martin Behaim's 'Erdapfel'.* (originally published as *Martin Behaim, his Life and his Globe,* London, George Philip, 1908). London: Greaves & Thomas, 1992.

Reid, Anthony, *Southeast Asia in the Age of Commerce 1450-1680, Volume One: The Lands below the Winds.* New Haven and London: Yale University Press, 1988.

Richardson, W. A. R., 'Piloting a Toponymic Course through Sixteenth-Century Southeast Asian Waters,' in *Terrae Incognitae,* Vol 20, 1988, p1-20.

Rodrigues, Francisco, *The Book of Francisco Rodrigues,* translated and edited by Armando Cortesão. Hakluyt Society, 1944 (Kraus reprint, 1967).

Roncière, Monique de la, 'Manuscript Charts by John Thornton, Hydrographer of the East Indian Company (1669-1701),' in *Imago Mundi,* Vol. XIX

Rooney, Dawn, 'Ayuthaya, the ancient capital of Siam on European maps,' in *The Map Collector,* issue 57, Winter, 1991.

— The Mapping of Thailand: an Introduction, (unpublished), paper presented in Singapore at the 1991 IMCoS Symposium.

Rubruquis, Friar William de, *The Journal of Friar William de Rubruquis.* (see Mandeville).

Quirino, Carlos, *Philippine Cartography.* Amsterdam: Nico Israel 1963 (1969).

SarDesai, D. R., *Southeast Asia Past & Present.* Boulder and San Francisco: Westview Press, 1989 (1994).

Schilder, Günter, 'Organization and Evolution of the Dutch East India Company's Hydrographic Office in the Seventeenth Century,' in *Imago Mundi,* Vol. 28, pp 61-78. Kent: 1976.

— *Australia Unveiled.* Theatrum Orbis Terrarum Ltd. Amsterdam: 1976.

— *Monumenta Cartographica Neerlandica II.* Alphen aan den Rijn: Canaletto, 1987.

Schouten, Joost (and François Caron), *A True Description of the Mighty Kingdoms of Japan and Siam.* London, 1671 [The Siam Society, 1986].

Skelton, R. A., *Explorers' Maps.* London: Routledge & Kegan Paul Limited, 1958 [Hamlyn Publishing Group Ltd., 1970].

Smithies, Michael (ed.), *Early Accounts of Phetchaburi.* Bangkok: The Siam Society, 1987.

— 'The Travels in France of the Siamese Ambassadors 1686-7,' in *Journal of the Siam Society,* Vol 77, Pt. 2, 1989, pp. 59-70.

— *The Siamese Embassy to the Sun King,* of unproven authorship but possibly a translation (from French and in turn from Thai) of the memoirs of Kosa Pan. Bangkok: Editions Duang Kamol, 1990.

— 'Jacques de Bourges (c. 1630-1714) and Siam,' in *Journal of the Siam Society,* Volume 81, Part 2 (1993).

Sunthorn Bhoo, *Phra Abhai Mani,* retold by M. L. Manich Jumsai, Bangkok, 1986.

Stahl, William Harris, (see Macrobius).

Steinberg *et al, In Search of Southeast Asia.* New York: Praeger, 1971.

Sternstein, Larry, "LOW' Maps of Siam,' in *The Journal of the Siam Society,* Vol. 73, part 1 & 2, Jan-July, 1985. Bangkok.

— 'Low's Description of the Siamese Empire in 1824,' in The *Journal of the Siam Society,* Vol. 78, Part 1, pp 8-34. Bangkok, 1990.

Stevenson, Edward Luther, *Marine World Chart of Nicolo de Canerio Januensis 1502 (circa) A Critical Study.* New York: The Hispanic Society of America, 1908.

Tachard, Guy, *A Relation of the Voyage to Siam Performed by Six Jesuits, Sent by the French King, to the Indies and China, in the Year, 1685.* London, A. Churchill, 1688 [Bangkok: Asia Books Co. Ltd., 1981].

Taylor, E.G.R., *The Haven-Finding Art.* New York: Abelard-Schuman Limited, 1957.

Taylor, Keith Weller, *The Birth of Vietnam,* University of California Press, Berkely and Los Angeles, 1983

Thrower, Norman (ed.), *The Compleat Plattmaker.* Berkeley, Los Angeles, and London: University of California Press, 1978.

— *Sir Francis Drake and the Famous Voyage, 1577-1580.* Berkeley Los Angeles, and London: University of California Press, 1984.

— *Maps and Civilization, Cartography on Culture and Society.* Chicago and London: The University of Chicago Press, 1996.

Tibbetts, G. R., *A Study of the Arabic Texts Containing Material on South-East Asia.* Published for the Royal Asiatic Society. Leiden & London: E. J. Brill, 1979.

Tuck, Patrick J. N., 'Auguste Pavie and the Exploration of the Mekong Valley, 1886-95', in *Terrae Incognitae XIV,* 1982 (pp. 41-60).

Tyache, Sarah, *London Map-Sellers* 1660-1720. Hertfordshire: Map Collector Publications Ltd., 1978.

Unno, Kazutaka, 'The Asian Lake Chiamay in the Early European Cartography', *Imago et Mensura Mundi,* Istituto della Enciclopedia Italiana, 1985, pp. 287-296.

Varadarajan, Lotika, 'Glimpses of Seventeenth Century Currency and Mensuration in Siam,' page 207, note 13, in *Journal of the Siam Society,* Volume 83, Parts 1&2, 1995

Varthema, Ludovico di, *The Itinerary of Ludovico di Varthema of Bologna from 1502 to 1508.* Translated from the original Italian edition of 1510 by John Winter Jones in 1863, for the Hakluyt Society; commentary by Sir Richard Carnac Temple. London: The Argonaut Press, 1928.

Vaughan, Thomas, E. A. P. Crownhart-Vaughan & Mercedes Palau de Iglesias, *Voyages of Enlightenment.* Portland: Oregon Historical Society, 1977.

Verner, Coolie, & R. A. Skelton, *Bibliographical Note* to the T.O.T. facsimile of John Thornton, *The English Pilot, The Third Book* (London, 1703). Amsterdam: Theatrum Orbis Terrarum Ltd. 1970.

Vrij, Marijke de, *The World on Paper.* Amsterdam: Theatrum Orbis Terrarum Ltd., 1967.

Wallis, Helen, *Sir Francis Drake.* London: British Museum Publications, 1977.

— 'The Influence of Father Ricci on Far Eastern Cartography,' in *Imago Mundi,* Vol. XIX.

Wheatley, Paul, *The Golden Khersonese.* Kuala Lumpur: University of Malaya Press, 1961 [reprint, 1980].

— *Lochac Revisited,* Center for South and Southeast Asia Studies, Reprint No. 351. Reprinted with permission from *Oriens Extremus,* 16 Jahrgang, Heft 1, Juni 1965.

— 'A Curious Feature on Early Maps of Malaya,' in *Imago Mundi,* Vol. XI [reprint edition]. Amsterdam: N. Israel, 1967.

White, John, *History of a Voyage to the China Sea.* Boston: 1827.

Williams, Lea, *Southeast Asia: A History.* New York: Oxford University Press, 1976.

Winichakul, Thongchai, *Siam Mapped a History of the Geo-body of a Nation.* Honolulu: University of Hawaii Press, 1994 [Chiang Mai, Silkworm Books].

Woodward, David (ed.), *Art and Cartography six historical essays.* Chicago and London: The University of Chicago Press, 1987.

— *The Maps and Prints of Paolo Forlani,* Chicago, The Newberry Library, 1990

Wright. John Kirtland, *The Geographical Lore of the Time of the Crusades.* New York: American Geographical Society, Research Series No. 15, 1925.

— *The Leardo Map of the World 1452 or 1453.* New York: American Geographical Society Library Series No. 4, 1928.

Wroth, Lawrence C., *The Early Cartography of the Pacific.* New York: The Papers of the Bibliographical Society of America, Volume Thirty-Eight Number Two, 1944.

Index

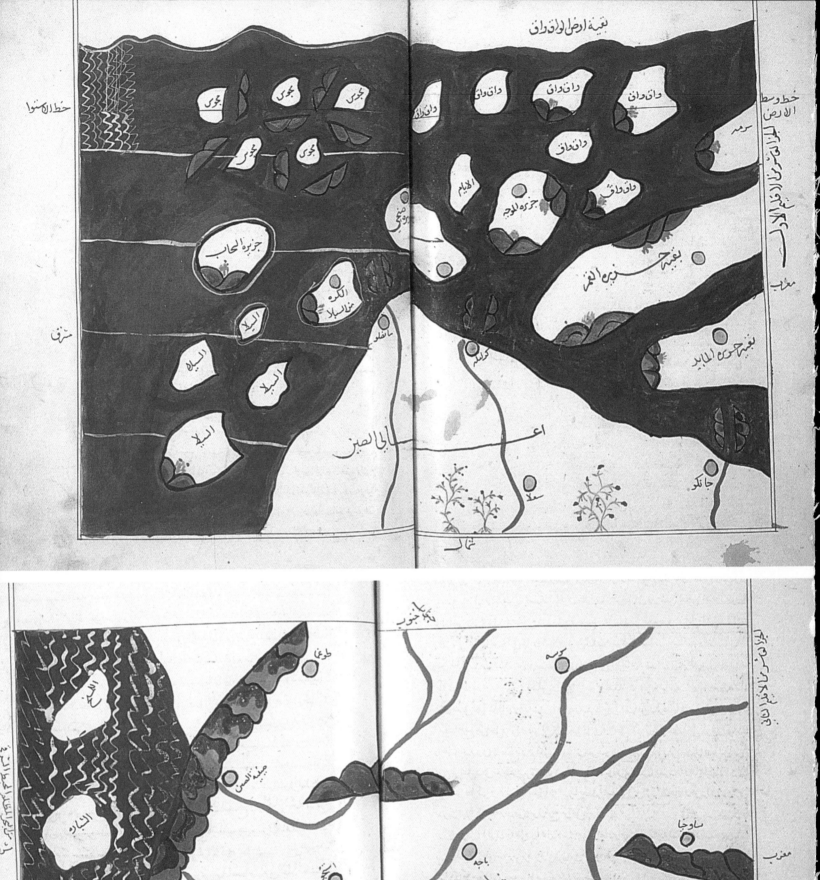